# Mathematics of Machine Learning

Master linear algebra, calculus, and probability for machine learning

**Tivadar Danka**

**‹packt›**

# Mathematics of Machine Learning

**Portfolio Director:** Sunith Shetty
**Relationship Lead:** Tushar Gupta
**Project Manager:** Amit Ramadas
**Content Engineer:** Deepayan Bhattacharjee
**Technical Editor:** Kushal Sharma
**Copy Editor:** Safis Editing
**Indexer:** Hemangini Bari
**Proofreader:** Deepayan Bhattacharjee
**Production Designer:** Ganesh Bhadwalkar
**Growth Leads:** Merlyn M Shelley & Bhavesh Amin
**Marketing Owner:** Ankur Mulasi

First published: May 2025

Production reference: 1210525

Published by Packt Publishing Ltd.
Grosvenor House
11 St Paul's Square
Birmingham
B3 1RB, UK.

ISBN 978-1-83702-787-3

www.packtpub.com

*This book is dedicated to my mother, whom I lost while making this book.*
*Thanks, Mom! You are inside every line I write.*
*– Tivadar Danka*

# Foreword

I met Tivadar during Covid. We were all stuck at home, unsure what to do with all the extra time, so we started talking about building something together.

I wanted to teach people Machine Learning. I had this idea about building a website that would ask random questions for people to answer. I wanted the site to do a hundred different things, but one thing was non-negotiable: I wanted people to leave feeling they had learned something different.

Tivadar was the answer to that.

Machine Learning is tough, and unfortunately, most educational content you find online suffers from chronic handwaving syndrome: overused buzzwords, skipped intuition, and more confusion than when you started.

At the time, Tivadar was already writing online about math. He wasn't the only one, but he was different. He was taking seemingly mundane topics and telling stories around them that were surprisingly effective.

There wasn't any handwaving or burying people under a mountain of theoretical ideas. The writing was different, sharp, and fresh.

I had never been excited about math before. I read every single one of Tivadar's posts. I wasn't just learning the rules, I was learning how to think. And, shockingly, I was entertained.

I had never seen that combination before.

I asked Tivadar to help me with the site, and he did – for a while – until he decided to move on to start writing this book. I remember telling him I understood, but I was secretly sad – really sad.

Today, I'm thrilled this happened the way it did.

Mathematics of Machine Learning is the inevitable consequence of those short posts that excited me about math for the first time. It's not just the best book I've read on the subject, it's the one I wish had existed when I started.

This book does something rare: it teaches you the math behind machine learning without boring you with vague concepts—or making you forget why you showed up in the first place.

The book is laser-focused on what you need and says nothing about what you don't. The explanations are vintage Tivadar: sharp, detailed, and entertaining. You can't just read or memorize them; you'll understand them.

I've been reading this book since it was an idea and a bunch of notes and sketches. I've watched it grow from online posts to something polished and powerful. And I've learned a lot – not just about math, but about how to explain math.

I'll leave you to it. You're in for a treat. Enjoy the journey – I know I did.

*Santiago Valdarrama,*
*Founder of ml.school*

# Contributors

## About the author

**Tivadar Danka** is an independent thinker, who believes that the truth value of any proposition is independent of the titles, awards, qualifications, and affiliations of the one asserting it. If you are looking for confirmation that you made a good purchase with this book, start at *Chapter 1*.

*Yes, that's really a reference to the first chapter in the author bio; that's where the important part begins.*

# About the reviewers

**Matthew Kehoe** earned a PhD in Computational Mathematics from the University of Illinois at Chicago, where he specialized in numerical partial differential equations and inverse electromagnetic scattering theory. Following two graduate internships with the National Science Foundation and several years in software development and technical consulting, he now serves as a senior researcher leading projects in radar signal processing, scientific machine learning, and natural language processing. In addition to his applied research, he maintains a strong interest in analytic number theory and has begun writing a book on computing zeros of the Riemann zeta function.

**Shravan Patankar**, PhD, is a researcher and software engineer with deep expertise in artificial intelligence, machine learning, and data science. He earned his PhD in mathematics from the University of Illinois at Chicago, where his research led to three peer-reviewed publications, including two in top mathematical journals and one in Scientific Reports (Nature) on COVID-19 death estimates. With a strong foundation in mathematical reasoning and statistical modeling, Shravan has applied these principles across both academic and industry contexts.

Professionally, he is a Software Engineer in AI/ML at KPIT Technologies, where he works on software-defined vehicles and contributes to the safety of autonomous driving systems. Shravan has also served as an instructor and teaching assistant at UIC, teaching subjects from introductory calculus and statistics to applied Python programming. His collaborative and interdisciplinary work has been showcased at national conferences and international seminars.

*I would like to thank my mentors and peers for shaping my journey. I am especially grateful to my family and friends for their unwavering encouragement and support during the review process.*

# Join our community on Discord

Read this book alongside other users, deep learning experts, and the author himself.

Ask questions, provide solutions to other readers, chat with the author via Ask Me Anything sessions, and much more.

Scan the QR code or visit the link to join the community:

https://packt.link/math

# Table of Contents

# Introduction

*Why do I have to learn mathematics?* - This is a question I am asked daily.

Well, you don't have to. But you should!

On the surface, advanced mathematics doesn't impact software engineering and machine learning in a production setting. You don't have to calculate gradients, solve linear equations, or find eigenvalues by hand. Basic and advanced algorithms are abstracted away into libraries and APIs, performing all the hard work for you.

Nowadays, implementing a state-of-the-art deep neural network is almost equivalent to instantiating an object in PyTorch, loading the pre-trained weights, and letting the data blaze through the model. Just like all technological advances, this is a double-edged sword. On the one hand, frameworks that accelerate prototyping and development enable machine learning in practice. Without them, we wouldn't have seen the explosion in deep learning that we witnessed in the last decade.

On the other hand, high-level abstractions are barriers between us and the underlying technology. User-level knowledge is only sufficient when one is treading on familiar paths. (Or until something breaks.)

If you are not convinced, let's do a thought experiment! Imagine moving to a new country without speaking the language and knowing the way of life. However, you have a smartphone and a reliable internet connection.

How do you start exploring?

With Google Maps and a credit card, you can do many awesome things there: explore the city, eat in excellent restaurants, and have a good time. You can do the groceries every day without speaking a word: just put the stuff in your basket and swipe your card at the cashier.

After a few months, you'll also start to pick up some language – simple things like saying greetings or introducing yourself. You are off to a good start!

There are built-in solutions for everyday tasks that just work – food ordering services, public transportation, etc. However, at some point, they will break down. For instance, you need to call the delivery person who dropped off your package at the wrong door. You need to call help if your rental car breaks down.

You may also want to do more. Get a job, or perhaps even start your own business. For that, you need to communicate with others effectively.

Learning the language when you plan to live somewhere for a few months is unnecessary. However, if you want to stay there for the rest of your life, it is one of the best investments you can make.

Now, replace the country with machine learning and the language with mathematics.

The fact is that algorithms are written in the language of mathematics. To get proficient with algorithms, you have to speak it.

# What is this book about?

> *"There is a similarity between knowing one's way about a town and mastering a field of knowledge; from any given point one should be able to reach any other point. One is even better informed if one can immediately take the most convenient and quickest path from one point to the other."*
>
> — *George Pólya and Gábor Szegő, in the introduction of the legendary book* Problems and Theorems in Analysis

The above quote is one of my all-time favorites. For me, it says that knowledge rests on many pillars. Like a chair has four legs, a well-rounded machine learning engineer also has a broad skill set that enables them to be effective in their job. Each of us focus on a balanced constellation of skills, and mathematics is a great addition for many. You can start machine learning without advanced mathematics, but at some point in your career, getting familiar with the mathematical background of machine learning can help you bring your skills to the next level.

There are two paths to mastery in deep learning. One starts from the practical parts and the other starts from theory. Both are perfectly viable, and eventually, they intertwine. This book is for those who started on the practical, application-oriented path, like data scientists, machine learning engineers, or even software developers interested in the topic.

This book is not a 100% pure mathematical treatise. At points, I will make some shortcuts to balance between clarity and mathematical correctness. My goal is to give you the "Eureka!" moments and help you understand the bigger picture instead of preparing you for a PhD in mathematics.

Most machine learning books I have read fall into one of two categories.

1. Focus on practical applications, but unclear and imprecise with mathematical concepts.
2. Focus on theory, involving heavy mathematics with almost no real applications.

I want this book to offer the best of both approaches: a sound introduction of basic and advanced mathematical concepts, keeping machine learning in sight at all times.

My goal is not only to cover the bare fundamentals but to give a breadth of knowledge. In my experience, to master a subject, one needs to go both deep and wide. Covering only the very essentials of mathematics would be like a tightrope walk. Instead of performing a balancing act every time you encounter a mathematical subject in the future, I want you to gain a stable footing. Such confidence can bring you very far and set you apart from others.

During our journey, we are going to follow a roadmap that takes us through

1. *linear algebra*,
2. *calculus*,
3. *multivariable calculus*,
4. and *probability theory*.

We are going to begin our journey with linear algebra. In machine learning, data is represented by vectors. Training a learning algorithm is the same as finding more descriptive representations of data through a series of transformations.

Linear algebra is the study of vector spaces and their transformations.

Simply put, a neural network is just a function that maps the data to a high-level representation. Linear transformations are the fundamental building blocks of these. Developing a good understanding of them will go a long way, as they are everywhere in machine learning.

While linear algebra shows how to describe predictive models, calculus has the tools to fit them to the data. When you train a neural network, you are almost certainly using gradient descent, a technique rooted in calculus and the study of differentiation.

Besides differentiation, its "inverse" is also a central part of calculus: integration. Integrals express essential quantities such as expected value, entropy, mean squared error, etc. They provide the foundations for probability and statistics.

However, when doing machine learning, we deal with functions with millions of variables. In higher dimensions, things work differently. This is where multivariable calculus comes in, where differentiation and integration are adapted to these spaces.

With linear algebra and calculus under our belt, we are ready to describe and train neural networks. However, we lack the understanding of extracting patterns from data. How do we draw conclusions from experiments and observations? How do we describe and discover patterns in them? These are answered by probability theory and statistics, the logic of scientific thinking. In the final chapter, we extend the classical binary logic and learn to deal with uncertainty in our predictions.

# How to read this book

Mathematics follows a definition-theorem-proof structure that might be difficult to follow at first. If you are unfamiliar with such a flow, don't worry. I'll give a gentle introduction right now.

In essence, mathematics is the study of abstract objects (such as functions) through their fundamental properties. Instead of empirical observations, mathematics is based on logic, making it universal. If we want to use the powerful tool of logic, the mathematical objects need to be precisely defined. Definitions are presented in boxes like this below.

**Definition 0.0.1 (An example definition)**

Definitions appear like this.

Given a definition, results are formulated as *if A, then B* statements, where $A$ is the premise, and $B$ is the conclusion. Such results are called theorems. For instance, *if* a function is differentiable, *then* it is also continuous. *If* a function is convex, *then* it has global minima. *If* we have a function, *then* we can approximate it with arbitrary precision using a single-layer neural network. You get the pattern. Theorems are the core of mathematics.

We must provide a sound logical argument to accept the validity of a proposition, one that deduces the conclusion from the premise. This is called a *proof,* responsible for the steep learning curve of mathematics. Contrary to other scientific disciplines, proofs in mathematics are indisputable statements, set in stone forever. On a practical note, look out for these boxes.

**Theorem 0.0.1** *(An example theorem)*

*Let x be a fancy mathematical object. The following two statements hold.*

*(a If A, then B.*

*(b) If C and D, then E.*

*Proof.* This is where the proof goes.

To enhance the learning experience, I'll often make good-to-know but not absolutely essential information into remarks.

> **Remark 0.0.1 (An exciting remark)**
>
> Mathematics is awesome. You'll be a better engineer because of it.

The most effective way of learning is building things and putting theory into practice. In mathematics, this is the only way to learn. What this means is that you need to read through the text carefully. Don't take anything for granted just because it is written down. Think through every sentence. Take every argument and calculation apart. Try to prove theorems by yourself before reading the proofs.

With that in mind, let's get to it! Buckle up for the ride; the road is long and full of twists and turns.

# Conventions used

There are a number of text conventions used throughout this book.

`CodeInText` indicates code words in text, database table names, folder names, filenames, file extensions, pathnames, or URLs. For example: "Slicing works by specifying the first and last elements with an optional step size, using the syntax `object[first:last:step]`."

A block of code is set as follows:

```
from sklearn.datasets import load_iris
data = load_iris()
X, y = data["data"], data["target"]
X[:10]
```

Any command-line input or output is written as follows:

```
(3.5, -2.71, 'a string')
```

*Italics* indicate new concepts or emphasis. For instance, words in menus or dialog boxes appear in the text like this. For example: "This is our first example of a *non-differentiable function*."

# What this book covers

*Chapter 1, Vectors and vector spaces* covers what vectors are and how to work with them. We'll travel from concrete examples through precise mathematical definitions to implementations, understanding vector spaces and NumPy arrays, which are used to represent vectors efficiently. Besides the fundamentals, we'll learn

_Chapter 2, The geometric structure of vector spaces_ moves forward by studying the concept of norms, distances, inner products, angles, and orthogonality, enhancing the algebraic definition of vector spaces with some much-needed geometric structure. These are not just tools for visualization; they play a crucial role in machine learning. We'll also encounter our first algorithm, the _Gram-Schmidt orthogonalization method,_ turning any set of vectors into an orthonormal basis.

In _Chapter 3, Linear algebra in practice,_ we break out NumPy once more, and implement _everything_ that we've learned so far. Here, we learn how to work with the high-performance NumPy arrays in practice: operations, broadcasting, functions, culminating in the from-scratch implementation of the Gram-Schmidt algorithm. This is also the first time we encounter _matrices,_ the workhorses of linear algebra.

_Chapter 4, Linear transformations_ is about the true nature of matrices; that is, structure-preserving transformations between vector spaces. This way, seemingly arcane things – such as the definition of matrix multiplication – suddenly make sense. Once more, we take the leap from algebraic structures to geometric ones, allowing us to study matrices as transformations that distort their underlying space. We'll also look at one of the most important descriptors of matrices: the _determinants,_ describing how the underlying linear transformations affect the _volume_ of the spaces.

_Chapter 5, Matrices and equations_ presents the third (and for us, the final) face of matrices as _systems of linear equations._ In this chapter, we first learn how to solve systems of linear equations by hand using the _Gaussian elimination,_ then supercharge it via our newfound knowledge of linear algebra, obtaining the mighty _LU decomposition._ With the help of the LU decomposition, we go hard and achieve a roughly 70000 × speedup on computing determinants.

_Chapter 6_ introduces two of the most important descriptors of matrices: _eigenvalues_ and _eigenvectors._ Why do we need them?

Because in _Chapter 7, Matrix factorizations,_ we are able to reach the pinnacle of linear algebra with their help. First, we show that real and symmetric matrices can be written in diagonal form by constructing a basis from their eigenvectors, known as the _spectral decomposition theorem._ In turn, a clever application of the spectral decomposition leads to the _singular value decomposition,_ the single most important result of linear algebra.

_Chapter 8, Matrices and graphs_ closes the linear algebra part of the book by studying the fruitful connection between linear algebra and graph theory. By representing matrices as graphs, we are able to show deep results such as _the Frobenius normal form,_ or even talk about the eigenvalues and eigenvectors of graphs.

In _Chapter 9, Functions,_ we take a detailed look at _functions,_ a concept that we have used intuitively so far. This time, we make the intuition mathematically precise, learning that functions are essentially arrows between dots.

_Chapter 10, Numbers, sequences, and series_ continues down the rabbit hole, looking at the concept of numbers.

Each step from natural numbers towards real numbers represents a conceptual jump, peaking at the study of *sequences* and *series*.

With *Chapter 11, Topology, limits, and continuity*, we are almost at the really interesting parts. However, in calculus, the objects, concepts, and tools are most often described in terms of limits and continuous functions. So, we take a detailed look at what they are.

*Chapter 12* is about the single most important concept in calculus: *Differentiation*. In this chapter, we learn that the derivative of a function describes 1) the slope of the tangent line, and 2) the best local linear approximation to a function. From a practical side, we also look at how derivatives behave with respect to operations, most importantly the function composition, yielding the essential *chain rule*, the bread and butter of backpropagation.

After all the setup, *Chapter 13, Optimization* introduces the algorithm that is used to train virtually every neural network: *gradient descent*. For that, we learn how the derivative describes the monotonicity of functions and how local extrema can be characterized with the first and second order derivatives.

*Chapter 14, Integration* wraps our study of univariate functions. Intuitively speaking, *integration* describes the (signed) area under the functions' graph, but upon closer inspection, it also turns out to be the inverse of differentiation. In machine learning (and throughout all of mathematics, really), integrals describe various probabilities, expected values, and other essential quantities.

Now that we understand how calculus is done in single variables, *Chapter 15* leads us to the world of *Multivariable functions*, where machine learning is done. There, we have an entire zoo of functions: *scalar-vector*, *vector-scalar*, and *vector-vector* ones.

In *Chapter 16, Derivatives and gradients*, we continue our journey, overcoming the difficulties of generalizing differentiation to multivariable functions. Here, we have three kinds of derivatives: *partial*, *total*, and *directional*; resulting in the gradient vector and the Jacobian and Hessian matrices.

As expected, optimization is also slightly more complicated in multiple variables. This issue is cleared up by *Chapter 17, Optimization in multiple variables*, where we learn the analogue of the univariate second-derivative test, and implement the almighty gradient descent in its final form, concluding our study of calculus.

Now that we have a mechanistic understanding of machine learning, *Chapter 18, What is probability?* shows us how to reason and model under uncertainty. In mathematical terms, *probability spaces* are defined by the *Kolmogorov axioms*, and we'll also learn the tools that allow us to work with probabilistic models.

*Chapter 19* introduces *Random variables and distributions*, allowing us not only to bring the tools of calculus into probability theory, but to compact probabilistic models into sequences or functions.

Finally, in *Chapter 20*, we learn the concept of *The expected value*, quantifying probabilistic models and distributions with averages, variances, covariances, and entropy.

# To get the most out of this book

The code for this book is provided in the form of Jupyter notebooks, hosted on GitHub at `https://github.c om/cosmic-cortex/mathematics-of-machine-learning-book`. To run the notebooks, you'll need to install the required packages.

The easiest way to install them is using Conda. Conda is a great package manager for Python. If you don't have Conda installed on your system, the installation instructions can be found here: `https://bit.ly/Insta llConda`.

Note that Conda's license might have some restrictions for commercial use. After installing Conda, follow the environment installation instructions in the book's repository **README.md**.

# Download the example code files

The code bundle for the book is hosted on GitHub at `https://github.com/cosmic-cortex/mathematic s-of-machine-learning-book`. We also have other code bundles from our rich catalog of books and videos available at `https://github.com/PacktPublishing/`. Check them out!

# Download the color images

We also provide a PDF file that has color images of the screenshots/diagrams used in this book. You can download it here: `https://packt.link/gbp/9781837027873`.

# Get in touch

Feedback from our readers is always welcome.

**General feedback**: Email `feedback@packtpub.com` and mention the book's title in the subject of your message. If you have questions about any aspect of this book, please email us at `questions@packtpub.com`.

**Errata**: Although we have taken every care to ensure the accuracy of our content, mistakes do happen. If you have found a mistake in this book, we would be grateful if you reported this to us. Please visit `http://www.packtpub.com/submit-errata`, click **Submit Errata**, and fill in the form.

**Piracy**: If you come across any illegal copies of our works in any form on the internet, we would be grateful if you would provide us with the location address or website name. Please contact us at `copyright@packtpub.com` with a link to the material.

**If you are interested in becoming an author**: If there is a topic that you have expertise in and you are interested in either writing or contributing to a book, please visit `http://authors.packtpub.com`.

# Share your thoughts

Once you've read *Mathematics of Machine Learning*, we'd love to hear your thoughts! Scan the QR code below to go straight to the Amazon review page for this book and share your feedback.

`https://packt.link/r/1837027870`

Your review is important to us and the tech community and will help us make sure we're delivering excellent quality content.

# Download a free PDF copy of this book

Thanks for purchasing this book!

Do you like to read on the go but are unable to carry your print books everywhere? Is your eBook purchase not compatible with the device of your choice?

Don't worry; with every Packt book, you now get a DRM-free PDF version of that book at no cost.

Read anywhere, on any device. Search, copy, and paste code from your favorite technical books directly into your application.

The perks don't stop there! You can get exclusive access to discounts, newsletters, and great free content in your inbox daily.

Follow these simple steps to get the benefits:

1.  Scan the QR code or visit the link below:

https://packt.link/free-ebook/9781837027873

2.  Submit your proof of purchase.
3.  That's it! We'll send your free PDF and other benefits to your email address directly.

# Part I
# Linear Algebra

This part comprises the following chapters:

# 1

# Vectors and Vector Spaces

*"I want to point out that the class of abstract linear spaces is no larger than the class of spaces whose elements are arrays. So what is gained by abstraction? First of all, the freedom to use a single symbol for an array; this way we can think of vectors as basic building blocks, unencumbered by components. The abstract view leads to simple, transparent proofs of results."*

*— Peter D. Lax, in Chapter 1 of his book Linear Algebra and its Applications*

The mathematics of machine learning rests upon three pillars: *linear algebra, calculus,* and *probability theory.* Linear algebra describes how to represent and manipulate data; calculus helps us fit the models; while probability theory helps interpret them.

These build on top of each other, and we will start at the beginning: *representing and manipulating data.*

To guide us throughout this section, we will look at the famous Iris dataset ( https://en.wikipedia.org/wiki/Iris_flower_data_set). This contains the measurements from three species of Iris: the lengths and widths of sepals and petals. Each data point includes these four measurements, for which we also have the corresponding species: Iris setosa, Iris virginica, or Iris versicolor. (Sepals are the typically green, leaf-like structures at the base of a flower that protect the developing bud before it opens. Petals are the colorful, soft parts of a flower that attract pollinators like insects or birds.)

The dataset can be loaded right away from scikit-learn (`https://scikit-learn.org/`), so let's take a look!

```
from sklearn.datasets import load_iris
data = load_iris()
X, y = data["data"], data["target"]
X[:10]
```

```
array([[5.1, 3.5, 1.4, 0.2],
       [4.9, 3. , 1.4, 0.2],
       [4.7, 3.2, 1.3, 0.2],
       [4.6, 3.1, 1.5, 0.2],
       [5. , 3.6, 1.4, 0.2],
       [5.4, 3.9, 1.7, 0.4],
       [4.6, 3.4, 1.4, 0.3],
       [5. , 3.4, 1.5, 0.2],
       [4.4, 2.9, 1.4, 0.2],
       [4.9, 3.1, 1.5, 0.1]])
```

Before going into the mathematical definitions, let's establish a common vocabulary first. The measurements themselves are stored in a tabular format. Rows represent samples, and columns represent measurements. A particular measurement type is often called a *feature*. As `X.shape` tells us, the Iris dataset has 150 data points and four features:

```
X.shape
```

```
(150, 4)
```

(Don't worry if you are not familiar with NumPy. We'll learn about the details in due time. For now, it's enough to understand that an array's shape describes its dimensions.)

For a given sample, the corresponding species is called the *label*. In our case, this is either Iris setosa, Iris virginica, or Iris versicolor. Here, the labels are encoded with the numbers 0, 1, and 2:

```
y
```

```
array([0, 0, 0, 0, 0, 0, 0, 0, 0, 0, 0, 0, 0, 0, 0, 0, 0, 0, 0, 0,
       0, 0, 0, 0, 0, 0, 0, 0, 0, 0, 0, 0, 0, 0, 0, 0, 0, 0, 0, 0,
       0, 0, 0, 0, 0, 0, 1, 1, 1, 1, 1, 1, 1, 1, 1, 1, 1, 1, 1, 1,
       1, 1, 1, 1, 1, 1, 1, 1, 1, 1, 1, 1, 1, 1, 1, 1, 1, 1, 1, 1,
       1, 1, 1, 1, 1, 1, 1, 1, 1, 1, 1, 1, 2, 2, 2, 2, 2, 2, 2, 2,
       2, 2, 2, 2, 2, 2, 2, 2, 2, 2, 2, 2, 2, 2, 2, 2, 2, 2, 2, 2,
```

```
2, 2, 2, 2, 2, 2, 2, 2, 2, 2, 2, 2, 2, 2, 2, 2, 2])
```

In mathematical terms, the Iris dataset forms a *matrix*, and the data points form *vectors*. Simply speaking, matrices are *tables*, while vectors are *tuples*. (Tuples are just finite and ordered sequences of objects, like $(1.297, -2.35, 32.3, 29.874)$.) However, this simplistic view doesn't show us the big picture. Vectors and matrices have a beautiful geometrical and algebraic structure, and exploring their mathematical theory allows us to see the patterns behind the data.

How so? Say, besides representing the data points in a compact form, we want to perform operations on them, like addition and scalar multiplication. Why do we need to add data points together? To give you a simple example, it is often beneficial if the features are on the same scale. If a given feature is distributed on a smaller scale than the others, it will have less influence on the predictions.

Think about this: if somebody is whispering to you something from the next room while speakers blast loud music right next to your ear, you won't hear anything of what the person is saying to you. Large-scale features are the blasting music, while the smaller ones are the whisper. You may obtain much more information from the whisper, but you need to quiet down the music first.

To see this phenomenon in action, let's take a look at the distribution of the features of our dataset!

```python
import pandas as pd
import seaborn as sns
import matplotlib.pyplot as plt
import numpy as np

sns.set_theme(style="white", rc={"axes.facecolor": (0, 0, 0, 0)})

# Create the data
x = X.ravel()
labels = ["sepal length", "sepal width", "petal length", "petal width"]
g = np.tile(labels, len(X))
df = pd.DataFrame(dict(x=x, g=g))

# Initialize the FacetGrid object
pal = sns.cubehelix_palette(10, rot=-.25, light=.7)
g = sns.FacetGrid(df, row="g", hue="g", aspect=10, height=1.5, palette=pal)

# Draw the densities
g.map(sns.kdeplot, "x", bw_adjust=.5, clip_on=False, fill=True, alpha=1, linewidth=1.5)
g.map(sns.kdeplot, "x", clip_on=False, color="w", lw=2, bw_adjust=.5)

# Add reference line
g.refline(y=0, linewidth=2, linestyle="-", color=None, clip_on=False)
```

```
# Label each plot
g.map(lambda x, color, label: plt.gca().text(0, .2, label, fontweight="bold",
color=color,ha="left", va="center", transform=plt.gca().transAxes), "x")

# Adjust subplots and aesthetics
g.figure.subplots_adjust(hspace=-.25)
g.set_titles("")
g.set(yticks=[], ylabel="")
g.despine(bottom=True, left=True)

plt.show()
```

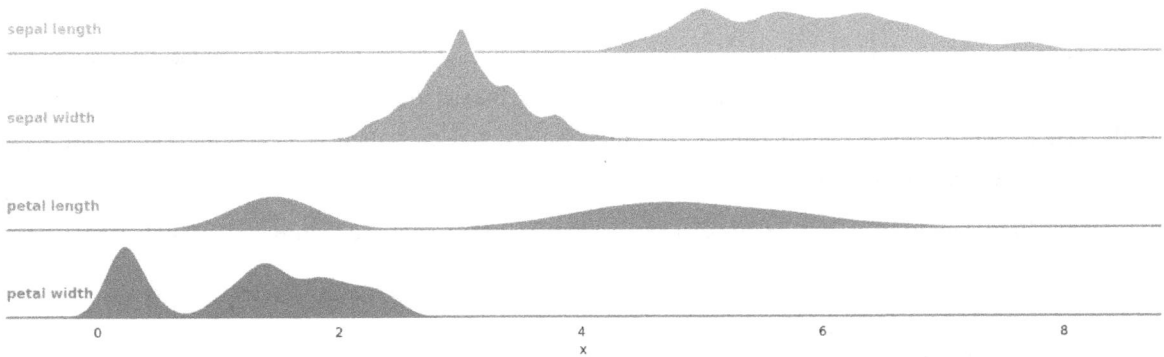

*Figure 1.1: The raw features of the Iris dataset*

You can see in the figure above that some are more stretched out (like sepal length), while others are narrower (like sepal width). In practical scenarios, this can hurt the predictive performance of our algorithms.

To solve it, we can remove the *mean* and the *standard deviation* of a dataset. If the dataset consists of the vectors $x_1, x_2, \ldots, x_{150}$ , we can calculate their mean by

$$\mu = \frac{1}{150} \sum_{i=1}^{150} x_i \in \mathbb{R}^4$$

and their standard deviation by

$$\sigma = \sqrt{\frac{1}{150} \sum_{i=1}^{150} (x_i - \mu)^2} \in \mathbb{R}^4,$$

where the subtraction and square operation in $(x_i - \mu)^2$ is taken elementwise.

The components of $\boldsymbol{\mu} = (\mu_1, \mu_2, \mu_3, \mu_4)$ and $\boldsymbol{\sigma} = (\sigma_1, \sigma_2, \sigma_3, \sigma_4)$ are the means and variances of the individual features. (Recall that the Iris dataset contains 150 samples and 4 features per sample.)

In other words, the mean describes the average of samples, while the standard deviation represents the average distance from the mean. The larger the standard deviation is, the more spread out the samples are.

With these quantities, the scaled dataset can be described as

$$\frac{x_1 - \boldsymbol{\mu}}{\sigma}, \frac{x_2 - \boldsymbol{\mu}}{\sigma}, \dots, \frac{x_{150} - \boldsymbol{\mu}}{\sigma},$$

where both the subtraction and the division are taken elementwise.

If you are familiar with Python and NumPy, this is how it is done. (Don't worry if you are not – everything you need to know about them will be explained in the next chapter, with example code.)

```python
X_scaled = (X - X.mean(axis=0))/X.std(axis=0)
X_scaled[:10]
```

```
array([[-0.90068117,  1.01900435, -1.34022653, -1.3154443 ],
       [-1.14301691, -0.13197948, -1.34022653, -1.3154443 ],
       [-1.38535265,  0.32841405, -1.39706395, -1.3154443 ],
       [-1.50652052,  0.09821729, -1.2833891 , -1.3154443 ],
       [-1.02184904,  1.24920112, -1.34022653, -1.3154443 ],
       [-0.53717756,  1.93979142, -1.16971425, -1.05217993],
       [-1.50652052,  0.78880759, -1.34022653, -1.18381211],
       [-1.02184904,  0.78880759, -1.2833891 , -1.3154443 ],
       [-1.74885626, -0.36217625, -1.34022653, -1.3154443 ],
       [-1.14301691,  0.09821729, -1.2833891 , -1.44707648]])
```

```python
# Create the data
x = X_scaled.ravel()
labels = ["sepal length", "sepal width", "petal length", "petal width"]
g = np.tile(labels, X_scaled.shape[0])
df = pd.DataFrame(dict(x=x, g=g))

# Initialize the FacetGrid object
pal = sns.cubehelix_palette(10, rot=-.25, light=.7)
grid = sns.FacetGrid(df, row="g", hue="g", aspect=10, height=1.5, palette=pal)

# Draw the densities
grid.map(sns.kdeplot, "x", bw_adjust=.5, clip_on=False, fill=True, alpha=1,
linewidth=1.5)
grid.map(sns.kdeplot, "x", clip_on=False, color="w", lw=2, bw_adjust=.5)
```

```
# Add reference line
grid.refline(y=0, linewidth=2, linestyle="-", color=None, clip_on=False)

# Add labels to each plot
grid.map(lambda x, color, label: plt.gca().text(0, .2, label, fontweight="bold",
color=color,ha="left", va="center", transform=plt.gca().transAxes), "x")

# Adjust subplots and aesthetics
grid.figure.subplots_adjust(hspace=-.25)
grid.set_titles("")
grid.set(yticks=[], ylabel="")
grid.despine(bottom=True, left=True)

plt.show()
```

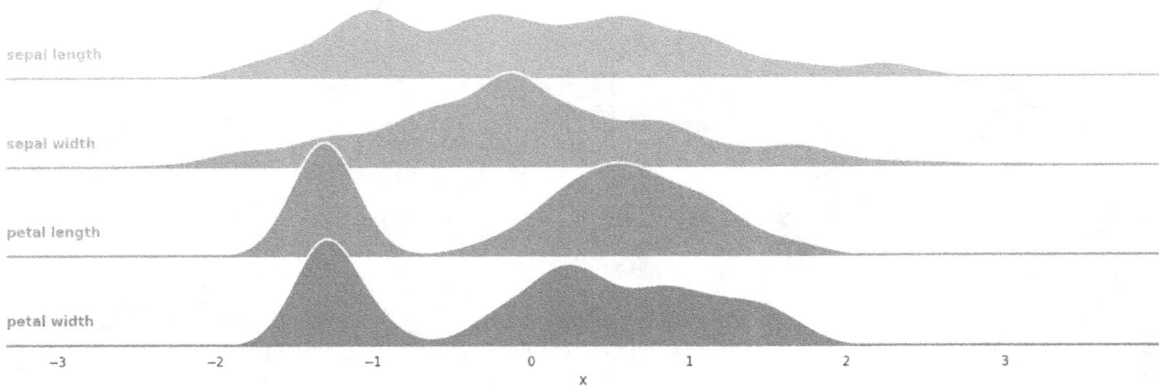

*Figure 1.2: The scaled features of the Iris dataset*

If you compare the modified version to the original, you can see that its features are on the same scale. In other words, we transformed the dataset to a more expressive one. From a (very) abstract point of view, machine learning is nothing else but a series of learned data transformations, turning raw data into a form where prediction is simple.

In a mathematical setting, manipulating data and modeling its relations to the labels arise from the concept of *vector spaces* and transformations between them. Let's take the first steps by making the definition of vector spaces precise!

# 1.1   What is a vector space?

Representing multiple measurements as a tuple $(x_1, x_2, \ldots, x_n)$ is a natural idea that has a ton of merits. The tuple form suggests that the components belong together in a precise order, giving a clear and concise way to store information.

However, this comes at a cost: now we have to work with more complex objects. Despite dealing with tuples like $(x_1, \ldots, x_n)$ instead of numbers, there are similarities. For instance, any two tuple $x = (x_1, \ldots, x_n)$ and $y = (y_1, \ldots, y_n)$

- can be added together by $x + y = (x_1 + y_1, \ldots, x_n + y_n)$,
- and can be multiplied with scalars: if $c \in \mathbb{R}$, then $cx = (cx_1, \ldots, cx_n)$.

It's almost like using a number.

These operations have clear geometric interpretations as well. Addition is the same as *translation*, while multiplication with a scalar is a simple *stretching*. (Or squeezing, if $|c| < 1$.)

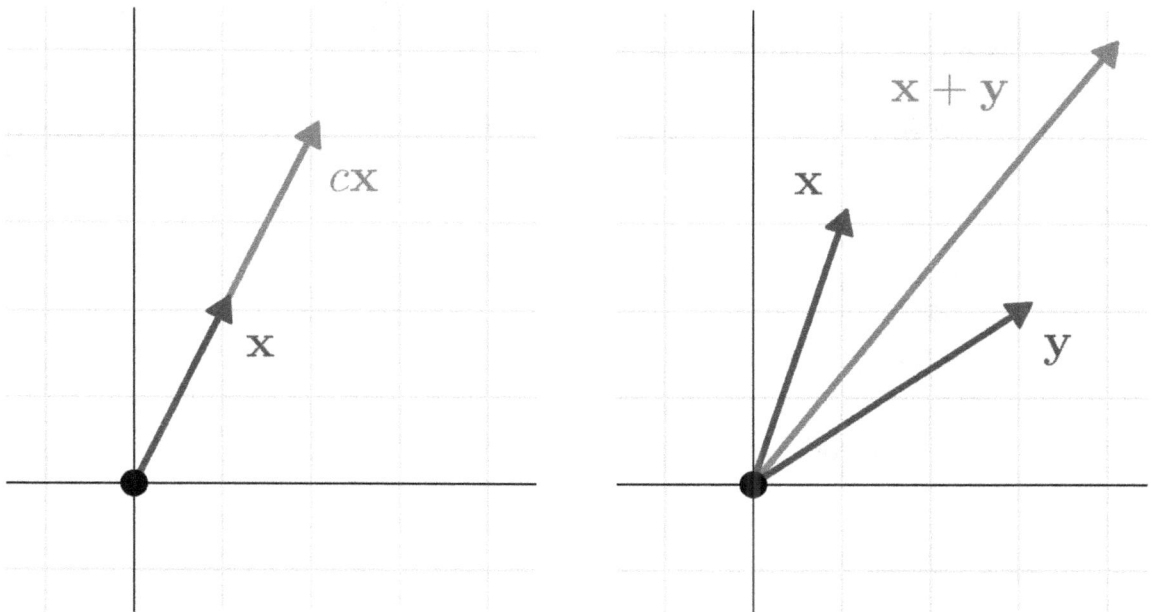

*Figure 1.3: Geometric interpretation of addition and scalar multiplication*

On the other hand, if we want to follow our geometric intuition (which we definitely do), it is unclear how to define vector multiplication. Even though the definition

$$xy = (x_1 y_1, \ldots, x_n y_n)$$

makes sense algebraically, we don't see what it means in a geometric sense.

When we think about vectors and vector spaces, we are thinking about a mathematical structure that fits our intuitive views and expectations. So, let's turn these into the definition!

---

**Definition 1.1.1 (Vector spaces)**

A *vector space* is a mathematical structure $(V, F, +, \cdot)$, where

*(a)* $V$ is the set of vectors,

*(b)* $F$ is a field of scalars (most commonly the real numbers $\mathbb{R}$ or the complex numbers $\mathbb{C}$),

*(c)* $+ : V \times V \to V$ is the addition operation, satisfying the following properties:

- $x + y = y + x$ (commutativity),
- $x + (y + z) = (x + y) + z$ (associativity),
- there is an element $0 \in V$ such that $x + 0 = x$ (existence of the null vector),
- and there is an inverse $-x \in V$ for each $x \in V$ such that $x + (-x) = 0$ (existence of additive inverses)

for all vectors $x, y, z \in V$,

*(d)* and $\cdot : F \times V \to V$ is the scalar multiplication operation, satisfying

- $a(bx) = (ab)x$ (associativity),
- $a(x + y) = ax + ay$ (distributivity),
- and $1x = x$

for all scalars $a, b \in F$ and vectors $x, y \in V$.

---

This definition is overloaded with new concepts, so let's unpack it.

First, looking at operations like addition and scalar multiplication as *functions* might be unusual for you, but this is a perfectly natural representation. (We'll learn about functions later in detail, but for now, feel free to think about them intuitively.) In writing, we use the notation $x + y$, but when thinking about $+$ as a function of two variables, we might as well write $+(x, y)$. The form $x + y$ is called *infix* notation, while $+(x, y)$ is called *prefix* notation.

In vector spaces, the inputs of addition are two vectors and the result is a single vector, thus $+$ is a function that maps the Cartesian product $V \times V$ to $V$.

Similarly, scalar multiplication takes a scalar and a vector, resulting in a vector; meaning a function that maps $F \times V$ to $V$.

(The Cartesian product $V \times V$ is just a set of ordered pairs:

$$V \times V = \{(\boldsymbol{u}, \boldsymbol{v}) \,:\, \boldsymbol{u}, \boldsymbol{v} \in V\}.$$

Feel free to check out the set theory appendix (*Appendix C*) for more details, but for now, the intuitive understanding is enough.)

This is also good place to note that mathematical definitions are always formalized in hindsight, after the objects themselves are somewhat crystallized and familiar to the users. Mathematics is often presented as definitions first, theorems second. This is not how it is done in practice. Examples motivate definitions, not the other way around.

In general, the field of scalars can be something other than real or complex numbers. The term *field* refers to a well-defined mathematical structure, which makes a natural notion mathematically precise. Without going into the technical details, we will think about fields as "a set of numbers where addition and multiplication work just as for real numbers".

Since we are not concerned with the most general case, we will use $\mathbb{R}$ or $\mathbb{C}$ to avoid unnecessary difficulty. If you are not familiar with the exact mathematical definition of a field, don't worry – just think of $\mathbb{R}$ each time you read the word "field".

When everything is clear from the context, $(V, \mathbb{R}, +, \cdot)$ will often be referred to as $V$ for notational simplicity. So, if the field $F$ is not specified, it is implicitly assumed to be $\mathbb{R}$. When we want to emphasize this, we'll call these *real* vector spaces.

At first sight, *Definition 1.1.1* is certainly too complex to comprehend. It seems like just a bunch of sets, operations, and properties thrown together. However, to help us build a mental model, we can imagine a vector as an arrow, starting from the null vector. (Recall that the null vector 0 is that special one for which $x + 0 = x$ holds for all $x$. Thus, it can be considered as an arrow with zero length; the origin.)

To further familiarize ourselves with the concept, let's see some examples of vector spaces!

## 1.1.1   Examples of vector spaces

Examples are one of the best ways of building insight into seemingly difficult concepts like vector spaces. We humans, usually think in terms of *models* instead of abstractions. (Yes, this includes pure mathematicians. Even though they might deny it.)

**Example 1.** The most ubiquitous instance of the vector space is $(\mathbb{R}^n, \mathbb{R}, +, \cdot)$, the same one we used to motivate the definition itself. ($\mathbb{R}^n$ refers to the $n$-fold Cartesian product of the set of real numbers. If you are unfamiliar with this notion, check the set theory tutorial in *Appendix C.*)

$(\mathbb{R}^n, \mathbb{R}, +, \cdot)$ is the canonical model, the one we use to guide us throughout our studies. If $n = 2$, we are simply talking about the familiar Euclidean plane.

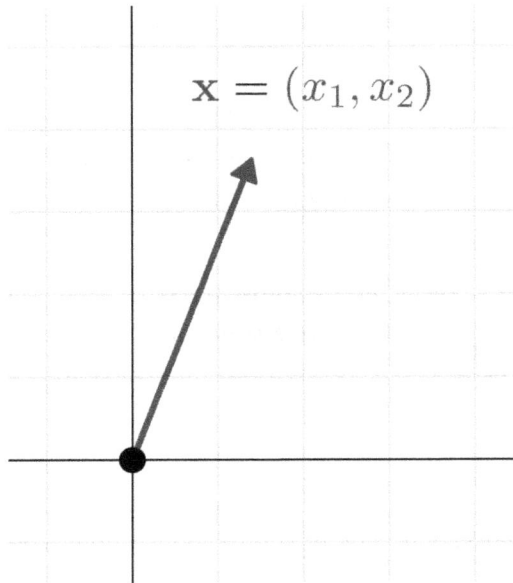

Figure 1.4: *The Euclidean plane as a vector space*

Using $\mathbb{R}^2$ or $\mathbb{R}^3$ for visualization can help a lot. What works here will usually work in the general case, although sometimes this can be dangerous. Math relies on both intuition and logic. We develop ideas using our intuition, but we confirm them with our logic.

**Example 2.** Vector spaces are not just a collection of finite tuples. An example is the space of polynomial functions with real coefficients, defined by

$$\mathbb{R}[x] = \left\{ \sum_{i=0}^{n} p_i x^i \ : \ p_i \in \mathbb{R}, n = 0, 1, \ldots \right\}.$$

Two polynomials $p(x)$ and $q(x)$ can be added together by

$$p(x) + q(x) := \sum_{k=1}^{n} (p_i + q_i)x^i,$$

and can be multiplied with a real scalar by

$$cp(x) = \sum_{k=1}^{n} cp_i x^i.$$

With these operations, $(\mathbb{R}[x], \mathbb{R}, +, \cdot)$ is a vector space. Although most of the time we percieve polynomials as functions, they can be represented as tuples of coefficients as well:

$$\sum_{i=0}^{n} p_i x^i \longleftrightarrow (p_0, \ldots, p_n).$$

Note that $n$ – the degree of the polynomial – is unbounded. As a consequence, this vector space has a significantly richer structure than $\mathbb{R}^n$.

**Example 3.** The previous example can be further generalized. Let $C([0, 1])$ denote the set of all continuous real functions $f : [0, 1] \to \mathbb{R}$. Then $(C(\mathbb{R}), \mathbb{R}, +, \cdot)$ is a vector space, where the addition and scalar multiplication are defined elementwise:

$$(f + g)(x) := f(x) + g(x), \quad (cf)(x) = cf(x)$$

for all $f, g \in C(\mathbb{R})$ and $c \in \mathbb{R}$. (Although continuity is a concept that we haven't defined yet, feel free to think of a continuous function as one whose graph can be drawn without lifting your pen.)

Yes, that is right: functions can be thought of as vectors as well. Spaces of functions play a significant role in mathematics, and they come in several different forms. We often restrict the space to continuous functions, differentiable functions, or basically any subset that is closed under the given operations.

(In fact, $\mathbb{R}^n$ can be also thought of as a function space. From an abstract viewpoint, each vector $x = (x_1, \ldots, x_n)$ is a mapping from $\{1, 2, \ldots, n\}$ to $\mathbb{R}$.)

Function spaces are encountered in more advanced topics, such as inverting ResNet architectures, which we won't deal with in this book. However, it is worth seeing examples that are different (and not as straightforward) as $\mathbb{R}^n$.

# 1.2   The basis

Although our vector spaces contain infinitely many vectors, we can reduce the complexity by finding special subsets that can express *any* other vector.

To make this idea precise, let's consider our recurring example $\mathbb{R}^n$. There, we have a special vector set

$$\boldsymbol{e}_1 = (1, 0, \dots, 0)$$
$$\boldsymbol{e}_2 = (0, 1, \dots, 0)$$
$$\vdots$$
$$\boldsymbol{e}_n = (0, 0, \dots, 1)$$

which can be used to express each vector $\boldsymbol{x} = (x_1, \dots, x_n)$ as

$$\boldsymbol{x} = \sum_{i=1}^{n} x_i \boldsymbol{e}_i, \quad x_i \in \mathbb{R}, \quad \boldsymbol{e}_i \in \mathbb{R}^n$$

For instance, $\boldsymbol{e}_1 = (1, 0)$ and $\boldsymbol{e}_2 = (0, 1)$ in $\mathbb{R}^2$.

What we have just seen feels extremely trivial and it seems to only complicate things. Why would we need to write vectors in the form of $\boldsymbol{x} = \sum_{i=1}^{n} x_i \boldsymbol{e}_i$, instead of simply using the coordinates $(x_1, \dots, x_n)$ ? Because, in fact, the coordinate notation depends on the underlying vector set ($\{\boldsymbol{e}_1, \dots, \boldsymbol{e}_n\}$ in our case) used to express other vectors.

A vector is *not* the same as its coordinates! A single vector can have multiple different coordinates in different systems, and switching between these is a useful tool.

Thus, the set $E = \{\boldsymbol{e}_1, \dots, \boldsymbol{e}_n\} \subseteq \mathbb{R}^n$ is rather special, as it significantly reduces the complexity of representing vectors. With the vector addition and scalar multiplication operations, it *spans* our vector space entirely. $E$ is an instance of a vector space *basis*, a set that serves as a skeleton of $\mathbb{R}^n$.

In this section, we are going to introduce and study the concept of vector space basis in detail.

## 1.2.1   Linear combinations and independence

Let's zoom out from the special case $\mathbb{R}^n$ and start talking about general vector spaces. From our motivating example regarding bases, we have seen that sums of the form

$$\sum_{i=1}^{n} x_i \boldsymbol{v}_i,$$

where the $\boldsymbol{v}_i$-s are vectors and the $x_i$ coefficients are scalars, play a crucial role. These are called *linear combinations*. A linear combination is called *trivial* if all of the coefficients are zero.

Given a set of vectors, the same vector can potentially be expressed as a linear combination in multiple ways. For example, if $\boldsymbol{v}_1 = (1,0), \boldsymbol{v}_2 = (0,1)$, and $\boldsymbol{v}_3 = (1,1)$, then

$$(2,1) = 2\boldsymbol{v}_1 + \boldsymbol{v}_2$$
$$= \boldsymbol{v}_1 + \boldsymbol{v}_3.$$

This suggests that the set $S = \{\boldsymbol{v}_1, \boldsymbol{v}_2, \boldsymbol{v}_3\}$ is redundant, as it contains duplicate information. The concept of *linear dependence and independence* makes this precise.

> **Definition 1.2.1 (Linear dependence and independence)**
>
> Let $V$ be a vector space and $S = \{\boldsymbol{v}_1, \ldots, \boldsymbol{v}_n\}$ be a subset of its vectors. $S$ is said to be *linearly dependent* if it only contains the zero vector, or there is a nonzero $\boldsymbol{v}_k$ that can be expressed as a linear combination of the other vectors $\boldsymbol{v}_1, \ldots, \boldsymbol{v}_{k-1}, \boldsymbol{v}_{k+1}, \ldots, \boldsymbol{v}_n$.
>
> $S$ is said to be *linearly independent* if it is not linearly dependent.

Linear dependence and independence can be looked at from a different angle. If

$$\boldsymbol{v}_k = \sum_{i=1}^{k-1} x_i \boldsymbol{v}_i + \sum_{i=k+1}^{n} x_i \boldsymbol{v}_i,$$

for some nonzero $\boldsymbol{v}_k$, then by subtracting $\boldsymbol{v}_k$, we obtain that the null vector can be obtained as a nontrivial linear combination

$$0 = \sum_{i=1}^{n} x_i \boldsymbol{v}_i$$

for some scalars $x_i$, where $x_k = -1$. This is an equivalent definition of linear dependence. With this, we have proved the following theorem.

**Theorem 1.2.1** *Let $V$ be a vector space and $S = \{\boldsymbol{v}_1, \dots, \boldsymbol{v}_n\}$ be a subset of its vectors.*

*(a) $S$ is linearly dependent if and only if the null vector $\boldsymbol{0}$ can be obtained as a nontrivial linear combination.*

*(b) $S$ is linearly independent if and only if whenever $0 = \sum_{i=1}^{n} x_i \boldsymbol{v}_i$, all coefficients $x_i$ are zero.*

## 1.2.2 Spans of vector sets

Linear combinations provide a way to take a small set of vectors and generate a whole lot of others from them. For a set of vectors $S$, taking all of its possible linear combinations is called *spanning*, and the generated set is called the *span*. Formally, it is defined by

$$\mathrm{span}(S) = \left\{ \sum_{i=1}^{n} x_i \boldsymbol{v}_i \; : \; n \in \mathbb{N}, \boldsymbol{v}_i \in S, x_i \text{ is a scalar} \right\}.$$

Note that the vector set $S$ is not necessarily finite. To help illustrate the concept of span, we can visualize the process in three dimensions. The span of two linearly independent vectors is a *plane*.

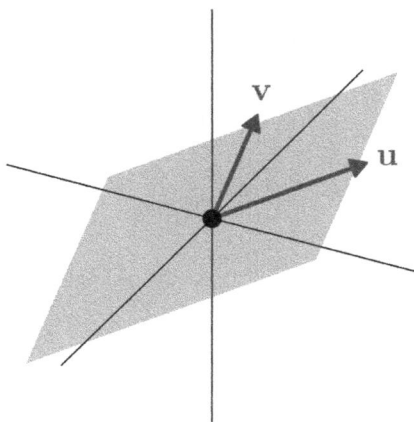

*Figure 1.5: The span of two linearly independent vectors $\boldsymbol{u}, \boldsymbol{v} \in \mathbb{R}^3$*

When we talk about the span of a finite set $\{v_1, \dots, v_n\}$, we denote the span as

$$\text{span}(v_1, \dots, v_n).$$

This helps us avoid overcomplicating notations by naming every set.

> **Proposition 1.2.1** *Let $V$ be a vector space and $S, S_1, S_2 \subseteq V$ be subsets of its vectors.*
>
> *(a) If $S_1 \subseteq S_2$, then $\text{span}(S_1) \subseteq \text{span}(S_2)$.*
>
> *(b) $\text{span}(\text{span}(S)) = \text{span}(S)$.*

This is our very first proof! Give it a read, and if it's too difficult, move on and revisit it later. Just make sure that you understand what the proposition says.

> *Proof.* The property *(a)* follows directly from the definition. To prove *(b)*, we have to show that $\text{span}(S) \subseteq \text{span}(\text{span}(S))$ and $\text{span}(\text{span}(S)) \subseteq \text{span}(S)$.
>
> (This is one of those *steep learning curve* moments, but think about it for a second: two sets $A$ and $B$ are equal if and only if $A \subseteq B$ and $B \subseteq A$.)
>
> The former follows from the definition. For the latter, let $x \in \text{span}(\text{span}(S))$. Then
>
> $$x = \sum_{i=1}^{n} x_i v_i$$
>
> for some $v_i \in \text{span}(S)$. Because of $v_i$ being in the span of $S$, we have
>
> $$v_i = \sum_{j=1}^{m} v_{i,j} u_j$$
>
> for some $u_j \in S$. Thus,
>
> $$x = \sum_{i=1}^{n} x_i v_i = \sum_{i=1}^{n} x_i \sum_{j=1}^{m} v_{i,j} u_j = \sum_{j=1}^{m} \left( \sum_{i=1}^{n} x_i v_{i,j} \right) u_j,$$
>
> implying that $x \in \text{span}(S)$ as well.

Because of $\text{span}(\text{span}(S)) = \text{span}(S)$, if $S$ is linearly dependent, we can remove the redundant vectors and still keep the span the same.

Think about it: if $S = \{v_1, \ldots, v_n\}$ and, say, $v_n = \sum_{i=1}^{n-1} x_i v_i$ , then $v_n \in \text{span}(S \setminus \{v_n\})$. So,

$$\text{span}(S \setminus \{v_n\}) = \text{span}\big(\text{span}(S \setminus \{v_n\})\big) = \text{span}(S).$$

(The operation $A \setminus B$ is the set difference, containing all that are elements of $A$, but not elements of $B$. Feel free to check out *Appendix C* for more details.)

Among sets of vectors, those that generate the entire vector space are special. After all this setup, we are ready to make a formal definition. Any set of vectors $S$ that have the property $\text{span}(S) = V$ is called a *generating set* for $V$.

$S$ can be thought of as a "lossless compression" of $V$, as it contains all the information needed to reconstruct any element in $V$, yet it is smaller than the entire space. Thus, we want to reduce the size of the generating set as much as possible. This leads us to one of the most important concepts in linear algebra: minimal generating sets, or *bases*, as we prefer to call them.

## 1.2.3  Bases, the minimal generating sets

With all the intuition we have built so far, let's jump into the definition right away!

> **Definition 1.2.2  (Basis)**
>
> Let $V$ be a vector space and $S$ be a subset of its vectors. $S$ is a *basis* of $V$ if:
>
> *(a)* $S$ is linearly independent,
>
> *(b)* and $\text{span}(S) = V$.
>
> The elements of a basis set are called *basis vectors*.

It can be shown that these defining properties mean that every vector $x$ can be *uniquely* written as a linear combination of $S$. (This is left as an exercise for the reader.)

Let's see some examples! In $\mathbb{R}^3$, the set $\{(1,0,0), (0,1,0), (0,0,1)\}$ is a basis, but so is $\{(1,1,1), (1,1,0), (0,1,1)\}$. So, there can be more than one basis for the same vector space.

For $\mathbb{R}^n$, the most commonly used basis is $\{e_1, \ldots, e_n\}$, where $e_i$ is a vector whose all coordinates are 0, except the $i$-th one, which is 1. This is called the *standard basis*.

In terms of the "information" contained in a set of vectors, bases hit the sweet spot. Adding any new vector to a basis set would introduce redundancy; removing any of its elements would cause the set to be incomplete.

These notions are formalized in the two theorems below.

> **Theorem 1.2.2** *Let $V$ be a vector space and $S = \{v_1, \dots, v_n\}$ be a subset of vectors. The following are equivalent:*
>
> *(a) $S$ is a basis.*
>
> *(b) $S$ is linearly independent and for any $x \in V \setminus S$, the vector set $S \cup \{x\}$ is linearly dependent. In other words, $S$ is a maximal linearly independent set.*

*Proof.* To show the equivalence of two propositions, we have to prove two things: that *(a)* implies *(b)*; and that *(b)* implies *(a)*. Let's start with the first one!

*(a)* $\implies$ *(b)* If $S$ is a basis, then any $x \in V$ can be written as

$$x = \sum_{i=1}^{n} x_i v_i$$

for some $x_i \in \mathbb{R}$. Thus, by definition, $S \cup \{x\}$ is linearly dependent.

*(b)* $\implies$ *(a)* Our goal is to show that any $x$ can be written as a linear combination of the vectors in $S$. By our assumption, $S \cup \{x\}$ is linearly dependent, so $\mathbf{0}$ can be written as a nontrivial linear combination:

$$\mathbf{0} = \alpha x + \sum_{i=1}^{n} x_i v_i,$$

where not all coefficients are zero. Because $S$ is linearly independent, $\alpha$ cannot be zero (as it would imply the linear dependence of $S$, which would go against our assumptions). Thus,

$$x = -\sum_{i=1}^{n} \frac{x_i}{\alpha} v_i,$$

showing that $S$ is a basis.

Next, we are going to show that every vector of a basis is essential.

> **Theorem 1.2.3** *Let $V$ be a vector space and $S = \{v_1, \dots, v_n\}$ a basis. Then, for any $v_i \in S$,*
>
> $$\mathrm{span}(S \setminus \{v_i\}) \subset V,$$
>
> *that is, the span of $S \setminus \{v_i\}$ is a proper subset of $V$.*

*Proof.* We are going to prove this by contradiction. Without loss of generality, we can assume that $i = 1$. If

$$\text{span}(S \setminus \{\boldsymbol{v}_1\}) = V,$$

then

$$\boldsymbol{v}_1 = \sum_{i=2}^{n} x_i \boldsymbol{v}_i.$$

This means that $S = \{\boldsymbol{v}_1, \dots, \boldsymbol{v}_n\}$ is *not* linearly independent, contradicting our assumptions.

In other words, the above results mean that a basis is a *maximal linearly independent* and a *minimal generating* set at the same time.

Given a basis $S = \{\boldsymbol{v}_1, \dots, \boldsymbol{v}_n\}$, we implictly write the vector $\boldsymbol{x} = \sum_{i=1}^{n} x_i \boldsymbol{v}_i$ as $\boldsymbol{x} = (x_1, \dots, x_n)$. Since this decomposition is unique, we can do this without issues. The coefficients $x_i$ are also called *coordinates*. (Note that the coordinates strongly depend on the basis. Given two different bases, the coordinates of the same vector can be different.)

## 1.2.4  Finite dimensional vector spaces

As we have seen previously, a single vector space can have many different bases, so bases are not unique. A very natural question that arises in this context is the following. If $S_1$ and $S_2$ are two bases for $V$, then does $|S_1| = |S_2|$ hold? (Where $|S|$ denotes the *cardinality* of the set $S$, that is, its "size".)

In other words, can we do better if we select our basis more cleverly? It turns out that we cannot, and the sizes of *any* two basis sets is equal. We are not going to prove this, but here is the theorem in its entirety.

**Theorem 1.2.4** *Let $V$ be a vector space, and let $S_1$ and $S_2$ be two bases of $V$. Then, $|S_1| = |S_2|$.*

This gives us a way to define the *dimension* of a vector space, which is simply the cardinality of its basis. We'll denote the dimension of $V$ as $\dim(V)$. For example, $\mathbb{R}^n$ is $n$-dimensional, as shown by the standard basis $\{(1, 0, \dots, 0), \dots, (0, 0, \dots, 1)\}$.

If you recall the previous theorems, we assumed that a basis is finite. You might ask the question: is this always true? The answer is no. Examples 2 and 3 show that this is not the case. For instance, the countably infinite set $\{1, x, x^2, x^3, \dots\}$ is a basis for $\mathbb{R}[x]$. So, according to the theorem above, no finite basis can exist there.

This marks an important distinction between vector spaces: those with finite bases are called *finite-dimensional*. I have some good news: *all* finite-dimensional real vector spaces are essentially $\mathbb{R}^n$. (Recall that we call a vector space *real* if its scalars are the real numbers.)

To see why, suppose that $V$ is an $n$-dimensional real vector space with basis $\{v_1, \ldots, v_n\}$, and define the mapping $\varphi : V \to \mathbb{R}^n$ by

$$\varphi : \sum_{i=1}^{n} x_i v_i \to (x_1, \ldots, x_n).$$

$\varphi$ is invertible and preserves the structure of $V$, that is, the addition and scalar multiplication operations. Indeed, if $u, v \in V$ and $\alpha, \beta \in \mathbb{R}$, then $\varphi(\alpha u + \beta v) = \alpha \varphi(x) + \beta \varphi(y)$. Such mappings are called *isomorphisms*. The word itself is derived from ancient Greek, with *isos* meaning *same* and *morphe* meaning *shape*. Even though this sounds abstract, the existence of an isomorphism between two vector spaces mean that they have the same structure. So, $\mathbb{R}^n$ is not just an example of finite dimensional real vector spaces, it is a universal model of them. Note that if the scalars are not the real numbers, the isomorphism to $\mathbb{R}^n$ is not true. (We'll talk more about transformations like this in later chapters.)

Considering that we'll almost exclusively deal with finite dimensional real vector spaces, this is good news. Using $\mathbb{R}^n$ is not just a heuristic, it is a good mental model.

## 1.2.5 Why are bases so important?

If every finite-dimensional real vector space is essentially the same as $\mathbb{R}^n$, what do we gain from abstraction? Sure, we can just work with $\mathbb{R}^n$ without talking about bases, but to develop a deep understanding of the core mathematical concepts in machine learning, we need the abstraction.

Let's look ahead briefly and see an example. If you have some experience with neural networks, you know that matrices play an essential role there. Without any context, matrices are just a table of numbers with seemingly arbitrary rules of computation. Have you ever wondered why matrix multiplication is defined the way it is?

Although we haven't precisely defined matrices yet, you have probably encountered them previously. We'll learn all about them in *Chapter 3* and *Chapter 4*, but for the two matrices

$$A = \begin{bmatrix} a_{1,1} & a_{1,2} & \ldots & a_{1,n} \\ a_{2,1} & a_{2,2} & \ldots & a_{2,n} \\ \vdots & \vdots & \ddots & \vdots \\ a_{n,1} & a_{n,2} & \ldots & a_{n,n} \end{bmatrix}, \quad B = \begin{bmatrix} b_{1,1} & b_{1,2} & \ldots & b_{1,n} \\ b_{2,1} & b_{2,2} & \ldots & b_{2,n} \\ \vdots & \vdots & \ddots & \vdots \\ b_{n,1} & b_{n,2} & \ldots & b_{n,n} \end{bmatrix},$$

their product $AB$ is defined by

$$AB = \begin{bmatrix} \sum_{k=1}^{n} a_{1,k} b_{k,1} & \sum_{k=1}^{n} a_{1,k} b_{k,2} & \cdots & \sum_{k=1}^{n} a_{1,k} b_{k,n} \\ \sum_{k=1}^{n} a_{2,k} b_{k,1} & \sum_{k=1}^{n} a_{2,k} b_{k,2} & \cdots & \sum_{k=1}^{n} a_{2,k} b_{k,n} \\ \vdots & \vdots & \ddots & \vdots \\ \sum_{k=1}^{n} a_{n,k} b_{k,1} & \sum_{k=1}^{n} a_{n,k} b_{k,2} & \cdots & \sum_{k=1}^{n} a_{n,k} b_{k,n} \end{bmatrix},$$

that is, the $(i, j)$-th element of $AB$ is defined by

$$\sum_{k=1}^{n} a_{i,k} b_{k,j}.$$

This definition feels random. Why not just take the componentwise product $(a_{i,j} b_{i,j})_{i,j=1}^{n}$? The definition becomes crystal clear once we look at a matrix as a tool to describe linear transformations between vector spaces, as the elements of the matrix describe the images of basis vectors. In this context, multiplication of matrices is just the composition of linear transformations.

Instead of just putting out the definition and telling you how to use it, I want you to *understand* why it is defined that way. In the next chapters, we are going to learn every nook and cranny of matrix multiplication.

## 1.2.6   The existence of bases

At this point, you might ask the question: for a given vector space, are we guaranteed to find a basis? Without such a guarantee, the previous setup might be wasted. (As there might not be a basis to work with.)

Fortunately, this is not the case. As the proof is extremely difficult, we will not show this, but this is so important that we should at least state the theorem. If you are interested in how this can be done, I included a proof sketch. Feel free to skip this, as it is not going to be essential for our purposes.

**Theorem 1.2.5** *Every vector space has a basis.*

*Proof.* (Sketch.) The proof of this uses an advanced technique called *transfinite induction*, which is way beyond our scope. (Check out *Naive Set Theory* by Paul Halmos for details.) Instead of being precise, let's just focus on building intuition about how to construct a basis for any vector space.

For our vector space $V$, we will build a basis one by one. Given any non-null vector $v_1$, if $\text{span}(S_1) \neq V$, the set $S_1 = \{v_1\}$ is not yet a basis. Thus, we can find a vector $v_2 \in V \setminus \text{span}(S_1)$ so that $S_2 := S_1 \cup \{v_2\}$ is still linearly independent.

Is $S_2$ a basis? If not, we can continue the process. In case the process stops in finitely many steps, we are done. However, this is not guaranteed. Think about $\mathbb{R}[x]$, the vector space of polynomials, which is not finite-dimensional, as we saw in *Section 1.2.4*.

This is where we need to employ some set-theoretical heavy machinery (which we don't have).

If the process doesn't stop, we need to find a set $S_{\aleph_0}$ that contains all $S_i$ as a subset. (Finding this $S_{\aleph_0}$ set is the tricky part.) Is $S_{\aleph_0}$ a basis? If not, we continue the process.

This is difficult to show, but the process eventually stops, and we can't add any more vectors to our linearly independent vector set without destroying the independence property. When this happens, we have found a *maximal linearly independent* set — that is, a basis.

For finite dimensional vector spaces, the above process is easy to describe. In fact, one of the pillars of linear algebra is the so-called Gram-Schmidt process, used to explicitly construct special bases for vector spaces. As several quintessential results rely on this, we are going to study it in detail during the next chapters.

## 1.2.7 Subspaces

Before we get our hands dirty with vectors in Python, there is one more subject we need to talk about, one that will come in handy when talking about linear transformations. (But again, linear transformations are at the heart of machine learning. Everything we learn is to get to know them better.) For a given vector space $V$, we are often interested in one of its subsets that is a vector space in its entirety. This is described by the concept of *subspaces*.

> **Definition 1.2.3 (Subspaces)**
>
> Let $V$ be a vector space. The set $U \subseteq V$ is a *subspace* of $V$ if it is closed under addition and scalar multiplication.
>
> $U$ is a *proper subspace* if it is a subspace and $U \subset V$.

By definition, subspaces are vector spaces themselves, so we can define their dimension as well. There are at least two subspaces of each vector space: itself and $\{0\}$. These are called *trivial* subspaces. Besides those, the span of a set of vectors is always a subspace. One such example is illustrated in *Figure 1.5*.

One of the most important aspects of subspaces is that we can use them to create more subspaces. This notion is made precise below.

---

**Definition 1.2.4 (Direct sum of subspaces)**

Let $V$ be a vector space and $U_1, U_2$ be two of its subspaces. The *direct sum* of $U_1$ and $U_2$ is defined by

$$U_1 + U_2 = \{u_1 + u_2 \,:\, u_1 \in U_1, u_2 \in U_2\}.$$

---

You can easily verify that $U_1 + U_2$ is a subspace indeed, moreover $U_1 + U_2 = \mathrm{span}(U_1 \cup U_2)$. Subspaces and their direct sum play an essential role in several topics, such as matrix decompositions. For example, we'll see later that many of them are equivalent to decomposing a linear space into a sum of vector spaces.

The ability to select a basis whose subsets span certain given subspaces often comes in handy. This is formalized by the next result.

---

**Theorem 1.2.6** *Let $V$ be a vector space and $U_1, U_2$ be two of its subspaces such that $U_1 + U_2 = V$. Moreover, let $\{p_1, \ldots, p_k\} \subseteq U_1$ be a basis of $U_1$ and $\{q_1, \ldots, q_l\} \subseteq U_2$ be a basis of $U_2$. Then the union*

$$\{p_1, \ldots, p_k\} \cup \{q_1, \ldots, q_l\}$$

*is a basis in $V$.*

---

*Proof.* This follows directly from the direct sum's definition. If $V = U_1 + U_2$, then any $x \in V$ can be written in the form $x = a + b$, where $a \in U_1$ and $b \in U_2$.

In turn, since $p_1, \ldots, p_k$ form a basis in $U_1$ and $q_1, \ldots, q_l$ form a basis in $U_2$, the vectors $a$ and $b$ can be written as

$$a = \sum_{i=1}^{k} a_i p_i, \quad b = \sum_{i=1}^{l} b_i q_i.$$

Thus, any $x$ takes the form

$$x = \sum_{i=1}^{k} a_i p_i + \sum_{i=1}^{l} b_i q_i,$$

which is the definition of the basis.

We are barely scratching the surface. Bases are essential, but they only provide the skeleton for the vector spaces encountered in practice. To properly represent and manipulate data, we need to build a geometric structure around this skeleton. How can we measure the "distance" between two measurements? What about their similarity?

Besides all that, there is an even more crucial question: how on earth will we represent vectors inside a computer? In the next section, we will take a look at the data structures of Python, laying the foundation for the data manipulations and transformations we'll do later.

# 1.3 Vectors in practice

So far, we have mostly talked about the theory of vectors and vector spaces. However, our ultimate goal is to build computational models for discovering and analyzing patterns in data. To put theory into practice, we will take a look at how vectors are represented in computations.

In computer science, there is a stark contrast between how we think about mathematical structures and how we represent them inside a computer. Until this point, our goal was to develop a mathematical framework that enables us to reason about the structure of data and its transformations. We want a language that is

- expressive,
- easy to speak,
- as compact as possible.

However, our goals change when we aim to do computations instead of pure logical reasoning. We want implementations that are

- easy to work with,
- memory-efficient,
- fast to access, manipulate and transform.

These are often contradicting requirements, and particular situations might prefer one over the other. For instance, if we have plenty of memory but want to perform lots of computations, we can sacrifice size for speed. Because of all the potential use-cases, there are multiple formats to represent the same mathematical concepts. These are called *data structures*.

Different programming languages implement vectors differently. Because Python is ubiquitous in data science and machine learning, it'll be our language of choice. In this chapter, we are going to study all the possible data structures in Python to see which one is suitable to represent vectors for high performance computations.

# 1.3.1  Tuples

In standard Python, there are (at least) two built-in data structures that can be used to represent vectors: *tuples* and *lists*. Let's start with tuples! They can be simply defined by enumerating their elements between two parentheses, separating them with commas.

```
v_tuple = (1, 3.5, -2.71, "a string", 42)
v_tuple
```

```
(1, 3.5, -2.71, 'a string', 42)
```

```
type(v_tuple)
```

```
tuple
```

A single tuple can hold elements of various types. Even though we'll exclusively deal with floats in computational linear algebra, this property is extremely useful for general-purpose programming.

We can access the elements of a tuple by indexing. Just like in several other programming languages, indexing starts from zero. This is in stark contrast with mathematics, where we often start indexing from one. Accordingly, in most languages designed for scientific computing, such as Fortran, Matlab, or Julia, indexing starts from one.

(Don't tell this to anybody else, but indexing from zero used to drive me crazy. I am a mathematician by training.)

```
v_tuple[0]
```

```
1
```

The size of a tuple can be accessed by calling the built-in len function.

```
len(v_tuple)
```

```
5
```

Besides indexing, we can also access multiple elements by *slicing*.

```
v_tuple[1:4]
```

```
(3.5, -2.71, 'a string')
```

Slicing works by specifying the first and last elements with an optional step size, using the syntax
`object[first:last:step]`.

Tuples are rather inflexible, as you cannot change their components. Attempting to do so results in a `TypeError`,
Python's standard way of telling you that the object does not support the method you are trying to call. (In
our case, item assignment.)

```
v_tuple[0] = 2
```

```
---------------------------------------------------------------------
TypeError                               Traceback (most recent call last)
Cell In[22], line 1
----> 1 v_tuple[0] = 2

TypeError: 'tuple' object does not support item assignment
```

Besides that, extending the tuple with additional elements is also not supported. As we cannot change the state
of a `tuple` object in any way after it has been instantiated, they are *immutable*. Depending on the use-case,
immutability can be an advantage and a disadvantage as well. Immutable objects eliminate accidental changes,
but each operation requires the creation of a new object, resulting in a computational overhead. Thus, tuples
are not going to be optimal to represent large amounts of data in complex computations.

This issue is solved by *lists*. Let's take a look at them, and the new problems they introduce!

## 1.3.2  Lists

Lists are the workhorses of Python. In contrast with tuples, lists are extremely flexible and easy to use, albeit
this comes at the cost of runtime. Similarly to tuples, a `list` object can be created by enumerating its objects
between square brackets, separated by commas.

```
v_list = [1, 3.5, -2.71, "qwerty"]
type(v_list)
```

```
list
```

Just like tuples, accessing the elements of a list is done by indexing or slicing. We can do all kinds of operations on a list: overwrite its elements, append items, or even remove others.

```
v_list[0] = "this is a string"
v_list
```

```
['this is a string', 3.5, -2.71, 'qwerty']
```

This example illustrates that lists can hold elements of various types as well. Adding and removing elements can be done with methods like append, push, pop, and remove.

Before trying that, let's quickly take note of the memory address of our example list, accessed by calling the id function.

```
v_list_addr = id(v_list)
v_list_addr
```

```
126433407319488
```

This number simply refers to an address in my computer's memory, where the v_list object is located. Quite literally, as this book is compiled on my personal computer.

Now, we are going to perform a few simple operations on our list and show that the memory address doesn't change. Thus, no new object is created.

```
v_list.append([42])     # adding the list [42] to the end of our list
v_list
```

```
['this is a string', 3.5, -2.71, 'qwerty', [42]]
```

```
id(v_list) == v_list_addr     # adding elements doesn't create any new objects
```

```
True
```

```
v_list.pop(1)     # removing the element at the index "1"
v_list
```

```
['this is a string', -2.71, 'qwerty', [42]]
```

```
id(v_list) == v_list_addr    # removing elements still doesn't create any new objects
```

```
True
```

Unfortunately, adding lists together achieves a result that is completely different from our expectations.

```
[1, 2, 3] + [4, 5, 6]
```

```
[1, 2, 3, 4, 5, 6]
```

Instead of adding the corresponding elements together, like we want vectors to behave, the lists are concatenated. This feature is handy when writing general-purpose applications. However, this is not well-suited for scientific computations. "Scalar multiplication" also has strange results.

```
3*[1, 2, 3]
```

```
[1, 2, 3, 1, 2, 3, 1, 2, 3]
```

Multiplying a list with an integer repeats the list by the specified number of times. Given the behavior of the + operator on lists, this seems logical as multiplication with an integer is repeated addition:

$$a \cdot b = \underbrace{b + \cdots + b}_{a \text{ times}}.$$

Overall, lists can do much more than we need to represent vectors. Although we potentially want to change elements of our vectors, we don't need to add or remove elements from them, and we also don't need to store objects other than floats. Can we sacrifice these extra features and obtain an implementation that's suitable for our purposes yet has lightning-fast computational performance? Yes. Enter NumPy arrays.

## 1.3.3 NumPy arrays

Even though Python's built-in data structures are amazing, they are optimized for ease of use, not for scientific computation. This problem was realized early on in the language's development and was addressed by the NumPy library.

One of the main selling points of Python is how fast and straightforward it is to write code, even for complex tasks. This comes at the price of speed. However, in machine learning, speed is crucial for us. When training a neural network, a small set of operations are repeated millions of times. Even a small percentage of improvement in performance can save hours, days, or even weeks in the case of extremely large models.

The C language is at the other end of the spectrum. While C code is hard to write, it executes blazingly fast when done correctly. As Python is written in C, a tried and true method for achieving fast performance is to call functions written in C from Python. In a nutshell, this is what NumPy provides: C arrays and operations, all in Python.

To get a glimpse into the deep underlying issues with Python's built-in data structures, we should put numbers and arrays under our magnifying glass. Inside a computer's memory, objects are represented as fixed-length 0-1 sequences. Each component is called a *bit*. Bits are usually grouped into 8-, 16-, 32-, 64-, or even 128 sized chunks. Depending on what we want to represent, identical sequences can mean different things. For instance, the 8-bit sequence *00100110* can represent the integer 38 or the ASCII character "&".

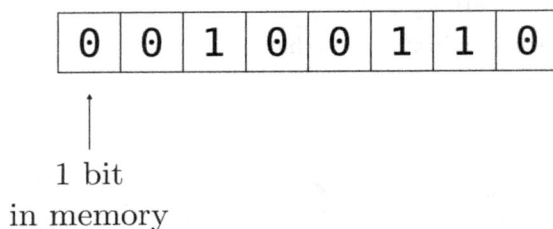

Figure 1.6: An 8-bit object in memory

By specifying the *data type*, we can decode binary objects. 32-bit integers are called `int32` types, 64-bit floats are `float64`, and so on.

Since a single bit contains very little information, memory is addressed by dividing it into 32- or 64-bit sized chunks and numbering them consecutively. This address is a *hexadecimal* number, starting from 0. (For simplicity, let's assume that the memory is addressed by 64 bits. This is customary in modern computers.)

A natural way to store a sequence of related objects (with matching data type) is to place them next to each other in the memory. This data structure is called an *array*.

array of `int64` objects

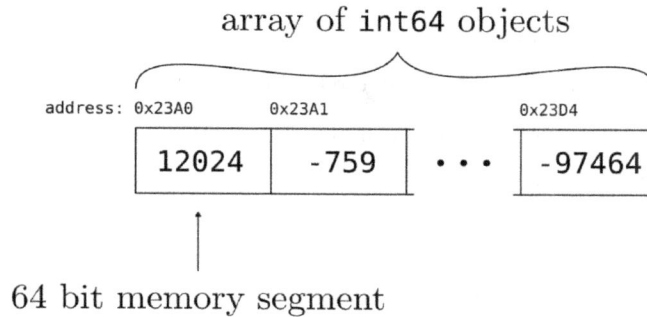

64 bit memory segment

*Figure 1.7: An array of int64 objects*

By storing the memory address of the first object, say `0x23A0`, we can instantly retrieve the *k*-th element by accessing the memory at `0x23A0 + k`.

We call this the static array or often the C array because this is how it is done in the magnificent C language. Although this implementation of arrays is lightning fast, it is relatively inflexible. First, you can only store objects of a single type. Second, you have to know the size of your array in advance, as you cannot use memory addresses that overextend the pre-allocated part. Thus, before you start working with your array, you have to allocate memory for it. (That is, reserve space so that other programs won't overwrite it.)

However, in Python, you can store arbitrarily large and different objects in the same list, with the option of removing and adding elements to it.

```
l = [2**142 + 1, "a string"]
l.append(lambda x: x)
l
```

```
[5575186299632655785383929568162090376495105,
 'a string',
 <function __main__.<lambda>(x)>]
```

In the example above, `l[0]` is an integer so large that it doesn't fit into 128 bits. Also, there are all kinds of objects in our list, including a function. How is this possible?

Python's `list` provides a flexible data structure by

1. Overallocating the memory, and
2. Keeping memory addresses to the objects in the list instead of the objects themselves.

(At least in the most widespread CPython implementation (`https://docs.python.org/3/faq/design.html#how-are-lists-implemented-in-cpython`).)

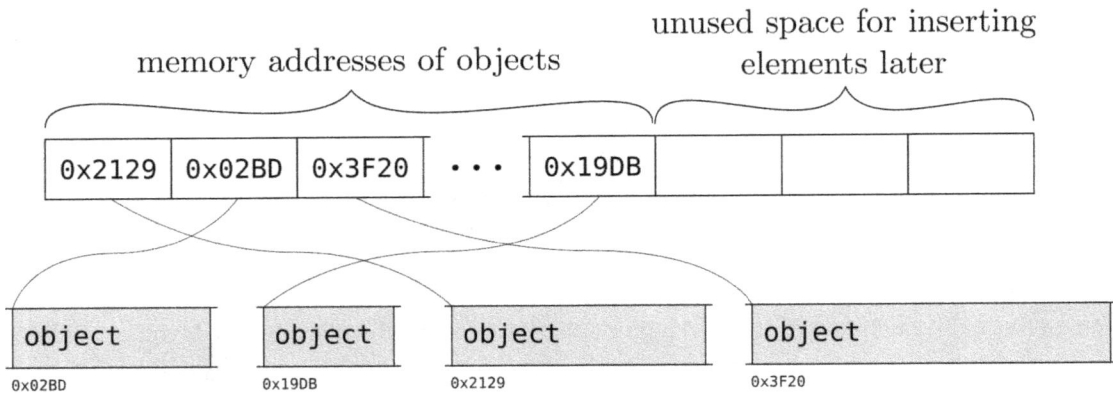

*Figure 1.8: CPython implementation of lists*

By checking the memory addresses of each object in our list 1, we can see that they are all over the memory.

```
[id(x) for x in l]
```

```
[126433412959232, 126433407528240, 126433410174944]
```

Due to the overallocation, deletion or insertion can always be done simply by shifting the remaining elements. Since the list stores the memory address of its elements, all types of objects can be stored within a single structure.

However, this comes at a cost. Because the objects are not contiguous in memory, we lose locality of reference (`https://en.wikipedia.org/wiki/Locality_of_reference`), meaning that since we frequently access distant locations of the memory, our reads are much slower. Thus, looping over a Python list is not efficient.

So, NumPy arrays are essentially the good old C arrays in Python, with the user-friendly interface of Python lists. (If you have ever worked with C, you know how big of a blessing this is.) Let's see how to work with them!

First, we import the numpy library. (To save on the characters, it is customary to import it as np.)

```python
import numpy as np
```

The main data structure is np.ndarray, short for *n*-dimensional array. We can use the np.array function to create NumPy arrays from standard Python containers or initialize from scratch. (Yes, I know. This is confusing, but you'll get used to it. Just take a mental note that np.ndarray is the class, and np.array is the function you use to create NumPy arrays from Python objects.)

```python
X = np.array([87.7, 4.5, -4.1, 42.1414, -3.14, 2.001])   # creating a NumPy array from a
Python list
X
```

```
array([87.7   ,  4.5   , -4.1   , 42.1414, -3.14  ,  2.001 ])
```

```python
np.ones(shape=7)    # initializing a NumPy array from scratch using ones
```

```
array([1., 1., 1., 1., 1., 1., 1.])
```

```python
np.zeros(shape=5)    # initializing a NumPy array from scratch using zeros
```

```
array([0., 0., 0., 0., 0.])
```

We can even initialize NumPy arrays using random numbers.

```python
np.random.rand(10)
```

```
array([0.92428404, 0.37719596, 0.92071695, 0.56905245, 0.12024811,
       0.02868856, 0.53215047, 0.51749348, 0.21022765, 0.96749756])
```

Most importantly, when we have a given array, we can initialize another one with the same dimensions using the np.zeros_like, np.ones_like, and np.empty_like functions.

```
np.zeros_like(X)
```

```
array([0., 0., 0., 0., 0., 0.])
```

Just like Python lists, NumPy arrays support item assignments and slicing.

```
X[0] = 1545.215
X
```

```
array([1545.215 ,    4.5  ,   -4.1  ,   42.1414,   -3.14 ,    2.001 ])
```

```
X[1:4]
```

```
array([ 4.5   ,  -4.1   ,  42.1414])
```

However, as expected, you can only store a single data type within each ndarray. When trying to assign a string as the first element, we get an error message.

```
X[0] = "str"
```

```
---------------------------------------------------------------
ValueError                                Traceback (most recent call last)
Cell In[48], line 1
----> 1 X[0] = "str"

ValueError: could not convert string to float: 'str'
```

As you might have guessed, every ndarray has a data type attribute that can be accessed at ndarray.dtype. If a conversion can be made between the value to be assigned and the data type, it is automatically performed, making the item assignment successful.

```
X.dtype
```

```
dtype('float64')
```

```
val = 23
type(val)
```

```
int
```

```
X[0] = val
X
```

```
array([23.    ,    4.5  ,   -4.1  , 42.1414, -3.14  ,   2.001 ])
```

NumPy arrays are iterable, just like other container types in Python.

```
for x in X:
    print(x)
```

```
23.0
4.5
-4.1
42.1414
-3.14
2.001
```

Are these suitable to represent vectors? Yes. We'll see why!

## 1.3.4   NumPy arrays as vectors

Let's talk about vectors once more. From now on, we are going to use NumPy ndarray-s to model vectors.

```
v_1 = np.array([-4.0, 1.0, 2.3])
v_2 = np.array([-8.3, -9.6, -7.7])
```

The addition and scalar multiplication operations are supported by default and perform as expected.

```
v_1 + v_2     # adding v_1 and v_2 together as vectors
```

```
array([-12.3,  -8.6,  -5.4])
```

```
10.0*v_1     # multiplying v_1 with a scalar
```

```
array([-40.,  10.,  23.])
```

```
v_1 * v_2     # the elementwise product of v_1 and v_2
```

```
array([ 33.2 ,  -9.6 , -17.71])
```

```
np.zeros(shape=3) + 1
```

```
array([1., 1., 1.])
```

Because of the dynamic typing of Python, we can (often) plug NumPy arrays into functions intended for scalars.

```
def f(x):
    return 3*x**2 - x**4
f(v_1)
```

```
array([-208.    ,    2.    ,  -12.1141])
```

So far, NumPy arrays satisfy almost everything we require to represent vectors. There is only one box to be checked: performance. To investigate this, we measure the execution time with Python's built-in `timeit` tool.

In its first argument, `timeit` (https://docs.python.org/3/library/timeit.html) takes a function to be executed and timed. Instead of passing a function object, it also accepts executable statements as a string. Since function calls have a significant computational overhead in Python, we are passing code rather than a function object in order to be more precise with the time measurements.

Below, we compare adding together two NumPy arrays vs. Python lists containing a thousand zeros.

```
from timeit import timeit
```

```
n_runs = 100000
size = 1000
```

```
t_add_builtin = timeit(
    "[x + y for x, y in zip(v_1, v_2)]",
    setup=f"size={size}; v_1 = [0 for _ in range(size)]; v_2 = [0 for _ in range(size)]",
    number=n_runs
)

t_add_numpy = timeit(
    "v_1 + v_2",
    setup=f"import numpy as np; size={size}; v_1 = np.zeros(shape=size);
    v_2 = np.zeros(shape=size)",
    number=n_runs
)


print(f"Built-in addition:      \t{t_add_builtin} s")
print(f"NumPy addition:         \t{t_add_numpy} s")
print(f"Performance improvement: \t{t_add_builtin/t_add_numpy:.3f} times faster")
```

```
Built-in addition:          3.3522969299992837 s
NumPy addition:             0.09616518099937821 s
Performance improvement:    34.860 times faster
```

NumPy arrays are much-much faster. This is because they are

- contiguous in memory,
- homogeneous in type,
- with operations implemented in C.

This is just the tip of the iceberg. We have only seen a small part of it, but NumPy provides much more than a fast data structure. As we progress in the book, we'll slowly dig deeper and deeper, eventually discovering the vast array of functionalities it provides.

## 1.3.5   Is NumPy really faster than Python?

NumPy is designed to be faster than vanilla Python. Is this really the case? Not all the time. If you use it wrong, it might even hurt performance! To know when it is beneficial to use NumPy, we will look at why exactly it is faster in practice.

To simplify the investigation, our toy problem will be random number generation. Suppose that we need just a single random number. Should we use NumPy? Let's test it! We are going to compare it with the built-in random number generator by running both ten million times, measuring the execution time.

```python
from numpy.random import random as random_np
from random import random as random_py


n_runs = 10000000
t_builtin = timeit(random_py, number=n_runs)
t_numpy = timeit(random_np, number=n_runs)

print(f"Built-in random:\t{t_builtin} s")
print(f"NumPy random:   \t{t_numpy} s")
```

```
Built-in random:        0.47474874800172984 s
NumPy random:           5.1664929229991685 s
```

For generating a single random number, NumPy is significantly slower. Why is this the case? What if we need an array instead of a single number? Will this also be slower?

This time, let's generate a list/array of a thousand elements.

```python
size = 1000
n_runs = 10000

t_builtin_list = timeit(
    "[random_py() for _ in range(size)]",
    setup=f"from random import random as random_py; size={size}",
    number=n_runs
)

t_numpy_array = timeit(
    "random_np(size)",
    setup=f"from numpy.random import random as random_np; size={size}",
    number=n_runs
)

print(f"Built-in random with lists:\t{t_builtin_list}s")
print(f"NumPy random with arrays:  \t{t_numpy_array}s")
```

```
Built-in random with lists:     0.5773125300001993s
NumPy random with arrays:       0.08449692800058983s
```

(Again, I don't want to wrap the timed expressions in lambdas since function calls have an overhead in Python. I want to be as precise as possible, so I pass them as strings to the `timeit` function.)

Things are looking much different now. When generating an array of random numbers, NumPy wins hands down.

There are some curious things about this result as well. First, we generated a single random number 10000000 times. Second, we generated an array of 1000 random numbers 10000 times. In both cases, we have 10000000 random numbers in the end. Using the built-in method, it took ~2x time when we put them in a list. However, with NumPy, we see a ~30x speedup compared to itself when working with arrays! (The actual numbers might be different on your computer.)

To see what happens behind the scenes, we are going to profile the code using cProfiler (`https://docs.pytho n.org/3/library/profile.html`). With this, we'll see exactly how many times a given function was called and how much time we spent inside it.

Let's take a look at the built-in function first. In the following function, we create 10000000 random numbers, just as before.

```python
def builtin_random_single(n_runs):
    for _ in range(n_runs):
        random_py()
```

From Jupyter Notebooks, where this book is written, cProfiler can be called with the magic command %prun.

```python
n_runs = 10000000

%prun builtin_random_single(n_runs)
```

```
10000558 function calls (10000539 primitive calls) in 2.082 seconds

   Ordered by: internal time

   ncalls  tottime  percall  cumtime  percall filename:lineno(function)
        1    0.937    0.937    1.671    1.671 2471337341.py:1(builtin_random_single)
 10000000    0.911    0.000    0.911    0.000 {method 'random' of '_random.Random'
objects}
      4/0    0.213    0.053    0.000          {method 'poll' of 'select.epoll' objects}
       10    0.009    0.001    0.016    0.002 socket.py:626(send)
        2    0.009    0.004    0.015    0.008 {method '__exit__' of 'sqlite3.Connection'
objects}
```

There are two important columns here for our purposes. `ncalls` shows how many times a function was called, while `tottime` is the total time spent in a function, excluding time spent in subfunctions.

The built-in function `random.random()` was called 10000000 times as expected. Take note of the total time spent in the function. (I can't give you an exact figure here, as it depends on the machine this book is built on.) What about the NumPy version? The results are surprising.

```
def numpy_random_single(n_runs):
    for _ in range(n_runs):
        random_np()

%prun numpy_random_single(n_runs)
```

```
448 function calls (444 primitive calls) in 7.203 seconds

   Ordered by: internal time

   ncalls  tottime  percall  cumtime  percall filename:lineno(function)
        1    7.029    7.029    7.029    7.029 2015715881.py:1(numpy_random_single)
        2    0.136    0.068    0.136    0.068 {method 'poll' of 'select.epoll' objects}
        2    0.015    0.007    0.015    0.007 {method '__exit__' of 'sqlite3.Connection'
   objects}
        1    0.011    0.011    0.011    0.011 {method 'execute' of 'sqlite3.Connection'
   objects}
        3    0.010    0.003    7.339    2.446 base_events.py:1910(_run_once)
        7    0.000    0.000    0.000    0.000 socket.py:626(send)
        1    0.000    0.000    0.000    0.000 {method 'disable' of '_lsprof.Profiler'
   objects}
        1    0.000    0.000    0.026    0.026 history.py:833(_writeout_input_cache)
        1    0.000    0.000    0.000    0.000 inspect.py:3102(_bind)
    88/84    0.000    0.000    0.000    0.000 {built-in method builtins.isinstance}
```

Similarly, as before, the `numpy.random.random()` function was indeed called 10000000 times, as expected. Yet, the script spent significantly more time in this function than in the Python built-in random before. Thus, it is more costly per call.

When we start working with large arrays and lists, things change dramatically. Next, we generate a list/array of 1000 random numbers, while measuring the execution time.

```
def numpy_random_single(n_runs):
    for _ in range(n_runs):
        random_np()
%prun numpy_random_single(n_runs)
```

```
448 function calls (444 primitive calls) in 7.203 seconds

   Ordered by: internal time

   ncalls  tottime  percall  cumtime  percall filename:lineno(function)
        1    7.029    7.029    7.029    7.029 2015715881.py:1(numpy_random_single)
        2    0.136    0.068    0.136    0.068 {method 'poll' of 'select.epoll' objects}
        2    0.015    0.007    0.015    0.007 {method '__exit__' of 'sqlite3.Connection'
   objects}
        1    0.011    0.011    0.011    0.011 {method 'execute' of 'sqlite3.Connection'
   objects}
        3    0.010    0.003    7.339    2.446 base_events.py:1910(_run_once)
        7    0.000    0.000    0.000    0.000 socket.py:626(send)
        1    0.000    0.000    0.000    0.000 {method 'disable' of '_lsprof.Profiler'
   objects}
        1    0.000    0.000    0.026    0.026 history.py:833(_writeout_input_cache)
        1    0.000    0.000    0.000    0.000 inspect.py:3102(_bind)
    88/84    0.000    0.000    0.000    0.000 {built-in method builtins.isinstance}
```

As we see, about 60% of the time was spent on the list comprehensions. (Note that tottime doesn't count subfunction calls like calls to random.random() here.)

Now we are ready to see why NumPy is faster when used right.

```python
def numpy_random_array(size, n_runs):
    for _ in range(n_runs):
        random_np(size)
%prun numpy_random_array(size, n_runs)
```

```
149 function calls (148 primitive calls) in 0.132 seconds

   Ordered by: internal time

   ncalls  tottime  percall  cumtime  percall filename:lineno(function)
        1    0.122    0.122    0.122    0.122 1681905588.py:1(numpy_random_array)
        2    0.009    0.004    0.009    0.004 {method '__exit__' of 'sqlite3.Connection'
   objects}
      2/1    0.000    0.000    0.122    0.122 {built-in method builtins.exec}
```

With each of the 10000 function calls, we get a numpy.ndarray of 1000 random numbers. The reason why NumPy is fast when used right is that its arrays are extremely efficient to work with. They are like C arrays instead of Python lists.

As we have seen, there are two significant differences between them.

- Python lists are dynamic, so for instance, you can append and remove elements. NumPy arrays have fixed lengths, so you cannot add or delete without creating a new one.
- Python lists can hold several data types simultaneously, while a NumPy array can only contain one.

So, NumPy arrays are less flexible but significantly more performant. When this additional flexibility is not needed, NumPy outperforms Python.

To see precisely at which size does NumPy overtakes Python in random number generation, we can compare the two by measuring the execution times for several sizes.

```python
sizes = list(range(1, 100))

runtime_builtin = [
    timeit(
        "[random_py() for _ in range(size)]",
        setup=f"from random import random as random_py; size={size}",
        number=100000
    )
    for size in sizes
]


runtime_numpy = [
    timeit(
        "random_np(size)",
        setup=f"from numpy.random import random as random_np; size={size}",
        number=100000
    )
    for size in sizes
]


import matplotlib.pyplot as plt


with plt.style.context("seaborn-v0_8"):
    plt.figure(figsize=(10, 5))
    plt.plot(sizes, runtime_builtin, label="built-in")
    plt.plot(sizes, runtime_numpy, label="NumPy")
    plt.xlabel("array size")
    plt.ylabel("time (seconds)")
    plt.title("Runtime of random array generation")
    plt.legend()
    plt.show()
```

Runtime of random array generation

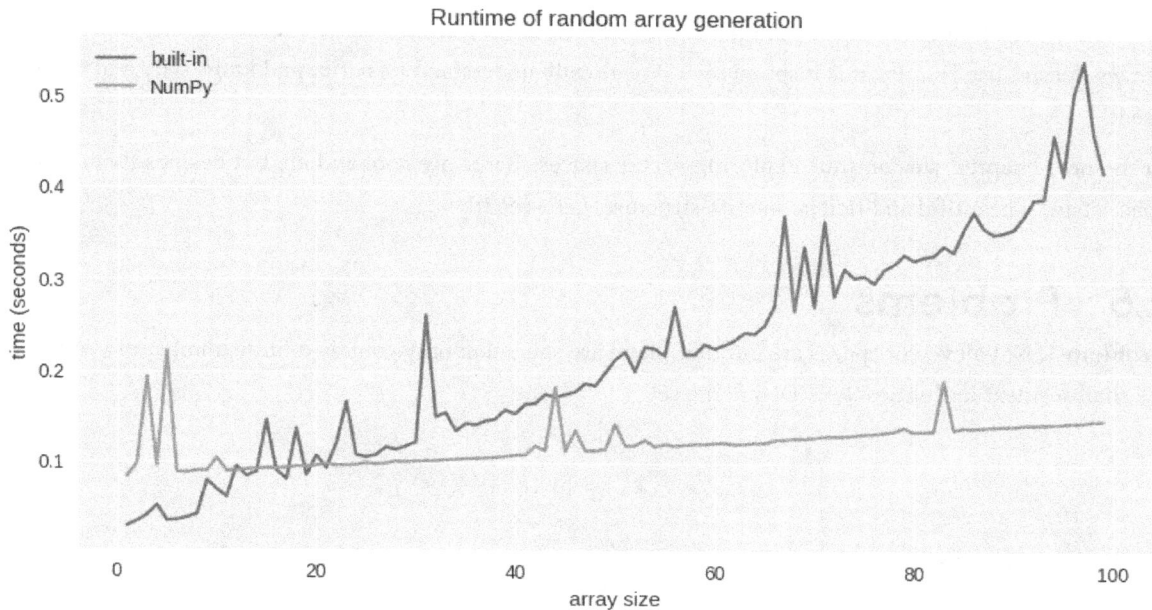

*Figure 1.9: Runtime of random array generation*

Around 20, NumPy starts to beat Python in performance. Of course, this number might be different for other operations like calculating the sine or adding numbers together, but the tendency will be the same. Python will slightly outperform NumPy for small input sizes, but NumPy wins by a large margin as the size grows.

# 1.4 Summary

In this chapter, we have learned what vectors are and why we must use them in data science and machine learning. Vectors are not just a bunch of numbers bundled together but a mathematical structure that allows us to reason about data more effectively, both in theory and in practice. Contrary to popular belief, vectors are vectors not because they have direction and magnitude but because you can add them together.

This is formalized by the concept of *vector spaces*, providing the mathematical framework for our studies. Vector spaces are best described by *bases*, that is, minimal and linearly independent generating sets. Understanding vector spaces and their bases will pay enormous dividends when we study linear transformations, the most important building block of predictive models.

Besides the leap of abstraction provided by vectors, we reap significant benefits in practice by vectorizing our code, compressing complex logic into one-liners such as data scaling:

```
X_scaled = (X - X.mean(axis=0)) / X.std(axis=0)
```

Besides the conceptual jump from scalars to vectors and matrices, efficient data processing is made possible by NumPy (short for Numerical Python), the number one library in the machine learning toolkit. If a tensor library doesn't use NumPy, it is inspired by it. We already understand its basics and know why and when to use it.

In the next chapter, we continue exploring vector spaces. Bases are cool and all, but besides them, vector spaces have a beautiful and rich geometric structure. Let's see it!

# 1.5   Problems

**Problem 1.** Not all vector spaces are infinite. There are some that only contain a finite number of vectors, as we shall see next in this problem. Define the set

$$\mathbb{Z}_2 := \{0, 1\},$$

where the operations $+, \cdot$ are defined by the rules

$$0 + 0 = 0$$
$$0 + 1 = 1$$
$$1 + 0 = 1$$
$$1 + 1 = 0$$

and

$$0 \cdot 0 = 0$$
$$0 \cdot 1 = 0$$
$$1 \cdot 0 = 0$$
$$1 \cdot 1 = 1.$$

This is called binary (or modulo-2) arithmetic.

*(a)* Show that $(\mathbb{Z}_2, \mathbb{Z}_2, +, \cdot)$ is a vector space.

*(b)* Show that $(\mathbb{Z}_2^n, \mathbb{Z}_2, +, \cdot)$ is also a vector space, where $\mathbb{Z}_2^n$ is the n-fold Cartesian product

$$\mathbb{Z}_2^n = \underbrace{\mathbb{Z}_2 \times \cdots \times \mathbb{Z}_2}_{n \text{ times}},$$

and the addition and scalar multiplication are defined elementwise:

$$x + y = (x_1 + y_1, \ldots, x_n + y_n), \quad x, y \in \mathbb{Z}_2^n,$$
$$cx = (cx_1, \ldots, cx_n), \quad c \in \mathbb{Z}_2.$$

**Problem 2.** Are the following vector sets linearly independent?

*(a)* $S_1 = \{(1, 0, 0), (1, 1, 0), (1, 1, 1)\} \subseteq \mathbb{R}^3$

*(b)* $S_2 = \{(1, 1, 1), (1, 2, 4), (1, 3, 9)\} \subseteq \mathbb{R}^3$

*(c)* $S_3 = \{(1, 1, 1), (1, 1, -1), (1, -1, -1)\} \subseteq \mathbb{R}^3$

*(d)* $S_4 = \{(\pi, e), (-42, 13/6), (\pi^3, -2)\} \subseteq \mathbb{R}^2$

**Problem 3.** Let $V$ be a finite $n$-dimensional vector space and let $S = \{v_1, \ldots, v_m\}$ be a linearly independent set of vectors, $m < n$. Show that there is a basis set $B$ such that $S \subset B$.

**Problem 4.** Let $V$ be a vector space and $S = \{v_1, \ldots, v_n\}$ be its basis. Show that every vector $x \in V$ can be uniquely written as a linear combination of vectors in $S$. (That is, if $x = \sum_{i=1}^n \alpha_i v_i = \sum_{i=1}^n \beta_i v_i$, then $\alpha_i = \beta_i$ for all $i = 1, \ldots, n$.)

**Problem 5.** Let $V$ be an arbitrary vector space and $U_1, U_2 \subseteq V$ be two of its subspaces. Show that $U_1 + U_2 = \text{span}(U_1 \cup U_2)$.

Hint: to prove the equality of these two sets, you need to show two things: 1) if $x \in U_1 + U_2$, then $x \in \text{span}(U_1 \cup U_2)$ as well,

2) if $x \in \text{span}(U_1 \cup U_2)$, then $x \in U_1 + U_2$ as well.

**Problem 6.** Consider the vector space of polynomials with real coefficients, defined by

$$\mathbb{R}[x] = \left\{ p(x) = \sum_{i=0}^{n} p_i x^i \ : \ p_i \in \mathbb{R}, n = 0, 1, \dots \right\}.$$

*(a)* Show that

$$x\mathbb{R}[x] := \left\{ p(x) = \sum_{i=1}^{n} p_i x^i \ : \ p_i \in \mathbb{R}, n = 1, 2, \dots \right\}$$

is a proper subspace of $\mathbb{R}[x]$.

*(b)* Show that

$$f \ : \ \mathbb{R}[x] \to x\mathbb{R}[x], \quad p(x) \mapsto xp(x)$$

is a bijective and linear. (A function $f : X \to Y$ is bijective if every $y \in Y$ has exactly one $x \in X$ for which $f(x) = y$. If you are not comfortable with this notion, feel free to revisit this problem after Chapter 9.)

In general, a linear and bijective function $f : U \to V$ between vector spaces is called an *isomorphism*. Given the existence of such a function, we call the vector spaces $U$ and $V$ *isomorphic*, meaning that they have an identical algebraic structure.

Combining *(a)* and *(b)*, we obtain that $\mathbb{R}[X]$ is isomorphic with its proper subspace $x\mathbb{R}[X]$. This is quite an interesting phenomenon: a vector space that is algebraically identical to its proper subspace. (Note that this cannot happen in finite dimensions, such as $\mathbb{R}^n$.)

# Join our community on Discord

Read this book alongside other users, Machine Learning experts, and the author himself.

Ask questions, provide solutions to other readers, chat with the author via Ask Me Anything sessions, and much more.

Scan the QR code or visit the link to join the community.

`https://packt.link/math`

# 2
# The Geometric Structure of Vector Spaces

Let's revisit the Iris dataset introduced in the previous chapter! I want to test your intuition. I plotted the petal widths against the petal lengths while hiding the class labels in *Figure 2.1*:

```python
import matplotlib.pyplot as plt
from sklearn.datasets import load_iris

# Load the iris dataset
iris = load_iris()
data = iris.data

# Extract petal length (3rd column) and petal width (4th column)
petal_length = data[:, 2]
petal_width = data[:, 3]

with plt.style.context("seaborn-v0_8"):
    # Create the scatter plot
    plt.figure(figsize=(7, 7))
    plt.scatter(petal_length, petal_width, color='indigo', alpha=0.8, edgecolor='none',
    s=70, marker='o')
    plt.xlabel('petal length (cm)')
    plt.ylabel('petal width (cm)')
```

```
plt.show()
```

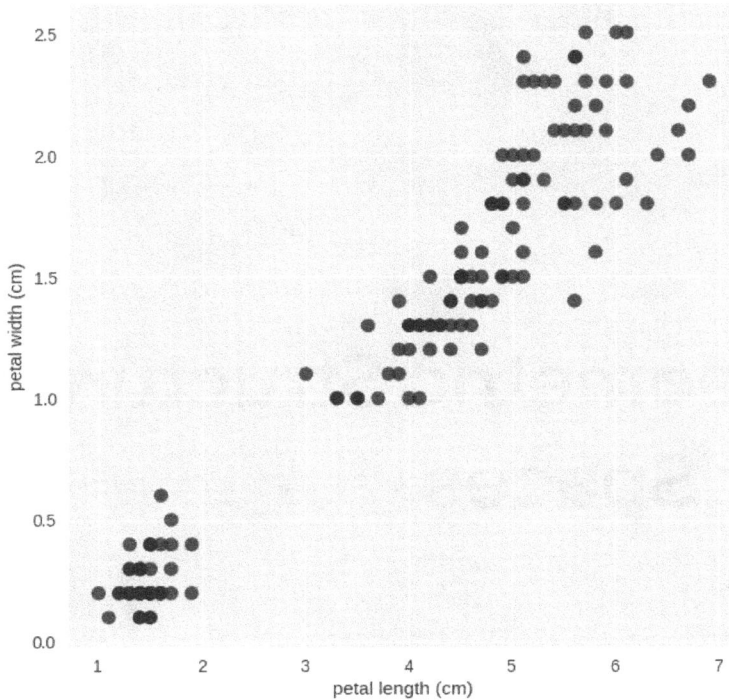

*Figure 2.1: The "petal width" and "petal length" features of the Iris dataset*

Even without knowing any labels, we can intuitively point out that there are probably at least two classes. Can you summarize your reasoning in a single sentence?

There are many valid arguments, but the most prevalent one is that the two clusters are *far* away from each other. As this example illustrates, the concept of distance plays an essential role in machine learning. In this chapter, we will translate the notion of distance into the language of mathematics and put it into the context of vector spaces.

# 2.1 Norms and distances

Previously, we saw that vectors are essentially arrows, starting from the null vector. In addition to their direction, vectors also have *magnitude*. For example, as we have learned in high school mathematics, the magnitude in the Euclidean plane is defined by

$$\|\mathbf{x}\| = \sqrt{x_1^2 + x_2^2}, \quad \mathbf{x} = (x_1, x_2),$$

while we can calculate the distance between $\mathbf{x}$ and $\mathbf{y}$ as

$$d(\mathbf{x}, \mathbf{y}) = \sqrt{(x_1 - y_1)^2 + (x_2 - y_2)^2}.$$

(The function $\|\cdot\|$ simply denotes the magnitude of a vector.)

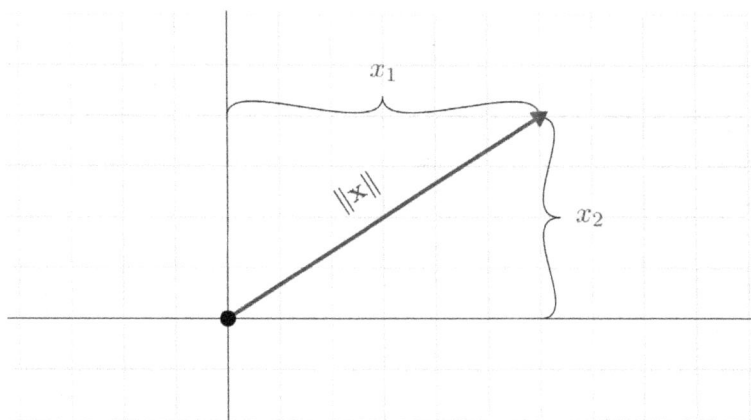

*Figure 2.2: Magnitude in the Euclidean plane*

The magnitude formula $\sqrt{x_1^2 + x_2^2}$ can be simply generalized to higher dimensions by

$$\|\mathbf{x}\| = \sqrt{x_1^2 + \cdots + x_n^2}, \quad \mathbf{x} = (x_1, \dots, x_n) \in \mathbb{R}^n.$$

However, just from looking at this formula, it is not clear why it is defined this way. What does the square root of a sum of squares have to do with distance and magnitude? Behind the scenes, it is just the Pythagorean theorem.

Recall that the Pythagorean theorem states that in right triangles, the squared length of the hypotenuse equals the sum of the squared lengths of the other sides, as illustrated by *Figure 2.3*.

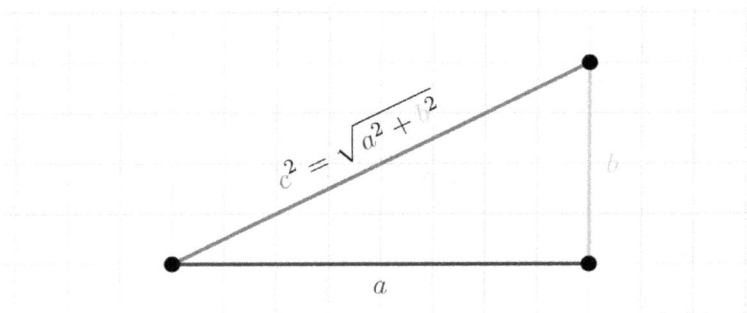

*Figure 2.3: The Pythagorean theorem*

To put this into an algebraic form, it states that $a^2 + b^2 = c^2$, when $c$ is the hypotenuse of the right triangle, and $a$ and $b$ are its two other sides. If we apply this to a two-dimensional vector $\mathbf{x} = (x_1, x_2)$, we can see that the Pythagorean theorem gives its magnitude $\|\mathbf{x}\| = \sqrt{x_1^2 + x_2^2}$ .

This can be generalized to higher dimensions. To see what is happening, we are going to check the three-dimensional case, as illustrated by *Figure 2.4*. Here, we can apply the Pythagorean theorem twice to obtain the magnitude!

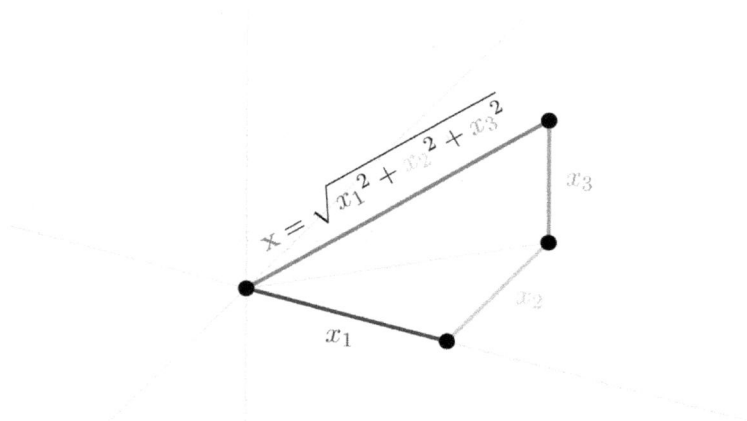

*Figure 2.4: The Pythagorean theorem in three dimensions*

For each vector $\mathbf{x} = (x_1, x_2, x_3)$, we can take a look at the triangle determined by $(0, 0, 0), (x_1, 0, 0)$, and $(x_1, x_2, 0)$ first. The length of the hypotenuse can be calculated by $\sqrt{x_1^2 + x_2^2}$. However, the points $(0, 0, 0), (x_1, x_2, 0)$, and $(x_1, x_2, x_3)$ form a right triangle. Applying the Pythagorean theorem once again, we obtain

$$\|\mathbf{x}\| = \sqrt{x_1^2 + x_2^2 + x_3^2},$$

which is called the *Euclidean norm*. This is exactly what is going on in the general *n*-dimensional case.

The notions of *magnitude* and *distance* are critical in machine learning, as we can use them to determine the similarity between data points, measure and control the complexity of neural networks, and much more.

Is the Pythagorean theorem the only viable way to measure magnitude and distance? Certainly not.

Because Manhattan's street layout is essentially a rectangular grid, its residents are famed for measuring distances in blocks. If something is two blocks to the north and three blocks east, it means that you have to travel two intersections to the north and three to the east to find it. This gives rise to a mathematically perfectly valid notion of measurement called *Manhattan distance*, defined by

$$d(\mathbf{x}, \mathbf{y}) = |x_1 - y_1| + |x_2 - y_2|.$$

When using the Manhattan distance, the shortest path between two points is not unique.

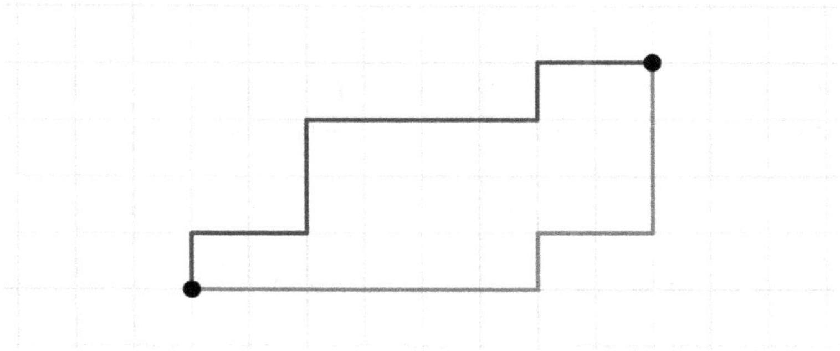

*Figure 2.5: For the Manhattan distance, the shortest path between two points is not unique*

Besides the Euclidean and Manhattan distances, there are several other metrics. Once again, we are going to step away from the concrete examples to take an abstract viewpoint.

If we talk about measurements and metrics in general, what are the properties that we expect from all of them? What makes a measurement *distance*? Essentially, there are three such traits:

- the distance should be nonnegative,

- it should preserve scaling (that is, $d(c\mathbf{x}, c\mathbf{y}) = c\, d(\mathbf{x}, \mathbf{y})$ for all scalars $c$),
- the distance straight from point $\mathbf{x}$ to $\mathbf{y}$ is always equal to or smaller than touching any other point $\mathbf{z}$.

These are formalized by the notion of *norms*.

---

**Definition 2.1.1 (Norms)**

Let $V$ be a vector space. A function $\|\cdot\| : V \to [0, \infty)$ is said to be a **norm** if for all $\mathbf{x}, \mathbf{y} \in V$, the following properties hold:

1. **Positive definiteness**: $\|\mathbf{x}\| \geq 0$ and $\|\mathbf{x}\| = 0$ if and only if $\mathbf{x} = 0$.
2. **Positive homogeneity**: $\|c\mathbf{x}\| = |c|\|\mathbf{x}\|$ for all $c \in \mathbb{R}$.
3. **Triangle inequality**: $\|\mathbf{x} + \mathbf{y}\| \leq \|\mathbf{x}\| + \|\mathbf{y}\|$ for all $\mathbf{x}, \mathbf{y} \in V$.

A vector space equipped with a norm is called a **normed space**.

---

Let's see some examples!

**Example 1.** Let $p \in [1, \infty)$ and define

$$\|\mathbf{x}\|_p = \left( \sum_{i=1}^{n} |x_i|^p \right)^{1/p}, \quad \mathbf{x} = (x_1, \dots, x_n)$$

on $\mathbb{R}^n$. The function $\|\cdot\|_p$ is called the $p$-norm. Showing that $\|\cdot\|_p$ is indeed a norm is a bit technical. Thus, we won't go into the details. (The triangle inequality requires some work, but the other two properties are easy to see.)

We have already seen two special cases: the Euclidean norm ($p = 2$) and the Manhattan norm ($p = 1$). Both of them frequently appear in machine learning. For instance, the familiar mean squared error is just the scaled Euclidean distance between prediction and ground truth:

$$\text{MSE}(\mathbf{y}, \hat{\mathbf{y}}) = \frac{1}{n}\|\mathbf{y} - \hat{\mathbf{y}}\|_2^2 = \frac{1}{n} \sum_{i=1}^{n} (y_i - \hat{y}_i)^2$$

As mentioned before, the 2-norm, along with the 1-norm, is commonly used to control the complexity of models during training. To give a concrete example, suppose that we are fitting a polynomial $f(x) = \sum_{i=0}^{m} q_i x^i$ to the data $\{(x_1, y_1), \dots, (x_n, y_n)\}$. To obtain a model that generalizes well to new data, we prefer our models to be as simple as possible. Thus, instead of using the plain mean squared error, we might consider minimizing the loss:

$$\text{Loss}(\mathbf{y}, \hat{\mathbf{y}}, \mathbf{q}) = \text{MSE}(\mathbf{y}, \hat{\mathbf{y}}) + \lambda\|\mathbf{q}\|_p, \quad \mathbf{q} = (q_0, q_1, \dots, q_m), \quad \lambda \in [0, \infty)$$

where the term $\|\mathbf{q}\|_p$ is responsible for keeping the coefficients of the polynomial $f(x)$ small, while $\lambda$ controls the strength of regularization. Usually, $p$ is either 1 or 2, but other values from $[1, \infty)$ are also valid.

**Example 2.** Let's stay in $\mathbb{R}^n$ for a bit longer! The so-called $\infty$-norm is defined by

$$\|\mathbf{x}\|_\infty = \max\{|x_1|, \dots, |x_n|\}.$$

Showing that $\|\cdot\|_\infty$ is indeed a norm is a simple task and left to you for practice. (This is perhaps one of the most notorious sentences written in mathematical textbooks, but trust me, this is truly easy. Give it a shot! If you don't see it, try the special case $\mathbb{R}^2$.)

This is called the $\infty$-norm, and is strongly related to the $p$-norm that we have just seen. In fact, if we let the value $p$ grow infinitely, $\|\mathbf{x}\|_p$ will be very close to $\|\mathbf{x}\|_\infty$, ultimately reaching it at the limit.

**Remark 2.1.1 (The $\infty$-norm as the limit of $p$-norm)**

If you are already familiar with convergent sequences and limits, you can see that this is called the $\infty$-norm because

$$\lim_{p \to \infty} \|\mathbf{x}\|_p = \|\mathbf{x}\|_\infty.$$

To show this, consider that

$$\lim_{p \to \infty} \|\mathbf{x}\|_p = \lim_{p \to \infty} \left( \sum_{i=1}^n |x_i|^p \right)^{1/p} = \lim_{p \to \infty} \|\mathbf{x}\|_\infty \left( \sum_{i=1}^n \left( \frac{|x_i|}{\|\mathbf{x}\|_\infty} \right)^p \right)^{1/p}.$$

Since $\frac{|x_i|}{\|\mathbf{x}\|_\infty} \leq 1$ by definition,

$$1 \leq \left( \sum_{i=1}^n \left( \frac{|x_i|}{\|\mathbf{x}\|_\infty} \right)^p \right)^{1/p} \leq n^{1/p}$$

holds. Because

$$\lim_{p \to \infty} n^{1/p} = 1,$$

we can conclude that

$$\lim_{p \to \infty} \|\mathbf{x}\|_p = \|\mathbf{x}\|_\infty.$$

This is the reason why the $\infty$-norm is considered a $p$-norm with $p = \infty$.

If you are not familiar with taking limits of sequences, don't worry. We'll cover everything in detail when studying single-variable calculus.

**Example 3.** $\infty$-norms can be generalized for function spaces. Remember $C([0, 1])$, the vector space of functions continuous on $[0, 1]$? We introduced this when talking about examples of vector spaces in *Section 1.1.1*. There, $\|\cdot\|_\infty$ can be defined as

$$\|f\|_\infty = \sup_{x\in[0,1]} |f(x)|.$$

This norm can be defined on other function spaces, like $C(\mathbb{R})$, the space of continuous real functions. Since the maximum is not guaranteed to exist (as for the sigmoid function in $C(\mathbb{R})$), the maximum is replaced with supremum. Hence, the $\infty$-norm is often called the *supremum norm.*

If you imagine the function as a landscape, the supremum norm is the height of the highest peak or the depth of the deepest trench (whichever is larger in absolute value).

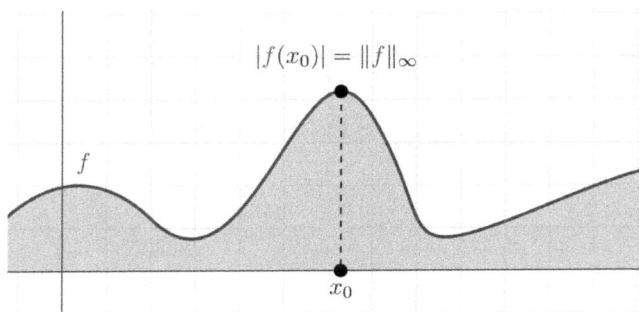

*Figure 2.6: The supremum norm*

When encountering this norm for the first time, it might seem challenging to understand what this has to do with any notion of magnitude. However, $\|f - g\|_\infty$ is a natural way to measure the distance between two functions $f$ and $g$, and in general, magnitude is just the distance from 0.

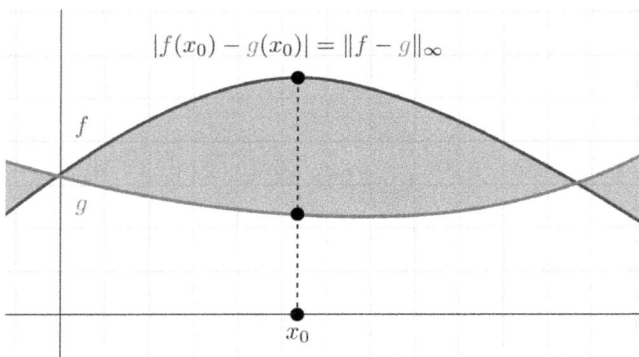

*Figure 2.7: The distance between two functions, given by the supremum norm*

## 2.1.1   Defining distances from norms

Besides measuring the magnitude of vectors, we are also interested in measuring the distance between them. If you are at the location $\mathbf{x}$ in some normed space, how far is $\mathbf{y}$? In normed vector spaces, we can define the distance between any $\mathbf{x}$ and $\mathbf{y}$ by

$$d(\mathbf{x}, \mathbf{y}) = \|\mathbf{x} - \mathbf{y}\|.$$

This is called the norm-induced metric. Thus, norms measure the distance from the zero vector, and the metric $d$ measures the norm of the difference.

In general, we say that a function $d : V \times V \to [0, \infty)$ is a *metric* if the following hold.

---

**Definition 2.1.2  (Metrics)**

Let $V$ be a vector space and $d : V \times V \to [0, \infty)$ be a function. $d$ is a *metric* if the following conditions hold for all $\mathbf{x}, \mathbf{y}, \mathbf{z} \in V$:

1. Whenever $d(\mathbf{x}, \mathbf{y}) = 0$, we have $\mathbf{x} = \mathbf{y}$ (*positive definiteness*).
2. $d(\mathbf{x}, \mathbf{y}) = d(\mathbf{y}, \mathbf{x})$ (*symmetry*).
3. $d(\mathbf{x}, \mathbf{z}) \leq d(\mathbf{x}, \mathbf{y}) + d(\mathbf{y}, \mathbf{z})$ (*triangle inequality*).

---

One of the immediate consequences of the definition is that if $\mathbf{x} \neq \mathbf{y}$, then $d(\mathbf{x}, \mathbf{y}) > 0$. (As the positive definiteness gives that $d(\mathbf{x}, \mathbf{y}) = 1$ implies $\mathbf{x} = \mathbf{y}$.)

Given the properties of norms, we can quickly check that $d(\mathbf{x}, \mathbf{y}) = \|\mathbf{x} - \mathbf{y}\|$ is indeed a metric. Due to the linear structure of vector spaces, the norm-generated metric is invariant to translation. That is, for any $\mathbf{x}, \mathbf{y}, \mathbf{z} \in V$, we have

$$d(\mathbf{x}, \mathbf{y}) = d(\mathbf{x} + \mathbf{z}, \mathbf{y} + \mathbf{z}).$$

In other words, it doesn't matter where you start: the distance only depends on your displacement. This is not true for any metric. Thus, norm-induced metrics are special. In our studies, we only deal with these special cases. Because of this, we won't even talk about metrics, just norms.

In itself, a vector space is just a skeleton that provides a way to represent data. On top of this, norms define a geometric structure that reveals properties such as magnitude and distance. Both of these are essential in machine learning. For instance, some unsupervised learning algorithms separate data points into clusters based on their mutual distances from each other.

There is yet another way to enhance the geometric structure of vector spaces: *inner products*, also called dot products. We are going to put this concept under our magnifying glass in the next section.

## 2.2   Inner products, angles, and lots of reasons to care about them

In the previous section, we imbued our vector spaces with *norms*, measuring the magnitude of vectors and the distance between points. In machine learning, these concepts can be used, for instance, to identify clusters in unlabeled datasets. However, without context, distance is often not enough. Following our geometric intuition, we can aspire to measure the *similarity* of data points. This is done by the *inner product* (also known as the dot product).

You can recall the inner product as a quantity that we used to measure the angle between two vectors in high school geometry classes. Given two vectors $\mathbf{x} = (x_1, x_2)$ and $\mathbf{y} = (y_1, y_2)$ from the plane, we defined their inner product by

$$\langle \mathbf{x}, \mathbf{y} \rangle = x_1 y_1 + x_2 y_2,$$

for which it can be shown that

$$\langle \mathbf{x}, \mathbf{y} \rangle = \|\mathbf{x}\|\|\mathbf{y}\| \cos \alpha \tag{2.1}$$

holds, where $\alpha$ is the angle between $\mathbf{x}$ and $\mathbf{y}$. (In fact, there are two such angles, but their cosine is equal.) Thus, the angle itself can be extracted by

$$\alpha = \arccos \frac{\langle \mathbf{x}, \mathbf{y} \rangle}{\|\mathbf{x}\|\|\mathbf{y}\|},$$

where $\arccos x$ is the inverse of the cosine function. We can use the inner products to determine whether two vectors are *orthogonal*, as this happens if and only if $\langle \mathbf{x}, \mathbf{y} \rangle = 0$ holds. During our earlier encounters with mathematics, geometric intuition (such as orthogonality) came first, on which we built tools such as the inner product. However, if we zoom out and take an abstract viewpoint, things are exactly the opposite. As we'll see soon, inner products emerge quite naturally, giving rise to the general concept of orthogonality.

In general, this is the formal definition of an inner product.

**Definition 2.2.1 (Inner products and inner product spaces)**

Let $V$ be a real vector space. The function $\langle \cdot, \cdot \rangle : V \times V \to \mathbb{R}$ is called an *inner product* if the following holds for all $\mathbf{x}, \mathbf{y}, \mathbf{z} \in V$ and $a \in \mathbb{R}$:

1. $\langle a\mathbf{x} + \mathbf{y}, \mathbf{z} \rangle = a\langle \mathbf{x}, \mathbf{z} \rangle + \langle \mathbf{y}, \mathbf{z} \rangle$ (*linearity of the first variable*).
2. $\langle \mathbf{x}, \mathbf{y} \rangle = \langle \mathbf{y}, \mathbf{x} \rangle$ (*symmetry*).
3. $\langle \mathbf{x}, \mathbf{x} \rangle > 0$ for all $\mathbf{x} \neq 0$ (*positive definiteness*).

Vector spaces with an inner product are called *inner product spaces.*

Right off the bat, we can immediately deduce two properties. First,

$$\langle 0, \mathbf{x} \rangle = \langle 0\mathbf{x}, \mathbf{x} \rangle = 0\langle \mathbf{x}, \mathbf{x} \rangle = 0. \tag{2.2}$$

As a special case, $\langle 0, 0 \rangle = 0$. Just like we have seen for norms, a bit more is true: if $\langle \mathbf{x}, \mathbf{x} \rangle = 0$, then $\mathbf{x} = 0$. This follows from positive definiteness and (2.2).

In addition, due to the symmetry and linearity of the first variable, inner products are also linear in the *second* variable. Because of this, they are called *bilinear.*

To familiarize ourselves with the concept, let's see some examples!

**Example 1.** As usual, the canonical and most prevalent example of inner product spaces is $\mathbb{R}^n$, where the inner product $\langle \cdot, \cdot \rangle$ is defined by

$$\langle \mathbf{x}, \mathbf{y} \rangle = \sum_{i=1}^{n} x_i y_i, \quad \mathbf{x} = (x_1, \dots, x_n), \quad \mathbf{y} = (y_1, \dots, y_n).$$

This bilinear function is often called the *dot product.* Equipped with this, $\mathbb{R}^n$ is called the n-dimensional *Euclidean space.* This is a central concept in machine learning, as data is most frequently represented in Euclidean spaces. Thus, we are going to explore the structure of this space in great detail throughout this book.

**Example 2.** Besides Euclidean spaces, there are other inner product spaces that play a significant role in mathematics and machine learning. If you are familiar with integration, in certain function spaces, the bilinear function

$$\langle f, g \rangle = \int_{-\infty}^{\infty} f(x)g(x)dx$$

defines an inner product space with a very rich and beautiful structure.

The symmetry and linearity of $\langle f, g \rangle$ is clear. Only the positive definiteness seems to be an issue.

For instance, if $f$ is defined by

$$f(x) = \begin{cases} 1 & \text{if } x = 0, \\ 0 & \text{otherwise,} \end{cases}$$

then $f \neq 0$, but $\langle f, f \rangle = 0$. This problem can be circumvented by "overloading" the equality operator and letting $f = g$ if and only if $\int_{-\infty}^{\infty} |f(x) - g(x)|^2 \, dx = 0$. Even though function spaces such as this play an important role in mathematics and machine learning, their study falls outside of our scope.

## 2.2.1   The generated norm

Recall that the 2-norm in $\mathbb{R}^n$ was defined by $\|\mathbf{x}\|_2 = \left( \sum_{i=1}^{n} x_i^2 \right)^{1/2}$, which, according to our definition of the inner product there, equals $\sqrt{\langle \mathbf{x}, \mathbf{x} \rangle}$. This is not a coincidence. Inner products can be used to define norms on vector spaces.

To show exactly how, we need a simple tool: the Cauchy-Schwarz inequality.

> **Theorem 2.2.1** *(Cauchy-Schwarz inequality)*
>
> *Let $V$ be an inner product space. Then, for any $\mathbf{x}, \mathbf{y} \in V$, the inequality*
>
> $$|\langle \mathbf{x}, \mathbf{y} \rangle|^2 \leq \langle \mathbf{x}, \mathbf{x} \rangle \langle \mathbf{y}, \mathbf{y} \rangle$$
>
> *holds.*

*Proof.* At this point, we don't know much about the inner product except its core defining properties. So, we are going to use a little trick. For any $\lambda \in \mathbb{R}$, the positive definiteness implies that

$$\langle \mathbf{x} + \lambda \mathbf{y}, \mathbf{x} + \lambda \mathbf{y} \rangle \geq 0.$$

On the other hand, because of bilinearity (that is, linearity in both variables) and symmetry, we have

$$\langle \mathbf{x} + \lambda \mathbf{y}, \mathbf{x} + \lambda \mathbf{y} \rangle = \langle \mathbf{x}, \mathbf{x} \rangle + 2\lambda \langle \mathbf{x}, \mathbf{y} \rangle + \lambda^2 \langle \mathbf{y}, \mathbf{y} \rangle, \tag{2.3}$$

which is a quadratic polynomial in $\lambda$. In general, we know that for any quadratic polynomial of the form $ax^2 + bx + c$, the roots are given by the formula

$$x_{1,2} = \frac{-b \pm \sqrt{b^2 - 4ac}}{2a}.$$

Since

$$\langle \mathbf{x} + \lambda \mathbf{y}, \mathbf{x} + \lambda \mathbf{y} \rangle \geq 0,$$

the polynomial defined by (2.3) must have at most one real root. Thus, the discriminant $b^2 - 4ac$ is non-positive. Plugging in the coefficients of the polynomial (2.3) into the discriminant formula, we obtain

$$|\langle \mathbf{x}, \mathbf{y} \rangle|^2 - \langle \mathbf{x}, \mathbf{x} \rangle \langle \mathbf{y}, \mathbf{y} \rangle \leq 0,$$

which completes the proof.

The Cauchy-Schwarz inequality is probably one of the most useful tools in studying inner product spaces. One application we are going to see next is to show how inner products define norms.

**Theorem 2.2.2** *(The norm generated by the inner product)*

*Let V be an inner product space. Then, the function $\| \cdot \| : V \to [0, \infty)$ defined by*

$$\|\mathbf{x}\| = \sqrt{\langle \mathbf{x}, \mathbf{x} \rangle}$$

*is a norm on V.*

*Proof.* According to the definition of norms, we have to show that three properties hold: positive definiteness, homogeneity, and the triangle inequality. The first two follow easily from the properties of inner products. The triangle inequality follows from the Cauchy-Schwarz inequality:

$$\begin{aligned}
\|\mathbf{x} + \mathbf{y}\|^2 &= \langle \mathbf{x} + \mathbf{y}, \mathbf{x} + \mathbf{y} \rangle \\
&= \|\mathbf{x}\|^2 + \|\mathbf{y}\|^2 + 2\langle \mathbf{x}, \mathbf{y} \rangle \\
&\leq \|\mathbf{x}\|^2 + \|\mathbf{y}\|^2 + 2\|\mathbf{x}\|\|\mathbf{y}\| \\
&= \left( \|\mathbf{x}\| + \|\mathbf{y}\| \right)^2,
\end{aligned}$$

from which the triangle inequality follows.

Thus, inner product spaces are normed spaces as well. They have the proper algebraic and geometric structure that we need to represent, manipulate, and transform data.

Most importantly, *Theorem 2.2.2* can be reversed! That is, given a norm $\| \cdot \|$, we can define a matching inner product.

---

**Theorem 2.2.3** *(The polarization identity)*

Let $V$ be an inner product space, and let $\| \cdot \|$ be the norm induced by the inner product. Then,

$$\langle \mathbf{x}, \mathbf{y} \rangle = \frac{1}{2} \left( \|\mathbf{x} + \mathbf{y}\|^2 - \|\mathbf{x}\|^2 - \|\mathbf{y}\|^2 \right). \tag{2.4}$$

---

In other words, one can generate an inner product from a norm, not just the other way around.

*Proof.* As the inner product is bilinear, we have

$$\langle \mathbf{x} + \mathbf{y}, \mathbf{x} + \mathbf{y} \rangle = \langle \mathbf{x}, \mathbf{x} \rangle + 2 \langle \mathbf{x}, \mathbf{y} \rangle + \langle \mathbf{y}, \mathbf{y} \rangle,$$

from which the polarization identity (2.4) follows.

## 2.2.2 Orthogonality

In vector spaces other than $\mathbb{R}^2$, the concept of enclosed angles is not clear at all. For instance, in spaces where vectors are functions, there is no intuitive way to define the angles between two functions. However, as (2.1) suggests, in the special case $\mathbb{R}^2$, these can be generalized.

---

**Definition 2.2.2 (Orthogonality of vectors)**

Let $V$ be an inner product space, and let $\mathbf{x}, \mathbf{y} \in V$. We say that $\mathbf{x}$ and $\mathbf{y}$ are *orthogonal* if

$$\langle \mathbf{x}, \mathbf{y} \rangle = 0.$$

Orthogonality is denoted as $\mathbf{x} \perp \mathbf{y}$.

---

To illustrate how inner products and orthogonality define geometry on vector spaces, let's see how the classic Pythagorean theorem looks in this new form. Recall that the "original" version states that in right triangles, $a^2 + b^2 = c^2$, where $c$ is the length of the hypotenuse, while $a$ and $b$ are the lengths of the other two sides.

In inner product spaces, this generalizes in the following way.

**Theorem 2.2.4** *(The Pythagorean theorem)*

*Let $V$ be an inner product space, and let $\mathbf{x}, \mathbf{y} \in V$. Then, $\mathbf{x}$ and $\mathbf{y}$ are orthogonal if and only if*

$$\langle \mathbf{x} + \mathbf{y}, \mathbf{x} + \mathbf{y} \rangle = \langle \mathbf{x}, \mathbf{x} \rangle + \langle \mathbf{y}, \mathbf{y} \rangle. \tag{2.5}$$

*Proof.* Given the definition of inner products and orthogonality, the proof is straightforward. Due to bilinearity, we have

$$\langle \mathbf{x} + \mathbf{y}, \mathbf{x} + \mathbf{y} \rangle = \langle \mathbf{x}, \mathbf{x} + \mathbf{y} \rangle + \langle \mathbf{y}, \mathbf{x} + \mathbf{y} \rangle = \langle \mathbf{x}, \mathbf{x} \rangle + 2\langle \mathbf{x}, \mathbf{y} \rangle + \langle \mathbf{y}, \mathbf{y} \rangle.$$

Since $\mathbf{x}$ and $\mathbf{y}$ are orthogonal, we have $\langle \mathbf{x}, \mathbf{y} \rangle = 0$. Thus, the equation simplifies to:

$$\langle \mathbf{x} + \mathbf{y}, \mathbf{x} + \mathbf{y} \rangle = \langle \mathbf{x}, \mathbf{x} \rangle + \langle \mathbf{y}, \mathbf{y} \rangle.$$

This completes the proof.

Why is this the Pythagorean theorem in another form? Because the norm and the inner product is related by $\langle \mathbf{x}, \mathbf{x} \rangle = \|\mathbf{x}\|^2$, (2.5) is equivalent to

$$\|\mathbf{x} + \mathbf{y}\|^2 = \|\mathbf{x}\|^2 + \|\mathbf{y}\|^2,$$

which is exactly the famous "$a^2 + b^2 = c^2$".

## 2.2.3 The geometric interpretation of inner products

Looking at the general definition, it is hard to get an insight into the inner product. However, by using the concept of orthogonality, we can visualize what $\langle \mathbf{x}, \mathbf{y} \rangle$ represents for any $\mathbf{x}$ and $\mathbf{y}$.

Intuitively, any $\mathbf{x}$ can be decomposed into the sum of two vectors $\mathbf{x}_o + \mathbf{x}_p$, where $\mathbf{x}_o$ is orthogonal to $\mathbf{y}$ and $\mathbf{x}_p$ is parallel to it.

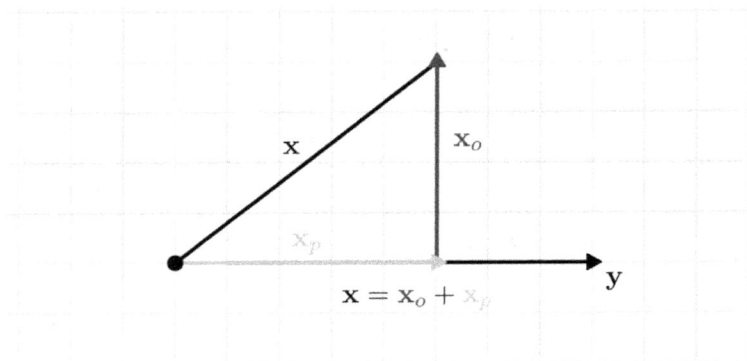

*Figure 2.8: Decomposition of x into components parallel and orthogonal to y*

Let's make the intuition precise. How can we find $\mathbf{x}_p$ and $\mathbf{x}_o$? Since $\mathbf{x}_p$ has the same direction as $\mathbf{y}$, it can be written in the form $\mathbf{x}_p = c\mathbf{y}$ for some scalar $c \in \mathbb{R}$. Because $\mathbf{x}_p$ and $\mathbf{x}_o$ sum up to $\mathbf{x}$, we also have $\mathbf{x}_o = \mathbf{x} - \mathbf{x}_p = \mathbf{x} - c\mathbf{y}$.

Since $\mathbf{x}_o$ is orthogonal to $\mathbf{y}$, the constant $c$ can be determined by solving the equation

$$\langle \mathbf{x} - c\mathbf{y}, \mathbf{y} \rangle = 0.$$

By using the bilinearity of the inner product, we can express $c$ from this equation. Thus, we have

$$c = \frac{\langle \mathbf{x}, \mathbf{y} \rangle}{\langle \mathbf{y}, \mathbf{y} \rangle}.$$

So,

$$\mathbf{x}_p = \frac{\langle \mathbf{x}, \mathbf{y} \rangle}{\langle \mathbf{y}, \mathbf{y} \rangle} \mathbf{y},$$
$$\mathbf{x}_o = \mathbf{x} - \frac{\langle \mathbf{x}, \mathbf{y} \rangle}{\langle \mathbf{y}, \mathbf{y} \rangle} \mathbf{y}. \tag{2.6}$$

We call $\mathbf{x}_p$ the *orthogonal projection* of $\mathbf{x}$ onto $\mathbf{y}$. This is a common transformation, so we are going to introduce the notation

$$\mathrm{proj}_{\mathbf{y}}(\mathbf{x}) = \frac{\langle \mathbf{x}, \mathbf{y} \rangle}{\langle \mathbf{y}, \mathbf{y} \rangle} \mathbf{y}. \tag{2.7}$$

From this, we can see that the scaling ratio between $\mathbf{y}$ and $\text{proj}_{\mathbf{y}}(\mathbf{x})$ can be described by inner products.

So far, we have seen that we can use inner products to define the orthogonality relation between two vectors. Can we use it to measure (and, in some cases, even define) the angle? The answer is yes! In the following, we are going to see how, arriving at the formula (2.1) already familiar from basic geometry.

To build our intuition, let's select two arbitrary $n$-dimensional vectors $\mathbf{x}, \mathbf{y} \in \mathbb{R}^n$. The inner product of the sum $\mathbf{x} + \mathbf{y}$ can be calculated using the bilinearity property.

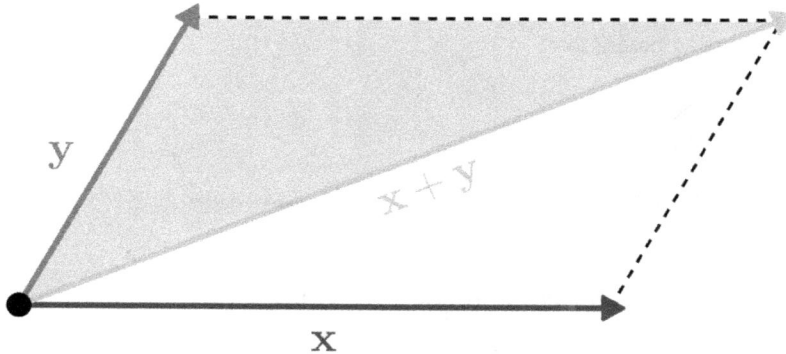

*Figure 2.9: The sum of $\mathbf{x}$ and $\mathbf{y}$*

With this, we obtain

$$\begin{aligned} \langle \mathbf{x} + \mathbf{y}, \mathbf{x} + \mathbf{y} \rangle &= \|\mathbf{x} + \mathbf{y}\|^2 \\ &= \|\mathbf{x}\|^2 + \|\mathbf{y}\|^2 + 2\langle \mathbf{x}, \mathbf{y} \rangle. \end{aligned} \tag{2.8}$$

On the other hand, considering that $\mathbf{x}$, $\mathbf{y}$, and $\mathbf{x} + \mathbf{y}$ form a triangle, we can use the law of cosines ( `https://en.wikipedia.org/wiki/Law_of_cosines`) to express $\langle \mathbf{x} + \mathbf{y}, \mathbf{x} + \mathbf{y} \rangle = \|\mathbf{x} + \mathbf{y}\|^2$ in a different form.

Here, the law of cosines implies

$$\|\mathbf{x} + \mathbf{y}\|^2 = \|\mathbf{x}\|^2 + \|\mathbf{y}\|^2 - 2\|\mathbf{x}\|\|\mathbf{y}\| \underbrace{\cos(\pi - \alpha)}_{= -\cos \alpha}. \tag{2.9}$$

By combining (2.8) and (2.9), we get

$$\langle \mathbf{x}, \mathbf{y} \rangle = \|\mathbf{x}\|\|\mathbf{y}\| \cos \alpha.$$

That is, in $\mathbb{R}^n$, the angle enclosed by **x** and **y** can be extracted by

$$\alpha = \arccos \frac{\langle \mathbf{x}, \mathbf{y} \rangle}{\|\mathbf{x}\| \|\mathbf{y}\|}. \tag{2.10}$$

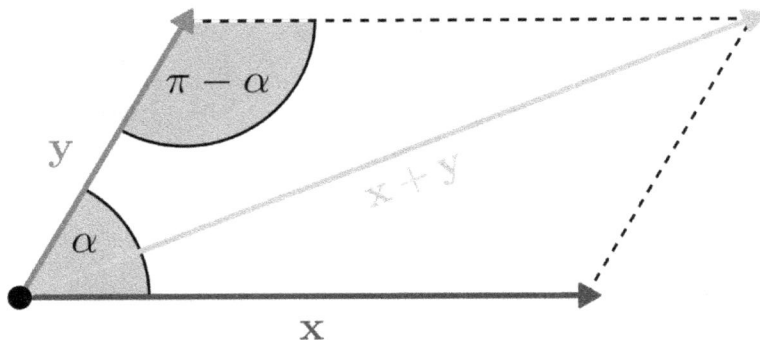

*Figure 2.10: The triangle formed by* **x**, **y**, *and* **x** + **y**

What about vector spaces where the angle between vectors is not defined? We have seen instances of vector spaces (*Section 1.1.1*) where the elements are polynomials, functions, and other mathematical objects. There, (2.10) can be used to *define* the angle!

Let's explore this idea further and see how to use inner products to measure *similarity*.

Given our geometric interpretation of inner products as orthogonal projections, let's focus on the case when both **x** and **y** have unit norms. In this special case, the orthogonal projection equals

$$\mathrm{proj}_{\mathbf{y}}(\mathbf{x}) = \langle \mathbf{x}, \mathbf{y} \rangle \mathbf{y} \quad (\|\mathbf{x}\| = \|\mathbf{y}\| = 1).$$

Thus, $\langle \mathbf{x}, \mathbf{y} \rangle$ precisely describes the signed magnitude of the orthogonal projection. (It can be negative when $\mathrm{proj}_{\mathbf{y}}(\mathbf{x})$ and **y** have an opposite direction.)

With this in mind, we can see that the inner product is equal to the cosine of the angle enclosed by the two vectors. Let's draw a picture to illustrate! (Recall that in right triangles, the cosine is the ratio of the length of the adjacent side and the hypotenuse. In this case, the adjacent side has a length of $\langle \mathbf{x}, \mathbf{y} \rangle$, while the hypotenuse is of unit length.)

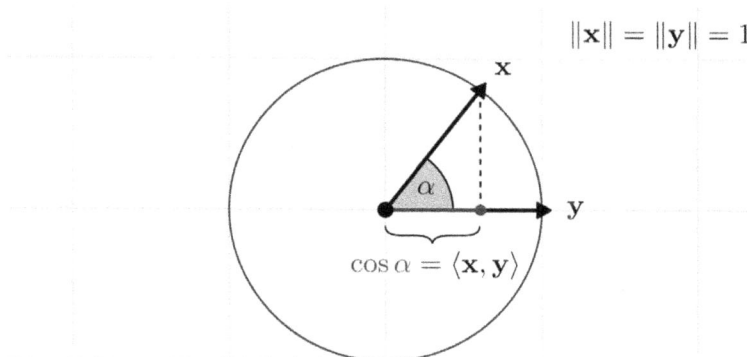

$$\|\mathbf{x}\| = \|\mathbf{y}\| = 1$$

$$\cos \alpha = \langle \mathbf{x}, \mathbf{y} \rangle$$

*Figure 2.11: The inner product of two unit vectors equals the cosine of their angle*

In machine learning, this quantity is frequently used to measure the similarity of two vectors.

Because any vector $\mathbf{x}$ can be scaled to unit norm with the transformation $\mathbf{x} \mapsto \mathbf{x}/\|\mathbf{x}\|$, we define the *cosine similarity* by

$$\cos(\mathbf{x}, \mathbf{y}) = \left\langle \frac{\mathbf{x}}{\|\mathbf{x}\|}, \frac{\mathbf{y}}{\|\mathbf{y}\|} \right\rangle. \tag{2.11}$$

If $\mathbf{x}$ and $\mathbf{y}$ represent the feature vectors of two data samples, $\cos(\mathbf{x}, \mathbf{y})$ tells us how much the features move together. Note that because of the scaling, two samples with a high cosine similarity can be far from each other. So, this reveals nothing about their relative positions in the feature space.

## 2.2.4 Orthogonal and orthonormal bases

Through the lens of similarity, orthogonality means that one vector does not contain "information" about the other. We will make this notion more precise when learning about correlation, but there are clear implications regarding the structure of inner product spaces. Recall that during the introduction of basis vectors (*Section 1.2*), our motivation was to find a *minimal* set of vectors that can be used to express any other vector. With the introduction of orthogonality, we can go one step further.

**Definition 2.2.3 (Orthogonal and orthonormal bases)**

Let $V$ be a vector space and $S = \{\mathbf{v}_1, ..., \mathbf{v}_n\}$ its basis. We say that $S$ is an *orthogonal basis* if $\langle \mathbf{v}_i, \mathbf{v}_j \rangle = 0$ whenever $i \neq j$.

Moreover, $S$ is called *orthonormal* if

$$\langle \mathbf{v}_i, \mathbf{v}_j \rangle = \begin{cases} 1, & \text{if } i = j, \\ 0, & \text{if } i \neq j. \end{cases}$$

In other words, $S$ is orthonormal if, in addition to being orthogonal, each vector has unit norm.

Orthogonal and orthonormal bases are extremely convenient to use. If a basis is orthogonal, we can easily obtain an orthonormal basis by simply scaling its vectors to unit norm. Thus, we'll use orthonormal basis vectors most of the time.

Why do we love orthonormal bases so much? To see this, let $\{\mathbf{v}_1, \ldots, \mathbf{v}_n\}$ be an arbitrary basis and let $\mathbf{x}$ be an arbitrary vector. We know that

$$\mathbf{x} = \sum_{i=1}^{n} x_i \mathbf{v}_i,$$

but how do we find the coefficients $x_i$? There is a general method involving linear equations that we will see later in *Chapter 5*, but if $\{\mathbf{v}_i\}_{i=1}^{n}$ is orthonormal, the situation is much simpler.

This is made more precise in the following theorem.

**Theorem 2.2.5** *Let $V$ be a vector space and $S = \{\mathbf{v}_1, \ldots, \mathbf{v}_n\}$ be an orthonormal basis of $V$. Then, for any $\mathbf{x} \in V$,*

$$\mathbf{x} = \sum_{i=1}^{n} \langle \mathbf{x}, \mathbf{v}_i \rangle \mathbf{v}_i \qquad (2.12)$$

*holds.*

*Proof.* Since $\mathbf{v}_1, \ldots, \mathbf{v}_n$ form a basis, we can express $\mathbf{x}$ as

$$\mathbf{x} = \sum_{i=1}^{n} x_i \mathbf{v}_i$$

for some scalars $x_i$. By the linearity of the inner product, we obtain

$$\langle \mathbf{x}, \mathbf{v}_j \rangle = \left\langle \sum_{i=1}^{n} x_i \mathbf{v}_i, \mathbf{v}_j \right\rangle = \sum_{i=1}^{n} x_i \langle \mathbf{v}_i, \mathbf{v}_j \rangle.$$

Since $\mathbf{v}_1, \ldots, \mathbf{v}_n$ form an orthonormal basis, we have

$$\langle \mathbf{v}_i, \mathbf{v}_j \rangle = \begin{cases} 1, & \text{if } i = j, \\ 0, & \text{if } i \neq j. \end{cases}$$

Thus, the sum reduces to

$$\langle \mathbf{x}, \mathbf{v}_j \rangle = x_j.$$

This proves the result.

Thus, the coefficients can be calculated by taking the inner product. In other words, for orthonormal bases, $x_j$ depends only on the $j$-th basis vector.

As another consequence of the orthonormality, calculating the norm is also easier, as we can always express it in terms of the coefficients. To be more precise, we have

$$\begin{aligned} \|\mathbf{x}\|^2 &= \langle \mathbf{x}, \mathbf{x} \rangle \\ &= \left\langle \sum_{i=1}^{n} x_i \mathbf{v}_i, \sum_{j=1}^{n} x_j \mathbf{v}_j \right\rangle \\ &= \sum_{i=1}^{n} \sum_{j=1}^{n} x_i x_j \langle \mathbf{v}_i, \mathbf{v}_j \rangle \\ &= \sum_{i=1}^{n} x_i^2. \end{aligned} \tag{2.13}$$

This is called *Parseval's identity*. In other words, if $\mathbf{x}$ is given in terms of an orthonormal basis, its norm is easy to find. It is not a coincidence that this formula resembles the Euclidean norm so much! (Note that, here, $\|\cdot\|$ is a general norm.) In fact, the squared Euclidean norm

$$\|\mathbf{x}\|_2^2 = \sum_{i=1}^{n} x_i^2, \quad \mathbf{x} = (x_1, \dots, x_n) \in \mathbb{R}^n$$

is just (2.13) using the standard basis.

## 2.2.5 The Gram-Schmidt orthogonalization process

Orthogonal bases are awesome and all, but how do we find them?

There is a general method called the *Gram-Schmidt orthogonalization process* that solves this problem. The algorithm takes any set of basis vectors $\{\mathbf{v}_1, \dots, \mathbf{v}_n\}$ and outputs an orthonormal basis $\{\mathbf{e}_1, \dots, \mathbf{e}_n\}$

such that

$$\text{span}(\mathbf{v}_1, \ldots, \mathbf{v}_k) = \text{span}(\mathbf{e}_1, \ldots, \mathbf{e}_k), \quad k = 1, \ldots, n,$$

that is, the subspaces generated by the first $k$ vectors of both sets match.

How do we do that? The process is straightforward. Let's focus on finding an *orthogonal* system first, which we can normalize later to achieve orthonormality. We are going to build our set $\{\mathbf{e}_1, \ldots, \mathbf{e}_n\}$ iteratively. It is clear that

$$\mathbf{e}_1 := \mathbf{v}_1$$

is a good choice. Now, our goal is to find $\mathbf{e}_2$ such that $\mathbf{e}_2 \perp \mathbf{e}_1$ and, together, they span the same subspace as $\{\mathbf{v}_1, \mathbf{v}_2\}$. Remember when we talked about the geometric interpretation of orthogonality in *Section 2.2.3*? The orthogonal component of $\mathbf{v}_2$ with respect to $\mathbf{e}_1$ will be a good choice for $\mathbf{e}_2$. Thus, let

$$\mathbf{e}_2 := \mathbf{v}_2 - \text{proj}_{\mathbf{e}_1}(\mathbf{v}_2) = \mathbf{v}_2 - \frac{\langle \mathbf{v}_2, \mathbf{e}_1 \rangle}{\langle \mathbf{e}_1, \mathbf{e}_1 \rangle} \mathbf{e}_1.$$

From the definition, it is clear that $\mathbf{e}_2 \perp \mathbf{e}_1$, and it is also clear that $\{\mathbf{e}_1, \mathbf{e}_2\}$ spans the same subspace as $\{\mathbf{v}_1, \mathbf{v}_2\}$.

In the next step, we perform the same process. We project $\mathbf{v}_3$ onto the subspace generated by $\mathbf{e}_1$ and $\mathbf{e}_2$, then define $\mathbf{e}_3$ as the difference of $\mathbf{v}_3$ and the projection. That is,

$$\mathbf{e}_3 := \mathbf{v}_3 - \text{proj}_{\mathbf{e}_1, \mathbf{e}_2}(\mathbf{v}_3) = \mathbf{v}_3 - \frac{\langle \mathbf{v}_3, \mathbf{e}_1 \rangle}{\langle \mathbf{e}_1, \mathbf{e}_1 \rangle} \mathbf{e}_1 - \frac{\langle \mathbf{v}_3, \mathbf{e}_2 \rangle}{\langle \mathbf{e}_2, \mathbf{e}_2 \rangle} \mathbf{e}_2.$$

With this, we essentially remove the "contributions" of $\mathbf{e}_1$ and $\mathbf{e}_2$ toward $\mathbf{v}_3$, thus obtaining an $\mathbf{e}_3$ that is orthogonal to the previous ones.

In general, if we have $\mathbf{e}_1, \ldots, \mathbf{e}_k$, the vector $\mathbf{e}_{k+1}$ can be found by

$$\mathbf{e}_{k+1} := \mathbf{v}_{k+1} - \text{proj}_{\mathbf{e}_1, \ldots, \mathbf{e}_k}(\mathbf{v}_{k+1}),$$

where

$$\text{proj}_{\mathbf{e}_1, \ldots, \mathbf{e}_k}(x) = \sum_{i=1}^{k} \frac{\langle \mathbf{x}, \mathbf{e}_i \rangle}{\langle \mathbf{e}_i, \mathbf{e}_i \rangle} \mathbf{e}_i \tag{2.14}$$

is the generalized orthogonal projection operator, projecting a vector to the subspace generated by $\{\mathbf{e}_1, \ldots, \mathbf{e}_k\}$.

To check that $\mathbf{e}_{k+1} \perp \mathbf{e}_1, \dots, \mathbf{e}_k$ , we have

$$
\begin{aligned}
\langle \mathbf{e}_{k+1}, \mathbf{e}_j \rangle &= \left\langle \mathbf{v}_{k+1} - \sum_{i=1}^{k} \frac{\langle \mathbf{v}_{k+1}, \mathbf{e}_i \rangle}{\langle \mathbf{e}_i, \mathbf{e}_i \rangle} \mathbf{e}_i, \mathbf{e}_j \right\rangle \\
&= \langle \mathbf{v}_{k+1}, \mathbf{e}_j \rangle - \sum_{i=1}^{k} \frac{\langle \mathbf{v}_{k+1}, \mathbf{e}_i \rangle}{\langle \mathbf{e}_i, \mathbf{e}_i \rangle} \langle \mathbf{e}_i, \mathbf{e}_j \rangle \\
&= \langle \mathbf{v}_{k+1}, \mathbf{e}_j \rangle - \langle \mathbf{v}_{k+1}, \mathbf{e}_j \rangle \\
&= 0.
\end{aligned}
\tag{2.15}
$$

due to the orthogonality of the $\mathbf{e}_i$-s and the linearity of the inner product. Since $\{\mathbf{e}_1, \dots, \mathbf{e}_k\}$ spans the same subspace as $\{\mathbf{v}_1, \dots, \mathbf{v}_k\}$ and $\mathbf{e}\{k+1\}$ is a linear combination of $\mathbf{v}_{k+1}$ and $\mathbf{e}_1, \dots, \mathbf{e}_k$ (where the coefficient of $\mathbf{v}_{k+1}$ is nonzero),

$$
\text{span}(\mathbf{v}_1, \dots, \mathbf{v}_{k+1}) = \text{span}(\mathbf{e}_1, \dots, \mathbf{e}_{k+1})
$$

also follows. This can be repeated until we run out of vectors and find $\{\mathbf{e}_1, \dots, \mathbf{e}_n\}$.

For the sake of further reference, mathematical correctness, and a tiny bit of perfectionism, let's summarize all the above in a single theorem.

**Theorem 2.2.6** *(Gram-Schmidt orthogonalization process)*

*Let $V$ be an inner product vector space and $\{\mathbf{v}_1, \dots, \mathbf{v}_n\} \subseteq V$ be a set of linearly independent vectors. Then, there exists an orthonormal set $\{\mathbf{e}_1, \dots, \mathbf{e}_n\} \subseteq V$ such that*

$$
\text{span}(\mathbf{e}_1, \dots, \mathbf{e}_k) = \text{span}(\mathbf{v}_1, \dots, \mathbf{v}_k)
$$

*holds for any $k = 1, \dots, n$.*

As a consequence, we can state that each finite inner product space has an orthonormal basis. We can even construct it explicitly via the Gram-Schmidt process.

**Corollary 2.2.1** *(Existence of orthonormal bases)*

*Let $V$ be a finite-dimensional inner product space. Then, there exists an orthonormal basis in $V$.*

Going one step further, we can view *Theorem 2.2.6* and its proof as an *algorithm*.

**Algorithm 2.2.1** *(Gram-Schmidt orthogonalization process)*

**Inputs**: *A set of linearly independent vectors* $\{v_1, \ldots, v_n\} \subseteq V$.

**Output**: *A set of orthonormal vectors* $\{e_1, \ldots, e_n\}$ *such that*

$$\text{span}(e_1, \ldots, e_k) = \text{span}(v_1, \ldots, v_k)$$

*holds for any* $k = 1, \ldots, n$.

**Remark 2.2.1 (Linearly dependent inputs in the Gram-Schmidt process)**

What happens if we apply the Gram-Schmidt orthogonalization process to a set of linearly *dependent* vectors?

To get a grip on the problem, let's consider a simple case of two vectors: $\{v_1, v_2 = cv_1\}$, where $c$ is an arbitrary scalar. $e_1$ is chosen to be $v_1$, and $e_2$ is defined by

$$e_2 = v_2 - \text{proj}_{e_1}(v_2).$$

By expanding the projection term, we obtain

$$e_2 = v_2 - \frac{\langle v_2, e_1 \rangle}{\langle e_1, e_1 \rangle} e_1.$$

Since $v_1 = e_1$ and $v_2 = cv_1$, we get that

$$e_2 = cv_1 - \frac{c\langle v_1, v_1 \rangle}{\langle v_1, v_1 \rangle} v_1$$

$$= cv_1 - cv_1$$

$$= 0.$$

This result generalizes: when the Gram-Schmidt process encounters an input vector that is linearly dependent from the previous ones, a zero vector is produced in the output.

## 2.2.6 The orthogonal complement

Earlier, we saw that given a fixed vector $\mathbf{y} \in V$, we can decompose any $\mathbf{x} \in V$ as $\mathbf{x} = \mathbf{x}_o + \mathbf{x}_p$, where $\mathbf{x}_o$ is orthogonal to $\mathbf{y}$, while $\mathbf{x}_p$ is parallel to it. (We used this to provide a geometric motivation for inner products in *Section 2.2.3*.)

This is an essential tool, and in this section, we will see that an analogue of this decomposition still holds true when $\mathbf{y}$ is replaced with an arbitrary subspace $S \subset V$. To see this, let's talk about the *orthogonality of subspaces*. (If you want to recall the definition of subspaces, check out *Definition 1.2.3*.)

---

**Definition 2.2.4 (Orthogonal subspaces)**

Let $V$ be an arbitrary inner product space. We say that the subspaces $S_1, S_2 \subseteq V$ are orthogonal if, for every pair of vectors $\mathbf{x} \in S_1$ and $\mathbf{y} \in S_2$, we have $\langle \mathbf{x}, \mathbf{y} \rangle = 0$. This is denoted as $S_1 \perp S_2$.

---

For example, the $x$-axis and the $y$-axis are orthogonal subspaces in $\mathbb{R}^2$. (Just as the $xy$ plane and the $z$-axis in $\mathbb{R}^3$.)

Similarly, we can talk about the orthogonality of a vector and a subspace: $\mathbf{x}$ is orthogonal to the subspace $S$, or $\mathbf{x} \perp S$ in symbols, if $\mathbf{x}$ is orthogonal to all vectors of $S$.

One of the most straightforward and essential ways to construct orthogonal subspaces is to take the *orthogonal complement*.

---

**Definition 2.2.5 (Orthogonal complement)**

Let $V$ be an arbitrary inner product space and let $S \subseteq V$ be a subspace. The set defined by

$$S^{\perp} := \{\mathbf{x} \in V \mid \mathbf{x} \perp S\} \tag{2.16}$$

is called the *orthogonal complement* of $S$.

---

$S^{\perp}$ is not just any set; it is a subspace, as we are about to see.

---

**Theorem 2.2.7** *Let $V$ be an arbitrary inner product space and $S \subseteq V$ one of its subspaces. Then*

1. *$S^{\perp}$ is orthogonal to $S$,*
2. *$S^{\perp}$ is a subspace of $V$,*
3. *and $S \cap S^{\perp} = \{0\}$.*

*Proof.* According to the definition of subspaces, we only have to show that $S^\perp$ is closed with respect to addition and scalar multiplication. As the inner product is bilinear, this is straightforward:

$$\langle a\mathbf{x} + b\mathbf{y}, \mathbf{z} \rangle = a\langle \mathbf{x}, \mathbf{z} \rangle + b\langle \mathbf{y}, \mathbf{z} \rangle = 0$$

holds for any vectors $\mathbf{x}, \mathbf{y} \in S^\perp$, $\mathbf{z} \in S$, and scalars $a, b$.

To see that $S \cap S^\perp = \{\mathbf{0}\}$, let's take an arbitrary $\mathbf{x} \in S \cap S^\perp$. By the definition of $S^\perp$, we have $\langle \mathbf{x}, \mathbf{x} \rangle = 0$. As the inner product is positive definite per definition, $\mathbf{x}$ must be the zero vector $\mathbf{0}$.

Recall the decomposition of any $\mathbf{x} \in V$ into a parallel and an orthogonal component with respect to a fixed vector $\mathbf{y}$? In terms of subspaces, we can restate this as

$$V = \mathrm{span}(\mathbf{y}) + \mathrm{span}(\mathbf{y})^\perp,$$

that is, $V$ can be written as the direct sum of the vector space spanned by $\mathbf{y}$, and its orthogonal complement. This is an extremely powerful result, as this allows us to decouple $\mathbf{x}$ from $\mathbf{y}$. For instance, if we think of vectors as a collection of features (just like the sepal and petal width and length measurements in our favorite Iris dataset), $\mathbf{y}$ can represent a certain trait that we want to exclude from our analysis.

With the notion of orthogonal complements, we can make this mathematically precise. We can also be more general. In fact, the decomposition

$$V = S + S^\perp$$

holds for *any* subspace $S$! Let's see the proof!

**Theorem 2.2.8** *Let $V$ be an arbitrary finite-dimensional inner product space and $S \subset V$ its subspace. Then,*

$$V = S + S^\perp$$

*holds.*

*Proof.* Let $\mathbf{v}_1, \ldots, \mathbf{v}_k \in S$ be a basis of $S$. As during the Gram-Schmidt process, we can define the generalized orthogonal projection (2.14), given by

$$\text{proj}_{\mathbf{v}_1,\ldots,\mathbf{v}_k}(\mathbf{x}) = \sum_{i=1}^{k} \langle \mathbf{x}, \mathbf{v}_i \rangle \mathbf{v}_i.$$

Using this, we can decompose any $\mathbf{x} \in V$ as

$$\mathbf{x} = \left( \mathbf{x} - \text{proj}_{\mathbf{v}_1,\ldots,\mathbf{v}_k}(\mathbf{x}) \right) + \text{proj}_{\mathbf{v}_1,\ldots,\mathbf{v}_k}(\mathbf{x}). \tag{2.17}$$

Since $\text{proj}_{\mathbf{v}_1,\ldots,\mathbf{v}_k}(\mathbf{x})$ is the linear combination of $\mathbf{v}_i$-s, it belongs to $S$. On the other hand, the bilinearity of the inner product gives that $\mathbf{x} - \text{proj}_{\mathbf{v}_1,\ldots,\mathbf{v}_k}(\mathbf{x}) \in S^{\perp}$. Indeed, as we have

$$\langle \mathbf{x} - \text{proj}_{\mathbf{v}_1,\ldots,\mathbf{v}_k}(x), \mathbf{v}_j \rangle = \left\langle \mathbf{x} - \sum_{i=1}^{k} \langle \mathbf{x}, \mathbf{v}_i \rangle \mathbf{v}_i, \mathbf{v}_j \right\rangle$$

$$= \langle \mathbf{x}, \mathbf{v}_j \rangle - \sum_{i=1}^{k} \langle \mathbf{x}, \mathbf{v}_i \rangle \langle \mathbf{v}_i, \mathbf{v}_j \rangle$$

$$= \langle \mathbf{x}, \mathbf{v}_j \rangle - \langle \mathbf{x}, \mathbf{v}_j \rangle$$

$$= 0,$$

the vector $\mathbf{x} - \text{proj}_{\mathbf{v}_1,\ldots,\mathbf{v}_k}(\mathbf{x})$ is orthogonal to each $\mathbf{v}_j$. Thus, since $\mathbf{v}_1, \ldots, \mathbf{v}_k$ is a basis of $S$, $\mathbf{x}$ is also orthogonal to $S$, hence $\mathbf{x} - \text{proj}_{\mathbf{e}_1,\ldots,\mathbf{v}_k}(\mathbf{x}) \in S^{\perp}$.

The fact that every $\mathbf{x} \in V$ can be decomposed as the sum of a vector from $S$ and a vector from $S^{\perp}$, as given by (2.17), means that $V = S + S^{\perp}$, which is what we had to prove.

## 2.3 Summary

In this chapter, we have learned that, besides the algebraic structure given by addition and scalar multiplication, vectors have a beautiful geometry that rises from the inner product. From the inner product, we have norms; from norms, we have metrics; and from metrics, we have geometry and topology.

Distance, similarity, angles, and orthogonality all arise from the simple concept of inner products. These are all extremely useful in both theory and practice. For instance, inner products give us a way to quantify the similarity of two vectors via the so-called cosine similarity, but they also provide a means to find optimal bases through the notion of orthogonality.

To summarize, we've learned what norms and distances are, the definition of the inner product, how inner products give angles and norms, and why all of these are useful in machine learning.

Besides the basic definitions and properties, we've encountered our very first algorithm: the Gram-Schmidt process, turning a set of vectors into an orthonormal basis. This is the best kind of basis.

In the next section, we'll take all that theory and put it into practice, taking our first steps in computational linear algebra. Let's go!

# 2.4  Problems

**Problem 1.** Let $V$ be a vector space and define the function $d : V \times V \to [0, \infty)$ by

$$
d(\mathbf{x}, \mathbf{y}) = \begin{cases} 0 & \text{if } \mathbf{x} = \mathbf{y}, \\ 1 & \text{otherwise.} \end{cases}
$$

*(a)* Show that $d$ is a metric (see *Definition 2.1.2*).

*(b)* Show that $d$ cannot come from a norm.

**Problem 2.** Let $S_n$ be the set of all ASCII strings of $n$ character length and define the Hamming distance $h(x, y)$ for any two $x, y \in S_n$ by the number of corresponding positions where $x$ and $y$ are different.

For instance,

$$h(\text{“001101”}, \text{“101110”}) = 2,$$

$$h(\text{“metal”}, \text{“petal”}) = 1.$$

Show that $h$ satisfies the three defining properties of a metric. (Note that $S_n$ is not a vector space so, technically, the Hamming distance is not a metric.)

**Problem 3.** Let $\| \cdot \|$ be a norm on the vector space $\mathbb{R}^n$, and define the mapping $f : \mathbb{R}^n \to \mathbb{R}^n$,

$$f : (x_1, x_2, \ldots, x_n) \mapsto (x_1, 2x_2, \ldots, nx_n).$$

Show that

$$\|\mathbf{x}\|_* := \|f(\mathbf{x})\|$$

is a norm on $\mathbb{R}^n$.

**Problem 4.** Let $a_1, \ldots, a_n > 0$ be arbitrary positive numbers. Show that

$$\langle \mathbf{x}, \mathbf{y} \rangle := \sum_{i=1}^{n} a_i x_i y_i, \quad \mathbf{x}, \mathbf{y} \in \mathbb{R}^n.$$

is an inner product, where $\mathbf{x} = (x_1, \ldots, x_n)$ and $\mathbf{y} = (y_1, \ldots, y_n)$.

**Problem 5.** Let $V$ be a finite-dimensional inner product space, let $\mathbf{v}_1, \ldots, \mathbf{v}_n \in V$ be a basis in $V$, and define

$$a_{i,j} := \langle \mathbf{v}_i, \mathbf{v}_j \rangle.$$

Show that for any $\mathbf{x}, \mathbf{y} \in V$,

$$\langle \mathbf{x}, \mathbf{y} \rangle = \sum_{i,j=1}^{n} a_{i,j} x_i y_j,$$

where $\mathbf{x} = \sum_{i=1}^{n} x_i \mathbf{v}_i$ and $\mathbf{y} = \sum_{i=1}^{n} y_i \mathbf{v}_i$.

**Problem 6.** Let $V$ be a finite-dimensional real inner product space.

*(a)* Let $\mathbf{y} \in V$ be an arbitrary vector. Show that

$$f : V \to \mathbb{R}, \quad \mathbf{x} \mapsto \langle \mathbf{x}, \mathbf{y} \rangle$$

is a linear function (that is, $f(\alpha \mathbf{u} + \beta \mathbf{v}) = \alpha f(\mathbf{u}) + \beta f(\mathbf{v})$ holds for all $\mathbf{u}, \mathbf{v} \in V$ and $\alpha, \beta \in \mathbb{R}$).

*(b)* Let $f : V \to \mathbb{R}$ an arbitrary linear function. Show that there exists a $\mathbf{y} \in V$ such that

$$f(\mathbf{x}) = \langle \mathbf{x}, \mathbf{y} \rangle.$$

(Note that *(b)* is the reverse of *(a)*, and a much more interesting result.)

**Problem 7.** Let $V$ be a real inner product space and let $\|\mathbf{x}\| = \sqrt{\langle \mathbf{x}, \mathbf{x} \rangle}$ be the generated norm. Show that

$$2\|\mathbf{x}\|^2 + 2\|\mathbf{y}\|^2 = \|\mathbf{x} + \mathbf{y}\|^2 + \|\mathbf{x} - \mathbf{y}\|^2. \tag{2.18}$$

This is called the *parallelogram law*, because if we think of $x$ and $y$ as the two sides determining a parallelogram, (2.18) relates the length of its sides to the length of its diagonals.

**Problem 8.** Let $V$ be a real inner product space and let $\mathbf{u}, \mathbf{v} \in V$. Show that if

$$\langle \mathbf{u}, \mathbf{x} \rangle = \langle \mathbf{v}, \mathbf{x} \rangle$$

holds for all $\mathbf{x} \in V$, then $\mathbf{u} = \mathbf{v}$.

**Problem 9.** Apply the Gram-Schmidt process to the input vectors

$$\mathbf{v}_1 = (2, 1, 1),$$
$$\mathbf{v}_2 = (1, 1, 1),$$
$$\mathbf{v}_3 = (1, 0, 1).$$

# Join our community on Discord

Read this book alongside other users, Machine Learning experts, and the author himself.

Ask questions, provide solutions to other readers, chat with the author via Ask Me Anything sessions, and much more.

Scan the QR code or visit the link to join the community.

`https://packt.link/math`

# 3

# Linear Algebra in Practice

Now that we understand the geometric structure of vector spaces, it's time to put the theory into practice once again. In this chapter, we'll take a hands-on look at norms, inner products, and NumPy array operations in general. Most importantly, we'll also meet matrices for the first time.

The last time we translated theory to code, we left off at finding an ideal representation for vectors: NumPy arrays. NumPy is built for linear algebra and handles computations much faster than the vanilla Python objects.

So, let's initialize two NumPy arrays to play around with!

```python
import numpy as np

x = np.array([1.8, -4.5, 9.2, 7.3])
y = np.array([-5.2, -1.1, 0.7, 5.1])
```

In linear algebra, and in most of machine learning, almost all operations involve looping through the vector components one by one. For instance, addition can be implemented like this.

```python
def add(x: np.ndarray, y: np.ndarray):
    x_plus_y = np.zeros(shape=len(x))

    for i in range(len(x_plus_y)):
```

```
        x_plus_y[i] = x[i] + y[i]

    return x_plus_y
```

```
add(x, y)
```

```
array([-3.4, -5.6,  9.9, 12.4])
```

Of course, this is far from optimal. (It may not even work if the vectors have different dimensions.)

For example, addition is massively parallelizable, and our implementation does not take advantage of that. With two threads, we can do two additions simultaneously. So, adding together two-dimensional vectors would require just one step, as one would compute x[0] + y[0], while the other x[1] + y[1]. Raw Python does not have access to such high-performance computing tools, but NumPy does, through functions implemented in C. In turn, C uses the LAPACK (Linear Algebra PACKage) library, which makes calls to BLAS (Basic Linear Algebra Subprograms). BLAS is optimized at the assembly level.

So, whenever it is possible, we should strive to work with vectors in a NumPythonic way. (Yes, I just made that term up.) For vector addition, this is simply the + operator, as we have seen earlier.

```
np.equal(x + y, add(x, y))
```

```
array([ True,  True,  True,  True])
```

By the way, you shouldn't ever compare floats with the == operator, as internal rounding errors can occur due to the float representation. The example below illustrates this.

```
1.0 == 0.3*3 + 0.1
```

```
False
```

```
0.3*3 + 0.1
```

```
0.999999999999999
```

To compare arrays, NumPy provides the functions `np.allclose` and `np.equal`. These compare arrays elementwise, returning a Boolean array. From this, the built-in `all` function can be used to see if all the elements match.

```
all(np.equal(x + y, add(x, y)))
```

```
True
```

In the following section, we'll briefly review how to work with NumPy arrays in practice.

# 3.1   Vectors in NumPy

There are two operations that we definitely want to do with our vectors: apply a function elementwise or take the sum/product of the elements. Since the +, *, and ** operators are implemented for our arrays, certain functions carry over from scalars, as the example below shows.

```
def just_a_quadratic_polynomial(x):
    return 3*x**2 + 1

x = np.array([1.8, -4.5, 9.2, 7.3])
just_a_quadratic_polynomial(x)
```

```
array([ 10.72,   61.75, 254.92, 160.87])
```

However, we can't just plug in `ndarrays` to every function. For instance, let's take a look at Python's built-in exp from its `math` module.

```
from math import exp

exp(x)
```

```
---------------------------------------------------------------
TypeError                                 Traceback (most recent call last)
Cell In[10], line 3
      1 from math import exp
----> 3 exp(x)

TypeError: only length-1 arrays can be converted to Python scalars
```

To overcome this problem, we could manually apply the function elementwise.

```python
def naive_exp(x: np.ndarray):
    x_exp = np.empty_like(x)

    for i in range(len(x)):
        x_exp[i] = exp(x[i])

    return x_exp
```

(Recall that np.empty_like(x) creates an uninitialized array that matches the dimensions of x.)

```python
naive_exp(x)
```

```
array([6.04964746e+00, 1.11089965e-02, 9.89712906e+03, 1.48029993e+03])
```

A bit less naive implementation would use a list comprehension to achieve the same effect.

```python
def bit_less_naive_exp(x: np.ndarray):
    return np.array([exp(x_i) for x_i in x])

bit_less_naive_exp(x)
```

```
array([   6. ,    0. , 9897.1, 1480.3])
```

Even though comprehensions are more concise and readable, they still don't avoid the core issue: for loops in Python.

This problem is solved by NumPy's famous ufuncs, that is, functions that operate element by element on whole arrays ( https://numpy.org/doc/stable/reference/generated/numpy.ufunc.html ). Since they are implemented in C, they are blazing fast. For instance, the exponential function $f(x) = e^x$ is given by np.exp.

```python
np.exp(x)
```

```
array([6.04964746e+00, 1.11089965e-02, 9.89712906e+03, 1.48029993e+03])
```

Not surprisingly, the results of our implementations match.

```
all(np.equal(naive_exp(x), np.exp(x)))
```

```
True
```

```
all(np.equal(bit_less_naive_exp(x), np.exp(x)))
```

```
True
```

Again, there are more advantages to using NumPy functions and operations than simplicity. In machine learning, we care a lot about speed, and as we are about to see, NumPy delivers once more.

```
from timeit import timeit


n_runs = 100000
size = 1000


t_naive_exp = timeit(
    "np.array([exp(x_i) for x_i in x])",
    setup=f"import numpy as np; from math import exp; x = np.ones({size})",
    number=n_runs
)

t_numpy_exp = timeit(
    "np.exp(x)",
    setup=f"import numpy as np; from math import exp; x = np.ones({size})",
    number=n_runs
)


print(f"Built-in exponential:     \t{t_naive_exp:.5f} s")
print(f"NumPy exponential:        \t{t_numpy_exp:.5f} s")
print(f"Performance improvement: \t{t_naive_exp/t_numpy_exp:.5f} times faster")
```

```
Built-in exponential:        18.35177 s
NumPy exponential:           0.87458 s
Performance improvement:     20.98356 times faster
```

For further reference, you can find the list of available ufuncs here: `https://numpy.org/doc/stable/refer`
`ence/ufuncs.html#available-ufuncs`.

What about operations that aggregate the elements and return a single value? Not surprisingly, these can be found within NumPy as well. For instance, let's take a look at the sum. In terms of mathematical formulas, we are looking to implement the function

$$\text{sum}(\mathbf{x}) = \sum_{i=1}^{n} x_i, \quad \mathbf{x} = (x_1, \dots, x_n) \in \mathbb{R}^n.$$

A basic approach would be something like this.

```python
def naive_sum(x: np.ndarray):
    val = 0

    for x_i in x:
        val += x_i

    return val

naive_sum(x)
```

```
np.float64(13.799999999999999)
```

Alternatively, we can use Python's built-in summing function.

```python
sum(x)
```

```
np.float64(13.799999999999999)
```

The story is the same: NumPy can do this better. We can either call the function `np.sum` or use the array method `np.ndarray.sum`.

```python
np.sum(x)
```

```
np.float64(13.799999999999999)
```

```
x.sum()
```

```
np.float64(13.799999999999999)
```

You know by now that I love timing functions, so let's compare the performances once more.

```
t_naive_sum = timeit(
    "sum(x)",
    setup=f"import numpy as np; x = np.ones({size})",
    number=n_runs
)

t_numpy_sum = timeit(
    "np.sum(x)",
    setup=f"import numpy as np; x = np.ones({size})",
    number=n_runs
)


print(f"Built-in sum:               \t{t_naive_sum:.5f} s")
print(f"NumPy sum:                  \t{t_numpy_sum:.5f} s")
print(f"Performance improvement: \t{t_naive_sum/t_numpy_sum:.5f} times faster")
```

```
Built-in sum:               5.52380 s
NumPy sum:                  0.35774 s
Performance improvement:    15.44076 times faster
```

Similarly, the product

$$\mathrm{prod}(\mathbf{x}) = \prod_{i=1}^{n} x_i, \quad \mathbf{x} = (x_1, \ldots, x_n) \in \mathbb{R}^n$$

is implemented by the np.prod function and the np.ndarray.prod method.

```
np.prod(x)
```

```
np.float64(-543.996)
```

On quite a few occasions, we need to find the maximum or minimum of an array. We can do this using the np.max and np.min functions. (Similarly to the others, these are also available as array methods.) The rule of thumb is if you want to perform any array operation, use NumPy functions.

## 3.1.1  Norms, distances, and dot products

Now that we have reviewed how to perform operations on our vectors efficiently, it's time to dive deep into the really interesting part: norms and distances.

Let's start with the most important one: the Euclidean norm, also known as the 2-norm, defined by

$$\|\mathbf{x}\|_2 = \left( \sum_{i=1}^{n} x_i^2 \right)^{1/2}, \quad \mathbf{x} = (x_1, \ldots, x_n) \in \mathbb{R}^n.$$

A straightforward implementation would be the following.

```
def euclidean_norm(x: np.ndarray):
    return np.sqrt(np.sum(x**2))
```

Note that our euclidean_norm function is dimension-agnostic; that is, it works for arrays of every dimension.

```
# a 1D array with 4 elements, which is a vector in 4-dimensional space
x = np.array([-3.0, 1.2, 1.2, 2.1])

# a 1D array with 2 elements, which is a vector in 2-dimensional space
y = np.array([8.1, 6.3])

euclidean_norm(x)
```

```
np.float64(4.036087214122113)
```

```
euclidean_norm(y)
```

```
np.float64(10.261578825892242)
```

But wait, didn't I just mention that we should use NumPy functions whenever possible? Norms are important enough to have their own functions: np.linalg.norm.

```
np.linalg.norm(x)
```

```
np.float64(4.036087214122113)
```

With a quick inspection, we can check that these match for our vector x.

```
np.equal(euclidean_norm(x), np.linalg.norm(x))
```

```
np.True_
```

However, the Euclidean norm is just a special case of $p$-norms. Recall that for any $p \in [0, \infty)$, we defined the $p$-norm by the formula

$$\|\mathbf{x}\|_p = \left( \sum_{i=1}^{n} |x_i|^p \right)^{1/p}, \quad \mathbf{x} = (x_1, \ldots, x_n) \in \mathbb{R}^n,$$

and

$$\|\mathbf{x}\|_\infty = \max\{|x_1|, \ldots, |x_n|\}, \quad \mathbf{x} = (x_1, \ldots, x_n) \in \mathbb{R}^n$$

for $p = \infty$. It is a good practice to keep the number of functions in a codebase minimal to reduce maintenance costs. Can we compact all $p$-norms into a single Python function that takes the value of $p$ as an argument? Sure. We only have a small issue: representing $\infty$. Python and NumPy both provide their own representations, but we will go with NumPy's np.inf. Surprisingly, this is a float type.

```
type(np.inf)
```

```
float
```

```python
def p_norm(x: np.ndarray, p: float):
    if np.isinf(p):
        return np.max(np.abs(x))
    elif p >= 1:
        return (np.sum(np.abs(x)**p))**(1/p)
    else:
        raise ValueError("p must be a float larger or equal than 1.0 or inf.")
```

Since $\infty$ can have multiple other representations, such as Python's built-in math.inf, we can make our function more robust by using the np.isinf function to check if an object represents $\infty$ or not.

A quick check shows that p_norm works as intended.

```
x = np.array([-3.0, 1.2, 1.2, 2.1])

for p in [1, 2, 42, np.inf]:
    print(f"p-norm for p = {p}: \t {p_norm(x, p=p):.5f}")
```

```
p-norm for p = 1:        7.50000
p-norm for p = 2:        4.03609
p-norm for p = 42:       3.00000
p-norm for p = inf:      3.00000
```

However, once again, NumPy is one step ahead of us. In fact, the familiar np.linalg.norm already does this out of the box. We can achieve the same with less code by passing the value of $p$ as the argument ord, short for *order*. For ord = 2, we obtain the good old 2-norm.

```
for p in [1, 2, 42, np.inf]:
    print(f"p-norm for p = {p}: \t {np.linalg.norm(x, ord=p):.5f}")
```

```
p-norm for p = 1:        7.50000
p-norm for p = 2:        4.03609
p-norm for p = 42:       3.00000
p-norm for p = inf:      3.00000
```

Somewhat surprisingly, distances don't have their own NumPy functions. However, as the most common distance metrics are generated from norms (*Section 2.1.1*), we can often write our own. For instance, here is the Euclidean distance.

```
def euclidean_distance(x: np.ndarray, y: np.ndarray):
    return np.linalg.norm(x - y, ord=2)
```

Besides norms and distances, the third component that defines the geometry of our vector spaces is the inner product. During our journey, we'll almost exclusively use the *dot product*, defined in the vector space $\mathbb{R}^n$ by

$$\langle \mathbf{x}, \mathbf{y} \rangle = \sum_{i=1}^{n} x_i y_i, \quad \mathbf{x}, \mathbf{y} \in \mathbb{R}^n.$$

By now, you can easily smash out a Python function that calculates this. In principle, the one-liner below should work.

```python
def dot_product(x: np.ndarray, y: np.ndarray):
    return np.sum(x*y)
```

Let's test this out!

```python
x = np.array([-3.0, 1.2, 1.2, 2.1])
y = np.array([1.9, 2.5, 3.9, 1.2])

dot_product(x, y)
```

```
np.float64(4.5)
```

When the dimension of the vectors doesn't match, the function throws an exception as we expect.

```python
x = np.array([-3.0, 1.2, 1.2, 2.1])
y = np.array([1.9, 2.5])

dot_product(x, y)
```

```
---------------------------------------------------------------------------
ValueError                                Traceback (most recent call last)
Cell In[39], line 4
      1 x = np.array([-3.0, 1.2, 1.2, 2.1])
      2 y = np.array([1.9, 2.5])
----> 4 dot_product(x, y)

Cell In[37], line 2, in dot_product(x, y)
      1 def dot_product(x: np.ndarray, y: np.ndarray):
----> 2     return np.sum(x*y)

ValueError: operands could not be broadcast together with shapes (4,) (2,)
```

However, upon further attempts to break the code, a strange thing occurs. Our function `dot_product` should fail when called with an *n*-dimensional and one-dimensional vector, and this is not what happens.

```python
x = np.array([-3.0, 1.2, 1.2, 2.1])
y = np.array([2.0])
```

```
dot_product(x, y)
```

```
np.float64(3.0)
```

I always advocate breaking solutions in advance to avoid later surprises, and the above example excellently illustrates the usefulness of this principle. If the previous phenomenon occurs in production, you would have code that executes properly but gives a totally wrong result. That's the worst kind of bug.

Behind the scenes, NumPy is doing something called *broadcasting*. When performing an operation on two arrays with mismatching shapes, it tries to guess the correct sizes and reshape them so that the operation can go through. Check out what takes place when calculating x*y.

```
x*y
```

```
array([-6. ,  2.4,  2.4,  4.2])
```

NumPy guessed that we want to multiply all elements of x by the scalar y[0], so it transforms y = np.array([2.0]) into np.array([2.0, 2.0, 2.0, 2.0]), then calculates the elementwise product.

Broadcasting is extremely useful because it allows us to write much simpler code by automagically performing transformations. Still, if you are unaware of how and when broadcasting is done, it can seriously come back to bite you. Just like in our case, as the inner product of a four-dimensional and one-dimensional vector is not defined.

To avoid writing excessive checks for edge cases (or missing them altogether), we calculate the inner product in practice using the np.dot function.

```
x = np.array([-3.0, 1.2, 1.2, 2.1])
y = np.array([1.9, 2.5, 3.9, 1.2])

np.dot(x, y)
```

```
np.float64(4.5)
```

When attempting to call np.dot with misaligned arrays, it fails as supposed to, even in cases when broadcasting bails out our custom implementation.

```
x = np.array([-3.0, 1.2, 1.2, 2.1])
y = np.array([2.0])

np.dot(x, y)
```

```
--------------------------------------------------------------
ValueError                          Traceback (most recent call last)
Cell In[43], line 4
      1 x = np.array([-3.0, 1.2, 1.2, 2.1])
      2 y = np.array([2.0])
----> 4 np.dot(x, y)

ValueError: shapes (4,) and (1,) not aligned: 4 (dim 0) != 1 (dim 0)
```

Now that we have a basic arsenal of array operations and functions, it is time to do something with them!

## 3.1.2  The Gram-Schmidt orthogonalization process

One of the most fundamental algorithms in linear algebra is the Gram-Schmidt orthogonalization process (*Theorem 2.2.6*), used to turn a set of linearly independent vectors into an orthonormal set.

To be more precise, for our input of a set of linearly independent vectors $v_1, \dots, v_n \in \mathbb{R}^n$, the Gram-Schmidt process finds the output set of vectors $e_1, \dots, e_n \in \mathbb{R}^n$ such that $\|e_i\| = 1$ and $\langle e_i, e_j \rangle = 0$ for all $i \neq j$ (that is, the vectors are orthonormal), and $\mathrm{span}(e_1, \dots, e_k) = \mathrm{span}(v_1, \dots, v_k)$ for all $k = 1, \dots, n$.

If you are having trouble recalling how this is done, feel free to revisit *Section 3.1.2*, where we first described the algorithm. The learning process is a spiral, where we keep revisiting old concepts from new perspectives. For the Gram-Schmidt process, this is our second iteration, where we put the mathematical formulation into code.

Since we are talking about a *sequence of vectors*, we need a suitable data structure for this purpose. There are several possibilities for this in Python. For now, we are going with the conceptually simplest, albeit computationally rather suboptimal, one: lists.

```
vectors = [np.random.rand(5) for _ in range(5)]    # randomly generated vectors in a list
vectors
```

```
[array([0.85885635, 0.05917163, 0.42449235, 0.39776749, 0.89750107]),
 array([0.49579437, 0.42797077, 0.21057023, 0.3091438 , 0.52590854]),
 array([0.73079791, 0.58140107, 0.09823772, 0.14323477, 0.63606972]),
 array([0.89495164, 0.40614454, 0.60637559, 0.61614928, 0.69006552]),
```

```
array([0.1996764 , 0.90298211, 0.70602567, 0.45721469, 0.02375226])]
```

The first component of the algorithm is the orthogonal projection operator, defined by

$$\text{proj}_{e_1,\dots,e_k}(x) = \sum_{i=1}^{k} \frac{\langle x, e_i \rangle}{\langle e_i, e_i \rangle} e_i.$$

With our NumPy tools, the implementation is straightforward by now.

```python
from typing import List

def projection(x: np.ndarray, to: List[np.ndarray]):
    """
    Computes the orthogonal projection of the vector `x`
    onto the subspace spanned by the set of vectors `to`.
    """
    p_x = np.zeros_like(x)

    for e in to:
        e_norm_square = np.dot(e, e)
        p_x += np.dot(x, e)*e/e_norm_square

    return p_x
```

To check if it works, let's look at a simple example and visualize the results. Since this book is written in Jupyter Notebooks, we can do it right here.

```python
x = np.array([1.0, 2.0])
e = np.array([2.0, 1.0])

x_to_e = projection(x, to=[e])
```

```python
import matplotlib.pyplot as plt

with plt.style.context("seaborn-v0_8"):
    plt.figure(figsize=(7, 7))
    plt.xlim([-0, 3])
    plt.ylim([-0, 3])
    plt.arrow(0, 0, x[0], x[1], head_width=0.1, color="r", label="x", linewidth=2)
    plt.arrow(0, 0, e[0], e[1], head_width=0.1, color="g", label="e", linewidth=2)
    plt.arrow(x_to_e[0], x_to_e[1], x[0] - x_to_e[0], x[1] - x_to_e[1], linestyle="--")
```

```
plt.arrow(0, 0, x_to_e[0], x_to_e[1], head_width=0.1, color="b",
label="projection(x, to=[e])")
plt.legend()
plt.show()
```

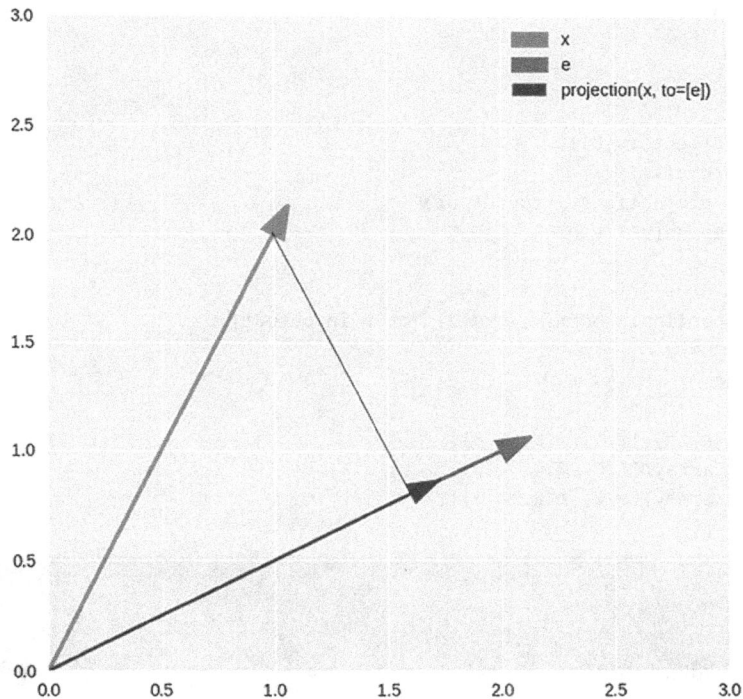

*Figure 3.1: The projection of x to e*

Checking the orthogonality of e and x - x to e provides another means of verification.

```
np.allclose(np.dot(e, x - x_to_e), 0.0)
```

```
True
```

When writing code for production, a couple of visualizations and ad hoc checks are not enough. An extensive set of unit tests is customarily written to ensure that a function works as intended. We are skipping this to keep our discussion on track, but feel free to add some of your tests. After all, mathematics and programming are not a spectator's sport.

With the `projection` function available to us, we are ready to knock the Gram-Schmidt algorithm out of the park.

```python
def gram_schmidt(vectors: List[np.ndarray]):
    """
    Creates an orthonormal set of vectors from the input
    that spans the same subspaces.
    """
    output = []

    # 1st step: finding an orthogonal set of vectors
    output.append(vectors[0])
    for v in vectors[1:]:
        v_proj = projection(v, to=output)
        output.append(v - v_proj)

    # 2nd step: normalizing the result
    output = [v/np.linalg.norm(v, ord=2) for v in output]

    return output

gram_schmidt([np.array([2.0, 1.0, 1.0]),
              np.array([1.0, 2.0, 1.0]),
              np.array([1.0, 1.0, 2.0])])
```

```
[array([0.81649658, 0.40824829, 0.40824829]),
 array([-0.49236596,  0.86164044,  0.12309149]),
 array([-0.30151134, -0.30151134,  0.90453403])]
```

Let's quickly test out this implementation with a simple example.

```python
test_vectors = [np.array([1.0, 0.0, 0.0]),
                np.array([1.0, 1.0, 0.0]),
                np.array([1.0, 1.0, 1.0])]

gram_schmidt(test_vectors)
```

```
[array([1., 0., 0.]), array([0., 1., 0.]), array([0., 0., 1.])]
```

So, we have just created our first algorithm from scratch. This is like the base camp for Mount Everest. We have come a long way, but there is much further to go before we can create a neural network from scratch. Until then, the journey is packed with some beautiful sections, and this is one of them. Take a while to appreciate this, then move on when you are ready.

**Remark 3.1.1 (Linearly dependent inputs of the Gram-Schmidt process)**

Recall that if the input vectors of the Gram-Schmidt are linearly dependent, some vectors of the output are zero (*Remark 2.2.1*). In practice, this causes a lot of problems.

For instance, we normalize the vectors in the end, using list comprehension:

```python
output = [v/np.linalg.norm(v, ord=2) for v in output]
```

This can cause numerical issues. If any v is approximately 0, its norm np.linalg.norm(v, ord=2) is going to be really small, and division with such small numbers is problematic.

This issue also affects the projection function. Take a look at the definition below:

```python
def projection(x: np.ndarray, to: List[np.ndarray]):
    """
    Computes the orthogonal projection of the vector `x`
    onto the subspace spanned by the set of vectors `to`.
    """

    p_x = np.zeros_like(x)

    for e in to:
        e_norm_square = np.dot(e, e)
        p_x += np.dot(x, e)*e / e_norm_square

    return p_x
```

If e is (close to) 0, which can happen if the input vectors are linearly dependent, then e_norm_square is small. One way to solve this is to add a small float, say, 1e-16.

```python
def projection(x: np.ndarray, to: List[np.ndarray]):
    p_x = np.zeros_like(x)

    for e in to:
        e_norm_square = np.dot(e, e)

        # note the change below:
        p_x += np.dot(x, e)*e / (e_norm_square + 1e-16)

    return p_x
```

Now, let's meet the single most important objects in machine learning: matrices.

## 3.2 Matrices, the workhorses of linear algebra

I am quite sure that you were already familiar with the notion of matrices before reading this book. Matrices are one of the most important data structures that can represent systems of equations, graphs, mappings between vector spaces, and many more. Matrices are the fundamental building blocks of machine learning.

At first look, we define a *matrix* as a table of numbers. If the matrix $A$ has, for instance, $n$ rows and $m$ columns of real numbers, we write

$$A = \begin{bmatrix} a_{1,1} & a_{1,2} & \dots & a_{1,m} \\ a_{2,1} & a_{2,2} & \dots & a_{2,m} \\ \vdots & \vdots & \ddots & \vdots \\ a_{n,1} & a_{n,2} & \dots & a_{n,m} \end{bmatrix} \in \mathbb{R}^{n \times m}. \tag{3.1}$$

When we don't want to write out the entire matrix as (3.1), we use the abbreviation $A = (a_{i,j})_{i=1,j=1}^{n,m}$.

The set of all $n \times m$ real matrices is denoted by $\mathbb{R}^{n \times m}$. We will exclusively talk about real matrices, but when referring to other types, this notation is modified accordingly. For instance, $\mathbb{Z}^{\{n \times m\}}$ denotes the set of integer matrices.

Matrices can be added and multiplied together, or multiplied by a scalar.

**Definition 3.2.1 (Matrix operations)**

*(a)* Let $A \in \mathbb{R}^{n \times m}$ be a matrix and $c \in \mathbb{R}$ a real number. The multiple of $A$ by the scalar $c$ is defined by

$$cA := (ca_{i,j})_{i,j=1}^{n,m} \in \mathbb{R}^{n \times m}.$$

*(b)* Let $A, B \in \mathbb{R}^{n \times m}$ be two matrices of matching dimensions. Their sum $A + B$ is defined by

$$A + B := (a_{i,j} + b_{i,j})_{i,j=1}^{n,m} \in \mathbb{R}^{n \times m}.$$

*(c)* Let $A \in \mathbb{R}^{n \times l}$ and $B \in \mathbb{R}^{l \times m}$ be two matrices. Their product $AB \in \mathbb{R}^{n \times m}$ is defined by

$$AB := \left( \sum_{k=1}^{l} a_{i,k} b_{k,j} \right)_{i,j=1}^{n,m} \in \mathbb{R}^{n \times m}.$$

Scalar multiplication and addition are clear, but matrix multiplication is not as simple to understand. Fortunately, visualization can help. In essence, the $(i, j)$-th element is the dot product of the $i$-th row of $A$ and the $j$-th column of $B$.

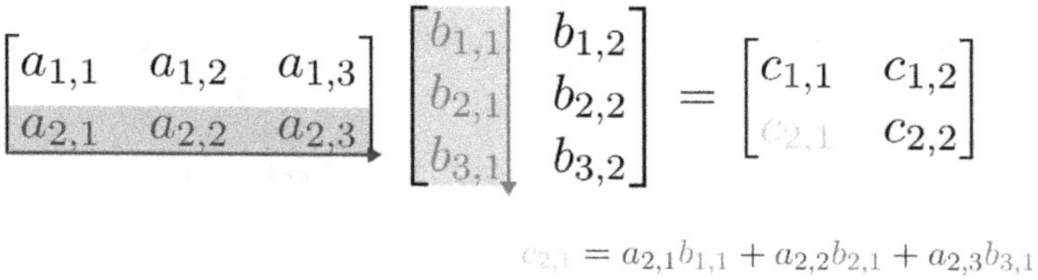

$$\begin{bmatrix} a_{1,1} & a_{1,2} & a_{1,3} \\ a_{2,1} & a_{2,2} & a_{2,3} \end{bmatrix} \begin{bmatrix} b_{1,1} & b_{1,2} \\ b_{2,1} & b_{2,2} \\ b_{3,1} & b_{3,2} \end{bmatrix} = \begin{bmatrix} c_{1,1} & c_{1,2} \\ c_{2,1} & c_{2,2} \end{bmatrix}$$

$$c_{2,1} = a_{2,1}b_{1,1} + a_{2,2}b_{2,1} + a_{2,3}b_{3,1}$$

*Figure 3.2: Visualizing matrix multiplication*

Besides addition and multiplication, there is another operation that is worth mentioning: *transposition*.

**Definition 3.2.2 (Matrix transposition)**

Let $A = (a_{i,j})_{i,j=1}^{\{n,m\}} \in \mathbb{R}^{\{n \times m\}}$ be a matrix. The matrix $A^T$, defined by

$$A^T = (a_{j,i})_{i,j=1}^{n,m} \in \mathbb{R}^{m \times n}$$

is called the *transpose* of $A$. The operation $A \mapsto A^T$ is called *transposition*.

Transposition simply means "flipping" the matrix, replacing rows with columns. For example,

$$A = \begin{bmatrix} a & b \\ c & d \end{bmatrix}, \quad A^T = \begin{bmatrix} a & c \\ b & d \end{bmatrix},$$

or

$$B = \begin{bmatrix} 0 & 1 \\ 2 & 3 \\ 4 & 5 \end{bmatrix}, \quad B^T = \begin{bmatrix} 0 & 2 & 4 \\ 1 & 3 & 5 \end{bmatrix}.$$

As opposed to addition and multiplication, transposition is a *unary* operation. (Unary means that it takes one argument. Binary operations take two arguments, and so on.)

Let's take another look at matrix multiplication, one of the most frequently used operations in computing. As it can be performed extremely fast on modern computers, it is common to vectorize certain algorithms just to express it in terms of matrix multiplications.

Thus, the more we know about it, the better. To get a grip on the operation itself, we can take a look at it from a few different angles. Let's start with a special case!

In machine learning, taking the product of a matrix and a column vector is a fundamental building block of certain models. For instance, this is linear regression in itself, or the famous fully connected layer in neural networks.

To see what happens in this case, let $A \in \mathbb{R}^{n \times m}$ be a matrix. If we treat $\mathbf{x} \in \mathbb{R}^m$ as a column vector $\mathbf{x} \in \mathbb{R}^{m \times 1}$, then $A\mathbf{x}$ can be written as

$$Ax = \begin{bmatrix} a_{1,1} & a_{1,2} & \cdots & a_{1,m} \\ a_{2,1} & a_{2,2} & \cdots & a_{2,m} \\ \vdots & \vdots & \ddots & \vdots \\ a_{n,1} & a_{n,2} & \cdots & a_{n,m} \end{bmatrix} \begin{bmatrix} x_1 \\ x_2 \\ \vdots \\ x_m \end{bmatrix} = \begin{bmatrix} \sum_{j=1}^{m} a_{1,j} x_j \\ \sum_{j=1}^{m} a_{2,j} x_j \\ \vdots \\ \sum_{j=1}^{m} a_{n,j} x_j \end{bmatrix}.$$

Based on this, the matrix $A$ describes a function that takes a piece of data $\mathbf{x}$, then transforms it into the form $A\mathbf{x}$.

This is the same as taking the linear combination of the columns of $A$, that is,

$$\begin{bmatrix} a_{1,1} & a_{1,2} & \cdots & a_{1,m} \\ a_{2,1} & a_{2,2} & \cdots & a_{2,m} \\ \vdots & \vdots & \ddots & \vdots \\ a_{n,1} & a_{n,2} & \cdots & a_{n,m} \end{bmatrix} \begin{bmatrix} x_1 \\ x_2 \\ \vdots \\ x_m \end{bmatrix} = x_1 \begin{bmatrix} a_{1,1} \\ a_{2,1} \\ \vdots \\ a_{n,1} \end{bmatrix} + \cdots + x_m \begin{bmatrix} a_{1,m} \\ a_{2,m} \\ \vdots \\ a_{n,m} \end{bmatrix}$$

With a bit more suggestive notation, by denoting the $i$-th column as $\mathbf{a}_i$, we can write

$$\begin{bmatrix} \mathbf{a}_1 & \mathbf{a}_2 & \cdots & \mathbf{a}_n \end{bmatrix} \begin{bmatrix} x_1 \\ \vdots \\ x_n \end{bmatrix} = \sum_{i=1}^{n} x_i \mathbf{a}_i, \quad \mathbf{a}_i = \begin{bmatrix} a_{1,i} \\ \vdots \\ a_{n,i} \end{bmatrix}. \tag{3.2}$$

If we replace the vector $\mathbf{x}$ with a matrix $B$, the columns in the product matrix $AB$ are linear combinations of the columns of $A$, where the coefficients are determined by $B$.

You should really appreciate that certain operations on the data can be written in the form $A\mathbf{x}$. Elevating this simple property to a higher level of abstraction, we can say that the data has the same representation as the function. If you are familiar with programming languages like Lisp, you know how beautiful this is.

There is one more way to think about the matrix product: taking the columnwise inner products. If $\mathbf{a}_i = (a_{i,1}, \dots, a_{i,n})$ denotes the $i$-th column of $A$, then $A\mathbf{x}$ can be written as

$$Ax = \begin{bmatrix} a_{1,1} & a_{1,2} & \dots & a_{1,m} \\ a_{2,1} & a_{2,2} & \dots & a_{2,m} \\ \vdots & \vdots & \ddots & \vdots \\ a_{n,1} & a_{n,2} & \dots & a_{n,m.} \end{bmatrix} \begin{bmatrix} x_1 \\ x_2 \\ \vdots \\ x_m \end{bmatrix} = \begin{bmatrix} \langle \mathbf{a}_1, \mathbf{x} \rangle \\ \langle \mathbf{a}_2, \mathbf{x} \rangle \\ \vdots \\ \langle \mathbf{a}_n, \mathbf{x} \rangle \end{bmatrix}, \tag{3.3}$$

That is, the transformation $\mathbf{x} \mapsto A\mathbf{x}$ projects the input $\mathbf{x}$ to the row vectors of $A$, then compacts the results in a vector.

## 3.2.1 Manipulating matrices

Because matrix operations are well defined, we can do algebra on matrices just as with numbers. However, there are some major differences. As manipulating matrix expressions is an essential skill, let's take a look at its fundamental rules!

> **Theorem 3.2.1** *(Properties of matrix addition and multiplication)*
>
> *(a) Let $A, B, C \in \mathbb{R}^{n \times l}$ be arbitrary matrices. Then,*
>
> $$A + (B + C) = (A + B) + C$$
>
> *holds. That is, matrix addition is associative.*
>
> *(b) Let $A \in \mathbb{R}^{n \times l}$, $B \in \mathbb{R}^{l \times k}$, $C \in \mathbb{R}^{k \times m}$ be arbitrary matrices. Then,*
>
> $$A(BC) = (AB)C$$
>
> *holds. That is, matrix multiplication is associative.*
>
> *(c) Let $A \in \mathbb{R}^{n \times l}$ and $B, C \in \mathbb{R}^{l \times m}$ be arbitrary matrices. Then,*
>
> $$A(B + C) = AB + AC$$
>
> *holds. That is, matrix multiplication is left-distributive with respect to addition.*
>
> *(d) Let $A, B \in \mathbb{R}^{n \times l}$ and $C \in \mathbb{R}^{l \times m}$ be arbitrary matrices. Then,*
>
> $$(A + B)C = AC + BC$$
>
> *holds. That is, matrix multiplication is right-distributive with respect to addition.*

As the proof is extremely technical and boring, we are going to skip it. However, there are a few things to note. Most importantly, matrix multiplication is *not* commutative; that is, $AB$ is not always equal to $BA$. (It might not even be defined.) For instance, consider

$$A = \begin{bmatrix} 1 & 1 \\ 1 & 1 \end{bmatrix}, \quad B = \begin{bmatrix} 1 & 0 \\ 0 & 2 \end{bmatrix}.$$

You can verify by hand that

$$AB = \begin{bmatrix} 1 & 2 \\ 1 & 2 \end{bmatrix}, \quad BA = \begin{bmatrix} 1 & 1 \\ 2 & 2 \end{bmatrix},$$

which are not equal.

In line with this, the algebraic identities that we use for scalars are quite different. For instance, if $A$ and $B$ are matrices, then

$$(A + B)(A + B) = A(A + B) + B(A + B) = A^2 + AB + BA + B^2.$$

or

$$(A + B)(A - B) = A(A - B) + B(A - B) = A^2 - AB + BA - B^2.$$

Transposition also behaves nicely with respect to addition and multiplication.

**Theorem 3.2.2** *(Properties of transposition)*

*(a) Let $A, B \in \mathbb{R}^{n \times m}$ be arbitrary matrices. Then,*

$$(A + B)^T = A^T + B^T$$

*holds.*

*(b) Let $A \in \mathbb{R}^{n \times l}$, $B \in \mathbb{R}^{l \times m}$ be arbitrary matrices. Then,*

$$(AB)^T = B^T A^T$$

*holds.*

We are not going to prove this either, but feel free to do so as an exercise.

## 3.2.2   Matrices as arrays

To perform computations with matrices inside a computer, we are looking for a data structure that represents a matrix A and supports

- accessing elements by A[i, j],
- assigning elements by A[i, j] = value
- addition and multiplication with the + and * operators,

and works lightning fast. These requirements only specify the interface of our matrix data structure, not the actual implementation. An obvious choice would be a list of lists, but as discussed when talking about representing vectors in computations (*Section 1.3*), this is highly suboptimal. Can we leverage the C array structure to store a matrix?

Yes, and this is precisely what NumPy does, providing a fast and convenient representation for matrices in the form of *multidimensional arrays*. Before learning how to use NumPy's machinery for our purposes, let's look a bit deeper into the heart of the issue.

At first glance, there seems to be a problem: a computer's memory is one-dimensional, thus addressed (indexed) by a single key, not two as we want. Thus, we can't just shove a matrix into the memory. The solution is to *flatten* the matrix and place each consecutive row next to each other, like *Figure 3.3* illustrates in the 3 × 3 case. This is called *row-major ordering*.

By storing the rows of any $n \times m$ matrix in a contiguous array, we get all the benefits of the array data structure at the low cost of a simple index transformation defined by

$$(i, j) \mapsto im + j.$$

(Note that for programming languages like Fortran or MATLAB that use column-major ordering — i.e., the columns are concatenated — this index transformation won't work. I leave figuring out the correct transformation as an exercise to check your understanding.)

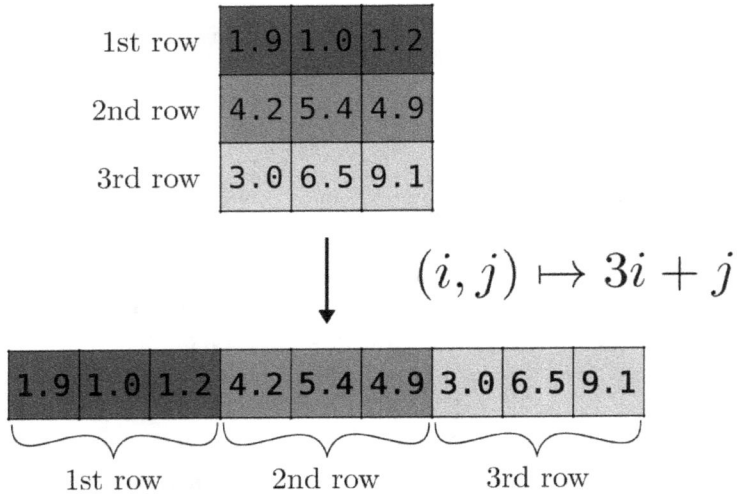

$$(i, j) \mapsto 3i + j$$

Figure 3.3: Flattening a matrix

To demonstrate what's happening, let's conjure up a prototypical `Matrix` class in Python that uses a single list to store all the values, yet supports accessing elements by row and column indices. For the sake of illustration, let's imagine that a Python list is actually a static array. (At least until this presentation is over.) This is for educational purposes only, as at the moment, we only want to understand the process, not to maximize performance.

Take a moment to review the code below. I'll explain everything line by line.

```python
from typing import Tuple


class Matrix:
    def __init__(self, shape: Tuple[int, int]):
        if len(shape) != 2:
            raise
            ValueError("The shape of a Matrix object must be a two-dimensional tuple.")

        self.shape = shape
        self.data = [0.0 for _ in range(shape[0]*shape[1])]

    def _linear_idx(self, i: int, j: int):
        return i*self.shape[1] + j

    def __getitem__(self, key: Tuple[int, int]):
        linear_idx = self._linear_idx(*key)
```

```
            return self.data[linear_idx]

    def __setitem__(self, key: Tuple[int, int], value):
        linear_idx = self._linear_idx(*key)
        self.data[linear_idx] = value

    def __repr__(self):
        array_form = [
            [self[i, j] for j in range(self.shape[1])]
            for i in range(self.shape[0])
        ]
        return "\n".join(["\t".join([f"{x}" for x in row]) for row in array_form])
```

The `Matrix` object is initialized with the `__init__` method. This is called when an object is created, like we are about to do now.

```
M = Matrix(shape=(3, 4))
```

Upon initialization, we supply the dimensions of the matrix in the form of a two-dimensional `tuple`, passed for the `shape` argument. In our concrete example, M is a 3 × 4 matrix, represented by an array of length 12. For simplicity, our simple `Matrix` is filled up with zeros by default.

Overall, the `__init__` method performs three main tasks:

1. **Validates** the shape parameter to ensure correctness
2. **Stores** the shape in an instance attribute for future reference
3. **Initializes** a list of size `shape[0] * shape[1]`, which serves as the primary data storage

The second method, suggestively named `_linear_idx`, is responsible for translating between the row-column indices of the matrix and the linear index for our internal one-dimensional representation. (In Python, it is customary to prefix methods with an underscore if they are not intended to be called externally. Many other languages, such as Java, support private methods. Python is not one of them, so we have to make do with such polite suggestions instead of strictly enforced rules.)

We can implement item retrieval via indexing by providing the `__getitem__` method, which expects a two-dimensional integer tuple as the key. For any `key = (i, j)`, the method:

1. **Calculates** the linear index using our `_linear_idx` method.
2. **Retrieves** the element located at the given linear index from the list.

Item assignment happens similarly, as given by the `__setitem__` magic method. Let's try these out to see if they work.

```
M[1, 2] = 3.14
M[1, 2]
```

```
3.14
```

By providing a `__repr__` method, we specify how a `Matrix` object is represented as a string. So, we can print it out to the standard output in a pretty form.

```
M
```

```
0.0      0.0      0.0      0.0
0.0      0.0      3.14     0.0
0.0      0.0      0.0      0.0
```

Pretty awesome. Now that we understand some of the internals, it is time to see how much we can achieve with NumPy.

## 3.2.3   Matrices in NumPy

As foreshadowed earlier, NumPy provides an excellent out-of-the-box representation for matrices in the form of *multidimensional arrays*. (These are often called *tensors*, but I'll just stick to the naming *array*.)

I have some fantastic news: these are the same `np.ndarray` objects we have been using! We can create one by simply providing a *list of lists* during initialization.

```
import numpy as np

A = np.array([[0, 1, 2, 3],
              [4, 5, 6, 7],
              [8, 9, 10, 11]])

B = np.array([[5, 5, 5, 5],
              [5, 5, 5, 5],
              [5, 5, 5, 5]])
A
```

```
array([[ 0,  1,  2,  3],
       [ 4,  5,  6,  7],
       [ 8,  9, 10, 11]])
```

Everything works the same as we have seen so far. Operations are performed elementwise, and you can plug them into functions like np.exp.

```
A + B          # pointwise addition
```

```
array([[ 5,  6,  7,  8],
       [ 9, 10, 11, 12],
       [13, 14, 15, 16]])
```

```
A*B            # pointwise multiplication
```

```
array([[ 0,  5, 10, 15],
       [20, 25, 30, 35],
       [40, 45, 50, 55]])
```

```
np.exp(A)   # pointwise application of the exponential function
```

```
array([[1.00000000e+00, 2.71828183e+00, 7.38905610e+00, 2.00855369e+01],
       [5.45981500e+01, 1.48413159e+02, 4.03428793e+02, 1.09663316e+03],
       [2.98095799e+03, 8.10308393e+03, 2.20264658e+04, 5.98741417e+04]])
```

Since we are working with multidimensional arrays, the transposition operator can be defined. Here, this is conveniently implemented as the np.transpose function, but can also be accessed at the np.ndarray.T attribute.

```
np.transpose(A)
```

```
array([[ 0,  4,  8],
       [ 1,  5,  9],
       [ 2,  6, 10],
       [ 3,  7, 11]])
```

```
A.T            # is the same as np.transpose(A)
```

```
array([[ 0,  4,  8],
       [ 1,  5,  9],
       [ 2,  6, 10],
       [ 3,  7, 11]])
```

As expected, we can get and set elements with the *indexing operator* []. The indexing starts from zero. (Don't even get me started.)

```
A[1, 2]    # 1st row, 2nd column (if we index rows and columns from zero)
```

```
np.int64(6)
```

Entire rows and columns can be accessed using slicing. Instead of giving the exact definitions, I'll just provide a few examples and let you figure it out with your internal pattern matching engine. (That is, your brain.)

```
A[:, 2]    # 2nd column
```

```
array([ 2,  6, 10])
```

```
A[1, :]    # 1st row
```

```
array([4, 5, 6, 7])
```

```
A[2, 1:4]    # 2nd row, 1st-4th elements
```

```
array([ 9, 10, 11])
```

```
A[1]    # 1st row
```

```
array([4, 5, 6, 7])
```

When used as an iterable, a two-dimensional array yields its rows at every step.

```
for row in A:
    print(row)
```

```
[0 1 2 3]
[4 5 6 7]
[ 8  9 10 11]
```

Initializing arrays can be done with the familiar np.zeros, np.ones, and other functions.

```
np.zeros(shape=(4, 5))
```

```
array([[0., 0., 0., 0., 0.],
       [0., 0., 0., 0., 0.],
       [0., 0., 0., 0., 0.],
       [0., 0., 0., 0., 0.]])
```

As you have guessed, that shape argument specifies the dimensions of the array. We are going to explore this property next. Let's initialize an example multidimensional array with three rows and four columns.

```
A = np.array([[0, 1, 2, 3],
              [4, 5, 6, 7],
              [8, 9, 10, 11]])
A
```

```
array([[ 0,  1,  2,  3],
       [ 4,  5,  6,  7],
       [ 8,  9, 10, 11]])
```

The shape of an array, stored inside the attribute np.ndarray.shape, is a tuple object describing its dimensions. In our example, since we have a $3 \times 4$ matrix, the shape equals (3, 4).

```
A.shape
```

```
(3, 4)
```

This innocent-looking attribute determines what kind of operations you can perform with your arrays. Let me tell you, as a machine learning engineer, shape mismatches will be the bane of your existence. You want to calculate the product of two matrices A and B? The second dimension of A must match the first dimension of B. Pointwise products? Matching or broadcastable shapes are required. Understanding shapes is vital.

However, we have just learned that multidimensional arrays are linear arrays in disguise. (See *Section 3.2.2.*) Because of this, we can reshape an array by slicing the linear view differently. For example, A can be reshaped into arrays with shapes (12, 1), (6, 2), (4, 3), (3, 4), (2, 6), and (1, 12).

```
A.reshape(6, 2)      # reshapes A into a 6 x 2 matrix
```

```
array([[ 0,  1],
       [ 2,  3],
       [ 4,  5],
       [ 6,  7],
       [ 8,  9],
       [10, 11]])
```

The np.ndarray.reshape method returns a newly constructed array object but doesn't change A. In other words, reshaping is not destructive in NumPy.

```
A
```

```
array([[ 0,  1,  2,  3],
       [ 4,  5,  6,  7],
       [ 8,  9, 10, 11]])
```

Reshaping is hard to wrap your head around for the first time. To help you visualize the process, *Figure 3.4* shows precisely what happens in our case.

If you are unaware of the exact dimension along a specific axis, you can get away by inputting -1 there during the reshaping. Since the product of dimensions is constant, NumPy is smart enough to figure out the missing one for you. This trick will get you out of trouble all the time, so it is worth taking note.

```
A.reshape(-1, 2)
```

```
array([[ 0,  1],
       [ 2,  3],
       [ 4,  5],
```

```
     [ 6,   7],
     [ 8,   9],
     [10, 11]])
```

```
A.reshape(-1, 4)
```

```
array([[ 0,  1,  2,  3],
       [ 4,  5,  6,  7],
       [ 8,  9, 10, 11]])
```

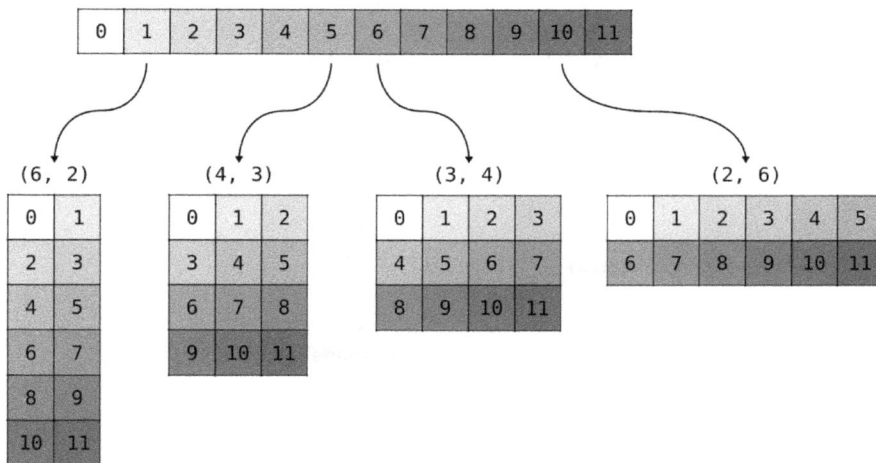

*Figure 3.4: Reshaping a one-dimensional array into multiple possible shapes*

We won't go into the details now, but as you probably guessed, multidimensional arrays can have more than two dimensions. The range of permitted shapes for the operations will be even more complicated, then. So, building a solid understanding now will provide a massive head start in the future.

## 3.2.4 Matrix multiplication, revisited

Without a doubt, one of the most important operations regarding matrices is multiplication. Computing determinants and eigenvalues? Matrix multiplication. Passing data through a fully connected layer? Matrix multiplication. Convolution? Matrix multiplication. We will see how these seemingly different things can be traced back to matrix multiplication; but first, let's discuss the operation itself from a computational perspective.

First, recap the mathematical definition. For any $A \in \mathbb{R}^{n \times m}$ and $B \in \mathbb{R}^{m \times l}$, their product is defined by the formula

$$AB = \left( \sum_{k=1}^{m} a_{i,k} b_{k,j} \right)_{i,j=1}^{n,l} \in \mathbb{R}^{n \times l}.$$

Notice that the element in the $i$-th row and $j$-th column of $AB$ is the dot product of $A$'s $i$-th row and $B$'s $j$-th column.

We can put this into code using the tools we have learned so far.

```python
from itertools import product


def matrix_multiplication(A: np.ndarray, B: np.ndarray):
    # checking if multiplication is possible
    if A.shape[1] != B.shape[0]:
        raise
        ValueError("The number of columns in A must match the number of rows in B.")

    # initializing an array for the product
    AB = np.zeros(shape=(A.shape[0], B.shape[1]))

    # calculating the elements of AB
    for i, j in product(range(A.shape[0]), range(B.shape[1])):
        AB[i, j] = np.sum(A[i, :]*B[:, j])

    return AB
```

Let's test our function with an example that is easy to verify by hand.

```python
A = np.ones(shape=(4, 6))
B = np.ones(shape=(6, 3))
matrix_multiplication(A, B)
```

```
array([[6., 6., 6.],
       [6., 6., 6.],
       [6., 6., 6.],
       [6., 6., 6.]])
```

The result is correct, as we expected.

Of course, matrix multiplication has its own NumPy function in the form of `numpy.matmul`.

```
np.matmul(A, B)
```

```
array([[6., 6., 6.],
       [6., 6., 6.],
       [6., 6., 6.],
       [6., 6., 6.]])
```

This yields the same result as our custom function. We can test it out by generating a bunch of random matrices and checking if the results match.

```
for _ in range(100):
    n, m, l = np.random.randint(1, 100), np.random.randint(1, 100), np.random.randint(1,
    100)
    A = np.random.rand(n, m)
    B = np.random.rand(m, l)

    if not np.allclose(np.matmul(A, B), matrix_multiplication(A, B)):
        print(f"Result mismatch for\n{A}\n and\n{B}")
        break
else:
    print("All good! Yay!")
```

```
All good! Yay!
```

According to this small test, our `matrix_multiplication` function yields the same result as NumPy's built-in one. We are happy, but don't forget: always use your chosen framework's implementations in practice, whether it be NumPy, TensorFlow, or PyTorch.

Since writing `np.matmul` is cumbersome when lots of multiplications are present, NumPy offers a way to abbreviate using the `@` operator.

```
A = np.ones(shape=(4, 6))
B = np.ones(shape=(6, 3))

np.allclose(A @ B, np.matmul(A, B))
```

```
True
```

## 3.2.5 Matrices and data

Now that we are familiar with matrix multiplication, it's time to make sense of them outside of linear algebra. Let's take a matrix $A \in \mathbb{R}^{n \times m}$ and a vector $\mathbf{x} \in \mathbb{R}^m$. By treating $\mathbf{x}$ as a column vector $\mathbf{x} \in \mathbb{R}^{m \times 1}$, the product of $A$ and $\mathbf{x}$ can be calculated by

$$A\mathbf{x} = \begin{bmatrix} a_{1,1} & a_{1,2} & \dots & a_{1,m} \\ a_{2,1} & a_{2,2} & \dots & a_{2,m} \\ \vdots & \vdots & \ddots & \vdots \\ a_{n,1} & a_{n,2} & \dots & a_{n,m\cdot} \end{bmatrix} \begin{bmatrix} x_1 \\ x_2 \\ \vdots \\ x_m \end{bmatrix} = \begin{bmatrix} \sum_{j=1}^{m} a_{1,j} x_j \\ \sum_{j=1}^{m} a_{2,j} x_j \\ \vdots \\ \sum_{j=1}^{m} a_{n,j} x_j \end{bmatrix}.$$

Mathematically speaking, looking at $\mathbf{x}$ as a column vector is perfectly natural. Think of it as extending $\mathbb{R}^m$ with a dummy dimension, thus obtaining $\mathbb{R}^{m \times 1}$. This form also comes naturally by considering that the columns of a matrix are images of the basis vectors by their very definition.

In practice, things are not as simple as they look. Implicitly, we have made a choice here: to represent datasets as a horizontal stack of column vectors. To elaborate further, let's consider two data points with four features and a matrix that maps these into a three-dimensional feature space. That is, let $\mathbf{x}_1, \mathbf{x}_2 \in \mathbb{R}^4$ and $A \in \mathbb{R}^{3 \times 4}$.

```python
x1 = np.array([2, 0, 0, 0])      # first data point
x2 = np.array([-1, 1, 0, 0])     # second data point

A = np.array([[0, 1, 2, 3],
              [4, 5, 6, 7],
              [8, 9, 10, 11]])    # a feature transformation
```

(I specifically selected these numbers so that the calculations would be easily verifiable by hand.) To be sure, we double-check the shapes.

```python
A.shape
```

```
(3, 4)
```

```python
x1.shape
```

```
(4,)
```

What happens when we call the `np.matmul` function?

```
np.matmul(A, x1)
```

```
array([ 0, 8, 16])
```

The result is correct. However, when we have a bunch of input data points, we prefer to calculate the images using a single operation. This way, we can take advantage of vectorized code, locality of reference, and all the juicy computational magic we have seen so far.

We can achieve this by horizontally stacking the column vectors, each one representing a data point. Mathematically speaking, we want to perform the calculation in code.

$$\begin{bmatrix} 0 & 1 & 2 & 3 \\ 4 & 5 & 6 & 7 \\ 8 & 9 & 10 & 11 \end{bmatrix} \begin{bmatrix} 2 & -1 \\ 0 & 1 \\ 0 & 0 \\ 0 & 0 \end{bmatrix} = \begin{bmatrix} 0 & 1 \\ 8 & 1 \\ 16 & 1 \end{bmatrix}$$

Upon looking up the NumPy documentation, we quickly find that the `np.hstack` function might be the tool for the job, at least according to its official documentation ( https://numpy.org/doc/stable/reference/gen erated/numpy.hstack.html). Yay!

```
np.hstack([x1, x2])    # np.hstack takes a list of np.ndarrays as its argument
```

```
array([ 2, 0, 0, 0, -1, 1, 0, 0])
```

Not yay. What happened? `np.hstack` treats one-dimensional arrays differently, and even though the math works out perfectly by creatively abusing the notation, we don't get away that easily in the trenches of real-life computations. Thus, we have to reshape our inputs manually. Meet the true skill gap between junior and senior machine learning engineers: correctly shaping multidimensional arrays.

```
# x.reshape(-1,1) turns x into a column vector
data = np.hstack([x1.reshape(-1, 1), x2.reshape(-1, 1)])
data
```

```
array([[ 2, -1],
       [ 0,  1],
       [ 0,  0],
       [ 0,  0]])
```

Let's try this one more time.

```
np.matmul(A, data)
```

```
array([[ 0,  1],
       [ 8,  1],
       [16,  1]])
```

Yay! (For real this time.)

Note that we made an extremely impactful choice in this chapter: **representing individual data points as column vectors**. I have written this in bold to emphasize its importance.

Why? Because we could have gone the other way and treated samples as row vectors. With our current choice, we ended up with a multidimensional array of shape

$$\text{number of dimensions} \times \text{number of samples},$$

as opposed to

$$\text{number of samples} \times \text{number of dimensions}.$$

The former is called *batch-last*, while the latter is called *batch-first* format. Popular frameworks like TensorFlow and PyTorch use batch-first, but we are going with batch-last. The reasons go back to the very definition of matrices, where columns are the images of basis vectors under the given linear transformation. This way, we can write multiplication from left to right, like $A\mathbf{x}$ and $AB$.

Should we define matrices as *rows* of basis vector images, everything turns upside down. This way, if $f$ and $g$ are linear transformations with "matrices" $A$ and $B$, the "matrix" of the composed transformation $f \circ g$ would be $BA$. This makes the math complicated and ugly.

On the other hand, batch-first makes the data easier to store and read. Think about a situation when you have thousands of data points in a single CSV file. Due to how input-output is implemented, files are read line by line, so it is natural and convenient to have a single line correspond to a single sample.

There are no good choices here; there are sacrifices either way. Since the math works out much easier for batch-last, we will use that format. However, in practice, you'll find that batch-first is more common. With this textbook, I don't intend to give you just a manual. My goal is to help you understand the internals of machine learning. If I succeed, you'll be able to apply your knowledge to translate between batch-first and batch-last seamlessly.

## 3.3  Summary

In this chapter, we finally dug into the trenches of practice instead of merely looking out from the towers of theory. Previously, we saw that NumPy arrays are the ideal tools for numeric computations, especially linear algebra. Now, we use them to provide fast and elegant implementations of what we learned in the previous chapter: norms, distances, dot products, and the Gram-Schmidt process.

Besides vectors, we also finally introduced matrices, one of the most important tools of machine learning. This time, we introduced, in a practical manner, viewing matrices as a table of numbers. Matrices can be transposed and added together, and unlike vectors, they can be multiplied with each other as well.

Speaking of our "from scratch" approach, before looking into how to actually work with matrices in practice, we created our very own `Matrix` implementation in vanilla Python. Closing the chapter, we dealt with the fundamentals and best practices of two-dimensional NumPy arrays, the prime matrix representation that Python can offer.

In the next chapter, we'll once more take a theoretical approach. This is how we do it in this book: looking at both aspects at once, supercharging our understanding of mathematics (and machine learning, along the way). We'll see that matrices are not just plain tables of numbers; they are data transformations as well. This property is beautiful beyond words: data and their transformations are represented by the same object.

Let's get to it!

## 3.4  Problems

**Problem 1.** Implement the mean squared error

$$\mathrm{MSE}(\mathbf{x}, \mathbf{y}) = \frac{1}{n} \sum_{i=1}^{n} (x_i - y_i)^2, \quad \mathbf{x}, \mathbf{y} \in \mathbb{R}^n$$

both with and without using NumPy functions and methods. (The vectors $\mathbf{x}$ and $\mathbf{y}$ should be represented by NumPy arrays in both cases.)

**Problem 2.** Compare the performances of the built-in maximum function `max` and NumPy's `np.max` using `timeit.timeit`, like we did above. Try running a different number of experiments and changing the array sizes to figure out the breakeven point between the two performances.

**Problem 3.** Instead of implementing the general $p$-norm as we did earlier in this chapter in *Section 3.1.1* , we can change things around to obtain the version below.

```
def p_norm(x: np.ndarray, p: float):
    if p >= 1:
        return (np.sum(np.abs(x)**p))**(1/p)
    elif np.isinf(p):
        return np.max(np.abs(x))
    else:
        raise ValueError("p must be a float larger or equal than 1.0 or inf.")
```

However, this doesn't work for $p = \infty$. What is the problem with it?

**Problem 4.** Let $\mathbf{w} \in \mathbb{R}^n$ be a vector with nonnegative elements. Use NumPy to implement the weighted $p$-norm by

$$\|\mathbf{x}\|_p^w = \left( \sum_{i=1}^n w_i |x_i|^p \right)^{1/p}, \quad \mathbf{x} = (x_1, \dots, x_n) \in \mathbb{R}^n.$$

Can you come up with a scenario where this can be useful in machine learning?

**Problem 5.** Implement the cosine similarity function, defined by the formula

$$\cos(\mathbf{x}, \mathbf{y}) = \left\langle \frac{\mathbf{x}}{\|\mathbf{x}\|}, \frac{\mathbf{y}}{\|\mathbf{y}\|} \right\rangle, \quad \mathbf{x}, \mathbf{y} \in \mathbb{R}^n.$$

(Whenever possible, use built-in NumPy functions.)

**Problem 6.** Calculate the product of the following matrices.

*(a)*

$$A = \begin{bmatrix} -1 & 2 \\ 1 & 5 \end{bmatrix}, \quad B = \begin{bmatrix} 6 & -2 \\ 2 & -6 \\ -3 & 2 \end{bmatrix}.$$

*(b)*

$$A = \begin{bmatrix} 1 & 2 & 3 \\ 4 & 5 & 6 \end{bmatrix}, \quad B = \begin{bmatrix} 7 & 8 \\ 9 & 10 \end{bmatrix}.$$

**Problem 7.** The famous Fibonacci numbers are defined by the recursive sequence

$$F_0 = 0,$$
$$F_1 = 1,$$
$$F_n = F_{n-1} + F_{n-2}.$$

*(a)* Write a recursive function that computes the $n$-th Fibonacci number. (Expect it to be *really* slow.)

*(b)* Show that

$$\begin{bmatrix} 1 & 1 \\ 1 & 0 \end{bmatrix}^n = \begin{bmatrix} F_{n+1} & F_n \\ F_n & F_{n-1} \end{bmatrix},$$

and use this identity to write a non-recursive function that computes the $n$-th Fibonacci number.

Use Python's built-in `timeit` function to measure the execution of both functions. Which one is faster?

**Problem 8.** Let $A, B \in \mathbb{R}^{n \times m}$ be two matrices. Their *Hadamard product* is defined by

$$A \odot B = \begin{bmatrix} a_{1,1}b_{1,1} & a_{1,2}b_{1,2} & \cdots & a_{1,n}b_{1,n} \\ a_{2,1}b_{2,1} & a_{2,2}b_{2,2} & \cdots & a_{2,n}b_{2,n} \\ \vdots & \vdots & \ddots & \vdots \\ a_{n,1}b_{n,1} & a_{n,2}b_{n,2} & \cdots & a_{n,n}b_{n,n} \end{bmatrix}.$$

Implement a function that takes two identically shaped NumPy arrays, then performs the Hadamard product on them. (There are two ways to do this: with `for` loops and with NumPy operations. It is instructive to implement both.)

**Problem 9.** Let $A \in \mathbb{R}^{n \times n}$ be a square matrix. Functions of the form

$$B(\mathbf{x}, \mathbf{y}) = \mathbf{x}^T A \mathbf{y}, \quad \mathbf{x}, \mathbf{y} \in \mathbb{R}^n$$

are called *bilinear forms*. Implement a function that takes two vectors and a matrix (all represented by NumPy arrays), then calculates the corresponding bilinear form.

# Join our community on Discord

Read this book alongside other users, Machine Learning experts, and the author himself.

Ask questions, provide solutions to other readers, chat with the author via Ask Me Anything sessions, and much more.

Scan the QR code or visit the link to join the community.

https://packt.link/math

# 4

# Linear Transformations

*"Why do my eyes hurt?"*
*"You've never used them before."*

— *Morpheus to Neo, when waking up from the Matrix for the first time*

In most linear algebra courses, the curriculum is all about matrices. In machine learning, we work with them all the time. Here is the thing: matrices don't tell the whole story. It is hard to understand the patterns by looking only at matrices. For instance, why is matrix multiplication defined in such a complex way as it is? Why are relations like $B = T^{-1}AT$ important? Why are some matrices invertible and some are not?

To *really* understand what is going on, we have to look at what gives rise to matrices: linear transformations. Like for Neo, this might hurt a bit, but it will greatly reward us later down the line. Let's get to it!

## 4.1   What is a linear transformation?

With the introduction of inner products, orthogonality, and orthogonal/orthonormal bases, we know everything about the structure of our feature spaces. However, in machine learning, our interest mainly lies in *transforming* the data.

From this viewpoint, a neural network is just a function composed of smaller parts (known as *layers*), transforming the data to a new feature space in every step. One of the key components of models in machine learning are *linear transformations*.

You probably encountered them as functions of the form $f(\mathbf{x}) = A\mathbf{x}$, but this is only one way to look at them. This section will start from a geometric viewpoint, then move towards the algebraic representation that you are probably already familiar with. To understand how neural networks can learn powerful high-level representations of the data, looking at the geometry of transforms is essential.

So, what linear transformations are? Let's not hesitate a moment further, and jump into the definition right away!

---

**Definition 4.1.1 (Linear transformations)**

Let $U$ and $V$ be two vector spaces (over the same scalar field), and let $f : U \to V$ be a function between them. We say that $f$ is linear if

$$f(a\mathbf{x} + b\mathbf{y}) = af(\mathbf{x}) + bf(\mathbf{y}) \tag{4.1}$$

holds for all vectors $\mathbf{x}, \mathbf{y} \in U$ and all scalars $a, b$.

---

This is why linear algebra is called *linear* algebra. In essence, a linear transformation is a mapping between two vector spaces that preserves the algebraic structure: addition and scalar multiplication. (Functions between vector spaces are often called *transformations*, so we will use this terminology.)

**Remark 4.1.1** Linearity is essentially combining two properties in one: $f(\mathbf{x} + \mathbf{y}) = f(\mathbf{x}) + f(\mathbf{y})$ and $f(a\mathbf{x}) = af(\mathbf{x})$ for all vectors $\mathbf{x}, \mathbf{y}$ and all scalars $a$. From these two, (4.1) follows by

$$f(a\mathbf{x} + b\mathbf{y}) = f(a\mathbf{x}) + f(b\mathbf{y}) = af(\mathbf{x}) + bf(\mathbf{y}).$$

Two properties immediately jump out from the definition. First, since

$$f(\mathbf{x}) = f(\mathbf{x} + \mathbf{0})$$
$$= f(\mathbf{x}) + f(\mathbf{0}),$$

$f(\mathbf{0}) = \mathbf{0}$ holds for every linear transformation. In addition, the composition of linear transformations is still linear, as

$$f\big(g(a\mathbf{x} + b\mathbf{y})\big) = f\big(ag(\mathbf{x}) + bg(\mathbf{y})\big)$$
$$= af(g(\mathbf{x})) + bf(g(\mathbf{y}))$$

shows for any linear $f$ and $g$ and scalars $a$ and $b$.

As usual, let's see some examples to build intuition.

**Example 1.** For any scalar $c$, the scaling transformation $f(\mathbf{x}) = c\mathbf{x}$ is linear.

This is probably the simplest example out there, and it can be defined in all vector spaces.

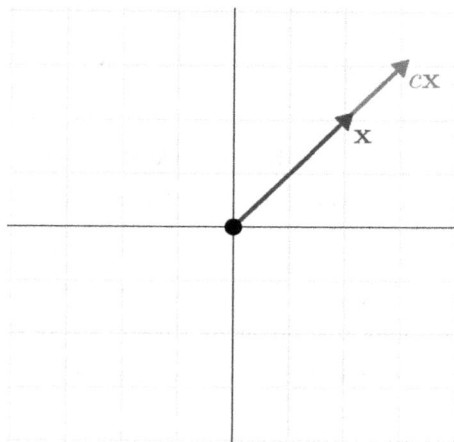

*Figure 4.1: Scaling as a linear transformation*

It's easy to see that scaling is linear:

$$c(a\mathbf{x} + b\mathbf{y}) = c(a\mathbf{x}) + c(b\mathbf{y})$$
$$= a(c\mathbf{x}) + b(c\mathbf{y}).$$

**Example 2.** In $\mathbb{R}^2$, rotations around the origin by an angle $\alpha$ are also linear.

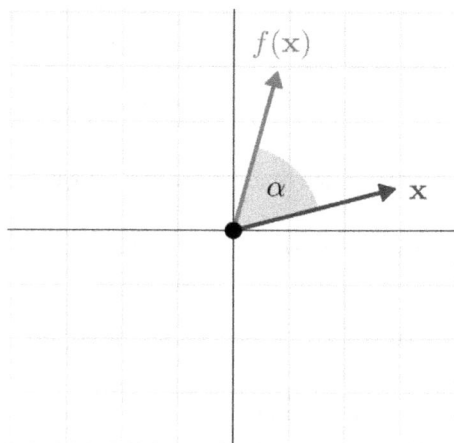

*Figure 4.2: Rotation in the Euclidean plane as a linear transformation*

To show that rotations are indeed linear, I pull the definition out from the hat: the rotation of a planar vector $\mathbf{x} = (x_1, x_2)$ with the angle $\alpha$ is described by

$$f(\mathbf{x}) = (x_1 \cos \alpha - x_2 \sin \alpha, x_1 \sin \alpha + x_2 \cos \alpha),$$

from which (4.1) is easily confirmed. I know that this looks like magic, but trust me, the rotation formula will be explained in detail. You can sweat it out with some basic trigonometry, or wait until we do this later with matrices.

In general, linear transformations have a strong connection with the geometry of the space. Later, we are going to study the linear transformations of $\mathbb{R}^2$ in detail, with an emphasis on geometric ones such as this. (Note that rotations are slightly more complicated in higher dimensions, as they will require an axis to rotate around.)

**Example 3.** In any vector space $V$ and a nonzero vector $\mathbf{v} \in V$, the translation defined by $f(\mathbf{x}) = \mathbf{x} + \mathbf{v}$ is *not* linear, as $f(\mathbf{0}) = \mathbf{v} \neq \mathbf{0}$.

We'll see more examples later in the section. For now, let's move on to some general properties of linear transformations. For any linear transformation $f : U \to V$, the image

$$\text{im}(f) = \{\mathbf{v} \in V : \mathbf{v} = f(\mathbf{u}) \text{ for some } \mathbf{u} \in U\}$$

is always a subspace *Section 1.2.7* of $\mathbf{V}$. This is easy to check: if $\mathbf{v}_1, \mathbf{v}_2 \in \text{im } f$, then there exist $\mathbf{u}_1, \mathbf{u}_2 \in U$ such that $f(\mathbf{u}_1) = \mathbf{v}_1$ and $f(\mathbf{u}_2) = \mathbf{v}_2$, as

$$a\mathbf{v}_1 + b\mathbf{v}_2 = af(\mathbf{u}_1) + bf(\mathbf{u}_2) = f(a\mathbf{u}_1 + b\mathbf{u}_2) \in \text{im } f.$$

To add one more level of abstraction, we will see that the set of all linear transformations form a vector space.

**Theorem 4.1.1** *Let U and V be two vector spaces over the same field F. Then the set of all linear transformations*

$$L(U, V) = \{f : U \to V \mid f \text{ is linear}\} \tag{4.2}$$

*is also a vector space over F, with the usual definitions for function addition and scalar multiplication.*

The proof of this is just a boring checklist, going through the items of the definition of vector spaces (*Definition 1.1.1*). I recommend you walk through it at least once to solidify your understanding of vector spaces, but there is really nothing special there.

## 4.1.1 Linear transformations and matrices

The definition of linear transformations, as we saw, is a bit abstract. However, there is a simple and expressive way to characterize them.

To see this, let $f : U \to V$ be a linear transformation between two vector spaces $U$ and $V$. Suppose that $\{\mathbf{u}_1, \dots, \mathbf{u}_m\}$ is a basis in $U$, while $\{\mathbf{v}_1, \dots, \mathbf{v}_n\}$ is a basis in $V$. Since every $\mathbf{x} \in U$ can be written in the form

$$\mathbf{x} = \sum_{i=1}^{m} x_i \mathbf{u}_i,$$

the linearity of $f$ implies

$$f\left( \sum_{j=1}^{m} x_j \mathbf{u}_j \right) = \sum_{j=1}^{m} x_j f(\mathbf{u}_j), \tag{4.3}$$

meaning that $f(\mathbf{x})$ is a linear combination of $f(\mathbf{u}_1), \dots, f(\mathbf{u}_m)$. In other words, every linear transformation is completely determined by the images of basis vectors. To expand this idea, suppose that for every $\mathbf{u}_j$, we have

$$f(\mathbf{u}_j) = \sum_{i=1}^{n} a_{i,j} \mathbf{v}_i$$

for some scalars $a_{i,j}$.

These $n \times m$ numbers completely describe $f$. For notational simplicity, we store these in a $n \times m$-sized table called a *matrix*, which we'll denote by $A_f$:

$$f \leftrightarrow A_f = \begin{bmatrix} a_{1,1} & a_{1,2} & \dots & a_{1,m} \\ a_{2,1} & a_{2,2} & \dots & a_{2,m} \\ \vdots & \vdots & \ddots & \vdots \\ a_{n,1} & a_{n,2} & \dots & a_{n,m} \end{bmatrix},$$

meaning that linear transformations are represented by matrices. This connection is heavily utilized throughout machine learning.

Expanding (4.3) further, for every $\mathbf{x} = \sum_{j=1}^{m} x_j \mathbf{u}_j$ we have

$$
\begin{aligned}
f(\mathbf{x}) &= \sum_{j=1}^{m} x_j f(\mathbf{u}_j) \\
&= \sum_{j=1}^{m} x_j \sum_{i=1}^{n} a_{i,j} \mathbf{v}_i \\
&= \sum_{i=1}^{n} \left( \sum_{j=1}^{m} a_{i,j} x_j \right) \mathbf{v}_i.
\end{aligned}
$$

Thus, the image of $\mathbf{x}$ can be expressed as $A_f \mathbf{x}$:

$$
f(\mathbf{x}) = \begin{bmatrix} a_{1,1} & a_{1,2} & \cdots & a_{1,m} \\ a_{2,1} & a_{2,2} & \cdots & a_{2,m} \\ \vdots & \vdots & \ddots & \vdots \\ a_{n,1} & a_{n,2} & \cdots & a_{n,m} \end{bmatrix} \begin{bmatrix} x_1 \\ x_2 \\ \vdots \\ x_m \end{bmatrix} = \begin{bmatrix} \sum_{j=1}^{m} a_{1,j} x_j \\ \sum_{j=1}^{m} a_{2,j} x_j \\ \vdots \\ \sum_{j=1}^{m} a_{n,j} x_j \end{bmatrix}.
$$

Two things to note here. First, we implicitly chose to represent vectors as columns instead of rows. This is a seriously impactful decision and will affect many of the computations later in this book. We'll keep pointing it out.

Second, the matrix representation depends on the choice of the basis! If, say, $P = \{\mathbf{p}_1, \dots, \mathbf{p}_n\} \subset U$ is the basis of our matrix, we denote this dependence in the subscript, writing $A_{f,P}$.

To avoid confusion, we'll almost exclusively define linear transformations by giving their matrices in the standard orthonormal basis. In practical scenarios, this makes it much easier to understand what is going on. So, whenever I write something like "*let A be the matrix of a linear transformation f*", it is implictly assumed that $A$ is written in the basis $\mathbf{e}_1 = (1, 0, \dots, 0), \mathbf{e}_2 = (0, 1, \dots, 0), \dots, \mathbf{e}_n = (0, 0, \dots, 1)$.

On a philosophical note, have you heard about Plato's allegory of the cave? In this thought experiment, people are assumed to be living in a cave constantly facing a single wall, only observing their shadows projected by a fire behind them. What they observe and use to build an internal representation of the world is very different from reality. Applying this analogy to linear algebra, matrices are the shadows that we observe and use in practical scenarios. In many introductory courses, linear transformations are hidden, and only matrix calculus is taught. My first exposition into the subject was similar: the first linear algebra course I took talked exclusively about matrices. It was as complicated and confusing as a math course can be. (Which, I can assure you, can be *very* complicated and confusing.) Later in my studies, everything clicked when I discovered that you could look at matrices from the perspective of linear transformations.

Without seeing what is behind matrices, it is impossible to master linear algebra. If my approach feels too abstract for you, keep this in mind: years later, when you are a practicing data scientist/machine learning engineer/researcher or whatever, going below the surface will pay huge dividends.

Let's get back on track and continue our discussion about linear transformations. The most commonly used matrix is the matrix of the identity transformation id : $\mathbf{x} \mapsto \mathbf{x}$. We'll denote this by $I$. It is easy to see that

$$
I = \begin{bmatrix} 1 & 0 & \dots & 0 \\ 0 & 1 & \dots & 0 \\ \vdots & \vdots & \ddots & \vdots \\ 0 & 0 & \dots & 1 \end{bmatrix}. \tag{4.4}
$$

To summarize, for a matrix $A$, a linear transformation can be given by $\mathbf{x} \mapsto A\mathbf{x}$. In fact, the mapping

$$
f \mapsto A_{f,P}
$$

defines a one-to-one correspondence between the space of linear transformations $L(U, V)$ defined by (4.2) and the set of $n \times m$ matrices, where $n$ and $m$ are the corresponding dimensions.

## 4.1.2   Matrix operations revisited

Functions can be added and composed. Because of the connection between linear transformations and matrices, matrix operations are inherited from the corresponding function operations.

With this principle in mind, we defined matrix addition so that the the matrix of the sum of two linear transformations is the sum of the corresponding matrices.

Mathematically speaking, if $f, g : U \to V$ are two linear transformations with matrices, $f \leftrightarrow A$ and $g \leftrightarrow B$, then

$$
(f + g)(\mathbf{u}_j) = f(\mathbf{u}_j) + g(\mathbf{u}_j) = \sum_{i=1}^{n} (a_{i,j} + b_{i,j})\mathbf{v}_i.
$$

Thus, the corresponding matrices can be added together elementwise:

$$
A + B = (a_{i,j} + b_{i,j})_{i,j=1}^{n,m}.
$$

Multiplication between matrices is defined by the composition of the corresponding transformations.

To see how, we study a special case first. (In general, it is a good idea to look at special cases first, as they often reduce the complexity and allow you to see patterns without information overload.) So, let $f, g : U \rightarrow U$ be two linear transformations, mapping $U$ onto itself. To determine the elements of the matrix corresponding to $f \circ g$, we have to express $f(g(\mathbf{u}_j))$ in terms of all the basis vectors $\mathbf{u}_1, \dots, \mathbf{u}_n$. For this, we have

$$(fg)(\mathbf{u}_j) = f(g(\mathbf{u}_j)) = f\left( \sum_{k=1}^{n} b_{k,j} \mathbf{u}_k \right)$$

$$= \sum_{k=1}^{n} b_{k,j} f(\mathbf{u}_k)$$

$$= \sum_{k=1}^{n} b_{k,j} \sum_{i=1}^{n} a_{i,k} \mathbf{u}_i$$

$$= \sum_{i=1}^{n} \left( \sum_{k=1}^{n} a_{i,k} b_{k,j} \right) \mathbf{u}_i.$$

By considering how we defined a transformation's matrix, the scalar $\left( \sum_{k=1}^{n} a_{i,k} b_{k,j} \right)$ is the element in the $i$-th row and $j$-th column of the matrix of $f \circ g$. Thus, matrix multiplication can be defined by

$$AB = \left( \sum_{k=1}^{n} a_{i,k} b_{k,j} \right)_{i,j=1}^{n}.$$

In the general case, we can only define the product of matrices if the corresponding linear transformations can be composed. That is, if $f : U \rightarrow V$, then $g$ must start from $V$. Translating this into the language of the matrices, the number of columns in $A$ must match the number of rows in $B$. So, for any $A \in \mathbb{R}^{n \times m}$ and $B \in \mathbb{R}^{m \times l}$, their product is defined by

$$AB = \left( \sum_{k=1}^{m} a_{i,k} b_{k,j} \right)_{i,j=1}^{n,l} \in \mathbb{R}^{n \times l}.$$

## 4.1.3  Inverting linear transformations

Regarding linear transformations, the question of invertibility is extremely important. For example, have you encountered a system of equations like this?

$$2x_1 + x_2 = 5$$

$$x_1 - 3x_2 = -8$$

If we define

$$A = \begin{bmatrix} 2 & 1 \\ 1 & -3 \end{bmatrix}, \quad \mathbf{b} = \begin{bmatrix} 5 \\ -8 \end{bmatrix}, \quad \mathbf{x} = \begin{bmatrix} x_1 \\ x_2 \end{bmatrix},$$

the above system can be written in the form $A\mathbf{x} = \mathbf{b}$. These are called *linear equations*, modeling various processes from finance to biology.

How would you write the solution of such an equation? If there would be a matrix $A^{-1}$ such that $A^{-1}A$ is the identity matrix $I$ (defined by (4.4)), then multiplying the equation $A\mathbf{x} = \mathbf{b}$ from the left by $A^{-1}$ would yield the solution in the form $\mathbf{x} = A^{-1}\mathbf{b}$.

The matrix $A^{-1}$ is called the *inverse matrix* of $A$. It might not always exist, but when it does, it is extremely important for several reasons. We'll talk about linear equations later, but first, let's study the fundamentals of invertibility! Here is the general definition.

**Definition 4.1.2 (Inverse of a linear transformation)**

Let $f : U \to V$ be a linear transformation between the vector spaces $U$ and $V$. We say that $f$ is invertible if there is a linear transformation $f^{-1}$ such that $f^{-1} \circ f$ and $f \circ f^{-1}$ are the identity functions; that is,

$$f^{-1}\big(f(\mathbf{u})\big) = \mathbf{u},$$
$$f\big(f^{-1}(\mathbf{v})\big) = \mathbf{v}$$

holds for all $\mathbf{u} \in U, \mathbf{v} \in V$. $f^{-1}$ is called the *inverse* of $f$.

Not all linear transformations are invertible. For instance, if $f$ maps all vectors to the zero vector, you cannot define an inverse.

There are certain conditions that guarantee the existence of the inverse. One of the most important ones connects the concept of basis with invertibility.

**Theorem 4.1.2 *(Invertibility of linear transformations)***

Let $f : U \to V$ *be a linear transformation and let* $\mathbf{u}_1, \ldots, \mathbf{u}_n$ *be a basis in* $U$. *Then* $f$ *is invertible if and only if* $f(\mathbf{u}_1), \ldots, f(\mathbf{u}_n)$ *is a basis in* $V$.

The following proof is straightforward, but can be a bit overwhelming. Feel free to skip this at the first reading, you can always revisit it later.

*Proof.* As usual, the proof of the *if and only if* type theorems consist of two parts, as these statements involve two implications.

*(a)* First, we prove that $f$ is invertible, then $f(\mathbf{u}_1), \dots, f(\mathbf{u}_n)$ is a basis. That is, we need to show that $f(\mathbf{u}_1), \dots, f(\mathbf{u}_n)$ is linearly independent and every $\mathbf{y} \in V$ can be written as their linear combination.

Since $f$ is invertible, $f(\mathbf{0}) = \mathbf{0}$, moreover there are no nonzero vectors $\mathbf{x} \in U$ such that $f(\mathbf{x}) = \mathbf{0}$. In other words, $\mathbf{0}$ cannot be written as the nontrivial linear combination of $f(\mathbf{u}_1), \dots, f(\mathbf{u}_n)$, from which *Theorem 1.2.2* implies the linear independence.

On the other hand, invertibility implies that every $\mathbf{y} \in V$ can be obtained as $\mathbf{y} = f(\mathbf{x})$ for some $\mathbf{x} \in U$. (With the choice $\mathbf{x} = f^{-1}(\mathbf{y})$.) As $\mathbf{u}_1, \dots, \mathbf{u}_n$ is a basis, $\mathbf{x} = \sum_{i=1}^{n} x_i \mathbf{u}_i$. Thus,

$$\mathbf{y} = f(\mathbf{x})$$
$$= f\left( \sum_{i=1}^{n} x_i \mathbf{u}_i \right)$$
$$= \sum_{i=1}^{n} x_i f(\mathbf{u}_i),$$

showing that $\mathrm{span}(f(\mathbf{u}_1), \dots, f(\mathbf{u}_n)) = V$.

The linear independence $f(\mathbf{u}_1), \dots, f(\mathbf{u}_n)$ and the fact that it spans $V$ gives that it is indeed a basis.

*(b)* Now we prove the other implication: if $f(\mathbf{u}_1), \dots, f(\mathbf{u}_n)$ is a basis, then $f$ is invertible.

If $f(\mathbf{u}_1), \dots, f(\mathbf{u}_n)$ is indeed a basis, then every $\mathbf{y} \in V$ can be written as

$$\mathbf{y} = \sum_{i=1}^{n} y_i f(\mathbf{u}_i) = f\left( \sum_{i=1}^{n} y_i \mathbf{u}_i \right),$$

which shows the surjectivity. Regarding the injectivity, if $\mathbf{y} = f(\mathbf{a}) = f(\mathbf{b})$ for some $\mathbf{a}, \mathbf{b} \in U$, then, since both $\mathbf{a}$ and $\mathbf{b}$ can be written as a linear combination of the $\mathbf{u}_i$ basis vectors, we would have

$$\mathbf{y} = f(\mathbf{a}) = f\left( \sum_{i=1}^{n} a_i \mathbf{u}_i \right) = \sum_{i=1}^{n} a_i f(\mathbf{u}_i)$$

and

$$\mathbf{y} = f(\mathbf{b}) = f\left( \sum_{i=1}^{n} b_i \mathbf{u}_i \right) = \sum_{i=1}^{n} y_i f(\mathbf{u}_i).$$

Thus, $\mathbf{0} = \sum_{i=1}^{n}(a_i - b_i)\mathbf{u}_i$, and since $\mathbf{u}_1, \dots, \mathbf{u}_n$ is a basis in U, $a_i = b_i$ must hold. Hence $f$ is injective.

A consequence of this theorem is that a linear transformation $f : U \to V$ is not invertible if the dimensions of $U$ and $V$ are different. We can look at invertibility from the aspect of matrices as well. For any $A \in \mathbb{R}^{n \times n}$, if the corresponding linear transformation is invertible, there exists a matrix $A^{-1} \in \mathbb{R}^{n \times n}$ such that $A^{-1}A = AA^{-1} = I$. If a matrix is not square, it is not invertible in the classical sense.

## 4.1.4 The kernel and the image

Regarding the invertibility of a linear transformation, two special sets play an essential role: the *kernel* and the *image*. Let's see them!

---

**Definition 4.1.3 (Kernel and image of linear transformations)**

Let $f : U \to V$ be a linear transformation. Its image and kernel is defined by

$$\operatorname{im} f := \{f(\mathbf{u}) : \mathbf{u} \in U\} \subseteq V$$

and

$$\ker f := \{\mathbf{u} \in U : f(\mathbf{u}) = 0\} \subseteq U.$$

---

Often, we write im $A$ and ker $A$ for some matrix $A$, referring to the linear transformation defined by $\mathbf{x} \mapsto A\mathbf{x}$. Due to the linearity of $f$, it is easy to see that im $f$ is a subspace of $V$ and ker $f$ is a subspace of $U$. As mentioned, they are closely connected with invertibility, as we shall see next.

---

**Theorem 4.1.3 *(Invertibility in terms of linear transformations)***

*Let $f : U \to V$ be a linear transformation.*

*(a) A is injective if and only if* ker $f = \{0\}$.

*(b) A is surjective if and only if* im $f = V$.

*(c) A is bijective (that is, invertible) if and only if* ker $f = \{0\}$ *and* im $f = V$.

---

*Proof.* *(a)* If $f$ is injective, there can only be one vector in $U$ that is mapped to $0$. Since $f(0) = 0$ for any linear transformation, ker $f = \{0\}$.

On the other hand, if there are two different vectors $\mathbf{x}, \mathbf{y} \in U$ such that $f(\mathbf{x}) = f(\mathbf{y})$, then $f(\mathbf{x} - \mathbf{y}) = f(\mathbf{x}) - f(\mathbf{y}) = 0$, so $\mathbf{x} - \mathbf{y} \in \ker f$. Thus, ker $f = \{0\}$ implies $\mathbf{x} = \mathbf{y}$, which gives the injectivity.

*(b)* This is just the definition of surjectivity.

*(c)* This immediately follows from combining *(a)* and *(b)* above.

Because matrices define linear transformations, it makes sense to talk about the inverse of a matrix.

Algebraically speaking, the inverse of an $A \in \mathbb{R}^{n \times n}$ is the matrix $A^{-1} \in \mathbb{R}^{n \times n}$ such that $A^{-1}A = AA^{-1} = I$ holds. The connection between linear transforms and matrices imply that $A^{-1}$ is the matrix of $f^{-1}$, so no surprise here.

Don't worry if this section about invertibility feels like a bit too much algebra. Later, when talking about the determinant of a transformation, we are going to study invertibility from a geometric perspective later in this chapter. In terms of matrices, later we are going to see a general method to calculate the inverse matrix in *Section 5.1.6*. We'll be there soon, but first, we take a look at how the choice of basis determines the matrix representation.

## 4.2  Change of basis

Previously in this section, we have seen that any linear transformation can be described with the images of the basis vectors (see *Section 4.1.1*). This gave us the matrix representation that we use all the time. However, this very much depends on the choice of basis. Different bases yield different matrices for the same transformation.

For instance, let's take a look at $f : \mathbb{R}^2 \to \mathbb{R}^2$, which maps $\mathbf{e}_1 = (1, 0)$ to the vector $(2, 1)$ and $\mathbf{e}_2 = (0, 1)$ to $(1, 2)$. Its matrix in the standard orthonormal basis $E = \{\mathbf{e}_1, \mathbf{e}_2\}$ is given by

$$A_{f,E} = \begin{bmatrix} 2 & 1 \\ 1 & 2 \end{bmatrix}. \tag{4.5}$$

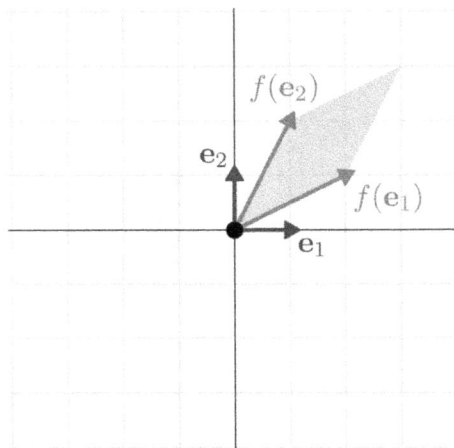

*Figure 4.3: The linear transformation $f$, defined by (4.5)*

The effect of $f$ is visualized in *Figure 4.3*.

What if we select a different basis, say $P = \{\mathbf{p}_1 = (1,1), \mathbf{p}_2 = (-1,1)\}$? With a quick calculation, we can check that

$$\begin{bmatrix} 2 & 1 \\ 1 & 2 \end{bmatrix} \begin{bmatrix} 1 \\ 1 \end{bmatrix} = \begin{bmatrix} 3 \\ 3 \end{bmatrix}, \quad \begin{bmatrix} 2 & 1 \\ 1 & 2 \end{bmatrix} \begin{bmatrix} -1 \\ 1 \end{bmatrix} = \begin{bmatrix} -1 \\ 1 \end{bmatrix}.$$

In other words, $f(\mathbf{p}_1) = 3\mathbf{p}_1 + 0\mathbf{p}_2$ and $f(\mathbf{p}_2) = 0\mathbf{p}_1 + \mathbf{p}_2$. This is visualized by *Figure 4.4.*

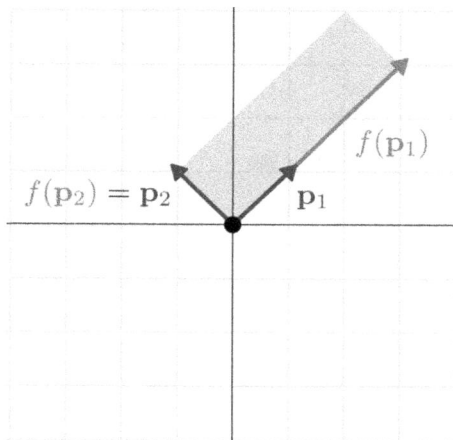

*Figure 4.4: The effect of $f$ on $p_1 = (1,1)$ and $p_2 = (-1,1)$*

This means that if $P = \{\mathbf{p}_1, \mathbf{p}_2\}$ is our basis (thus, if writing $(a, b)$ means $a\mathbf{p}_1 + b\mathbf{p}_2$), the matrix of $f$ becomes

$$A_{f,P} = \begin{bmatrix} 3 & 0 \\ 0 & 1 \end{bmatrix}.$$

In this form, $A_{f,P}$ is a diagonal matrix. (That is, its elements below and above the diagonal are zero.) As you can see, having the right basis can significantly simplify the linear transformation. For instance, in $n$ dimensions, applying a transformation in diagonal form requires only $n$ operations, as

$$\begin{bmatrix} d_1 & 0 & \dots & 0 \\ 0 & d_2 & \dots & 0 \\ \vdots & \vdots & \ddots & \vdots \\ 0 & 0 & \dots & d_n \end{bmatrix} \begin{bmatrix} x_1 \\ x_2 \\ \vdots \\ x_n \end{bmatrix} = \begin{bmatrix} d_1 x_1 \\ d_2 x_2 \\ \vdots \\ d_n x_n \end{bmatrix}$$

Otherwise, $n^2$ operations are needed. So, we can save a lot there.

## 4.2.1 The transformation matrix

We have just seen that the matrix of a linear transformation depends on our choice of basis. However, there is a special relation between matrices of the same transformation. We'll explore this next. Let $f : U \to U$ be a linear transformation, and let $P = \{\mathbf{p}_1, \ldots, \mathbf{p}_n\}$ and $Q = \{\mathbf{q}_1, \ldots, \mathbf{q}_n$ be two bases. As before, $A_{f,S}$ denotes the matrix of $f$ in some basis $S$.

Suppose that we know $A_{f,P}$, but we have our vectors represented in terms of the other basis $Q$. How do we calculate the images our vectors under the linear transformation? A natural idea is to first *transform our vector representations* from $Q$ to $P$, apply $A_{f,P}$, then transform the representations back. In the following, we are going to make this precise.

Let $t : U \to U$ be a transformation defined by $\mathbf{p}_i \mapsto \mathbf{q}_i$ for all $i \in \{1, \ldots, n\}$. (In other words, $t$ maps one set of basis vectors to another.) Since $P$ and $Q$ are bases (so the sets are linearly independent), $t$ is invertible. Suppose that the matrix $A_{f,Q} = (a_{i,j}^Q)_{i,j=1}^n$ is known to us, that is,

$$f(\mathbf{q}_j) = A_{f,Q}\mathbf{q}_j = \sum_{i=1}^n a_{i,j}^Q \mathbf{q}_i$$

holds for all $j$. So, we have

$$(t^{-1}ft)(\mathbf{p}_j) = t^{-1}f(\mathbf{q}_j)$$
$$= t^{-1}\left( \sum_{i=1}^n a_{i,j}^Q \mathbf{q}_i \right)$$
$$= \sum_{i=1}^n a_{i,j}^Q t^{-1}(\mathbf{q}_i)$$
$$= \sum_{i=1}^n a_{i,j}^Q \mathbf{p}_i.$$

In other words, the matrix of the composed transformation $t^{-1}ft$ in the basis $P$ is the same as the matrix of $f$ in $Q$. In terms of formulas,

$$T^{-1}A_{f,P}T = A_{f,Q}, \tag{4.6}$$

where $T$ denotes the matrix of $t$ in $P$. (For notational simplicity, we omit the subscript. Most often, we don't care what base it is in.)

We'll call $T$ the change of basis matrix. These types of relations are prevalent in linear algebra, so we'll take the time to introduce a definition formally.

> **Definition 4.2.1 (Similar matrices)**
>
> Let $A, B \in \mathbb{R}^{n \times n}$ be two arbitrary matrices. $A$ and $B$ are called *similar* if there exists a matrix $T \in \mathbb{R}^{n \times n}$ such that
>
> $$B = T^{-1}AT$$
>
> holds. We call mappings of the form $A \mapsto T^{-1}AT$ *similarity transformations*.

In these terms, (4.6) says that the matrices of a given linear transformation are all similar to each other.

With this under our belt, we can finish up with the example (4.5). In this case, $T$ and $T^{-1}$ can be written as

$$T = \begin{bmatrix} 1 & -1 \\ 1 & 1 \end{bmatrix}, \quad T^{-1} = \begin{bmatrix} 1/2 & 1/2 \\ -1/2 & 1/2 \end{bmatrix}.$$

(Later, we'll see a general method to compute the inverse of any matrix, but for now, you can verify this by hand.) Thus,

$$\begin{bmatrix} 1/2 & 1/2 \\ -1/2 & 1/2 \end{bmatrix} \begin{bmatrix} 2 & 1 \\ 1 & 2 \end{bmatrix} \begin{bmatrix} 1 & -1 \\ 1 & 1 \end{bmatrix} = \begin{bmatrix} 3 & 0 \\ 0 & 1 \end{bmatrix}, \tag{4.7}$$

or equivalently,

$$\begin{bmatrix} 2 & 1 \\ 1 & 2 \end{bmatrix} = \begin{bmatrix} 1 & -1 \\ 1 & 1 \end{bmatrix} \begin{bmatrix} 3 & 0 \\ 0 & 1 \end{bmatrix} \begin{bmatrix} 1/2 & 1/2 \\ -1/2 & 1/2 \end{bmatrix}. \tag{4.8}$$

*Figure 4.5* shows what (4.8) looks like in geometric terms.

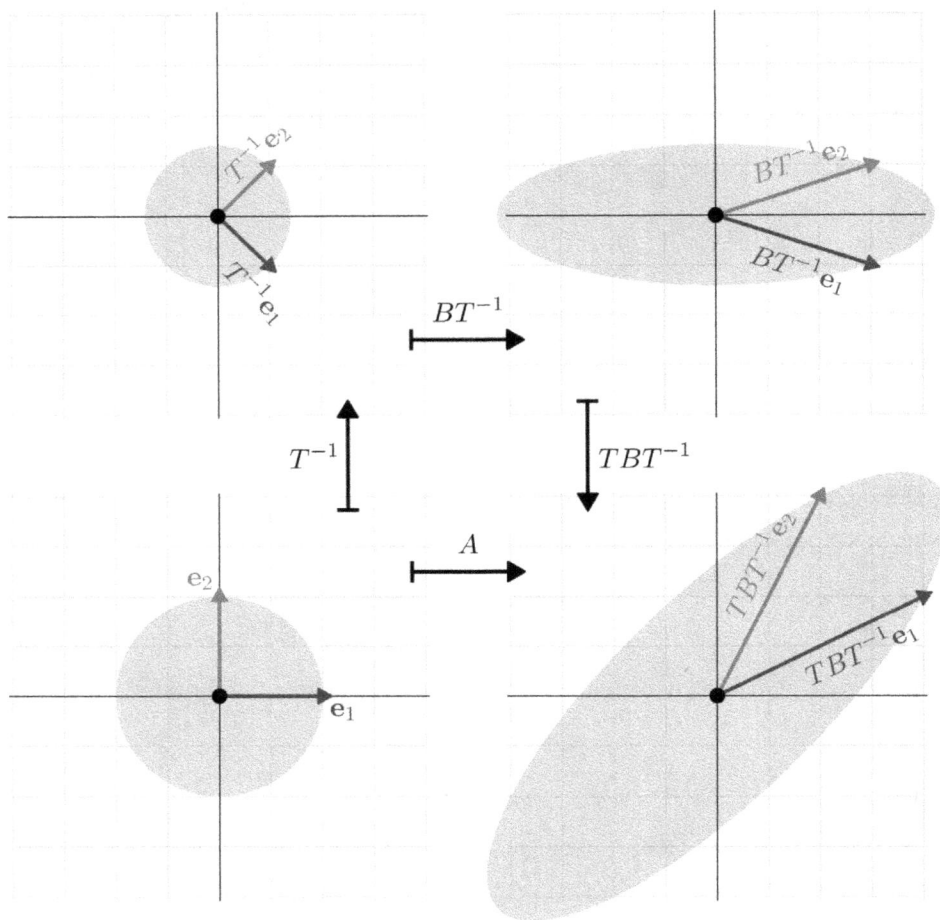

Figure 4.5: Change of basis, illustrated

From this example, we can see that a properly selected similarity transformation can diagonalize certain matrices. Is this a coincidence? Spoiler alert: no. In *Chapter 7*, we will see exactly when and how this can be done.

I know, this is a bit too abstract. As always, examples illustrate a concept best, so let's see some!

# 4.3    Linear transformations in the Euclidean plane

We have just seen that a linear transformation can be described by the image of a basis set. From a geometric viewpoint, they are functions mapping parallelepipeds to parallelepipeds.

Because of the linearity, you can imagine this as distorting the *grid* determined by the bases.

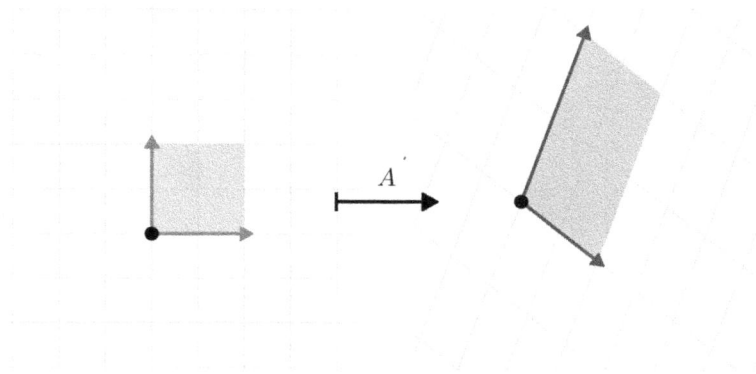

*Figure 4.6: How linear transforms distort the grid determined by the basis vectors*

In two dimensions, we have seen a few examples of geometric maps such as scaling and rotation as linear transformations. Now we can put them into matrix form. There are five of them in particular that we will study: stretching, shearing, rotation, reflection, and projection.

These simple transformations are not only essential to build intuition, but they are also frequently applied in computer vision. Flipping, rotating, and stretching are essential parts of image augmentation pipelines, greatly enhancing the performance of models.

## 4.3.1   Stretching

The simplest one is a generalization of scaling. We have seen a variant of this in Example 1 above (see *Section 4.1*). In matrix form, this is given by

$$A = \begin{bmatrix} c_1 & 0 \\ 0 & c_2 \end{bmatrix}, \quad c_1, c_2 \in \mathbb{R}.$$

Linear transformations such as this can be visualized by plotting the image of the unit square determined by the standard basis $\mathbf{e}_1 = (1, 0), \mathbf{e}_2 = (0, 1)$.

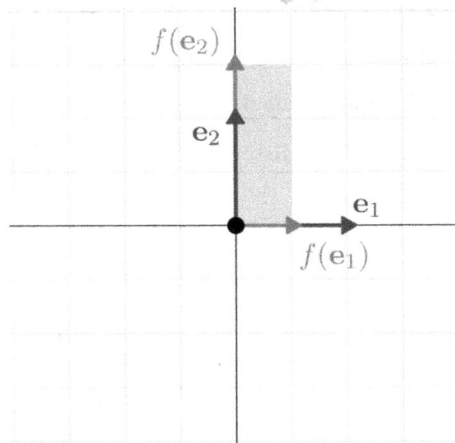

*Figure 4.7: Stretching*

## 4.3.2 Rotations

Rotations are given by the matrix

$$R_\alpha = \begin{bmatrix} \cos \alpha & -\sin \alpha \\ \sin \alpha & \cos \alpha \end{bmatrix}.$$

To see why, recall that each column of the transformation's matrix describes the *image* of the basis vectors. The rotation of $(1, 0)$ is given by $(\cos \alpha, \sin \alpha)$, while the rotation of $(0, 1)$ is $(\cos(\alpha + \pi/2), \sin(\alpha + \pi/2))$. This is illustrated by *Figure 4.8*.

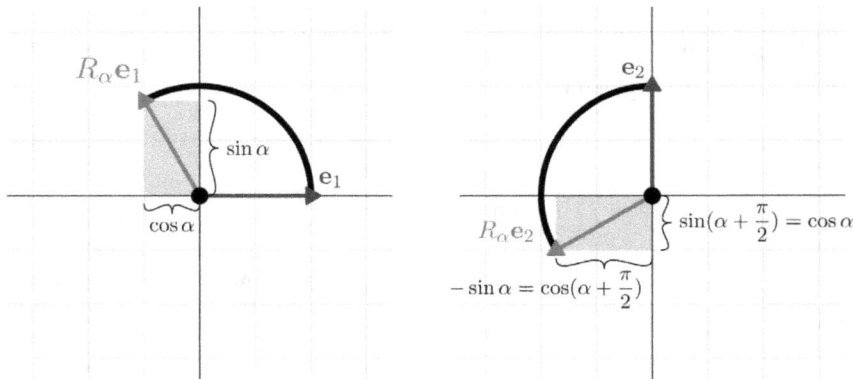

*Figure 4.8: The rotation matrix explained*

Like above, we can visualize the image of the unit square to gain a geometric insight into what is happening.

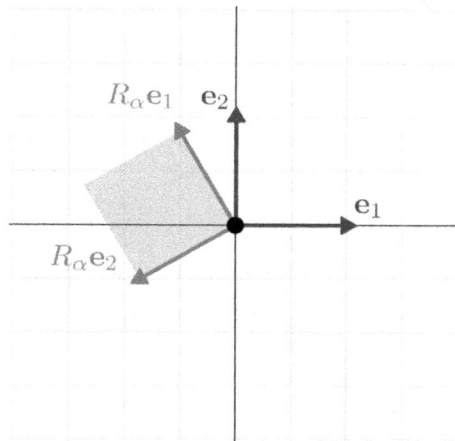

*Figure 4.9: Rotation*

## 4.3.3 Shearing

Another essential geometric transform is *shearing*, which is frequently applied in physics. A shearing force (https://en.wikipedia.org/wiki/Shear_force) is a pair of forces with opposite directions, acting on the same body.

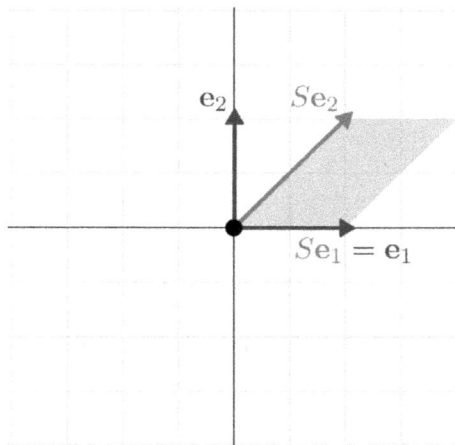

*Figure 4.10: Shearing*

Its matrix is given in the form

$$S_x = \begin{bmatrix} 1 & k_x \\ 0 & 1 \end{bmatrix}, \quad S_y = \begin{bmatrix} 1 & 0 \\ k_y & 1 \end{bmatrix}, \quad S = \begin{bmatrix} 1 & k_x \\ k_y & 1 \end{bmatrix},$$

where $S_x$, $S_y$, and $S$ represent shearing transformations in the $x$, $y$, and in both directions.

## 4.3.4  Reflection

Until this point, all the transformations we have seen in the Euclidean plane preserved the "orientation" of the space. However, this is not always the case. The transformation given by the matrices

$$R_1 = \begin{bmatrix} -1 & 0 \\ 0 & 1 \end{bmatrix}, \quad R_2 = \begin{bmatrix} 1 & 0 \\ 0 & -1 \end{bmatrix}$$

act as reflections with respect to the $x$ and the $y$ axes.

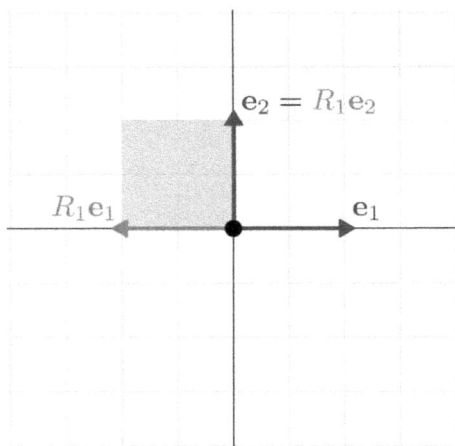

*Figure 4.11: Reflection*

When combined with a rotation, we can use reflections to flip bases. For instance, the transformation maps $\mathbf{e}_1$ to $\mathbf{e}_2$ and $\mathbf{e}_2$ to $\mathbf{e}_1$.

$$R = \underbrace{\begin{bmatrix} 0 & -1 \\ 1 & 0 \end{bmatrix}}_{\text{rotation with } \pi/2} \underbrace{\begin{bmatrix} 1 & 0 \\ 0 & -1 \end{bmatrix}}_{=R_2} = \begin{bmatrix} 0 & 1 \\ 1 & 0 \end{bmatrix}$$

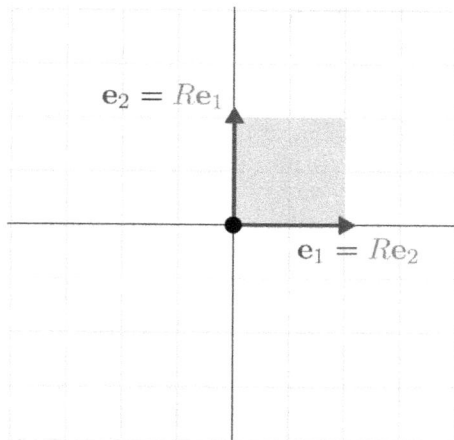

*Figure 4.12: Swapping $\mathbf{e}_1$ and $\mathbf{e}_2$ is a reflection and rotation*

These types of transformations play an essential role in understanding determinants, as we will soon see in the next chapter.

In general, reflections can be easily defined in higher dimensional spaces. For instance,

$$R = \begin{bmatrix} 1 & 0 & 0 \\ 0 & 1 & 0 \\ 0 & 0 & -1 \end{bmatrix}$$

is a reflection in $\mathbb{R}^3$ that flips $\mathbf{e}_3$ to the opposite direction. It is just like looking in the mirror: it turns left to right and right to left.

Reflections can flip orientations multiple times. The transformation given by

$$R = \begin{bmatrix} 1 & 0 & 0 \\ 0 & -1 & 0 \\ 0 & 0 & -1 \end{bmatrix}$$

flips $\mathbf{e}_2$ and $\mathbf{e}_3$, changing the orientation twice. Later, we'll see that the "number of changes in orientation" of a given transformation is one of its essential descriptors.

## 4.3.5 Orthogonal projection

One of the most important transformations (not only in two dimensions) is the *orthogonal projection*. We have seen this already when talking about inner products and their geometric representation in *Section 2.2.3*. By taking a closer look, it turns out that they are linear transformations.

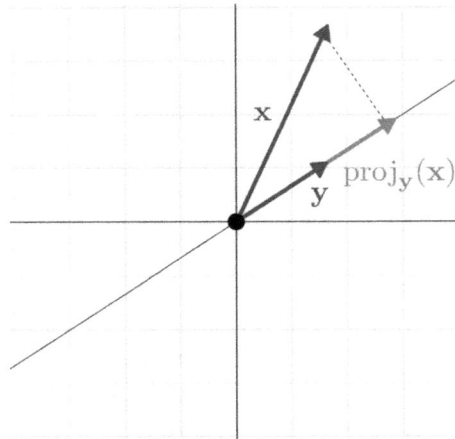

*Figure 4.13: Orthogonal projection*

Recall from (2.7) that the orthogonal projection of $x$ to some $y$ can be written as

$$\text{proj}_\mathbf{y}(\mathbf{x}) = \frac{\langle \mathbf{x}, \mathbf{y}\rangle}{\langle \mathbf{y}, \mathbf{y}\rangle}\mathbf{y}.$$

The bilinearity of $\langle \cdot, \cdot\rangle$ immediately implies that $\text{proj}_\mathbf{y}(\mathbf{x})$ is also linear. With a bit of algebra, we can rewrite this in terms of matrices. We have

$$\begin{aligned}
\text{proj}_\mathbf{y}(\mathbf{x}) &= \frac{\langle \mathbf{x}, \mathbf{y}\rangle}{\langle \mathbf{y}, \mathbf{y}\rangle}\mathbf{y}\\
&= \frac{x_1 y_1 + x_2 y_2}{\|\mathbf{y}\|^2}\begin{bmatrix} y_1 \\ y_2 \end{bmatrix}\\
&= \frac{1}{\|\mathbf{y}\|^2}\begin{bmatrix} y_1^2 & y_1 y_2 \\ y_1 y_2 & y_2^2 \end{bmatrix}\begin{bmatrix} x_1 \\ x_2 \end{bmatrix}\\
&= \frac{1}{\|\mathbf{y}\|^2}\mathbf{y}\mathbf{y}^T\mathbf{x},
\end{aligned}$$

thus,

$$\text{proj}_{\mathbf{y}} = \frac{1}{\|\mathbf{y}\|^2} \begin{bmatrix} y_1^2 & y_1 y_2 \\ y_1 y_2 & y_2^2 \end{bmatrix}$$

$$= \frac{1}{\|\mathbf{y}\|^2} \mathbf{y}\mathbf{y}^T \mathbf{x}$$

Notice that

$$\text{proj}_{\mathbf{y}}(\mathbf{e}_2) = \frac{y_2}{y_1} \text{proj}_{\mathbf{y}}(\mathbf{e}_1),$$

so the images of the standard basis vectors are not linearly independent. As a consequence, the image of the plane under $\text{proj}_{\mathbf{y}}$ is span($\mathbf{y}$), which is a one-dimensional subspace. From this example, we can see that the image of a vector space under a linear transformation is not necessarily of the same dimension as the starting space.

With these examples and knowledge under our belt, we have a basic understanding of linear transformations, the most basic building blocks of neural networks. In the next section, we will study how linear transformations affect the geometric structure of the vector space.

## 4.4 Determinants, or how linear transformations affect volume

In the previous sections, we have seen that linear transformations (*Definition 4.1.1*) can be thought of as distorting the grid determined by the basis vectors.

Following our geometric intuition, we suspect that measuring *how much* a transformation distorts volume and distance can provide some valuable insight. As we will see in this chapter, this is exactly the case. Transformations that preserve distance or norm are special, giving rise to methods such as Principal Component Analysis.

## 4.4.1  How linear transformations scale the area

Let's go back to the Euclidean plane one more time. Consider any linear transformation $A$, mapping the unit square to a parallelogram.

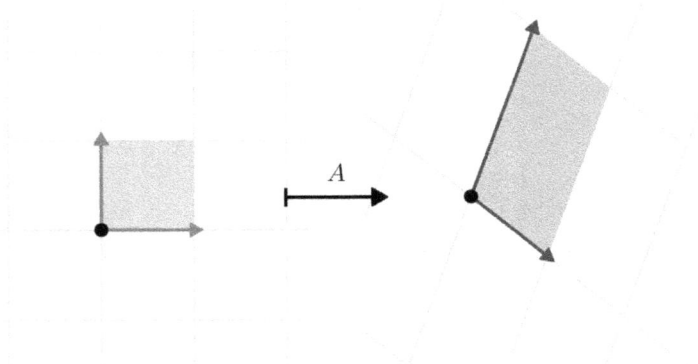

*Figure 4.14: Image of the unit square under a linear transformation*

The area of this parallelogram describes how $A$ scales the unit square. Let's call it $\lambda$ for now; that is,

$$\text{area}\big(A(C)\big) = \lambda \cdot \text{area}(C),$$

where $C = [0, 1] \times [0, 1]$ is the unit square, and $A(C)$ is its image

$$A(C) := \{A\mathbf{x} : \mathbf{x} \in C\}.$$

Due to linearity, $\lambda$ also matches the scaling ratio between the area of any rectangle (with parallel sides to the coordinate axes) and its image under $A$. As *Figure 4.15* shows, we can approximate any planar object as the union of rectangles.

If all rectangles are scaled by $\lambda$, then unions of rectangles also scale by that factor. Thus, it follows that $\lambda$ is also the scaling ratio between *any* planar object $E$ and its image $A(E) = \{A\mathbf{x} : \mathbf{x} \in E\}$.

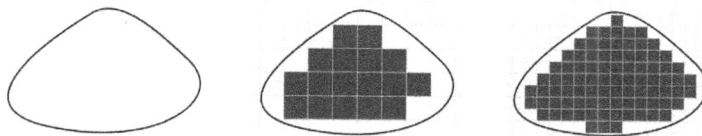

*Figure 4.15: Approximating planar objects with a union of rectangles*

This quantity $\lambda$ reveals a lot about the transformation itself, but there is a question remaining: how can we calculate it?

Suppose that our linear transformation is given by

$$A = \begin{bmatrix} x_1 & y_1 \\ x_2 & y_2 \end{bmatrix},$$

thus its columns $\mathbf{x} = (x_1, x_2)$ and $\mathbf{y} = (y_1, y_2)$ describe the two sides of the parallelogram. This is the image of the unit square.

*Figure 4.16: Image of the unit square under a linear transformation*

Our area scaling factor $\lambda$ equals the area of this parallelogram, so our goal is to calculate this.

The area of any parallelogram can be calculated by multiplying the length of the base ($\|\mathbf{x}\|$ in this case) with the height $h$. (You can easily see this by cutting off a triangle at the right side of the parallelogram and putting it to the left side, rearranging it as a rectangle.) $h$ is unknown, but with basic trigonometry, we can see that $h = \sin \alpha \|\mathbf{y}\|$, where $\alpha$ is the angle between $\mathbf{x}$ and $\mathbf{y}$.

Thus,

$$\text{area} = \sin \alpha \|\mathbf{y}\| \|\mathbf{x}\|.$$

This is almost the dot product of $\mathbf{x}$ and $\mathbf{y}$. (Recall that the dot product can be written as $\langle \mathbf{x}, \mathbf{y} \rangle = \|\mathbf{x}\| \|\mathbf{y}\| \cos \alpha$.) However, the $\sin \alpha$ part is not a match.

Fortunately, there is a clever trick we can use to turn this into a dot product! Since $\sin \alpha = \cos \left( \alpha - \frac{\pi}{2} \right)$, we have

$$\text{area} = \cos \left( \alpha - \frac{\pi}{2} \right) \|\mathbf{x}\| \|\mathbf{y}\|.$$

The issue is the angle between $\mathbf{x}$ and $\mathbf{y}$ is not $\alpha - \frac{\pi}{2}$. However, we can solve this easily by applying a rotation (*Section 4.3.2*). Applying the transformation

$$R = \begin{bmatrix} 0 & 1 \\ -1 & 0 \end{bmatrix},$$

we obtain

$$\mathbf{y}_{\text{rot}} = R\mathbf{y} = (y_2, -y_1).$$

Since $\|\mathbf{y}_{\text{rot}}\| = \|\mathbf{y}\|$, we have

$$\text{area} = \sin \alpha \|\mathbf{y}\| \|\mathbf{x}\|$$
$$= \cos \left( \alpha - \frac{\pi}{2} \right) \|\mathbf{x}\| \|\mathbf{y}\|$$
$$= \cos \left( \alpha - \frac{\pi}{2} \right) \|\mathbf{x}\| \|\mathbf{y}_{\text{rot}}\|$$
$$= \langle \mathbf{x}, \mathbf{y}_{\text{rot}} \rangle.$$

The quantity $\langle \mathbf{x}, \mathbf{y}_{\text{rot}} \rangle$ can be calculated using only the elements of the matrix $A$:

$$\langle \mathbf{x}, \mathbf{y}_{\text{rot}} \rangle = x_1 y_2 - x_2 y_1.$$

Notice that $\langle \mathbf{x}, \mathbf{y}_{\text{rot}} \rangle$ can be negative! This happens when the angle between $\mathbf{y} = A\mathbf{e}_2$ and $\mathbf{x} = A\mathbf{e}_1$, measured from a counter-clockwise direction, is larger than $\pi$, as this implies $\cos \left( \alpha - \frac{\pi}{2} \right) < 0$.

Hence, the quantity $\langle \mathbf{x}, \mathbf{y}_{\text{rot}} \rangle$ is called the *signed area* of the parallelogram.

In two dimensions, we call this the *determinant* of the linear transformation. That is, for any given linear transformation/matrix $A \in \mathbb{R}^{2 \times 2}$, its determinant is defined by

$$\det A = ad - cb, \quad A = \begin{bmatrix} a & b \\ c & d \end{bmatrix}. \tag{4.9}$$

The determinant is often written as $|A|$, but we'll avoid this notation. We'll deal with determinants for any matrix $A \in \mathbb{R}^{n \times n}$, but let's stay with the $2 \times 2$ case just a bit to build intuition.

The determinant also reveals the orientation of the vectors: positive determinant means positive orientation, negative determinant means negative orientation. (Intuitively, positive orientation means that the angle measured from **x** to **y** in a counter-clockwise direction is between $0$ and $\pi$; equivalently, the angle measured from **x** to **y** in a clockwise direction is between $\pi$ and $2\pi$.) This is demonstrated in *Figure 4.17* below.

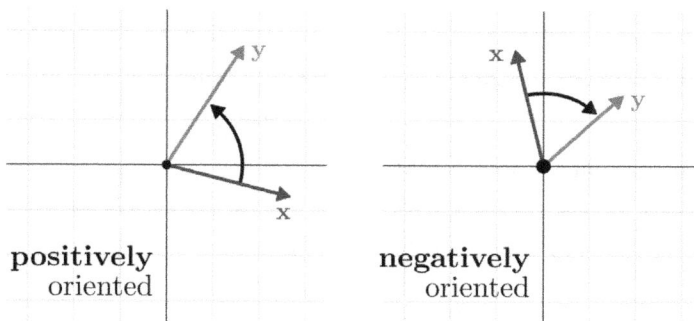

**positively** oriented    **negatively** oriented

Figure 4.17: Orientation of two vectors in the plane

Overall,

$$\text{area}\big(A(E)\big) = |\det A|\text{area}(E) \tag{4.10}$$

holds, where $E \subseteq \mathbb{R}^2$ is a planar object, and

$$A(E) = \{A\mathbf{x} \,:\, \mathbf{x} \in E\}$$

is the image of $E$ under the transformation $A$.

Even though we have only shown (4.10) in two dimensions, this holds in general. (Although we don't know how to define the determinant there yet.)

So, if $e_1$ and $e_2$ is a basis on the plane, equations (4.9) and (4.10) tell us that the determinant in two dimensions equals to

$$\det A = \text{orientation}(Ae_1, Ae_2) \times \text{area}(Ae_1, Ae_2)$$

Based on the example of the Euclidean plane, we have built enough geometric intuition on understanding how linear transformations distort volume and change the orientation of the space. These are described by the concept of *determinants*, which we have defined in the special case (4.9). We are going to move on to study the concept in its full generality.

To introduce the formal definition of the determinant, we will take a route that is different from the usual. Most commonly, the determinant of a linear transformation $A$ is defined straight away with a complicated formula, then all of its geometric properties are shown.

Instead of this, we will deduce the determinant formula by generalizing the geometric notion we have learned in the previous section. Here, we are roughly going to follow the outline of *Linear Algebra and Its Applications* by Peter D. Lax.

We set the foundations by introducing some key notations. Let

$$A = (a_{i,j})_{i,j=1}^{n} \in \mathbb{R}^{n \times n}$$

be a matrix with columns $a_1, \ldots, a_n$. When we introduced the notion of matrices as linear transformations in *Section 4.1.1*, we saw that the $i$-th column is the image of the $i$-th basis vector. For simplicity, let's assume that $e_1, e_2, \ldots, e_n$ is the standard orthonormal basis, that is, $e_i$ is the vector whose $i$-th coordinate is 1 and the rest is 0. Thus, $Ae_i = a_i$.

During our explorations in the Euclidean plane *Section 4.3*, we have seen that the determinant is the orientation of the images of basis vectors, times the area of the parallelogram defined by them. Following this logic, we could define the determinant for $n \times n$ matrices by

$$\det A = \text{orientation}(Ae_1, \ldots, Ae_n) \times \text{volume}(Ae_1, \ldots, Ae_n)$$

Two questions surface immediately. First, how do we define the orientation of multiple vectors in the $n$-dimensional space? Second, how can we even calculate the area?

Instead of finding the answers for these questions, we are going to add a twist into the story: first, we'll find a convenient formula for determinants, then use it to *define* orientation.

## 4.4.2 The multi-linearity of determinants

To make the relation between the determinant and the columns of the matrix $\mathbf{a}_i = A\mathbf{e}_i$ more explicit, we'll write

$$\det A = \det(\mathbf{a}_1, \dots, \mathbf{a}_n).$$

Thinking about determinants this way, det is just a function of multiple variables:

$$\det : \underbrace{\mathbb{R}^n \times \cdots \times \mathbb{R}^n}_{n \text{ times}} \to \mathbb{R}.$$

Good news: $\det(\mathbf{a}_1, \dots, \mathbf{a}_n)$ is linear in each variable. That is,

$$\det(\mathbf{a}_1, \dots, \alpha\mathbf{a}_i + \beta\mathbf{b}_i, \dots \mathbf{a}_n) = \alpha \det(\mathbf{a}_1, \dots, \mathbf{a}_i, \dots \mathbf{a}_n) + \beta \det(\mathbf{a}_1, \dots, \mathbf{b}_i, \dots \mathbf{a}_n)$$

holds. We are not going to prove this, but as the determinant represents the signed volume, you can convince yourself by checking out *Figure 4.18*.

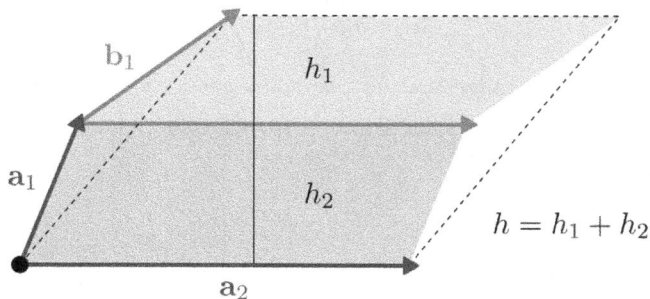

Figure 4.18: *The multilinearity of* $\det(\mathbf{a}_1, \mathbf{a}_2)$

A consequence of linearity is that we can express the determinant as a linear combination of determinants for the standard basis vectors $\mathbf{e}_1, \dots, \mathbf{e}_n$. For instance, consider the following. Since

$$A\mathbf{e}_1 = \mathbf{a}_1 = \sum_{i=1}^{n} a_{i,1}\mathbf{e}_i,$$

we have

$$\det(\mathbf{a}_1, \mathbf{a}_2, \dots, \mathbf{a}_n) = \sum_{i=1}^{n} a_{i,1} \det(\mathbf{e}_i, \mathbf{a}_2, \dots, \mathbf{a}_n).$$

Going one step further and using that

$$\mathbf{a}_2 = \sum_{j=1}^{n} a_{j,2} \mathbf{e}_j,$$

we start noticing a pattern. With the linearity, we have

$$\det(\mathbf{a}_1, \mathbf{a}_2, \dots, \mathbf{a}_n) = \sum_{i=1}^{n} \sum_{j=1}^{n} a_{i,1} a_{j,2} \det(\mathbf{e}_i, \mathbf{e}_j, \mathbf{a}_3, \dots, \mathbf{a}_n). \tag{4.11}$$

We can see that the row indices in the coefficients $a_{i,1} a_{j,2}$ match the indices of $\mathbf{e}_k$-s in $\det(\mathbf{e}_i, \mathbf{e}_j, \mathbf{a}_3, \dots, \mathbf{a}_n)$. In the general case, this pattern can be formalized in terms of permutations; that is, orderings of the set $\{1, 2, \dots, n\}$.

You can imagine a permutation as a function $\sigma$ mapping $\{1, 2, \dots, n\}$ to itself in a way that for every $j \in \{1, 2, \dots, n\}$, there is exactly one $i \in \{1, 2, \dots, n\}$ with $\sigma(i) = j$. In other words, you take every integer between 1 and $n$, and putting them in an order. The set of all possible permutations on $\{1, 2, \dots, n\}$ is denoted by $S_n$.

Continuing (4.11) and further expanding the determinant of $A$, we have

$$\det(\mathbf{a}_1, \dots, \mathbf{a}_n) = \sum_{\sigma \in S_n} \left[ \prod_{i=1}^{n} a_{\sigma(i),i} \right] \det(\mathbf{e}_{\sigma(1)}, \dots, \mathbf{e}_{\sigma(n)}).$$

This formula is not the easiest one to understand. You can think about each term $\prod_{i=1}^{n} \mathbf{a}_{\sigma(i),i}$ as placing $n$ chess rooks on an $n \times n$ board such that none of them can capture each other.

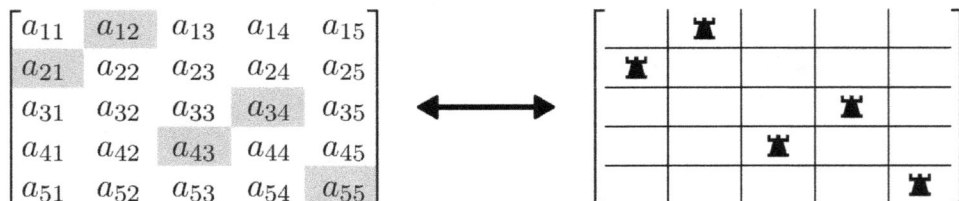

$$\sigma(1) = 2, \quad \sigma(2) = 1, \quad \sigma(3) = 4, \quad \sigma(4) = 3, \quad \sigma(5) = 5$$

*Figure 4.19: The anatomy of the term* $\mathbf{a}_{\sigma(1)1} \cdots \mathbf{a}_{\sigma(n)n}$

The formula

$$\sum_{\sigma \in S_n} \left[ \prod_{i=1}^{n} \mathbf{a}_{\sigma(i),i} \right] \det(\mathbf{e}_{\sigma(1)}, \dots, \mathbf{e}_{\sigma(n)})$$

combines all the possible ways we can do this.

There is only one thing left: calculating $\det(\mathbf{e}_{\sigma(1)}, \dots, \mathbf{e}_{\sigma(n)})$.

Remember when we discussed the combination of reflections and rotations in the Euclidean plane (*Section 4.3.4*)? The transformation determined by $\mathbf{e}_i \mapsto \mathbf{e}_{\sigma(i)}$ is similar to that. When talking about permutations, it's good to know that each one can be obtained by switching two elements at a time. The number of *transpositions* - that is, permutations affecting two elements - in a permutation is called the *sign* of $\sigma$. In the context of our linear transformation $\mathbf{e}_i \mapsto \mathbf{e}_{\sigma(i)}$, the number of transpositions in $\sigma$ is the number of reflections required and $\text{sign}(\sigma)$ is the *orientation* of $(\mathbf{e}_{\sigma(1)}, \dots, \mathbf{e}_{\sigma(n)})$.

Thus, with these, we can finally give a formal definition for determinants and the orientation.

---

**Definition 4.4.1 (Determinants and orientation)**

Let $A \in \mathbb{R}^{n \times n}$ be an arbitrary matrix and let $\mathbf{a}_i \in \mathbb{R}^n$ be its $i$-th column. The *determinant* of $A$ is defined by

$$\det A = \det(\mathbf{a}_1, \dots, \mathbf{a}_n) = \sum_{\sigma \in S_n} \text{sign}(\sigma) \left[ \prod_{i=1}^{n} \mathbf{a}_{\sigma(i),i} \right], \tag{4.12}$$

and the *orientation* of the vectors $\mathbf{a}_1, \dots, \mathbf{a}_n$ is

$$\text{orientation}(\mathbf{a}_1, \dots, \mathbf{a}_n) := \text{sign}(\det A).$$

---

When the det notation is not convenient, we denote determinants by putting the elements of the matrix inside a big absolute value sign:

$$\det A = \begin{vmatrix} a_{1,1} & a_{1,2} & \dots & a_{1,n} \\ a_{2,1} & a_{2,2} & \dots & a_{2,n} \\ \vdots & \vdots & \ddots & \vdots \\ a_{n,1} & a_{n,2} & \dots & a_{n,n} \end{vmatrix}.$$

When I was a young math student, the determinant formula (4.12) was presented as-is in my first linear algebra class. Without explaining the connection to volume and orientation, it took me years to properly understand it. I still think that the determinant is one of the most complex concepts in linear algebra, especially when presented without a geometric motivation for the definition.

Now that you have a basic understanding of the determinant, you might ask: how can we calculate it in practice? Summing over the set of all permutations and calculating their sign is not an easy operation from a computational perspective.

Good news: there is a recursive formula for the determinant. Bad news: for an $n \times n$ matrix, it involves $n$ pieces of $(n-1) \times (n-1)$ matrices. Still, it is a big step from the permutation formula. Let's see it!

**Theorem 4.4.1** *(Recursive formula for determinants)*

*Let $A \in \mathbb{R}^{n \times n}$ be an arbitrary square matrix. Then*

$$\det A = \sum_{j=1}^{n} (-1)^{j+1} a_{1,j} \det A_{1,j}, \tag{4.13}$$

*where $A_{i,j}$ is the $(n-1) \times (n-1)$ matrix obtained from A by removing its i-th row and j-th column.*

Instead of a proof, we are going to provide an example to demonstrate the formula. For $3 \times 3$ matrices, this is how it looks:

$$\begin{vmatrix} a & b & c \\ d & e & f \\ g & h & i \end{vmatrix} = a \begin{vmatrix} e & f \\ h & i \end{vmatrix} - b \begin{vmatrix} d & f \\ g & i \end{vmatrix} + c \begin{vmatrix} d & e \\ g & h \end{vmatrix}.$$

Now that we have both the geometric intuition and the recursive formula, let's see the most important properties of the determinants!

## 4.4.3   Fundamental properties of the determinants

When working with determinants, we prefer to create basic building blocks and rules for combining them. (As we have seen with this pattern so many times, even when deducing the (4.12).) These rules are manifested by the fundamental properties of determinants, which we will discuss now.

The first property is concerned with the relation of composition and the determinant.

**Theorem 4.4.2** *(The product of determinants)*

$\in \mathbb{R}^{n \times n}$ *be two matrices. Then*

$$\det AB = \det A \det B. \tag{4.14}$$

The equation (4.14) is called the *determinant product rule*, and its proof involves some heavy computations based on the formulas (4.12) and (4.13). Instead of providing a fully fleshed-out proof, I'll give an intuitive explanation. After all, we want to build algorithms *using* mathematics, not building mathematics.

So, the explanation of $\det AB = \det A \det B$ is quite simple. If we think about the matrices $A, B \in \mathbb{R}^{n \times n}$ as linear transformations, we have just seen that $\det A$ and $\det B$ determine how they scale the unit cube.

Since the composition of these linear transformations is the matrix product $AB$, the linear transformation $AB$ scales the unit cube to a parallelepiped with signed volume $\det A \det B$. (Because applying $AB$ is the same as applying $B$ first, then applying $A$ on the result.)

Thus, by our understanding of the determinant, as the scaling factor of $AB$ is also $\det AB$, (4.14) holds.

We can do the actual proof of this, for example, by induction based on the recursive formula (4.13), leading to a long and involved calculation.

An immediate corollary of the product rule is a special relation between the determinants of a matrix and its inverse.

**Theorem 4.4.3** *Let $A \in \mathbb{R}^{n \times n}$ be an arbitrary invertible matrix. Then*

$$\det A^{-1} = (\det A)^{-1}.$$

*Proof.* Using the product rule, we have

$$1 = \det I = \det AA^{-1} = (\det A)(\det A^{-1}),$$

from which the theorem follows.

Because of this, we can also conclude that the determinant is preserved by the similarity relation.

**Theorem 4.4.4** *Let $A, B \in \mathbb{R}^{n \times n}$ be two similar matrices with $B = T^{-1}AT$ for some $T \in \mathbb{R}^{n \times n}$. Then*

$$\det A = \det B.$$

*Proof.* This simply follows from

$$\det B = \det T^{-1}AT$$
$$= \det T^{-1} \det A \det T$$
$$= \det A,$$

which is what we had to show.

Another important consequence is that the determinant is independent of the basis the matrix is in. If $A : U \to U$ is a *linear transformation*, and $P = \{\mathbf{p}_1, \dots, \mathbf{p}_n\}$ and $R = \{\mathbf{r}_1, \dots, \mathbf{r}_n\}$ are two bases of $U$, then we know that the matrices of the transformation are related (*Section 4.2.1*) by

$$A_P = T^{-1}A_R T,$$

where $A_S$ is the matrix of the transformation $A$ in a basis $S$ and $T \in \mathbb{R}^{n \times n}$ is the change of basis matrix (*Section 4.2.1*). The previous theorem implies that $\det A_P = \det A_R$. Thus, the determinant is properly defined for *linear transformations*, not just matrices!

There is an essential duality relation regarding determinants: you can swap the rows and columns of a matrix, keeping all determinant-related identities true.

**Theorem 4.4.5** *Let $A \in \mathbb{R}^{n \times n}$ be an arbitrary matrix. Then*

$$\det A = \det A^T.$$

*Proof.* Suppose that $A = (a_{i,j})_{i,j=1}^n$. Let's denote the elements of its transpose by $a_{i,j}^t = a_{j,i}$. According to (4.12), we have

$$\det A^T = \sum_{\sigma \in S_n} \operatorname{sign}(\sigma) \prod_{i=1}^n a_{\sigma(i),i}^t$$
$$= \sum_{\sigma \in S_n} \operatorname{sign}(\sigma) \prod_{i=1}^n a_{i,\sigma(i)}.$$

Now comes the trick. Since the product $\prod_{i=1}^n a_{i,\sigma(i)}$ iterates through all $i$-s, and the order of the terms doesn't matter, we might as well order the terms as $i = \sigma^{-1}(1), \dots, \sigma^{-1}(n)$. Since

$$\operatorname{sign}(\sigma^{-1}) = \operatorname{sign}(\sigma),$$

by continuing the above calculation, we have

$$\sum_{\sigma \in S_n} \text{sign}(\sigma) \prod_{i=1}^{n} a_{i,\sigma(i)} = \sum_{\sigma \in S_n} \text{sign}(\sigma^{-1}) \prod_{j=1}^{n} a_{\sigma^{-1}(j),j}.$$

Because every permutation is invertible and $\sigma \mapsto \sigma^{-1}$ is a bijection, summing over $\sigma \in S_n$ is the same as summing over $\sigma^{-1} \in S_n$.

Combining all of the above, we obtain that

$$\det A = \sum_{\sigma \in S_n} \text{sign}(\sigma^{-1}) \prod_{j=1}^{n} a_{j,\sigma^{-1}(j)}$$

$$= \sum_{\sigma \in S_n} \text{sign}(\sigma) \prod_{j=1}^{n} a_{\sigma(j),j}$$

$$= \det A^T.$$

which is what we had to show.

**Theorem 4.4.6** *Let $A \in \mathbb{R}^{n \times n}$ be an arbitrary matrix and let $A^{i,j}$ denote the matrix which can be obtained by swapping the $i$-th and $j$-th column of $A$. Then*

$$\det A^{i,j} = -\det A,$$

*or in other words, swapping any two columns of $A$ will change the sign of the determinant. Similarly, swapping two rows also changes the sign of the determinant.*

*Proof.* This follows from a clever application of (4.14), noticing that $A^{i,j} = AI^{i,j}$, where $I^{i,j}$ is obtained from the identity matrix by swapping its $i$-th and $j$-th column. $\det I^{i,j}$ is a determinant of the form $\det(\mathbf{e}_{\sigma(1)}, \dots, \mathbf{e}_{\sigma(n)})$, where $\sigma$ is a permutation simply swapping $i$ and $j$. (That is, $\sigma$ is a transposition.) Thus,

$$\det A^{i,j} = \det A \det I^{i,j} = -\det A,$$

which is what we had to show.

Regarding swapping rows, we can apply the previous result because transposing a matrix preserves the determinant.

As a consequence, matrices with two matching rows have a zero determinant.

**Theorem 4.4.7** *Let $A \in \mathbb{R}^{n \times n}$ be a matrix that has two identical rows or columns. Then* $\det A = 0$.

*Proof.* Suppose that the $i$-th and the $j$-th columns are matching. Since the two columns are equal, $\det A^{i,j} = \det A$. However, applying the previous theorem (which states that swapping two columns changes the sign of the determinant), we obtain $\det A^{i,j} = -\det A$. This can only be true if $\det A = 0$.

Again, transposing the matrix gives the statement for rows.

As yet another consequence, we obtain an essential connection between linearly dependent vector systems and determinants.

**Theorem 4.4.8** *Let $A \in \mathbb{R}^{n \times n}$ be a matrix. Then its columns are linearly dependent if and only if $\det A = 0$. Similarly, the rows of $A$ are linearly dependent if and only if $\det A = 0$.*

*Proof.* (i) First, we are going to show that linearly dependent columns (or rows) imply $\det A = 0$. As usual, let's denote the columns of $A$ as $\mathbf{a}_1, \dots, \mathbf{a}_n$ and for the sake of simplicity, assume that

$$\mathbf{a}_1 = \sum_{i=2}^{n} \alpha_i \mathbf{a}_i.$$

Since the determinant is a linear function of the columns, we have

$$\det(\mathbf{a}_1, \mathbf{a}_2, \dots, \mathbf{a}_n) = \sum_{i=2}^{n} \alpha_i \det(\mathbf{a}_i, \mathbf{a}_2, \dots, \mathbf{a}_n).$$

Because of the previous theorem, all terms $\det(\mathbf{a}_i, \mathbf{a}_2, \dots, \mathbf{a}_n)$ are zero, implying $\det A = 0$, which is what we had to show. If the rows are linearly dependent, we apply the above to obtain that $\det A = \det A^T = 0$.

(ii) Now, let's show that $\det A = 0$ means linearly dependent columns. Instead of the exact proof, which is rather involved, we should have an intuitive explanation instead.

Recall that the determinant is orientation times volume of the parallelepiped given by the columns. Since the orientation is $\pm 1$, $\det A$ implies that the volume of the parallelepiped is 0. This can only happen if the $n$ columns lie in an $n - 1$-dimensional subspace, meaning that they are linearly dependent.

We can immediately apply this to get the following result.

> **Corollary 4.4.1** *Let $A \in \mathbb{R}^{n \times n}$ be a matrix with a constant zero column (or row). Then $\det A = 0$.*

As the determinant is the signed volume of the basis vectors' image, it can be zero in certain cases. These transformations are rather special. When can it happen? Let's go back to the Euclidean plane to build some intuition.

There, we have

$$\begin{vmatrix} x_1 & y_1 \\ x_2 & y_2 \end{vmatrix} = x_1 y_2 - x_2 y_1 = 0,$$

or in other words, $\frac{x_1}{y_1} = \frac{x_2}{y_2}$. There is one more interpretation of this: the vector $(y_1, y_2)$ is a scalar multiple of $(x_1, x_2)$; that is, they are *colinear*, meaning that they lie on the same line through the origin. Thinking in terms of linear transformations, this means that the images of $\mathbf{e}_1$ and $\mathbf{e}_2$ lie on a subspace of $\mathbb{R}^2$. As we shall see next, this is closely connected with the invertibility of the transformation.

> **Theorem 4.4.9** *(Invertibility and the determinants)*
>
> *The linear transformation $A \in \mathbb{R}^{n \times n}$ is invertible if and only if $\det A \neq 0$.*

*Proof.* When we introduced the concept of *invertibility* (*Definition 4.1.2*), we saw that $A$ is invertible if and only if its columns $\mathbf{a}_1, \dots, \mathbf{a}_n$ form a basis. Thus, they are linearly independent.

Since linear independence (*Definition 1.2.1*) of columns is equivalent to a nonzero determinant, the result follows.

# 4.5 Summary

So, do your eyes hurt after finally using them for the first time? Mine sure did when I first learned about matrices as *linear transformations*. Here's a part where the abstract viewpoint pays off for the first time, and trust me, it'll pay even more dividends later.

Let's recap this chapter quickly.

We've learned that besides a table of numbers, a matrix can represent linear transformations:

$$
A\mathbf{x} = \begin{bmatrix} a_{1,1} & a_{1,2} & \cdots & a_{1,m} \\ a_{2,1} & a_{2,2} & \cdots & a_{2,m} \\ \vdots & \vdots & \ddots & \vdots \\ a_{n,1} & a_{n,2} & \cdots & a_{n,m}, \end{bmatrix} \begin{bmatrix} x_1 \\ x_2 \\ \vdots \\ x_m \end{bmatrix} = \begin{bmatrix} \sum_{j=1}^{m} a_{1,j} x_j \\ \sum_{j=1}^{m} a_{2,j} x_j \\ \vdots \\ \sum_{j=1}^{m} a_{n,j} x_j \end{bmatrix},
$$

where the columns of $A$ describe the images of the basis vectors under the linear transformation $\mathbf{x} \to A\mathbf{x}$.

Why on Earth is this useful for us? Think of mathematics as a problem-solving tool. The crux of problem-solving is often finding the proper representation of our objects of interest. Looking at matrices as a way to transform data gives us the much-needed geometric perspective, opening up a whole new avenue of methods.

When looking at matrices this way, we quickly understand why matrix multiplication is defined as it is. The definition

$$
AB = \left( \sum_{k=1}^{l} a_{i,k} b_{k,j} \right)_{i,j=1}^{n,m}
$$

is daunting at first, but from the perspective of linear transformations, it's all revealed to be a simple composition: first, we apply the transformation $B$, then $A$.

Be careful, though: linear transformations and matrices are not exactly the same, as the matrix representation depends on the underlying basis of our vector space. (See, I told you that bases are going to be useful.)

Matrices also possess an important quantity called *determinants*, originally defined by the mind-boggingly complex formula

$$
\det(\mathbf{a}_1, \ldots, \mathbf{a}_n) = \sum_{\sigma \in S_n} \left[ \prod_{i=1}^{n} a_{\sigma(i),i} \right] \det(\mathbf{e}_{\sigma(1)}, \ldots, \mathbf{e}_{\sigma(n)}),
$$

but an investigation exploiting our newfound geometric perspective reveals that the determinant simply describes how much the linear transformation distorts the volume of the domain space, and how it changes the orientation of the basis vectors.

For us machine learning practitioners, making the conceptual jump from matrices to linear transformations is the more interesting one. (As opposed to the theory, where we often learn about linear transformations first, matrices second.) For instance, this allows us to see a layer in a neural network as stretching, rotating, shearing, and potentially reflecting the feature space.

In the next chapter, we'll revisit matrices from a slightly different perspective: systems of equations. Of course, everything is connected, and we'll end up where we started, looking at what we know from a higher perspective. This is because learning is a spiral, and we are ascending fast.

## 4.6 Problems

**Problem 1.** Show that if $A \in \mathbb{R}^{n \times n}$ is an invertible matrix, then

$$(A^{-1})^T = (A^T)^{-1}.$$

**Problem 2.** Let $R_\alpha$ be the two-dimensional rotation matrix defined by

$$R_\alpha = \begin{bmatrix} \cos \alpha & -\sin \alpha \\ \sin \alpha & \cos \alpha \end{bmatrix}.$$

Show that $R_\alpha R_\beta = R_{\alpha+\beta}$.

**Problem 3.** Let $A = (a_{i,j})_{i,j=1}^n \in \mathbb{R}^{n \times n}$ be a matrix and let $D \in \mathbb{R}^{n \times n}$ be a *diagonal* matrix defined by

$$D = \begin{bmatrix} d_1 & 0 & \ldots & 0 \\ 0 & d_2 & \ldots & 0 \\ 0 & 0 & \ldots & d_n \end{bmatrix},$$

where all of its elements are zero outside the diagonal. Show that

$$DA = \begin{bmatrix} d_1 a_{1,1} & d_2 a_{1,2} & \ldots & d_n a_{1,n} \\ d_1 a_{2,1} & d_2 a_{2,2} & \ldots & d_n a_{2,n} \\ \vdots & \vdots & \ddots & \vdots \\ d_1 a_{n,1} & d_2 a_{n,2} & \ldots & d_n a_{n,n} \end{bmatrix}$$

and

$$AD = \begin{bmatrix} d_1 a_{1,1} & d_1 a_{1,2} & \ldots & d_1 a_{1,n} \\ d_2 a_{2,1} & d_2 a_{2,2} & \ldots & d_2 a_{2,n} \\ \vdots & \vdots & \ddots & \vdots \\ d_n a_{n,1} & d_n a_{n,2} & \ldots & d_n a_{n,n} \end{bmatrix}.$$

**Problem 4.** Let $\| \cdot \|$ be a norm on $\mathbb{R}^n$, and let $A \in \mathbb{R}^{n \times n}$ be an arbitrary matrix.

Show that $A$ is invertible if and only if the function

$$\|\mathbf{x}\|_* := \|A\mathbf{x}\|$$

is a norm on $\mathbb{R}^n$.

**Problem 5.** Let $U$ be a normed space and $f : U \to U$ be a linear transformation.

If

$$\|\mathbf{x}\|_* := \|f(\mathbf{x})\|$$

is a norm, is $f$ necessarily invertible?

*Hint:* Consider the vector space $\mathbb{R}[x]$ with the norm

$$\|p\| = \left( \sum_{i=0}^{n} p_i^2 \right)^{1/2}, \quad p(x) = \sum_{i=0}^{n} p_i x^i$$

and the linear transformation $f : p(x) \mapsto x p(x)$.

**Problem 6.** Let $\langle \cdot, \cdot \rangle$ be an inner product on $\mathbb{R}^n$. Show that there is a matrix $A \in \mathbb{R}^{n \times n}$ such that

$$\langle \mathbf{x}, \mathbf{y} \rangle = \mathbf{x}^T A \mathbf{y}, \quad \mathbf{x}, \mathbf{y} \in \mathbb{R}^n.$$

(Recall that we treat vectors $\mathbf{x}, \mathbf{y} \in \mathbb{R}^n$ as column vectors.)

**Problem 7.** Let $A \in \mathbb{R}^{n \times n}$ be a matrix. $A$ is called *positive definite* if $\mathbf{x}^T A \mathbf{x} > 0$ for every nonzero $\mathbf{x} \in \mathbb{R}^n$.

Show that $A$ is positive definite if and only if

$$\langle \mathbf{x}, \mathbf{y} \rangle := \mathbf{x}^T A \mathbf{y}$$

is an inner product.

**Problem 8.** Let $A \in \mathbb{R}^{n \times m}$ be a matrix, and denote its columns by $\mathbf{a}_1, \dots, \mathbf{a}_n \in \mathbb{R}^n$.

*(a)* Show that for all $\mathbf{x} \in \mathbb{R}^m$, we have $A\mathbf{x} \in \mathrm{span}(\mathbf{a}_1, \dots, \mathbf{a}_n)$.

*(b)* Let $B \in \mathbb{R}^{m \times k}$, and denote the columns of $AB$ by $\mathbf{v}_1, \dots, \mathbf{v}_k \in \mathbb{R}^n$. Show that

$$\mathbf{v}_1, \dots, \mathbf{v}_k \in \operatorname{span}(\mathbf{a}_1, \dots, \mathbf{a}_n).$$

**Problem 9.** Let $A \in \mathbb{R}^{n \times n}$ be a matrix. Show that

$$\langle A\mathbf{x}, \mathbf{y} \rangle = \langle \mathbf{x}, A^T \mathbf{y} \rangle$$

holds for all $\mathbf{x}, \mathbf{y} \in \mathbb{R}^n$, where $\langle \cdot, \cdot \rangle$ is the Euclidean inner product.

**Problem 10.** Calculate the determinant of

$$A = \begin{bmatrix} 1 & 2 & 3 \\ 4 & 5 & 6 \\ 7 & 8 & 9 \end{bmatrix}.$$

**Problem 11.** Let $A \in \mathbb{R}^{n \times n}$ be a matrix and let $c \in \mathbb{R}$ be a constant.

*(a)* Show that

$$\begin{vmatrix} a_{1,1} & \dots & ca_{1,i} & \dots & a_{1,n} \\ a_{2,1} & \dots & ca_{2,i} & \dots & a_{2,n} \\ \vdots & \ddots & \vdots & \ddots & \vdots \\ a_{n,1} & \dots & ca_{n,i} & \dots & a_{n,n} \end{vmatrix} = c \det A$$

holds for all $i = 1, \dots, n$.

*(b)* Show that

$$\begin{vmatrix} a_{1,1} & a_{1,2} & \dots & a_{1,n} \\ \vdots & \vdots & \ddots & \vdots \\ ca_{i,1} & ca_{i,2} & \dots & ca_{i,n} \\ \vdots & \vdots & \ddots & \vdots \\ a_{n,1} & a_{n,2} & \dots & a_{n,n} \end{vmatrix} = c \det A$$

holds for all $i = 1, \dots, n$ and

*(c)* Show that

$$\det(cA) = c^n \det A.$$

**Problem 12.** Let $A \in \mathbb{R}^{n \times n}$ be an upper triangular matrix. (That is, all elements below the diagonal are zero.) Show that

$$\det A = \prod_{i=1}^{n} a_{i,i}.$$

Show that the same holds for lower triangular matrices. (That is, matrices where elements above the diagonal are zero.)

**Problem 13.** Let $M \in \mathbb{R}^{n \times m}$ be a matrix with the block structure

$$M = \begin{bmatrix} A & B \\ 0 & C \end{bmatrix},$$

where $A \in \mathbb{R}^{k \times k}$, $B \in \mathbb{R}^{k \times l}$, and $C \in \mathbb{R}^{l \times l}$.

Show that

$$\det M = \det A \det C.$$

# Join our community on Discord

Read this book alongside other users, Machine Learning experts, and the author himself.

Ask questions, provide solutions to other readers, chat with the author via Ask Me Anything sessions, and much more.

Scan the QR code or visit the link to join the community.

```
https://packt.link/math
```

# 5

# Matrices and Equations

So, matrices are not just tables of numbers but linear transformations; we have spent a lengthy chapter discovering this relationship.

Now, I want us to circle back to the good old tables of numbers once more, but representing systems of linear equations this time. Why? Simple. Because solving linear equations is the motivator behind key theoretical and technical innovations. In the previous chapter, we talked about inverse matrices but didn't compute one in practice. With what we're about to learn, we will be able to not only compute inverse matrices but do so blazing fast.

Let's get to work!

## 5.1  Linear equations

In practice, we can translate several problems into linear equations. For example, a cash dispenser has $900 in $20 and $50 bills. We know that there are twice as many $20 bills than $50. The question is, how many of each bill does the machine have?

If we denote the number of $20 bills by $x_1$ and the number of $50 bills by $x_2$, we obtain the equations

$$x_1 - 2x_2 = 0$$
$$20x_1 + 50x_2 = 900.$$

For two variables, as we have now, these are easily solvable by expressing one in terms of the other. Here, the first equation would imply $x_1 = 2x_2$. Plugging it back into the second equation, we obtain $90x_2 = 900$, which gives $x_2 = 10$. Coming full circle, we can substitute this into $x_1 = 2x_2$, yielding the solutions

$$x_1 = 20$$
$$x_2 = 10.$$

However, for thousands of variables like in real applications, we need a bit more craft. This is where linear algebra comes in. By introducing the matrix and vectors

$$A = \begin{bmatrix} 1 & -2 \\ 20 & 50 \end{bmatrix}, \quad \mathbf{x} = \begin{bmatrix} x_1 \\ x_2 \end{bmatrix}, \quad \mathbf{b} = \begin{bmatrix} 0 \\ 900 \end{bmatrix},$$

the equation can be written in the form $A\mathbf{x} = \mathbf{b}$. That is, in terms of linear transformations, we can reformulate the question: which vector $\mathbf{x}$ is mapped to $\mathbf{b}$ by the transformation $A$? This question is central in linear algebra, and we'll dedicate this section to solving it.

## 5.1.1 Gaussian elimination

Let's revisit our earlier example:

$$x_1 - 2x_2 = 0$$
$$20x_1 + 50x_2 = 900.$$

We can use the first equation $x_1 - 2x_2 = 0$ to get rid of the term $x_1$ in the second equation $20x_1 + 50x_2 = 900$. We can do this by multiplying it by 20 and subtracting it from the second row, obtaining $90x_2 = 900$, from which $x_2 = 10$ is obtained. This can be substituted back into the first row, yielding $x_1 = 20$.

What about the general case? Would this work for a general $A \in \mathbb{R}^{n \times n}$ and $\mathbf{x}, \mathbf{b} \in \mathbb{R}^n$? Absolutely. So far, we have used two rules for manipulating the equations in a linear system:

1. Multiplying an equation with a nonzero scalar won't change the solutions.
2. Adding a scalar multiple of one row to another won't change the solutions either.

Earlier, we applied these repeatedly to eliminate variables progressively in our simple example. We can easily do the same for $n$ variables! First, let's see what we are talking about!

**Definition 5.1.1 (System of linear equations)**

Let $A \in \mathbb{R}^{n \times n}$ be a matrix and $\mathbf{b} \in \mathbb{R}^n$ be a vector. The collection of equations

$$a_{11}x_1 + a_{12}x_2 + \cdots + a_{1n}x_n = b_1$$
$$a_{21}x_1 + a_{22}x_2 + \cdots + a_{2n}x_n = b_2$$
$$\vdots$$
$$a_{n1}x_1 + a_{n2}x_2 + \cdots + a_{nn}x_n = b_n$$

(5.1)

are called the *system of linear equations* determined by $A$ and $\mathbf{b}$.

A system of linear equations is often written in the short form $A\mathbf{x} = \mathbf{b}$, where $A$ is called its *coefficient matrix*. If the vector $\mathbf{x}$ satisfies $A\mathbf{x} = \mathbf{b}$, it is called a *solution*.

Speaking of solutions, are there even any, and if so, how can we find them?

If $a_{11}$ is nonzero, we can multiply the first equation of (5.1) by $\frac{a_{k1}}{a_{11}}$ and subtract it from the $k$-th equation. This way, $x_1$ will be eliminated from all but the first row, obtaining

$$a_{11}x_1 + a_{12}x_2 + \cdots + a_{1n}x_n = b_1$$
$$0x_1 + \left(a_{22} - a_{12}\frac{a_{21}}{a_{11}}\right)x_2 + \cdots + \left(a_{2n} - a_{1n}\frac{a_{21}}{a_{11}}\right)x_n = b_2 - b_1\frac{a_{21}}{a_{11}}$$
$$\vdots$$
$$0x_1 + \left(a_{n2} - a_{12}\frac{a_{n1}}{a_{11}}\right)x_2 + \cdots + \left(a_{nn} - a_{1n}\frac{a_{n1}}{a_{11}}\right)x_n = b_n - b_1\frac{a_{n1}}{a_{11}}.$$

(5.2)

To clear up this notation a bit, let's denote the new coefficients with $a_{ij}^{(1)}$ and $b_i^{(1)}$. So, we have

$$a_{11}x_1 + a_{12}x_2 + \cdots + a_{1n}x_n = b_1$$
$$0x_1 + a_{22}^{(1)}x_2 + \cdots + a_{2n}^{(1)}x_n = b_2^{(1)}$$
$$\vdots$$
$$0x_1 + a_{n2}^{(1)}x_2 + \cdots + a_{nn}^{(1)}x_n = b_n^{(1)}.$$

(5.3)

We can repeat the above process and use the second equation to get rid of the $x_2$ variable in the third equation, and so forth.

This can be done $n - 1$ times in total, ultimately leading to an equation system $A^{(n-1)}\mathbf{x} = \mathbf{b}^{(n-1)}$ where all coefficients below the diagonal of $A^{(n-1)}$ are 0:

$$A^{(n-1)} = \begin{bmatrix} a_{11} & a_{12} & a_{13} & \dots & a_{1n} \\ 0 & a_{22}^{(1)} & a_{23}^{(1)} & \dots & a_{2n}^{(1)} \\ 0 & 0 & a_{33}^{(2)} & \dots & a_{3n}^{(2)} \\ \vdots & \vdots & \vdots & \ddots & \vdots \\ 0 & 0 & 0 & \dots & a_{nn}^{(n-1)} \end{bmatrix}. \tag{5.4}$$

Notice that the $k$-th elimination step only affects the coefficients from the $(k + 1)$-th row. Now we can work backward: the last equation $a_{nn}^{(n-1)}x_n = b_n^{(n-1)}$ can be used to find $x_n$. This can be substituted to the $(n - 1)$-th equation, yielding $x_{n-1}$. Continuing like this, we can eventually find all $x_1, \dots, x_n$, obtaining a solution for our linear system.

This process is called *Gaussian elimination*, and it's kind of a big deal. It is not only useful for solving linear equations, but it can also be used to calculate determinants, factor matrices into the product of simpler ones, and much more. We'll talk about all of this in detail, but let's focus on equations a little more.

Unfortunately, not all linear equations can be solved. For instance, consider the system

$$x_1 + x_2 = 1$$
$$2x_1 + 2x_2 = -1.$$

Subtracting the first equation from the second one yields

$$x_1 + x_2 = 1$$
$$x_1 + x_2 = -2$$

in the very first step, making it apparent that the equation has no solutions.

Before we turn to the technical details, let's see a simple example of how Gaussian elimination is done in practice!

## 5.1.2 Gaussian elimination by hand

To build a deeper understanding of Gaussian elimination, let's consider the simple equation system

$$x_1 + 0x_2 - 3x_3 = 6$$
$$2x_1 + 1x_2 + 5x_3 = 2$$
$$-2x_1 - 3x_2 + 8x_3 = 2.$$

To keep track of our progress (and, since we are lazy, to avoid writing too much), we record the intermediate results as

$$\begin{array}{ccc|c} 1 & 0 & -3 & 6 \\ 2 & 1 & 5 & 2 \\ -2 & -3 & 8 & 2 \end{array}$$

with the coefficient matrix $A$ on the left side and $b$ on the other. To get a good grip on the method, I encourage you to follow along and do the calculations yourself by hand.

After eliminating the first variable from the second and third equations, we have

$$\begin{array}{ccc|c} 1 & 0 & -3 & 6 \\ 0 & 1 & 11 & -10 \\ 0 & -3 & 2 & 14 \end{array},$$

while the final step yields

$$\begin{array}{ccc|c} 1 & 0 & -3 & 6 \\ 0 & 1 & 11 & -10 \\ 0 & 0 & 35 & -16 \end{array}.$$

From this form, we can unravel the solutions one by one.

In the 21st century, your chances of having to solve a linear equation by hand are close to 0. (If you are reading this book during the 22nd century or later, I am incredibly honored and surprised at the same time. Or, at least, I would be if I were still alive.) Still, understanding the general principles behind solving linear equations can take you very far.

## 5.1.3   When can we perform Gaussian elimination?

If you followed the description of Gaussian elimination carefully, you might have noticed that the process can break down. We might accidentally divide by 0 during any elimination step!

For instance, after the first step given by equation (5.2), the new coefficients are of the form

$$\left( a_{ij} - a_{1j} \frac{a_{i1}}{a_{11}} \right),$$

which is invalid if $a_{11} = 0$. In general, the $k$-th step involves division by $a_{kk}^{(k-1)}$. Since $a_{kk}^{(k-1)}$ is defined recursively, describing it in terms of $A$ is not straightforward. For this, we introduce the concept of *principal minors*, the upper-left subdeterminants of a matrix.

> **Definition 5.1.2  (Principal minors)**
>
> Let $A = (a_{ij})_{i,j=1}^{n} \in \mathbb{R}^{n \times n}$ be an arbitrary square matrix. Define the submatrix $A_k \in \mathbb{R}^{k \times k}$ by omitting all rows and columns of $A$ with indices larger than $k$. For instance,
>
> $$A_1 = \begin{bmatrix} a_{11} \end{bmatrix}, \quad A_2 = \begin{bmatrix} a_{11} & a_{12} \\ a_{21} & a_{22} \end{bmatrix},$$
>
> and so on. The k-th principal minor of $A$, denoted by $M_k$, is defined by
>
> $$M_k := \det A_k.$$

The first and last principal minors are special, as $M_1 = a_{11}$ and $M_n = \det A$. With principal minors, we can describe when Gaussian elimination is possible. In fact, it turns out that

$$a_{11} = M_1, \quad a_{22}^{(1)} = \frac{M_2}{M_1}, \quad \dots, a_{nn}^{(n-1)} = \frac{M_n}{M_{n-1}}$$

and, in general, $a_{kk}^{(k-1)} = \frac{M_k}{M_{k-1}}$.

To summarize, we can state the following.

> **Theorem 5.1.1** *Let $A \in \mathbb{R}^{n \times n}$ be an arbitrary square matrix, and let $M_k$ be its $k$-th principal minor. If $M_k \neq 0$ for all $k = 1, 2, \dots, n - 1$, then Gaussian elimination can be successfully performed.*

As the proof is a bit involved, we are not going to do it here. (The difficult step is showing $a_{kk}^{(k-1)} = M_k / M_{k-1}$; the rest follows immediately.) The point is, if none of the principal minors are 0, the algorithm finishes.

We can simplify this requirement a bit and describe the Gaussian elimination in terms of the determinant, not the principal minors.

**Theorem 5.1.2** *Let $A \in \mathbb{R}^{n \times n}$ be an arbitrary square matrix. If $\det A \neq 0$, then all principal minors are nonzero as well.*

As a consequence, if the determinant is nonzero, the Gaussian elimination is successful. A simple and nice requirement.

## 5.1.4   The time complexity of Gaussian elimination

To get a handle on how fast the Gaussian elimination algorithm executes, let's do a little complexity analysis. As described by (5.2), the first elimination step involves an addition and a multiplication for each element, except for those in the first row. That is $2n(n-1)$ operations in total.

The next step is essentially the first step, done on the $(n-1) \times (n-1)$ matrix obtained from $A^{(1)}$ by removing its first row and column. This time, we have $2(n-1)(n-2)$ operations.

Following this train of thought, we quickly get that the total number of operations is

$$\sum_{i=1}^{n} 2(n-i+1)(n-i),$$

which doesn't look that friendly. Since we are looking for the order of complexity instead of an exact number, we can be generous and suppose that at each elimination step, we are performing $O(n^2)$ operations. So, we have a time complexity of

$$\sum_{i=1}^{n} O(n^2) = O(n^3),$$

meaning that we need around $cn^3$ operations for Gaussian elimination, where $c$ is an arbitrary positive constant. This might seem a lot, but in the beautiful domain of algorithms, this is good. $O(n^3)$ is polynomial time, and we can have much, much worse.

## 5.1.5   When can a system of linear equations be solved?

So, we have just seen that for any linear equation

$$A\mathbf{x} = \mathbf{b}, \quad A \in \mathbb{R}^{n \times n}, \quad \mathbf{x}, \mathbf{b} \in \mathbb{R}^n,$$

Gaussian elimination can be successfully performed if the principal minors $M_1, \ldots, M_{n-1}$ are nonzero. Notice one caveat about the result: $M_n = \det A$ can be zero as well. Turns out, this is quite an important detail.

If you have closely followed the discussion leading up to this point, you will see that we missed a crucial point: are there any solutions at all for a given linear equation? There are three options:

1. There are no solutions.
2. There is exactly one solution.
3. There are multiple solutions.

All of these are relevant to us from a certain perspective, but let's start with the most straightforward one: when do we have exactly one solution? The answer is simple: when $A$ is invertible, the solution can be explicitly written as $\mathbf{x} = A^{-1}\mathbf{b}$. Speaking in terms of linear transformations, we can find a unique vector $\mathbf{x}$ that is mapped to $\mathbf{b}$. We summarize this idea in the following theorem.

**Theorem 5.1.3** *Let $A \in \mathbb{R}^{n \times n}$ be an invertible matrix. Then, for any $\mathbf{b} \in \mathbb{R}^n$, the equation $A\mathbf{x} = \mathbf{b}$ has a unique solution that can be written as $\mathbf{x} = A^{-1}\mathbf{b}$.*

If $A$ is invertible, then $\det A$ is nonzero. Thus, using what we have learned previously, Gaussian elimination can be performed, yielding the unique solution. Nice and simple.

If $A$ is not invertible, the two remaining possibilities are in play: no vector is mapped to $\mathbf{b}$, which means there are no solutions, or multiple vectors are mapped to $\mathbf{b}$, giving infinite solutions.

Do you remember how we used the kernel of a linear transformation to describe its invertibility in *Theorem 4.1.3*? It turns out that ker $A$ can also be used to find *all* solutions for a linear system.

**Theorem 5.1.4** *Let $A \in \mathbb{R}^{n \times n}$ be an arbitrary matrix and let $\mathbf{x}_0 \in \mathbb{R}^n$ be a solution to the linear equation $A\mathbf{x} = \mathbf{b}$, where $\mathbf{b} \in \mathbb{R}^n$. Then, the set of all solutions can be written as*

$$\mathbf{x}_0 + \ker A := \{\mathbf{x}_0 + \mathbf{y} \: : \: \mathbf{y} \in \ker A\}.$$

*Proof.* We have to show two things: (a) if $\mathbf{x} \in \mathbf{x}_0 + \ker A$, then $\mathbf{x}$ is a solution; and (b) if $\mathbf{x}$ is a solution, then $\mathbf{x} \in \mathbf{x}_0 + \ker A$.

(a) Suppose that $\mathbf{x} \in \mathbf{x}_0 + \ker A$, that is, $\mathbf{x} = \mathbf{x}_0 + \mathbf{y}$ for some $\mathbf{y} \in \ker A$. Then,

$$A\mathbf{x} = A(\mathbf{x}_0 + \mathbf{y}) = \underbrace{A\mathbf{x}_0}_{=\mathbf{b}} + \underbrace{A\mathbf{y}}_{=0} = \mathbf{b},$$

which shows that $\mathbf{x}$ is indeed a solution.

(b) Now let $\mathbf{x}$ be an arbitrary solution. We have to show that $\mathbf{x} - \mathbf{x}_0 \in \ker A$. This is easy, since

$$A(\mathbf{x} - \mathbf{x}_0) = A\mathbf{x} - A\mathbf{x}_0 = \mathbf{b} - \mathbf{b} = \mathbf{0}.$$

Thus, (a) and (b) imply that $\mathbf{x}_0 + \ker A$ is the set of all solutions.

In theory, this theorem provides an excellent way of finding all solutions for linear equations, generalizing far beyond finite-dimensional vector spaces. (Note that the proof goes through verbatim for *all* vector spaces and linear transformations.) For instance, this exact result is used to describe all solutions of an inhomogeneous linear differential equation.

## 5.1.6 Inverting matrices

So far, we have seen that the invertibility of a matrix $A \in \mathbb{R}^{n \times n}$ is key to solving linear equations. However, we haven't found a way to compute the inverse of a matrix yet.

Let's recap what the inverse is in terms of linear transformations. If the columns of $A$ are denoted by the vectors $\mathbf{a}_1, \dots, \mathbf{a}_n \in \mathbb{R}^n$, then $A$ is the linear transformation that maps the standard basis vectors to these vectors:

$$A : \mathbf{e}_i \mapsto \mathbf{a}_i, \quad i = 1, \dots, n.$$

If the direction of the arrows can be reversed, that is,

$$A^{-1} : \mathbf{a}_i \mapsto \mathbf{e}_i, \quad i = 1, \dots, n$$

is a well-defined linear equation, then $A^{-1}$ is called the inverse of $A$.

In light of all that we have seen in this chapter, the method for finding the inverse is simple: solve $A\mathbf{x} = \mathbf{e}_i$ for each $i$ where $\mathbf{e}_1, \dots, \mathbf{e}_n \in \mathbb{R}^n$ is the standard basis.

Suppose that $A\mathbf{x}_i = \mathbf{e}_i$. Then, if $\mathbf{b}$ can be written as $\mathbf{b} = \sum_{i=1}^{n} b_i \mathbf{e}_i$, the vector $\mathbf{x} = \sum_{i=1}^{n} b_i \mathbf{x}_i$ is the solution of $A\mathbf{x} = \mathbf{b}$:

$$A\left( \sum_{i=1}^{n} b_i x_i \right) = \sum_{i=1}^{n} b_i A\mathbf{x}_i$$

$$= \sum_{i=1}^{n} b_i \mathbf{e}_i$$

$$= \mathbf{b}.$$

Thus, the inverse is the matrix whose $i$-th column is $\mathbf{x}_i$.

I know, this seems paradoxical: to find the solution of $A\mathbf{x} = \mathbf{b}$, we need the inverse $A^{-1}$. To find the inverse, we need to solve $n$ equations. The answer is Gaussian elimination (*Section 5.1.1*), which gives us an exact computational method to obtain $A^{-1}$. In the next section, we are going to put this into practice and write our matrix-inverting algorithm from scratch. Pretty awesome.

# 5.2    The LU decomposition

In the previous chapter, I promised that you'd never have to solve a linear equation by hand. As it turns out, this task is perfectly suitable for computers. In this chapter, we will dive deep into the art of solving linear equations, developing the tools from scratch.

We start by describing the process of Gaussian elimination in terms of matrices. Why would we even do that? Because matrix multiplication can be performed extremely fast in modern computers. Expressing *any* algorithm in terms of matrices is a sure way to accelerate.

At the start, our linear equation $A\mathbf{x} = \mathbf{b}$ is given by the coefficient matrix

$$A = \begin{bmatrix} a_{11} & a_{12} & \dots & a_{1n} \\ a_{21} & a_{22} & \dots & a_{2n} \\ \vdots & \vdots & \ddots & \vdots \\ a_{n1} & a_{n2} & \dots & a_{nn} \end{bmatrix} \in \mathbb{R}^{n \times n},$$

and at the end of the elimination process, $A$ is transformed into the form

$$A^{(n-1)} = \begin{bmatrix} a_{11} & a_{12} & a_{13} & \dots & a_{1n} \\ 0 & a_{22}^{(1)} & a_{23}^{(1)} & \dots & a_{2n}^{(1)} \\ 0 & 0 & a_{33}^{(2)} & \dots & a_{3n}^{(2)} \\ \vdots & \vdots & \vdots & \ddots & \vdots \\ 0 & 0 & 0 & \dots & a_{nn}^{(n-1)} \end{bmatrix}.$$

$A^{(n-1)}$ is *upper diagonal*; that is, all elements below its diagonal are 0.

Gaussian elimination performs this task one step at a time, focusing on consecutive columns. After the first elimination step, this is turned into the equation (5.3), described by the coefficient matrix

$$A^{(1)} = \begin{bmatrix} a_{11} & a_{12} & \dots & a_{1n} \\ 0 & a_{22}^{(1)} & \dots & a_{2n}^{(1)} \\ \vdots & \vdots & \ddots & \vdots \\ 0 & a_{n2}^{(1)} & \dots & a_{nn}^{(1)} \end{bmatrix} \in \mathbb{R}^{n \times n}.$$

Can we obtain $A^{(1)}$ from $A$ via multiplication with some matrix; that is, can we find $G_1 \in \mathbb{R}^{n \times n}$ such that $A^{(1)} = G_1 A$ holds?

Yes. By defining $G_1$ as

$$G_1 = \begin{bmatrix} 1 & 0 & 0 & \dots & 0 \\ -\frac{a_{21}}{a_{11}} & 1 & 0 & \dots & 0 \\ -\frac{a_{31}}{a_{11}} & 0 & 1 & \dots & 0 \\ \vdots & \vdots & & \ddots & \vdots \\ -\frac{a_{n1}}{a_{11}} & 0 & 0 & \dots & 1 \end{bmatrix}, \tag{5.5}$$

we can see that $A^{(1)} = G_1 A$ is the same as performing the first step of Gaussian elimination. (Pick up a pen and paper and verify this by hand. It's an excellent exercise.) $G_1$ is *lower diagonal*; that is, all elements above its diagonal are 0. In fact, except for the first column, all elements below the diagonal are 0 as well. (Note that $G_1$ depends on $A$.)

By analogously defining

$$G_2 = \begin{bmatrix} 1 & 0 & 0 & \dots & 0 \\ 0 & 1 & 0 & \dots & 0 \\ 0 & -\frac{a_{32}^{(1)}}{a_{22}^{(1)}} & 1 & \dots & 0 \\ \vdots & \vdots & & \ddots & \vdots \\ 0 & -\frac{a_{n2}^{(1)}}{a_{22}^{(1)}} & 0 & \dots & 1 \end{bmatrix}, \tag{5.6}$$

we obtain $A^{(2)} = G_2 A^{(1)} = G_2 G_1 A$, a matrix that is *upper diagonal* in the first two columns. (That is, all elements are 0 below the diagonal, but only in the first two columns.)

We can continue this process until we obtain the upper triangular matrix

$$A^{(n-1)} = G_{n-1} \dots G_1 A. \tag{5.7}$$

The algorithm is starting to shape up nicely. The $G_i$ matrices are invertible, with inverses

$$G_1^{-1} = \begin{bmatrix} 1 & 0 & 0 & \dots & 0 \\ \frac{a_{21}}{a_{11}} & 1 & 0 & \dots & 0 \\ \frac{a_{31}}{a_{11}} & 0 & 1 & \dots & 0 \\ \vdots & \vdots & & \ddots & \vdots \\ \frac{a_{n1}}{a_{11}} & 0 & 0 & \dots & 1 \end{bmatrix}, \quad G_2^{-1} = \begin{bmatrix} 1 & 0 & 0 & \dots & 0 \\ 0 & 1 & 0 & \dots & 0 \\ 0 & \frac{a_{32}^{(1)}}{a_{22}^{(1)}} & 1 & \dots & 0 \\ \vdots & \vdots & & \ddots & \vdots \\ 0 & \frac{a_{n2}^{(1)}}{a_{22}^{(1)}} & 0 & \dots & 1 \end{bmatrix}, \dots,$$

and so on. Thus, by multiplying by their inverses one by one, we can express $A$ as

$$A = G_1^{-1} \dots G_{n-1}^{-1} A^{(n-1)}.$$

Fortunately, we can calculate $L := G_1^{-1} \dots G_{n-1}^{-1}$ by hand. After a quick computation, we obtain

$$L = \begin{bmatrix} 1 & 0 & 0 & \dots & 0 \\ \frac{a_{21}}{a_{11}} & 1 & 0 & \dots & 0 \\ \frac{a_{31}}{a_{11}} & \frac{a_{32}^{(1)}}{a_{22}^{(1)}} & 1 & \dots & 0 \\ \vdots & \vdots & & \ddots & \vdots \\ \frac{a_{n1}}{a_{11}} & \frac{a_{n2}^{(1)}}{a_{22}^{(1)}} & \frac{a_{n3}^{(2)}}{a_{33}^{(2)}} & \dots & 1 \end{bmatrix}, \tag{5.8}$$

which is lower diagonal. (Again, don't be shy about verifying (5.8) by hand.) By defining the upper diagonal matrix $U := A^{(n-1)}$, we obtain the famous LU decomposition, factoring $A$ into a lower and upper diagonal matrix:

$$A = LU.$$

Notice that with this algorithm, we perform two tasks for the price of one: factorizing $A$ into the product of an upper diagonal and lower diagonal matrix, and performing Gaussian elimination.

From a computational standpoint, the LU decomposition is an extremely important tool. Since it is just a refashioned Gaussian elimination, its complexity is $O(n^3)$, just as we saw this earlier (*Section 5.1.4*).

Bad news: LU decomposition not always available. Since it is tied to Gaussian elimination, we can characterize its existence in similar terms. Recall that for the Gaussian elimination to successfully finish, the principal minors are required to be nonzero (*Theorem 5.1.2*). This is directly transferred to the LU decomposition.

> **Theorem 5.2.1** *(Existence of the LU decomposition)*
>
> *Let $A \in \mathbb{R}^{n \times n}$ be an arbitrary square matrix, and let $M_k$ be its k-th principal minor. If $M_k \neq 0$ for all $k = 1, 2, \ldots, n - 1$, then A can be written as*
>
> $$A = LU, \quad L, U \in \mathbb{R}^{n \times n},$$
>
> *where L is a lower diagonal and U is an upper diagonal matrix. Moreover, the elements along the diagonal of L are equal to 1.*

The gist is the same: everything is fine if we avoid division by 0 during the algorithm. Note that the LU algorithm doesn't require a nonzero $M_n = \det A$, that is, an invertible matrix!

After all this preparation, we are ready to put things into practice!

## 5.2.1 Implementing the LU decomposition

To summarize the LU decomposition, it's essentially the iteration of two steps:

1. Calculate the elimination matrices of the input.
2. Multiply the input by the elimination matrices, feeding the output back into the first step.

The plan is clear: first, we write a function that computes the elimination matrices and their inverses; then, we iteratively perform the elimination steps using matrix multiplication.

```python
import numpy as np


def elimination_matrix(
    A: np.ndarray,
    step: int,
):
    """
    Computes the step-th elimination matrix and its inverse.

    Args:
        A (np.ndarray): The matrix of shape (n, n) for which
            the LU decomposition is being computed.
        step (int): The current step of elimination, an integer
```

```
                   between 1 and n-1

         Returns:
             elim_mtx (np.ndarray): The step-th elimination matrix
                 of shape (n, n)
             elim_mtx_inv (np.ndarray): The inverse of the
                 elimination matrix of shape (n, n)
         """

         n = A.shape[0]
         elim_mtx = np.eye(n)
         elim_mtx_inv = np.eye(n)

         if 0 < step < n:
             a = A[:, step-1]/A[step-1, step-1]
             elim_mtx[step:, step-1] = -a[step:]
             elim_mtx_inv[step:, step-1] = a[step:]

         return elim_mtx, elim_mtx_inv
```

Now, we are ready to perform the elimination steps.

```
def LU(A: np.ndarray):
    """
    Computes the LU factorization of a square matrix A.

    Args:
        A (np.ndarray): A square matrix of shape (n, n) to be factorized.
            It must be non-singular (invertible) for the
            decomposition to work.

    Returns:
        L (np.ndarray): A lower triangular matrix of shape (n, n)
            with ones on the diagonal.
        U (np.ndarray): An upper triangular matrix of shape (n, n).
    """

    n = A.shape[0]
    L = np.eye(n)
    U = np.copy(A)

    for step in range(1, n):
        elim_mtx, elim_mtx_inv = elimination_matrix(U, step=step)
        U = np.matmul(elim_mtx, U)
        L = np.matmul(L, elim_mtx_inv)
```

```
    return L, U
```

Let's test our function on a small matrix.

```
A = 10*np.random.rand(4, 4) - 5
A
```

```
array([[-4.61990165, -3.97616553, -1.34258661,  0.50835913],
       [-2.39491833, -2.3919011 , -1.3266581 ,  2.8658852 ],
       [ 4.32658116,  0.43607725,  4.41630776, -4.46731714],
       [-0.68329877,  4.76659965, -1.13602896, -2.12305592]])
```

```
L, U = LU(A)

print(f"Lower:\n{L}\n\nUpper:\n{U}")
```

```
Lower:
[[  1.          0.          0.          0.        ]
 [  0.51839163  1.          0.          0.        ]
 [ -0.93650936  9.94174964  1.          0.        ]
 [  0.14790331 -16.19246049 -1.18248523  1.        ]]

Upper:
[[-4.61990165e+00 -3.97616553e+00 -1.34258661e+00  5.08359130e-01]
 [ 0.00000000e+00 -3.30690182e-01 -6.30672445e-01  2.60235608e+00]
 [ 0.00000000e+00  0.00000000e+00  9.42895038e+00 -2.98632067e+01]
 [ 1.11022302e-16 -8.88178420e-16  1.77635684e-15  4.62750317e+00]]
```

Is the result correct? Let's test it by multiplying L and U together to see if it gives A back.

```
np.allclose(np.matmul(L, U), A)
```

```
True
```

Overall, the LU decomposition is a highly versatile tool, used as a stepping stone in the implementation of essential algorithms. One of them is computing the inverse matrix, as we shall see next.

## 5.2.2   Inverting a matrix, for real

So far, we have talked a lot about the inverse matrix. We explored the question of invertibility from several angles, in terms of the kernel and the image, the determinant, and the solvability of linear equations.

However, we haven't yet talked about actually computing the inverse. With the LU decomposition, we obtain a tool that can be used for this purpose. How? By plugging in a lower triangular matrix into the Gaussian elimination process, we get its inverse as a side effect. So, we

1. calculate the LU decomposition $A = LU$,
2. invert the lower triangular matrices $L$ and $U^T$,
3. use the identity $(U_{-1})T = (U_T)-1$ to get $U^{-1}$,
4. multiply $L^{-1}$ and $U^{-1}$ to finally obtain $A^{-1} = U^{-1}L^{-1}$.

That's a plan! Let's start with inverting lower triangular matrices.

Let $L \in \mathbb{R}^{n \times n}$ be an arbitrary lower triangular matrix. Following the same process that led to (5.7), we obtain

$$D = G_{n-1} \dots G_1 L,$$

where $D$, the final result of Gaussian elimination, is a diagonal matrix

$$D = \text{diag}(d_1, \dots, d_n) = \begin{bmatrix} d_1 & 0 & \dots & 0 \\ 0 & d_2 & \dots & 0 \\ \vdots & \vdots & \ddots & \vdots \\ 0 & 0 & \dots & d_n \end{bmatrix},$$

and the $G_i$-s are the elimination matrices defined by (5.5), (5.6), and so on.

Since the inverse of $D$ is simply $D^{-1} = \text{diag}(d_1^{-1}, \dots, d_n^{-1})$, we can express $L^{-1}$ as

$$L^{-1} = D^{-1} G_{n-1} \dots G_1.$$

We can implement this very similarly to the LU decomposition; we can even reuse our `elimination_matrix` function.

```
def invert_lower_triangular_matrix(L: np.ndarray):
    """
    Computes the inverse of a lower triangular matrix.

    Args:
```

```
        L (np.ndarray): A square lower triangular matrix of shape (n, n).
                        It must have non-zero diagonal elements for the
                        inversion to succeed.

    Returns:
        np.ndarray: The inverse of the lower triangular matrix L, with
                    shape (n, n).
    """
    n = L.shape[0]
    G = np.eye(n)
    D = np.copy(L)

    for step in range(1, n):
        elim_mtx, _ = elimination_matrix(D, step=step)
        G = np.matmul(elim_mtx, G)
        D = np.matmul(elim_mtx, D)

    D_inv = np.eye(n)/np.diagonal(D)    # NumPy performs this operation elementwise

    return np.matmul(D_inv, G)
```

With this done, we are ready to invert any matrix (that is actually invertible).

We are almost at the finish line. Every component is ready; the only thing left to do is to put them together. We can do this with a few lines of code.

```
def invert(A: np.ndarray):
    """
    Computes the inverse of a square matrix using its LU decomposition.

    Args:
        A (np.ndarray): A square matrix of shape (n, n). The matrix must be
                        non-singular (invertible) for the inversion to succeed.

    Returns:
        np.ndarray: The inverse of the input matrix A, with shape (n, n).
    """
    L, U = LU(A)
    L_inv = invert_lower_triangular_matrix(L)
    U_inv = invert_lower_triangular_matrix(U.T).T
    return np.matmul(U_inv, L_inv)
```

Voilà! Witness the result with your own eyes.

```python
A = np.random.rand(3, 3)
A_inv = invert(A)

print(f"A:\n{A}\n\nA^{-1}:\n{A_inv}\n\nAA^{-1}:\n{np.matmul(A, A_inv)}")
```

```
A:
[[0.17180745 0.79269571 0.36879642]
 [0.37772174 0.94712553 0.55310582]
 [0.93418085 0.38813821 0.51581695]]

A^{-1}:
[[  9.14230036  -8.87123133   2.97602074]
 [ 10.74480189  -8.5427258    1.47801811]
 [-24.64252111  22.49459369  -4.56327945]]

AA^{-1}:
[[ 1.00000000e+00 -1.37050081e-15  1.56962305e-17]
 [-4.06841165e-16  1.00000000e+00 -4.50714074e-16]
 [ 1.26848123e-14 -1.04564970e-14  1.00000000e+00]]
```

To test the correctness of our `invert` function, we quickly check the results on a few randomly generated matrices.

```python
for _ in range(1000):
    n = np.random.randint(1, 10)
    A = np.random.rand(n, n)
    A_inv = invert(A)
    if not np.allclose(np.matmul(A, A_inv), np.eye(n), atol=1e-5):
        print("Test failed.")
```

Since there is no error message above, the function is (probably) correct.

What seemed complex and abstract a few chapters ago is now in our hands. We can invert any matrix, not with built-in functions, but with one that we wrote from scratch. I love these moments when the pieces are finally put together, and everything clicks. Sit back, relax, and appreciate the journey that got us here!

## 5.2.3 How to actually invert matrices

Of course, our LU decomposition!matrices, invertingimplementation is far from optimal. When working with NumPy arrays, we can turn to the built-in functions. In NumPy, this is `np.linalg.inv`.

```python
A = np.random.rand(3, 3)
A_inv = np.linalg.inv(A)

print(f"A:\n{A}\n\nNumPy's A^{-1}:\n{A_inv}\n\nAA^{-1}:\n{np.matmul(A, A_inv)}")
```

```
A:
[[0.08503998 0.31186637 0.71032538]
 [0.48973954 0.77358354 0.96303592]
 [0.31250848 0.14359491 0.05593863]]

NumPy's A^{-1}:
[[ 2.13348825 -1.89861153  5.59470678]
 [-6.14268693  4.87769374 -5.97239945]
 [ 3.84931433 -1.91423645  1.95236829]]

AA^{-1}:
[[ 1.00000000e+00 -1.86546922e-16  2.74435800e-16]
 [-1.62367293e-16  1.00000000e+00  9.21871975e-17]
 [-1.41854334e-18  3.64838601e-17  1.00000000e+00]]
```

Let's compare the runtime of our implementation and NumPy's.

```python
from timeit import timeit


n_runs = 100
size = 100
A = np.random.rand(size, size)

t_inv = timeit(lambda: invert(A), number=n_runs)
t_np_inv = timeit(lambda: np.linalg.inv(A), number=n_runs)


print(f"Our invert:             \t{t_inv} s")
print(f"NumPy's invert:         \t{t_np_inv} s")
print(f"Performance improvement: \t{t_inv/t_np_inv} times faster")
```

```
Our invert:                    14.586225221995846 s
NumPy's invert:                0.46499890399718424 s
Performance improvement:       31.368300218798304 times faster
```

A massive improvement. Nice! (Don't forget that the execution time isLU decomposition!matrices, inverting hardware-dependent.) Why is NumPy that much faster? There are two main reasons. First, it directly calls the SGETRI function from LAPACK, which is extremely fast. Second, according to its documentation ( https://www. netlib.org/lapack/explore-html/da/d28/group__getri_gaa3bf1bb1432917f0e5fdf4c48bd6998c.html), SGETRI uses a faster algorithm:

```
"""
SGETRI computes the inverse of a matrix using the LU factorization
computed by SGETRF.

This method inverts U and then computes inv(A) by solving the system
inv(A)*L = inv(U) for inv(A).
"""
```

So, NumPy calls the LAPACK function, which uses LU factorization in turn. (I am not particularly adept at digging through Fortran code that is older than I am, so let me know if I am wrong here. Nevertheless, the fact that state-of-the-art frameworks still make calls to this ancient library is a testament to its power. Never underestimate old technologies like LAPACK and Fortran.)

Are there any other applications of the powerful LU decomposition that we've just learned? Glad you asked! Of course there is; that's the beauty of math. There's always a new and unexpected application even for the oldest of tools. This time, we'll finally see how to compute determinants fast! (And also slow.)

# 5.3   Determinants in practice

In the theory and practice of mathematics, the development of concepts usually has a simple flow. Definitions first arise from vague geometric or algebraic intuitions, eventually crystallizing in mathematical formalism.

However, mathematical definitions often disregard practicalities. For a very good reason, mind you! Keeping practical considerations out of sight gives us the power to reason about structure effectively. This is the strength of abstraction. Eventually, if meaningful applications are found, the development flows toward computational questions, putting speed and efficiency onto the horizon.

The epitome of this is neural networks themselves. From theoretical constructs to state-of-the-art algorithms that run on your smartphone, machine learning research followed this same arc.

This is also what we experience in this book on a microscopic level. Among many other examples, think about determinants. We introduced the determinant as the orientation of column vectors and the parallele-piped volume defined by them. Still, we haven't really worked on computing them in practice. Sure, we gave a formula or two, but it is hard to decide which one is the most convoluted. All of them are.

On the other hand, the mathematical study of determinants yielded a ton of useful results: invertibility of linear transformations, characterization of Gaussian elimination, and many more (and even more to come.).

In this section, we are ready to pay off our debts and develop tools to actually compute determinants. As before, we will take a straightforward approach and use one of the previously derived determinant formulas. Spoiler alert: This is far from optimal, so we'll find a way to compute the determinant with high speed.

## 5.3.1 The lesser of two evils

Let's recall what we know about determinants (*Definition 4.4.1*) so far. Given a matrix $A \in \mathbb{R}^{n \times n}$, its determinant det $A$ quantifies the volume distortion of the linear transformation $\mathbf{x} \to A\mathbf{x}$. That is, if $\mathbf{e}_1, \dots, \mathbf{e}_n$ is the standard orthonormal basis, then informally speaking,

$$\det A = \left( \text{orientation of } A\mathbf{e}_1, A\mathbf{e}_2, \dots, A\mathbf{e}_n \right)$$
$$\times \left( \text{area of the parallelepiped determined by } A\mathbf{e}_1, A\mathbf{e}_2, \dots, A\mathbf{e}_n \right).$$

We have derived two formulas to compute this quantity. Initially, we described the determinant in terms of summing over all permutations:

$$\det A = \sum_{\sigma \in S_n} \text{sign}(\sigma) a_{\sigma(1),1} \dots a_{\sigma(n),n}.$$

This is difficult to understand, let alone to programmatically compute. So, a recursive formula is derived, which we can also use. It states that

$$\det A = \sum_{j=1}^{n} (-1)^{j+1} a_{1,j} \det A_{1,j},$$

where $A_{i,j}$ is the matrix obtained by deleting the $i$-th row and $j$-th column of $A$. Which one would you rather use? Take a few minutes to figure out your reasoning.

Unfortunately, there are no right choices here. With the permutation formula, one has to find a way to generate all permutations first, then calculate their signs. Moreover, there are $n!$ unique permutations in $S_n$, so this sum has a *lot* of terms. Using this formula seems extremely difficult, so we will go with the recursive version. Recursion has its issues (as we are about to see very soon), but it is easy to handle from a coding standpoint.

Let's get to work!

## 5.3.2  The recursive way

Let's put the formula

$$\det A = \sum_{j=1}^{n} (-1)^{j+1} a_{1,j} \det A_{1,j}$$

under our magnifying glass. If $A$ is an $n \times n$ matrix, then $A_{1,j}$ (obtained from $A$ by deleting its first row and $j$-th column) is of size $(n-1) \times (n-1)$. This is a recursive step. For each $n \times n$ determinant, we have to calculate $n$ pieces of $(n-1) \times (n-1)$ determinants, and so on.

By the end, we have a lots of $1 \times 1$ determinants, which are trivial to calculate. So, we have a boundary condition, and with that, we are ready to put these together.

```python
def det(A: np.ndarray):
    """
    Recursively computes the determinant of a square matrix A.

    Args:
        A (np.ndarray): A square matrix of shape (n, n) for which the
        determinant is to be calculated.

    Returns:
        float: The determinant of matrix A.

    Raises:
        ValueError: If A is not a square matrix.
    """

    n, m = A.shape

    # making sure that A is a square matrix
    if n != m:
        raise ValueError("A must be a square matrix.")

    if n == 1:
        return A[0, 0]

    else:
        return sum([(-1)**j*A[0, j]*det(np.delete(A[1:], j, axis=1)) for j in range(n)])
```

Let's test the det function out on a small example. For $2 \times 2$ matrices, we can easily calculate the determinants using the rule

$$\det \begin{bmatrix} a & b \\ c & d \end{bmatrix} = ad - bc.$$

```python
A = np.array([[1, 2],
              [3, 4]])
det(A)      # should be -2
```

```
np.int64(-2)
```

It seems to work. So far, so good. What is the issue? Recursion. Let's calculate the determinant of a small $10 \times 10$ matrix, measuring the time it takes.

```python
from timeit import timeit

A = np.random.rand(10, 10)
t_det = timeit(lambda: det(A), number=1)

print(f"The time it takes to compute the determinant of a 10 x 10 matrix: {t_det} seconds")
```

```
The time it takes to compute the determinant of a 10 x 10 matrix:
63.98369195000123 seconds
```

That was long and unbearable. For such a simple task, this feels like an eternity.

For $n \times n$ inputs, we call the det function recursively $n$ times, on $(n - 1) \times (n - 1)$ inputs. That is, if $a_n$ denotes the time complexity of our algorithm for an $n \times n$ matrix, then, due to the recursive step, we have

$$a_n = na_{n-1},$$

which explodes really fast. In fact, $a_n = O(n!)$, which is the dreaded factorial complexity. Unlike some other recursive algorithms, caching doesn't help either. There are two reasons for this: sub-matrices rarely match, and numpy.ndarray objects are mutable, thus not hashable.

In practice, $n$ can be in the millions, so this formula is utterly useless. What can we do? Simple: LU decomposition.

## 5.3.3   How to actually compute determinants

Besides the two formulas, we have seen lots of useful properties of matrices and determinants. Can we apply what we have learned so far to simplify the problem?

Let's consider the LU decomposition. According to this, if $\det A \neq 0$, then $A = LU$, where $L$ is lower triangular and $U$ is upper triangular. Since the determinant behaves nicely with respect to matrix multiplication (see equation (4.11)), we have

$$\det A = \det L \det U.$$

Seemingly, we made our situation worse: instead of one determinant, we have to deal with two. However, $L$ and $U$ are rather special, as they are triangular. It turns out that computing a triangular matrix's determinant is extremely easy. We just have to multiply the elements on the diagonal together!

> **Theorem 5.3.1** *(Determinant of a triangular matrix)*
>
> Let $A \in \mathbb{R}^{n \times n}$ be a triangular matrix. (That is, it is either lower or upper triangular.) Then,
>
> $$\det A = \prod_{i=1}^{n} a_{ii}.$$

*Proof.*Suppose that $A$ is lower triangular. (That is, all elements above its diagonal are 0.) According to the recursive formula for $\det A$, we have

$$\det A = \sum_{j=1}^{n} (-1)^{j+1} a_{1,j} \det A_{1,j}.$$

Because $A$ is lower triangular, $a_{1,j} = 0$ if $j > 1$. Thus,

$$\det A = a_{11} \det A_{1,1}.$$

$A_{1,1} = (a_{ij})_{i,j=2}^{n}$ is also lower triangular. By iterating the previous step, we obtain

$$\det A = a_{11} a_{22} \dots a_{nn},$$

which is what we had to show.

If $A$ is upper triangular, its transpose $A^T$ is lower triangular. Thus, we can apply the previous result, so

$$\det A = \det A^T = a_{11} a_{22} \dots a_{nn}$$

holds as well.

Back to our original problem. Since the diagonal $L$ is constant 1 (see (5.8)), as guaranteed by the LU decomposition, we have

$$\det A = \det U = \prod_{i=1}^{n} u_{ii}.$$

So, the algorithm to compute the determinant is quite simple: get the LU decomposition, then calculate the product of $U$'s diagonal. Let's put this into practice!

```python
def fast_det(A: np.ndarray):
    """
    Computes the determinant of a square matrix using LU decomposition.

    Args:
        A (np.ndarray): A square matrix of shape (n, n) for which the determinant
                        needs to be computed. The matrix must be non-singular (invertible).

    Returns:
        float: The determinant of the matrix A.
    """
    L, U = LU(A)
    return np.prod(np.diag(U))
```

Yes, that simple. Let's see how it performs!

```python
A = np.random.rand(10, 10)

t_fast_det = timeit(lambda : fast_det(A), number=1)
print(f"The time it takes to compute the determinant of a 10 x 10 matrix: {t_fast_det} seconds")
```

```
The time it takes to compute the determinant of a 10 x 10 matrix:
0.0008458310039713979 seconds
```

It's faster by a huge margin. How much faster?

```
print(f"Recursive determinant:    \t{t_det} s")
print(f"LU determinant:           \t{t_fast_det} s")
print(f"Performance improvement: \t{t_det/t_fast_det} times faster")
```

```
Recursive determinant:        63.98369195000123 s
LU determinant:               0.0008458310039713979 s
Performance improvement:      75645.95250065446 times faster
```

That's quite an insane improvement! This can be even faster if we use a better implementation of the LU decomposition algorithm (for instance, `scipy.linalg.lu`, which relies on our old friend LAPACK).

I get emotional just by looking at this result. See how far we can go with a bit of linear algebra? This is why understanding the fundamentals such as Gaussian elimination is essential. Machine learning and deep learning are still very new fields, and even though an insane amount of research power is being put into it, moments like these happen all the time. Simple ideas often give birth to new paradigms.

## 5.4 Summary

In this chapter, we have looked at matrices from the perspective of linear equation systems, i.e., equations of the form

$$a_{11}x_1 + a_{12}x_2 + \cdots + a_{1n}x_n = b_1$$
$$a_{21}x_1 + a_{22}x_2 + \cdots + a_{2n}x_n = b_2$$
$$\vdots$$
$$a_{n1}x_1 + a_{n2}x_2 + \cdots + a_{nn}x_n = b_n.$$

Not surprisingly, these are described by matrices, and the above is equivalent to the expression $Ax = \mathbf{b}$. Solving linear equations is an ancient art, so why are we talking about it in the age of AI?

Remember: It's only AI if you are talking to investors. Deep down, it's linear algebra, calculus, and probability theory.

We wanted to solve linear equations, which led us to Gaussian elimination (well, led *Gauss* to Gaussian elimination). Which led us to the LU decomposition. Which led us to fast matrix inversion, and a bunch of other innovations on which our current technology is built on. Let me tell you, fast matrix multiplication and inversion are the bread and butter of computational linear algebra, and they all stem from that aforementioned ancient art of solving linear equations.

Let's recap the feats that we've achieved in this chapter one by one:

- solving linear equations by Gaussian elimination,
- characterizing the invertibility of matrices in terms of linear equations,
- discovering a matrix factorization technique called LU decomposition,
- building a crazy-fast matrix-inverting algorithm using the LU decomposition,
- and building a crazy-fast determinant-computing algorithm using – drumroll! – the LU decomposition.

However, we are not done with matrices. Recall the relationship we established between them and linear transformations, viewing matrices as data transforms that distort the underlying feature space. As it turns out, if we are looking from the right perspective, this distortion is always a stretching. Well, almost always. Well, almost a stretching.

Well, that was too many *well*s, so let's clear all of it up in the next chapter, diving into *eigenvalues* and *eigenvectors*. Let's go!

# 5.5 Problems

**Problem 1.** Show that the product of upper triangular matrices is upper triangular. Similarly, show that the product of lower triangular matrices is lower triangular. (We have used these facts extensively in this section but didn't give a proof. So, this is an excellent time to convince yourself about this if you haven't already.)

**Problem 2.** Write a function that, given an invertible square matrix $A \in \mathbb{R}^{n \times n}$ and a vector $\mathbf{b} \in \mathbb{R}^n$, finds the solution of the linear equation $A\mathbf{x} = \mathbf{b}$. (This can be done with a one-liner if you use one of the tools we have built here.)

**Problem 3.** Before we wrap this chapter up, let's go back to the definition of determinants. Even though there are lots of reasons against using the determinant formula, we have one for it: it is a good exercise, and implementing it will deepen your understanding. So, in this problem, you are going to build

$$\det A = \sum_{\sigma \in S_n} \text{sign}(\sigma) a_{\sigma(1)1} \dots a_{\sigma(n)n},$$

one step at a time.

*(i)* Implement a function that, given an integer $n$, returns all permutations of the set $\{0, 1, \dots, n-1\}$. Represent each permutation $\sigma$ as a list. For example,

```
[2, 0, 1]
```

would represent the permutation $\sigma$, where $\sigma(0) = 2$, $\sigma(1) = 0$, and $\sigma(2) = 1$.

*(ii)* Let $\sigma \in S_n$ be a permutation of the set $\{0, 1, \ldots, n-1\}$. Its *inversion number* is defined by

$$\text{inversion}(\sigma) = \left|\{(i, j) : i < j \text{ and } \sigma(i) > \sigma(j)\}\right|,$$

where $|\cdot|$ denotes the number of elements in the set. Essentially, inversion describes the number of times a permutation reverses the order of a pair of numbers.

Turns out, the sign of $\sigma$ can be written as

$$\text{sign}(\sigma) = (-1)^{\text{inversion}(\sigma)}.$$

Implement a function that first calculates the inversion number, then the sign of an arbitrary permutation. (Permutations are represented like in the previous problem.)

*(iii)* Put the solutions for Problem 1. and Problem 2. together and calculate the determinant of a matrix using the permutation formula. What do you think the time complexity of this algorithm is?

# Join our community on Discord

Read this book alongside other users, Machine Learning experts, and the author himself.

Ask questions, provide solutions to other readers, chat with the author via Ask Me Anything sessions, and much more.

Scan the QR code or visit the link to join the community.

https://packt.link/math

# 6

# Eigenvalues and Eigenvectors

So far, we have seen three sides of linear transformations: functions, matrices, and transforms that distort the grid of the underlying vector space. In the Euclidean plane, we saw some examples (*Section 4.3*) that shed some light on the geometric nature of them.

Following this line of thought, let's consider the linear transformation given by the matrix

$$A = \begin{bmatrix} 2 & 1 \\ 1 & 2 \end{bmatrix}. \tag{6.1}$$

Since the columns of $A$ are the images of the standard basis vectors $\mathbf{e}_1 = (1, 0)$ and $\mathbf{e}_2 = (0, 1)$, we can visualize the effect of $A$ on *Figure 6.1*. (Check *Section 4.1.1* if you don't recall this fact.)

This seems to shear, stretch, and rotate the entire grid. However, there are special directions along which $A$ is simply a stretching. For instance, consider the vector $\mathbf{u}_1 = (1, 1)$. By a simple calculation, you can verify that $A\mathbf{u}_1 = 3\mathbf{u}_1$.

Because of the linearity, this means that if a vector $\mathbf{x}$ is in span($\mathbf{u}_1$), its image under $A$ is $3\mathbf{x}$.

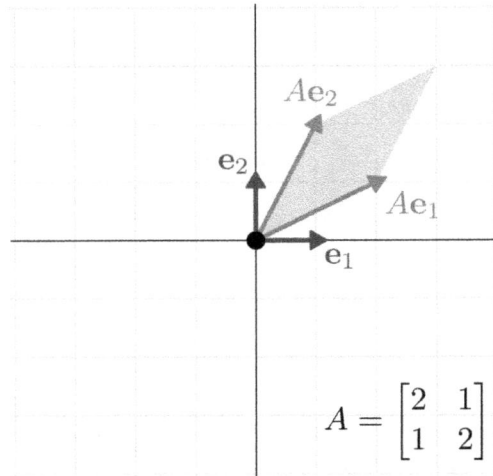

Figure 6.1: Images of the standard basis vectors under the linear transformation given by $A$

Another one is $\mathbf{u}_2 = (-1, 1)$, where we have $A\mathbf{u}_2 = \mathbf{u}_2$. Thus, any $\mathbf{x} \in$ span($\mathbf{u}_2$) is left in place. If we select $\mathbf{u}_1, \mathbf{u}_2$ as our base, the matrix of this transformation is

$$A_{\mathbf{u}_1,\mathbf{u}_2} = \begin{bmatrix} 3 & 0 \\ 0 & 1 \end{bmatrix},$$

that is, $A_{\mathbf{u}_1,\mathbf{u}_2}$ is diagonal.

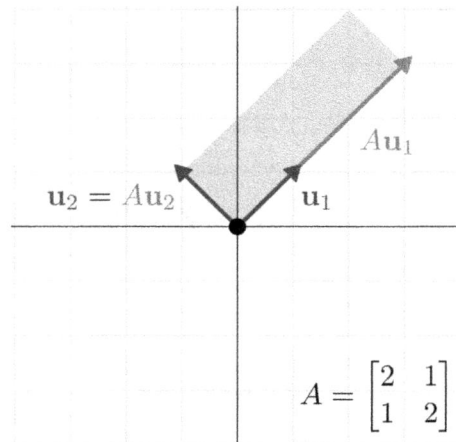

Figure 6.2: Images of $\mathbf{u}_1 = (1, 1)$ and $\mathbf{u}_2 = (-1, 1)$ under the linear transformation given by $A$

We love diagonal matrices in practice because multiplication with a diagonal matrix is much faster, as it requires $O(n)$ operations, opposed to $O(n^2)$.

Is this a general phenomena? Are these even useful? The answer is yes to both questions. What we have just seen is formalized by the concept of *eigenvalues* and *eigenvectors*. The terminology originates from the german word "eigen" meaning "own," resulting in one of the ugliest naming conventions in mathematics.

---

**Definition 6.0.1 (Eigenvalues and eigenvectors)**

Let $f : V \to V$ be an arbitrary linear transformation. We say that the $\lambda$ scalar and the $\mathbf{x} \in V \setminus \{\mathbf{0}\}$ nonzero vector is an eigenvalue-eigenvector pair of $f$ if $f(\mathbf{x}) = \lambda \mathbf{x}$ holds.

---

# 6.1 Eigenvalues of matrices

Although we have formally defined eigenvalues and eigenvectors for linear transformations, we often talk about them in context of matrices. (Because, as we have seen, matrices and linear transformations are two faces of the same coin.) Let's start by translating the definition into the language of matrices.

If $A \in \mathbb{R}^{n \times n}$ is a matrix, *Definition 6.0.1* translates to the following: the scalar $\lambda$ and the vector $\mathbf{x} \in \mathbb{R} \setminus \{\mathbf{0}\}$ is an eigenvalue-eigenvector pair of the *matrix* if

$$A\mathbf{x} = \lambda \mathbf{x} \tag{6.2}$$

holds. This can be simplified: as the linear transformation $\mathbf{x} \mapsto \lambda \mathbf{x}$ corresponds to the matrix $\lambda I$, (6.2) is equivalent to

$$(A - \lambda I)\mathbf{x} = \mathbf{0} \tag{6.3}$$

If you recall *Chapter 4, Section 4.1.1*, where we learned how matrices arise from linear transformations, you might ask the question: won't the eigenvalues depend on the choice of the matrix?

The following theorem states that this is not the case: the eigenvalues of a linear transformation and its matrices are the same.

---

**Theorem 6.1.1 *(Eigenvalues of similar matrices)***

Let $A, B \in \mathbb{R}^{n \times n}$ be two similar matrices, that is, suppose that there exists an invertible $T \in \mathbb{R}^{n \times n}$ such that $B = T^{-1}AT$. Then, if $A\mathbf{x} = \lambda \mathbf{x}$ holds for some scalar $\lambda$ and vector $\mathbf{x} \in \mathbb{R}^n$, *then*

$$B\mathbf{x}' = \lambda\mathbf{x}'$$

*holds for some* $\mathbf{x}' \in \mathbb{R}^n$ *as well.*

*Proof.* Let's massage the eigenvalue (6.3) a bit! We have

$$
\begin{aligned}
(A - \lambda I)\mathbf{x} &= (A - \lambda T T^{-1})\mathbf{x} \\
&= T(T^{-1}AT - \lambda I)T^{-1}\mathbf{x} \\
&= \mathbf{0}.
\end{aligned}
$$

Since $T$ is invertible, $T\big[(T^{-1}AT - \lambda I)T^{-1}\mathbf{x}\big] = \mathbf{0}$ can only happen if $(T^{-1}AT - \lambda I)T^{-1}\mathbf{x} = \mathbf{0}$. (Recall the relation of the kernel and invertibility in *Theorem 4.1.3.*) This looks almost like (6.3), just a bit more complicated. Let me use some suggestive parentheses to highlight the similarities:

$$\big[T^{-1}AT - \lambda I\big]\big[T^{-1}\mathbf{x}\big] = 0.$$

So, with the selection $\mathbf{x}' = T^{-1}\mathbf{x}$, we have

$$T^{-1}AT\mathbf{x}' = \lambda\mathbf{x}',$$

which is what we had to show.

In other words, the eigenvalues of similar matrices are the same. Consequently, we can talk about the eigenvalues of *matrices*, not just linear transformations. The above theorem implies that the eigenvalues of a transformation and its corresponding matrix are the same. Moreover, the eigenvalues of the matrix don't depend on the choice of basis.

To be more precise, suppose that $A : U \to U$ is a linear transformation and $P, Q$ are bases of $U$. The matrix of $A$ in some basis $S$ is denoted by $A_Q$. We know that there is a transformation matrix $T \in \mathbb{R}^{n \times n}$ such that

$$A_Q = T^{-1}A_P T.$$

So, the eigenvalues are the same.

All of the above begs the question: how do we actually find eigenvalues? Let's talk about this next.

## 6.2   Finding eigenvalue-eigenvector pairs

Even though the definition of eigenvalue-eigenvector pairs is easy to understand given the geometric interpretation we just saw, it does not give us any tools to find them in practice. Using them to get simpler representations of matrices is one thing, but we are stuck at square one without a method to find them.

First, let's focus on the eigenvalues. Suppose that for some $\lambda$, there is a nonzero vector $\mathbf{x}$ such that $A\mathbf{x} = \lambda\mathbf{x}$. The transformation defined by $\mathbf{x} \to \lambda\mathbf{x}$ is a linear one, and its matrix is diagonal:

$$\lambda\mathbf{x} = \begin{bmatrix} \lambda & 0 & \dots & 0 \\ 0 & \lambda & \dots & 0 \\ \vdots & \vdots & \ddots & \vdots \\ 0 & 0 & \dots & \lambda \end{bmatrix} \begin{bmatrix} x_1 \\ x_2 \\ \vdots \\ x_n \end{bmatrix},$$

where the matrix with $\lambda$-s in the diagonal is $\lambda I$, that is, $\lambda$ times the identity matrix.

Because linear transformations can be added and subtracted (as we saw in *Section 4.1.2*), the defining equation $A\mathbf{x} = \lambda\mathbf{x}$ is equivalent to

$$(A - \lambda I)\mathbf{x} = \mathbf{0},$$

where $I$ denotes the identity transformation, as defined by equation (4.3). In other words, the transformation $A - \lambda I$ maps a nonzero vector to $\mathbf{0}$, meaning that it is not invertible, as *Theorem 4.1.3* implies. We can characterize this with determinants: we need to find all $\lambda$-s such that

$$\det(A - \lambda I) = 0.$$

We can summarize the above findings in the following theorem.

> **Theorem 6.2.1** *Let $A : U \to U$ be an arbitrary linear transformation. Then $\lambda$ is its eigenvalue if and only if*
>
> $$\det(A - \lambda I) = 0.$$

Although we are one step closer, finding eigenvalues based on this still seems complicated. In the following, we are going to see what $\det(A - \lambda I)$ really is and how we can find the solutions of $\det(A - \lambda I) = 0$ in practice.

Before going into the generalities, let's revisit the example (6.1). There, we have

$$\det(A - \lambda I) = \begin{vmatrix} 2 - \lambda & 1 \\ 1 & 2 - \lambda \end{vmatrix}$$
$$= (2 - \lambda)^2 - 1$$
$$= \lambda^2 - 4\lambda + 3.$$

To find the eigenvalues, we have to solve the quadratic equation

$$\lambda^2 - 4\lambda + 3 = 0,$$

which we can do easily. Recall that the solutions of any quadratic equation $ax^2 + bx + c = 0$ are

$$x_{1,2} = \frac{-b \pm \sqrt{b^2 - 4ac}}{2a}.$$

Applying this, we have $\lambda_1 = 3$ and $\lambda_2 = 1$ as solutions. There are no other ones, so 1 and 3 are the only two eigenvalues for $A$.

Let's see what happens in the general case!

## 6.2.1   The characteristic polynomial

As the example above suggests, if the underlying vector space $U$ is $n$-dimensional, that is, $A$ is an $n \times n$ matrix, $\det(A - \lambda I)$ is an $n$-th degree polynomial in $\lambda$.

To see this, let's write $\det(A - \lambda I)$ explicitly in terms of matrices. With this in mind, we have

$$\det(A - \lambda I) = \begin{vmatrix} a_{11} - \lambda & a_{12} & \dots & a_{1n} \\ a_{21} & a_{22} - \lambda & \dots & a_{2n} \\ \vdots & \vdots & \ddots & \vdots \\ a_{n1} & a_{n2} & \dots & a_{nn} - \lambda \end{vmatrix}.$$

If you consider the formula to calculate the determinant given by (4.12), you can see that every term is a polynomial. Depending on how many fixed points $\sigma$ has (that is, points where $\sigma(i) = i$), the degree of this polynomial varies between 0 and $n$.

(Alternatively, you can see that $\det(A - \lambda I)$ is a polynomial of degree $n$ by using the recursive formula (4.13) and applying induction.)

**Definition 6.2.1 (Characteristic polynomial of matrices)**

Let $A \in \mathbb{R}^{n \times n}$ be an arbitrary matrix. The polynomial

$$p(\lambda) = \det(A - \lambda I)$$

is called the *characteristic polynomial* of $A$.

The roots of the characteristic polynomial are the eigenvalues. If $U$ is an $n$-dimensional complex vector space (that is, the set of scalars is $\mathbb{C}$), the fundamental theorem of algebra (*Theorem D.3.1*) guarantees that $\det(A - \lambda I) = 0$ has exactly $n$ roots.

As a consequence, every matrix $A \in \mathbb{C}^{n \times n}$ has at least one eigenvalue. Note that roots can have higher algebraic multiplicity. For instance, the characteristic polynomial for the matrix

$$B = \begin{bmatrix} 1 & 0 & 0 \\ 0 & 1 & 0 \\ 0 & 0 & 2 \end{bmatrix}$$

is $(1 - \lambda)^2(2 - \lambda)$. So, its roots are 1 (with algebraic multiplicity 2) and 2.

If we restrict ourselves to real matrices and real vector spaces, the existence of eigenvalues and eigenvectors are not guaranteed. For instance, consider

$$C = \begin{bmatrix} 0 & -1 \\ 1 & 0 \end{bmatrix}.$$

Its characteristic polynomial is $\lambda^2 + 1$, which doesn't have any real roots, only complex ones: $\lambda_1 = i$ and $\lambda_2 = -i$. Mathematically speaking, if we want to stay within the confines of real vector spaces, $C$ has no eigenvalues. However, we are here to do machine learning, not algebra. Thus, we are going to be a bit imprecise and treat real matrices as complex ones. We don't often need complex numbers to describe mathematical models of a dataset, but they frequently appear during the analysis of matrices.

## 6.2.2 Finding eigenvectors

When an eigenvalue $\lambda$ is identified, we can set out to find the corresponding eigenvectors; that is, vectors $\mathbf{x}$ where $(A - \lambda I)\mathbf{x} = \mathbf{0}$. In more precise terms, we are looking for $\ker(A - \lambda I)$.

As we have mentioned before in *Section 4.1.4*, the kernel of any linear transformation is a subspace. As it might be more than one-dimensional, identifying it often involves an implicit description like $x_1 + x_2 = 0$.

Let's check what happens with our recurring example

$$A = \begin{bmatrix} 2 & 1 \\ 1 & 2 \end{bmatrix}.$$

Previously, we have seen that $\lambda_1 = 3$ and $\lambda_2 = 1$ are the eigenvalues. To identify the corresponding eigenvectors for, say, $\lambda_1$, we have to find *all* solutions for the linear equation $(A - \lambda_1 I)\mathbf{x} = \mathbf{0}$. Expanding this, we have

$$-x_1 + x_2 = 0$$
$$x_1 - x_2 = 0.$$

Both equations imply that all $\mathbf{x} = (x_1, x_2)$ are solutions where $x_1 = x_2$.

# 6.3 Eigenvectors, eigenspaces, and their bases

**Definition 6.3.1 (Eigenspaces)**

Let $f : V \to V$ be an arbitrary linear transformation, and $\lambda$ its eigenvalue. The subspace of eigenvectors defined by

$$U_\lambda = \{\mathbf{x} : A\mathbf{x} = \lambda\mathbf{x}\}$$

is called the *eigenspace* of $\lambda$.

Eigenspaces play an important role in understanding the structure of linear transformations. First, we note that a linear transformation *keeps its eigenspaces invariant*. (That is, if $\mathbf{x}$ is in the $U_\lambda$ eigenspace, then $f(\mathbf{x}) \in U_\lambda$ as well.) This property makes it possible for us to restrict linear transformations to their eigenspaces.

To illustrate the concept of eigenspaces, let's revisit the already familiar matrix

$$A = \begin{bmatrix} 2 & 1 \\ 1 & 2 \end{bmatrix}$$

one more time. Its eigenvalues are $\lambda_1 = 3$ and $\lambda_2 = 1$, and by solving the equation $(A - \lambda_1 I)\mathbf{x} = \mathbf{0}$, we get that the eigenspace of $\lambda_1$ is

$$U_{\lambda_1} = \{\mathbf{x} \in \mathbb{R}^2 : x_1 = x_2\}.$$

Similarly, you can check that $U_{\lambda_2} = \{\mathbf{x} \in \mathbb{R}^2 \; : \; x_1 = -x_2\}$. (If you go back to *Figure 6.2*, you can visualize $U_{\lambda_1}$ and $U_{\lambda_2}$.)

Eigenspaces are not necessarily one-dimensional. For instance, consider one of the the previous examples

$$B = \begin{bmatrix} 1 & 0 & 0 \\ 0 & 1 & 0 \\ 0 & 0 & 2 \end{bmatrix},$$

with two eigenvalues $\lambda_1 = 1$ and $\lambda_2 = 2$. Substituting $\lambda_1$ back into the equation and solving for $(B - I)\mathbf{x} = 0$ for $\mathbf{x}$, we obtain that

$$U_{\lambda_1} = \{\mathbf{x} \in \mathbb{R}^3 \; : \; x_3 = 0\},$$

which is simply the plane determined by the first two axes.

The structure of eigenspaces determines whether or not we can diagonalize the matrix $A$ with a change of basis (*Section 4.2*) transformation $\Lambda = T^{-1}AU$. The following general theorem establishes this connection.

---

**Theorem 6.3.1 *(Diagonalization and eigenspaces)***

*Let $f : V \to V$ be a linear transformation, let $A \in \mathbb{R}^{n \times n}$ be its matrix in some basis, and let $U_{\lambda 1,...,U}\{\lambda_k\}$ be the eigenspaces of $f$. The following are equivalent.*

*(a) There is a matrix $T \in \mathbb{R}^{n \times n}$ such that*

$$\Lambda = T^{-1}AT,$$

*where $\Lambda$ is a diagonal matrix.*

*(b) There is a basis $\mathbf{u}_1, ..., \mathbf{u}_n$ for $V$ that can be selected from the eigenvectors of $f$.*

*(c) V can be written as the direct sum of the eigenspaces, that is,*

$$V = U_{\lambda_1} + \cdots + U_{\lambda_k}.$$

---

(Note that $k$, the number of eigenspaces, is not necessarily $n$.)

*Proof. (a)* $\implies$ *(b).* If $A$ is the matrix of $f$ in some basis, then a similarity transformation is equivalent to a change of basis.

That is, the new matrix $\Lambda = T^{-1}AT$ is the matrix of $f$ in a different basis, say $\mathbf{u}_1, ..., \mathbf{u}_n$.

If $\Lambda$ is diagonal, it can be written in the form

$$\Lambda = \begin{bmatrix} \lambda_1 & 0 & \cdots & 0 \\ 0 & \lambda_2 & \cdots & 0 \\ \vdots & \vdots & \ddots & \vdots \\ 0 & 0 & \cdots & \lambda_n \end{bmatrix}.$$

(Note that the $\lambda_i$-s are not necessarily mutually different.) Thus, $\Lambda\mathbf{u}_i = \lambda_i\mathbf{u}_i$, meaning that $\mathbf{u}_1, \ldots, \mathbf{u}_n$ is a basis from the eigenvectors of $f$.

*(b)* $\implies$ *(a).* If $\mathbf{u}_1, \ldots, \mathbf{u}_n$ is a basis from the eigenvectors of $f$, then its matrix $\Lambda$ in that basis is diagonal. Thus, $A$ is similar to $\Lambda$, which is what we had to show.

*(b)* $\implies$ *(c).* By definition, the direct sum (*Definition 1.2.4*) of the eigenspaces contains all linear combinations of the form

$$x = \sum_{i=1}^{n} x_i\mathbf{u}_i.$$

Since $\mathbf{u}_1, \ldots, \mathbf{u}_n$ is a basis, $V = U_{\lambda_1} + \cdots + U_{\lambda_k}$ holds.

*(c)* $\implies$ *(b).* From each eigenspace $U_{\lambda_i}$, we can select a basis. Due to the construction of $U_{\lambda_i}$, its basis will consist of eigenvectors.

Since $V = U_{\lambda_1} + \cdots + U_{\lambda_k}$, the union of of such bases $\mathbf{u}_1, \ldots, \mathbf{u}_n$ will be a basis for $V$.

Even though this theorem does not give us any useful recipes on how to diagonalize a matrix, it provides us with an extremely valuable insight: diagonalization is equivalent to finding an eigenvector basis. This is not always possible, but when it is, we are cooking with gas.

In the next chapter, we will take a deep dive into this topic, providing multiple ways to simplify matrices. If our journey in linear algebra is akin to a mountain climb, we will reach the peak soon.

# 6.4  Summary

In this chapter, we've veered into the theory side of math once again. This time, it was about *eigenvalues* and *eigenvectors* of a matrix, that is, scalars $\lambda$ and vectors $\mathbf{x}$ for which

$$A\mathbf{x} = \lambda\mathbf{x}, \quad A \in \mathbb{R}^{n \times n}$$

hold.

Just like most mathematical objects, this might seem daunting at first, but geometrically, this means that in the direction **x**, the linear transformation $A$ is the same as a stretching by $\lambda$. In practice, we can find eigenvectors by solving the so-called *characteristic equation*

$$\det(A - \lambda I) = 0$$

for $\lambda$.

What are eigenvalues used for? There are tons of applications, but one stands out: according to *Theorem 6.3.1*, if you can build a basis from the eigenvectors of the matrix $A \in \mathbb{R}^{n \times n}$, then you can find a $T \in \mathbb{R}^{n \times n}$ such that $T^{-1}AT$ is diagonal. This process is extremely useful. For one, multiplication with diagonal matrices is fast and simple, and we prefer to do it whenever we can. For another, diagonalization reveals a ton about the internal structure of the underlying linear transformation.

We've left this chapter with a multitude of questions. How do we find eigenvalues? What kind of matrices are diagonalizable? If a matrix is diagonalizable, how can we find such a form?

We'll answer all of these in the next chapter. Be warned: we are approaching the pinnacle of linear algebra. The next chapter might be our heaviest one yet. Just like the final stretch before reaching the peaks of Mount Everest. However, you have my total confidence. If you are here, you can climb it.

Let's go!

# 6.5 Problems

**Problem 1.** Compute the eigenvalues of the matrices

$$A = \begin{bmatrix} 4 & 1 & -1 \\ 1 & 3 & 1 \\ -1 & 1 & 2 \end{bmatrix}, \quad B = \begin{bmatrix} 2 & 1 & 1 \\ 1 & 2 & 1 \\ 1 & 1 & 2 \end{bmatrix},$$

and find an eigenvector for every eigenvalue.

**Problem 2.** Let $A \in \mathbb{R}^{n \times n}$ be an upper or lower triangular matrix. Show that the eigenvalues of $A$ are its diagonal elements.

**Problem 3.** Let $A \in \mathbb{R}^{n \times n}$ be a square matrix. Show that

$$\det(A - \lambda I)$$

is a polynomial of degree $n$ in $\lambda$.

This is the characteristic polynomial that we have talked about, and we have even mentioned this fact. However, we omitted the proof, so here's your chance to fill the gap.

**Problem 4.** Let $A \in \mathbb{R}^{n \times n}$, $B \in \mathbb{R}^{n \times m}$, and $C \in \mathbb{R}^{m \times m}$ arbitrary matrices, and we define the so-called *block matrix*

$$D = \begin{bmatrix} A & B \\ 0 & C \end{bmatrix} \in \mathbb{R}^{(n+m) \times (n+m)}.$$

Show that if $\lambda$ is an eigenvalue of $A$ or an eigenvalue of $B$, then it's also an eigenvalue of $C$.

# Join our community on Discord

Read this book alongside other users, Machine Learning experts, and the author himself.

Ask questions, provide solutions to other readers, chat with the author via Ask Me Anything sessions, and much more.

Scan the QR code or visit the link to join the community.

https://packt.link/math

# 7

# Matrix Factorizations

One of the recurring thoughts in this book is that problem-solving is about finding the best representations of your objects of study. Say linear transformations of a vector space are represented by matrices. Studying one is the same as studying the other, but each perspective comes with its own set of tools. Linear transformations are geometric, while matrices are algebraic sides of the same coin.

This thought can be applied on a smaller scale as well. Recall the LU decomposition from *Chapter 5*. You can think of this as another view of matrices.

Guess what: It's not the only one. This chapter is dedicated to the three most important ones:

- the spectral decomposition,
- the singular value decomposition,
- and the QR decomposition.

Buckle up. It's our most challenging adventure yet.

## 7.1  Special transformations

So far, we have aspired to develop a geometric view of linear algebra. Vectors are mathematical objects defined by their direction and magnitude. In the spaces of vectors, the concept of distance and orthogonality gives rise to a geometric structure.

Linear transformations, the building blocks of machine learning, are just mappings that distort this structure: rotating, stretching, and skewing the geometry. However, there are types of transformations that *preserve* some of the structure. In practice, these provide valuable insights, and additionally, they are much easier to work with. In this section, we will take a look at the most important ones, those that we'll encounter in machine learning.

## 7.1.1  The adjoint transformation

In machine learning, the most important stage is the Euclidean space $\mathbb{R}^n$. This is where data is represented and manipulated. There, the entire geometric structure is defined by the inner product

$$\langle \mathbf{x}, \mathbf{y} \rangle = \sum_{i=1}^{n} x_i y_i,$$

giving rise to the notion of magnitude, direction (in the form of *angles*), and orthogonality. Because of this, transformations that can be related to the inner product are special. For instance, if $\langle f(\mathbf{x}), f(\mathbf{x}) \rangle = \langle \mathbf{x}, \mathbf{x} \rangle$ holds for all $\mathbf{x} \in \mathbb{R}^n$ and the linear transformation $f : \mathbb{R}^n \to \mathbb{R}^n$, we know that $f$ leaves the norm invariant. That is, distance in the original and the transformed feature space have the same meaning.

First, we will establish a general relation between images of vectors under a transform and their inner product. This is going to be the foundation for our discussions in this chapter.

**Theorem 7.1.1** *(The adjoint transformation)*

*Let $f : \mathbb{R}^n \to \mathbb{R}^n$ be a linear transformation. Then, there exists a linear transformation $f^* : \mathbb{R}^n \to \mathbb{R}^n$ for which*

$$\langle f(\mathbf{x}), \mathbf{y} \rangle = \langle \mathbf{x}, f^*(\mathbf{y}) \rangle \tag{7.1}$$

*holds for all $\mathbf{x}, \mathbf{y} \in \mathbb{R}^n$. $f^*$ is called the adjoint transformation\* of $f$.*

*Moreover, if $A \in \mathbb{R}^{n \times n}$ is the matrix of $f$ in the standard orthonormal basis, then the matrix of $f^*$ is $A^T$. That is,*

$$\langle A\mathbf{x}, \mathbf{y} \rangle = \langle \mathbf{x}, A^T \mathbf{y} \rangle. \tag{7.2}$$

*Proof.* Suppose that $A \in \mathbb{R}^{n \times n}$ is the matrix of $f$ in the standard orthonormal basis. For any $\mathbf{x} = (x_1, \dots, x_n)$ and $\mathbf{y} = (y_1, \dots, y_n)$, the inner product is defined by

$$\langle \mathbf{x}, \mathbf{y} \rangle = \sum_{i=1}^{n} x_i y_i.$$

and $A\mathbf{x}$ can be written as

$$\begin{bmatrix} a_{11} & a_{12} & \cdots & a_{1n} \\ a_{21} & a_{22} & \cdots & a_{2n} \\ \vdots & \vdots & \ddots & \vdots \\ a_{n1} & a_{n2} & \cdots & a_{nn} \end{bmatrix} \begin{bmatrix} x_1 \\ x_2 \\ \vdots \\ x_n \end{bmatrix} = \begin{bmatrix} \sum_{j=1}^{n} a_{1j}x_j \\ \sum_{j=1}^{n} a_{2j}x_j \\ \vdots \\ \sum_{j=1}^{n} a_{nj}x_j \end{bmatrix}.$$

Using this form, we can express $\langle A\mathbf{x}, \mathbf{y} \rangle$ in terms of $a_{ij}$-s, $x_i$-s, and $y_i$-s. For this, we have

$$\langle A\mathbf{x}, \mathbf{y} \rangle = \sum_{i=1}^{n} \left( \sum_{j=1}^{n} a_{ij}x_j \right) y_i$$

$$= \sum_{j=1}^{n} \underbrace{\left( \sum_{i=1}^{n} a_{ij}y_i \right)}_{j\text{-th component of } A^T\mathbf{y}} x_j$$

$$= \langle \mathbf{x}, A^T\mathbf{y} \rangle.$$

This shows that the transformation given by $f^* : x \mapsto A^T\mathbf{x}$ satisfies (7.1) and (7.2), which is what we had to show.

Why is the quantity $\langle A\mathbf{x}, \mathbf{y} \rangle$ that important to us? Because inner products define the geometric structure of a vector space. Recall the equation (2.12), allowing us to fully describe any vector using only the inner products with respect to an orthonormal basis. In addition, $\langle \mathbf{x}, \mathbf{x} \rangle = \|\mathbf{x}\|^2$ defines the notion of distance and magnitude. Because of this, (7.1) and (7.2) will be quite useful for us.

As we are about to see, transformations that preserve the inner product are rather special, and these relationships provide us with a way to characterize them both algebraically and geometrically.

## 7.1.2 Orthogonal transformations

Let's jump straight into the definition.

**Definition 7.1.1 (Orthogonal transformations)**

Let $f : \mathbb{R}^n \to \mathbb{R}^n$ be an arbitrary linear transformation. $f$ is called *orthogonal* if

$$\langle f(\mathbf{x}), f(\mathbf{y}) \rangle = \langle \mathbf{x}, \mathbf{y} \rangle$$

holds for all $\mathbf{x}, \mathbf{y} \in \mathbb{R}^n$.

As a consequence, an orthogonal $f$ preserves the norm: $\|f(\mathbf{x})\|^2 = \langle f(\mathbf{x}), f(\mathbf{x}) \rangle = \langle \mathbf{x}, \mathbf{x} \rangle = \|\mathbf{x}\|^2$. Because the angle enclosed by two vectors is defined by their inner product, see equation (2.9), the property $\langle f(\mathbf{x}), f(\mathbf{y}) \rangle = \langle \mathbf{x}, \mathbf{y} \rangle$ means that an orthogonal transform also preserves angles.

We can translate the definition to the language of matrices as well. In practice, we are always going to work with matrices, so this characterization is essential.

**Theorem 7.1.2** *(Matrices of orthogonal transformations)*

*Let $f : \mathbb{R}^n \to \mathbb{R}^n$ be a linear transformation and $A \in \mathbb{R}^{n \times n}$ be its matrix in the standard orthonormal basis. Then, $f$ is orthogonal if, and only if, $A^T = A^{-1}$.*

*Proof.* As usual, we have to show the implication in both ways.

*(a)* Suppose that $f$ is orthogonal. Then, (7.2) gives

$$\langle \mathbf{x}, \mathbf{y} \rangle = \langle A\mathbf{x}, A\mathbf{y} \rangle = \langle \mathbf{x}, A^T A \mathbf{y} \rangle.$$

Thus, for any given $\mathbf{y}$,

$$\langle \mathbf{x}, (A^T A - I)\mathbf{y} \rangle = 0$$

holds for all $\mathbf{x}$. By letting $\mathbf{x} = (A^T A - I)\mathbf{y}$, the positive definiteness of the inner product implies that $(A^T A - I)\mathbf{y} = \mathbf{0}$ for all $\mathbf{y}$. Thus, $A^T A = I$, which means that $A^T$ is the inverse of $A$.

*(b)* If $A^T = A^{-1}$, we have

$$\begin{aligned} \langle A\mathbf{x}, A\mathbf{y} \rangle &= \langle \mathbf{x}, A^T A\mathbf{y} \rangle \\ &= \langle \mathbf{x}, A^{-1} A\mathbf{y} \rangle \\ &= \langle \mathbf{x}, \mathbf{y} \rangle, \end{aligned}$$

showing that $f$ is orthogonal.

The fact that $A^T = A^{-1}$ has a profound implication regarding the columns of $A$. If you think back to the definition of matrix multiplication in *Section 4.1.2*, the element in the $i$-th row and $j$-th column of $AB$ is the inner product of the $i$-th row of $A$ and the $j$-th column of $B$.

To be more precise, if the $i$-th column is denoted by $\mathbf{a}_i = (a_{1,i}, a_{2,i}, \dots, a_{n,i})$, then we have

$$A^T A = \left( \langle \mathbf{a}_i, \mathbf{a}_j \rangle \right)_{i,j=1}^{n} = I,$$

that is,

$$\langle \mathbf{a}_i, \mathbf{a}_j \rangle = \begin{cases} 1 & \text{if } i = j, \\ 0 & \text{otherwise.} \end{cases}$$

In other words, the columns of $A$ form an orthonormal system. This fact should not come as a surprise since orthogonal transformations preserve magnitude and orthogonality, and the columns of $A$ are the images of the standard orthonormal basis $\mathbf{e}_1, \dots, \mathbf{e}_n$.

In machine learning, performing an orthogonal transformation on our features is equivalent to looking at them from another perspective, without distortion. You might know it already, but this is what Principal Component Analysis (PCA) is doing.

## 7.2 Self-adjoint transformations and the spectral decomposition theorem

Besides orthogonal transformations, there is another important class: transformations whose adjoints are themselves. Bear with me a bit, and we'll see an example soon.

---

**Definition 7.2.1 (Self-adjoint transformations)**

Let $f : \mathbb{R}^n \to \mathbb{R}^n$ be a linear transformation. $f$ is *self-adjoint* if $f^* = f$, that is,

$$\langle f(\mathbf{x}), \mathbf{y} \rangle = \langle \mathbf{x}, f(\mathbf{y}) \rangle \tag{7.3}$$

holds for all $\mathbf{x}, \mathbf{y} \in \mathbb{R}^n$.

---

As always, we are going to translate this into the language of matrices. If $A$ is the matrix of $f$ in the standard orthonormal basis, we know that $A^T$ is the matrix of the adjoint. For self-adjoint transformations, it implies that $A^T = A$. Matrices such as these are called *symmetric*, and they have a lot of pleasant properties.

For us, the most important one is that symmetric matrices can be diagonalized! (That is, they can be transformed into a diagonal matrix with a check reference) The following theorem makes this precise.

**Theorem 7.2.1** *(Spectral decomposition of real symmetric matrices)*

*Let $A \in \mathbb{R}^{n \times n}$ be a real symmetric matrix. Then, $A$ has exactly $n$ real eigenvalues $\lambda_1 \geq \cdots \geq \lambda_n$, and the corresponding eigenvectors $\mathbf{u}_1, \ldots, \mathbf{u}_n$ can be selected such that they form an orthonormal basis.*

*Moreover, if we let $\Lambda = \mathrm{diag}(\lambda_1, \ldots, \lambda_n)$ and $U$ be the orthogonal matrix whose columns are $\mathbf{u}_1, \ldots, \mathbf{u}_n$, then*

$$A = U\Lambda U^T \tag{7.4}$$

*holds.*

Note that the eigenvalues $\lambda_1 \geq \cdots \geq \lambda_n$ are not necessarily distinct from each other.

*Proof.* (Sketch) Since the proof is pretty involved, we are better off getting to know the main ideas behind it, without all the mathematical details.

The main steps are the following.

1. If the matrix $A$ is symmetric, all of its eigenvalues are real.
2. Using this, it can be shown that an orthonormal basis can be formed from the *eigenvectors* of $A$.
3. Writing the matrix of the transformation $\mathbf{x} \rightarrow A\mathbf{x}$ in this orthonormal basis yields a diagonal matrix. Hence, a change of basis yields (7.4).

Showing that the eigenvalues are real requires some complex number magic (which is beyond the scope of this chapter). The tough part is the second step. Once that has been done, moving to the third one is straightforward, as we have seen when talking about eigenspaces and their bases (*Section 6.3*).

We still don't have a hands-on way to diagonalize matrices, but this theorem gets us one step closer: at least we know it is possible for symmetric matrices. This is an important stepping stone, as we'll be able to reduce the general case to the symmetric one.

The requirement for a matrix to be symmetric seems like a very special one. However, in practice, we can *symmetrize* matrices in several different ways. For any matrix $A \in \mathbb{R}^{n \times m}$, the products $AA^T$ and $A^T A$ will be symmetric. For square matrices, the average $\frac{A + A^T}{2}$ also works. So, symmetric matrices are more common than you think.

The orthogonal matrix $U$ and the corresponding orthonormal basis $\{\mathbf{u}_1, \ldots, \mathbf{u}_n\}$ that diagonalizes a symmetric matrix $A$ has a special property that is going to be very important in machine learning for the principal component analysis of data samples.

Before looking at the next theorem, we introduce the argmax notation. Recall that the expression $\max_{x \in A} f(x)$ denotes the maximum value of the function over the set $A$. Often, we would like to know *where* that maximum is attained, which is defined by the argmax:

$$x^* = \mathrm{argmax}_{x \in A} f(x),$$

$$f(x^*) = \max_{x \in A} f(x).$$

Now, let's see the fundamentals of PCA!

**Theorem 7.2.2** *Let $A \in \mathbb{R}^{n \times n}$ be a real symmetric matrix and let $\lambda_1 \geq \cdots \geq \lambda_n$ be its real eigenvalues in decreasing order. Moreover, let $U \in \mathbb{R}^{n \times n}$ be the orthogonal matrix that diagonalizes $A$, with the corresponding orthonormal basis $\{\mathbf{u}_1, \ldots, \mathbf{u}_n\}$.*

*Then,*

$$\arg \max_{\|\mathbf{x}\|=1} \mathbf{x}^T A \mathbf{x} = \mathbf{u}_1,$$

*and*

$$\max_{\|\mathbf{x}\|=1} \mathbf{x}^T A \mathbf{x} = \lambda_1.$$

*Proof.* Since $\{\mathbf{u}_1, \ldots, \mathbf{u}_n\}$ is an orthonormal basis, any $\mathbf{x}$ can be expressed as a linear combination of them:

$$\mathbf{x} = \sum_{i=1}^{n} x_i \mathbf{u}_i, \quad x_i \in \mathbb{R}.$$

Thus, since the $\mathbf{u}_i$ are eigenvectors of $A$,

$$A\mathbf{x} = A\left( \sum_{i=1}^{n} x_i \mathbf{u}_i \right) = \sum_{i=1}^{n} x_i A\mathbf{u}_i = \sum_{i=1}^{n} x_i \lambda_i \mathbf{u}_i.$$

Plugging it back into $\mathbf{x}^T A \mathbf{x}$, we have

$$\mathbf{x}^T A \mathbf{x} = \left( \sum_{j=1}^{n} x_j \mathbf{u}_j \right)^T A\left( \sum_{i=1}^{n} x_i \mathbf{u}_i \right)$$

$$= \left( \sum_{j=1}^{n} x_j \mathbf{u}_j^T \right) A\left( \sum_{i=1}^{n} x_i \mathbf{u}_i \right)$$

$$= \left( \sum_{j=1}^{n} x_j \mathbf{u}_j^T \right)\left( \sum_{i=1}^{n} x_i \lambda_i \mathbf{u}_i \right)$$

$$= \sum_{i,j=1}^{n} x_i x_j \lambda_i \mathbf{u}_j^T \mathbf{u}_i.$$

Since the $\mathbf{u}_i$-s form an orthonormal basis,

$$\mathbf{u}_j^T \mathbf{u}_i = \langle \mathbf{u}_i, \mathbf{u}_j \rangle = \begin{cases} 1 & \text{if } i = j, \\ 0 & \text{otherwise.} \end{cases}$$

In other words, $\mathbf{u}_j^T \mathbf{u}_i$ vanishes when $i \neq j$. Continuing the above calculation with this observation,

$$\mathbf{x}^T A \mathbf{x} = \sum_{i,j=1}^{n} x_i x_j \lambda_i \mathbf{u}_j^T \mathbf{u}_i = \sum_{i=1}^{n} x_i^2 \lambda_i.$$

When $\|\mathbf{x}\|^2 = \sum_{i=1}^{n} x_i^2 = 1$, the sum $\sum_{i=1}^{n} x_i^2 \lambda_i$ is a weighted average of the eigenvalues $\lambda_i$. So,

$$\sum_{i=1}^{n} x_i^2 \lambda_i \leq \sum_{i=1}^{n} x_i^2 \max_{k=1,\ldots,n} \lambda_k = \max_{k=1,\ldots,n} \lambda_k = \lambda_1,$$

from which $\mathbf{x}^T A \mathbf{x} \leq \lambda_1$ follows. (Recall that we can assume without loss in generality that the eigenvalues are decreasing.) On the other hand, by plugging in $\mathbf{x} = \mathbf{u}_1$, we can see that $\mathbf{u}_1^T A \mathbf{u}_1 = \lambda_1$, so the maximum is indeed attained. From these two, the theorem follows.

**Remark 7.2.1** In other words, *Theorem 7.2.2* gives that the function $\mathbf{x} \mapsto \mathbf{x}^T A \mathbf{x}$ assumes its maximum value at $\mathbf{u}_1$, and that maximum value is $\mathbf{u}_1^T A \mathbf{u}_1 = \lambda_1$. The quantity $\mathbf{x}^T A \mathbf{x}$ seems quite mysterious as well, so let's clarify this a bit. If we think in terms of features, the vectors $\mathbf{u}_1, \ldots, \mathbf{u}_n$ can be thought of as mixtures of the "old" features $\mathbf{e}_1, \ldots, \mathbf{e}_n$. When we have actual observations (that is, data), we can use the above process to diagonalize the covariance matrix. So, if $A$ denotes this covariance matrix, $\mathbf{u}_1^T A \mathbf{u}_1$ is the *variance* of the new feature $\mathbf{u}_1$.

Thus, this theorem says that $\mathbf{u}_1$ is the unique feature that maximizes the variance. So, among all the possible choices for new features, $\mathbf{u}_1$ conveys the most information about the data.

At this point, we don't have all the tools to see, but in connection to the principal component analysis, this says that the first principal vector is the one that maximizes variance.

*Theorem 7.2.2* is just a special case of the following general theorem.

> **Theorem 7.2.3** *Let $A \in \mathbb{R}^{n \times n}$ be a real symmetric matrix, let $\lambda_1 \geq \cdots \geq \lambda_n$ be its real eigenvalues in decreasing order. Moreover, let $U \in \mathbb{R}^{n \times n}$ be the orthogonal matrix that diagonalizes $A$, with the corresponding orthonormal basis $\{\mathbf{u}_1, \ldots, \mathbf{u}_n\}$.*
>
> *Then, for all $k = 1, \ldots, n$, we have*
>
> $$\mathbf{u}_k = \text{argmax}\left\{ \mathbf{x}^T A \mathbf{x} \ : \ \|\mathbf{x}\| = 1, \mathbf{x} \perp \{\mathbf{u}_1, \ldots, \mathbf{u}_{k-1}\} \right\},$$
>
> *and*
>
> $$\lambda_k \max \left\{ \mathbf{x}^T A \mathbf{x} \ : \ \|\mathbf{x}\| = 1, \mathbf{x} \perp \{\mathbf{u}_1, \ldots, \mathbf{u}_{k-1}\} \right\}.$$

(Sometimes, when the conditions are too complicated, we write $\max\{f(x) \ : \ x \in A\}$ instead of $\max_{x \in A} f(x)$.)

> *Proof.* The proof is almost identical to the previous one. Since $\mathbf{x}$ is required to be orthogonal to $\mathbf{u}_1, \ldots, \mathbf{u}_{k-1}$, it can be expressed as
>
> $$\mathbf{x} = \sum_{i=k}^{n} x_i \mathbf{u}_i.$$
>
> Following the calculations in the proof of the previous theorem, we have
>
> $$\mathbf{x}^T A \mathbf{x} = \sum_{i=k}^{n} x_i^2 \lambda_i \leq \lambda_k.$$
>
> On the other hand, similar to before, $\mathbf{u}_k^T A \mathbf{u}_k = \lambda_k$, so the theorem follows.

# 7.3 The singular value decomposition

So, we can diagonalize any real symmetric matrix with an orthogonal transformation. That's great, but what if our matrix is not symmetric? After all, this is a rather special case.

How can we do the same for a general matrix? We'll use a very strong tool, straight from the mathematician's toolkit: wishful thinking. We pretend to have the solution, then reverse engineer it. To be specific, let $A \in \mathbb{R}^{n \times m}$ be any real matrix. (It might not be square.) Since $A$ is not symmetric, we have to relax our wishes for factoring it into the form $U \Lambda U^T$. The most straightforward way is to assume that the orthogonal matrices to the left and to the right are not each other's transposes.

Thus, we are looking for orthogonal matrices $U \in \mathbb{R}^{n \times n}$ and $V \in \mathbb{R}^{m \times m}$ such that

$$A = U\Sigma V^T$$

holds for some diagonal $\Sigma \in \mathbb{R}^{n \times m}$. (A non-square matrix $\Sigma = (\sigma_{i,j})_{i,j=1}^{n,m}$ is diagonal if $\sigma_{i,j}$ is 0 when $i \neq j$.)

You might be wondering about the notational switch from $\Lambda$ to $\Sigma$. This is because $\Sigma$ will not necessarily contain eigenvalues, but *singular values*. We'll explain soon.

Here comes the reverse-engineering part. First, as we discussed earlier, $AA^T$ and $A^TA$ are symmetric matrices. Second, we can simplify them by using the orthogonality of $U$ and $V$, obtaining

$$AA^T = (U\Sigma V^T)(V\Sigma U^T)$$
$$= U\Sigma^2 U^T.$$

Similarly, we have $A^TA = V\Sigma^2 V^T$. Good news: We can actually find $U$ and $V$ by applying the spectral decomposition theorem (*Theorem 7.2.1*) to $AA^T$ and $A^TA$, respectively. Thus, the factorization $A = U\Sigma V^T$ is valid! This form is called the singular value decomposition (SVD), one of the pinnacle achievements of linear algebra.

Of course, we are not done yet; we just know where to look. Let's make this mathematically precise!

**Theorem 7.3.1** *(Singular value decomposition)*

*Let $A \in \mathbb{R}^{n \times m}$ be an arbitrary matrix. Then, there exists a diagonal matrix $\Sigma \in \mathbb{R}^{n \times m}$ and orthogonal matrices $U \in \mathbb{R}^{n \times n}$ and $V \in \mathbb{R}^{m \times m}$, such that*

$$A = U\Sigma V^T.$$

What is a non-square diagonal matrix? Let me give you two examples, and you'll immediately get the gist:

$$\begin{bmatrix} 1 & 0 \\ 0 & 2 \\ 0 & 0 \end{bmatrix}, \quad \begin{bmatrix} 1 & 0 & 0 \\ 0 & 2 & 0 \end{bmatrix}.$$

We can always write rectangular matrices $M \in \mathbb{R}^{n \times m}$ in the forms

$$M = \begin{bmatrix} M_1 \\ M_2 \end{bmatrix}, \quad M_1 \in \mathbb{R}^{m \times m}, \quad M_2 \in \mathbb{R}^{(n-m) \times m}$$

if $m < n$, and

$$M = \begin{bmatrix} M_1 & M_2 \end{bmatrix}, \quad M_1 \in \mathbb{R}^{n \times n}, \quad M_2 \in \mathbb{R}^{n \times (m-n)}$$

otherwise. Now, let's see the proof of the singular value decomposition!

*Proof.* (Sketch.) To illustrate the main ideas of the proof, we assume that 1) $A$ is square, and 2) $A$ is invertible; that is, 0 is not an eigenvalue of $A$,

Since $A^T A \in \mathbb{R}^{m \times m}$ is a real symmetric matrix, we can apply the spectral decomposition theorem to obtain a diagonal $\Sigma \in \mathbb{R}^{m \times m}$ and orthogonal $V \in \mathbb{R}^{m \times m}$ such that

$$A^T A = V \Sigma^2 V^T$$

holds. (Recall that the eigenvalues of a symmetric matrix are nonnegative, thus we can write the eigenvalues of $A^T A$ in the form $\Sigma^2$.)

As $A$ is invertible, $A^T A$ is invertible as well; thus, 0 is not an eigenvalue of $A^T A$. As a consequence, $\Sigma A^{-1}$ is well defined. Now, by defining $U := A V \Sigma A^{-1}$, the orthogonality of $V$ gives that

$$\begin{aligned} U \Sigma V^T &= (A V \Sigma^{-1}) \Sigma V^T \\ &= A V V^T \\ &= A, \end{aligned}$$

Thus, we are almost finished. The only thing left to show is that $U$ is indeed orthogonal, that is, $U^T U = I$. Here we go:

$$\begin{aligned} U^T U &= (A V \Sigma^{-1})^T A V \Sigma^{-1} \\ &= \Sigma^{-1} V^T \underbrace{A^T A}_{=V \Sigma^2 V^T} V \Sigma^{-1} \\ &= \Sigma^{-1} (V^T V) \Sigma^2 (V^T V) \Sigma^{-1} \\ &= \Sigma^{-1} \Sigma^2 \Sigma^{-1} \\ &= I, \end{aligned}$$

With that, have the singular value decomposition for the special case of square and invertible $A$.

To keep the complexity bearable, we won't work the rest of the details out; I'll leave that to you as an exercise. You have all the tools by now.

Let's take a moment to appreciate the power of the singular value decomposition. The columns of $U$ and $V$ are orthogonal matrices, which are rather special transformations. As they leave the inner products and the norm invariant, the structure of the underlying vector spaces is preserved. The diagonal $\Sigma$ is also special, as it is just a stretching in the direction of the bases. It is very surprising that *any* linear transformation is the composition of these three special ones.

Besides mapping out the fine structure of linear transformations, SVD offers a lot more. For instance, it generalizes the notion of eigenvectors, a concept that was defined only for square matrices. With this, we have

$$AV = U\Sigma,$$

which we can take a look at column-wise. Here, $\Sigma$ is diagonal, but its number of elements depends on the smaller one of $n$ or $m$.

So, if $\mathbf{u}_i$ is the $i$-th column of $U$, and $\mathbf{v}_i$ is the $i$-th column of $V$, the identity $AV = U\Sigma$ is translated to

$$A\mathbf{v}_i = \sigma_i\mathbf{u}_i, \quad 0 \leq i \leq \min(n, m).$$

This closely resembles the definition of eigenvalue-eigenvector pairs, except that instead of one vector, we have *two*. The $\mathbf{u}_i$ and $\mathbf{v}_i$ are the so-called left and right singular vectors, while the scalars $\sigma_i = \sqrt{\lambda_i}$ are called singular values.

To sum up, orthogonal transformations give us the singular value decomposition, but is that all? Are there any other special transformations and matrix decompositions? You bet there is.

Enter *orthogonal projections*.

# 7.4   Orthogonal projections

Linear transformations are essentially manipulations of data, revealing other (hopefully more useful) representations. Intuitively, we think about them as one-to-one mappings, faithfully preserving all the "information" from the input.

This is often not the case, to such an extent that sometimes a lossy compression of the data is highly beneficial. To give you a concrete example, consider a dataset with a million features, out of which only a couple hundred are useful. What we can do is identify the important features and throw away the rest, obtaining a representation that is more compact, thus easier to work with.

This notion is formalized by the concept of orthogonal projections. We already met them upon our first encounter with the inner products (see (2.7)).

Projections also play a fundamental role in the Gram-Schmidt process (*Theorem 2.2.6*), used to orthogonalize an arbitrary basis. Because we are already somewhat familiar with orthogonal projections, a formal definition is due.

---

**Definition 7.4.1 (Projections and orthogonal projections)**

Let $V$ be an arbitrary inner product space and $P : V \to V$ be a linear transformation. $P$ is a *projection* if $P^2 = P$.

A projection $P$ is *orthogonal* if the subspaces $\ker P$ and $\operatorname{im} P$ are orthogonal to each other. (That is, for every pair of $\mathbf{x} \in \ker P$ and $\mathbf{y} \in \operatorname{im} P$, we have $\langle \mathbf{x}, \mathbf{y} \rangle = 0$.)

---

Let's revisit the examples we have seen so far to get a grip on the definition!

**Example 1.** The simplest one is the orthogonal projection to a single vector. That is, if $u \in \mathbb{R}^n$ is an arbitrary vector, the transformation

$$\operatorname{proj}_{\mathbf{u}}(\mathbf{x}) = \frac{\langle \mathbf{x}, \mathbf{u} \rangle}{\langle \mathbf{u}, \mathbf{u} \rangle} \mathbf{u}$$

is the orthogonal projection to (the subspace spanned by) $u$. (We talked about this when discussing the geometric interpretation of inner products in *Section 2.2.3*, where this definition was deduced from a geometric intuition.) Applying this transformation repeatedly, we get

$$
\begin{aligned}
\operatorname{proj}_{\mathbf{u}}\big(\operatorname{proj}_{\mathbf{u}}(\mathbf{x})\big) &= \frac{\left\langle \frac{\langle \mathbf{x}, \mathbf{u} \rangle}{\langle \mathbf{u}, \mathbf{u} \rangle} \mathbf{u}, \mathbf{u} \right\rangle}{\langle \mathbf{u}, \mathbf{u} \rangle} \mathbf{u} \\
&= \frac{\frac{\langle \mathbf{x}, \mathbf{u} \rangle}{\langle \mathbf{u}, \mathbf{u} \rangle} \langle \mathbf{u}, \mathbf{u} \rangle}{\langle \mathbf{u}, \mathbf{u} \rangle} \mathbf{u} \\
&= \frac{\langle \mathbf{x}, \mathbf{u} \rangle}{\langle \mathbf{u}, \mathbf{u} \rangle} \mathbf{u} \\
&= \operatorname{proj}_{\mathbf{u}}(\mathbf{x}).
\end{aligned}
$$

Thus, faithfully to its name, $\operatorname{proj}_{\mathbf{u}}$ is indeed a projection. To see that it is orthogonal, let's examine its kernel and image! Since the value of $\operatorname{proj}_{\mathbf{u}}(\mathbf{x})$ is a scalar multiple of $\mathbf{u}$, its image is

$$\operatorname{im}(\operatorname{proj}_{\mathbf{u}}) = \operatorname{span}(\mathbf{u}).$$

Its kernel, the set of vectors mapped to $\mathbf{0}$ by $\operatorname{proj}_{\mathbf{u}}$, is also easy to find, as $\frac{\langle \mathbf{x}, \mathbf{u} \rangle}{\langle \mathbf{u}, \mathbf{u} \rangle} \mathbf{u} = \mathbf{0}$ can only happen if $\langle \mathbf{x}, \mathbf{u} \rangle = 0$, that is, if $\mathbf{x} \perp \mathbf{u}$.

In other words,

$$\ker(\text{proj}_{\mathbf{u}}) = \text{span}(\mathbf{u})^{\perp},$$

where $\text{span}(\mathbf{u})^{\perp}$ denotes the orthogonal complement (*Definition 2.2.5*) of $\text{span}(\mathbf{u})$. This means that $\text{proj}_{\mathbf{u}}$ is indeed an orthogonal projection.

We can also describe $\text{proj}_{\mathbf{u}}(\mathbf{x})$ in terms of matrices. By writing out $\text{proj}_{\mathbf{u}}(\mathbf{x})$ component-wise, we have

$$\text{proj}_{\mathbf{u}}(\mathbf{x}) = \frac{\langle \mathbf{x}, \mathbf{u} \rangle}{\langle \mathbf{u}, \mathbf{u} \rangle}\mathbf{u} = \frac{1}{\|\mathbf{u}\|^2} \begin{bmatrix} \langle \mathbf{x}, \mathbf{u} \rangle u_1 \\ \langle \mathbf{x}, \mathbf{u} \rangle u_2 \\ \vdots \\ \langle \mathbf{x}, \mathbf{u} \rangle u_n \end{bmatrix},$$

where $\mathbf{u} = (u_1, \ldots, u_n)$. This looks like some kind of matrix multiplication! As we saw earlier, multiplying a matrix and a vector can be described in terms of rowwise dot products. (See (3.3).)

So, according to this interpretation of matrix multiplication, we have

$$\text{proj}_{\mathbf{u}}(\mathbf{x}) = \frac{1}{\|\mathbf{u}\|^2} \begin{bmatrix} \langle \mathbf{x}, \mathbf{u} \rangle u_1 \\ \langle \mathbf{x}, \mathbf{u} \rangle u_2 \\ \vdots \\ \langle \mathbf{x}, \mathbf{u} \rangle u_n \end{bmatrix} = \frac{\mathbf{u}\mathbf{u}^T}{\|\mathbf{u}\|^2}\mathbf{x}. \tag{7.5}$$

Note that the scaling with $\|\mathbf{u}\|^2$ can be incorporated into the "matrix" product by writing

$$\frac{\mathbf{u}\mathbf{u}^T}{\|\mathbf{u}\|^2} = \frac{\mathbf{u}}{\|\mathbf{u}\|} \cdot \frac{\mathbf{u}^T}{\|\mathbf{u}\|},$$

The matrix $\mathbf{u}\mathbf{u}^T \in \mathbb{R}^{n \times n}$, obtained from the product of the vector $\mathbf{u} \in \mathbb{R}^{n(\times 1)}$ and its transpose $\mathbf{u}^T \in \mathbb{R}^{1 \times n}$, is a rather special one. They are called rank-1 projection matrices, and they frequently appear in mathematics.

(In general, the matrix $\mathbf{u}\mathbf{v}^T$ is called the *outer product* of the vectors $\mathbf{u}$ and $\mathbf{v}$. We won't use this extensively, but it appears frequently throughout linear algebra.)

**Example 2.** As we saw when introducing the Gram-Schmidt orthogonalization process (*Theorem 2.2.6*), the previous example can be generalized by projecting to multiple vectors. If $u_1, \ldots, u_k \in \mathbb{R}^n$ is a set of linearly independent and pairwise orthogonal vectors, then the linear transformation

$$\text{proj}_{u_1, \ldots, u_k}(x) = \sum_{i=1}^{k} \frac{\langle x, u_i \rangle}{\langle u_i, u_i \rangle} u_i$$

is an orthogonal projection onto the subspace $\text{span}(u_1, \ldots, u_k)$. This is easy to see, and I recommend the reader to do this as an exercise. (This can be found in the problems section as well.)

From (7.5), we can determine the matrix form of $\text{proj}_{u_1, \ldots, u_k}$ as well:

$$\text{proj}_{u_1, \ldots, u_k}(x) = \underbrace{\left( \sum_{i=1}^{k} \frac{u_i u_i^T}{\|u_i\|^2} \right)}_{\in \mathbb{R}^{n \times n}} x.$$

This is good to know, as projection matrices are often needed in the implementation of certain algorithms.

## 7.4.1 Properties of orthogonal projections

Now that we have seen a few examples, it is time to discuss orthogonal projections in more general terms. There are lots of reasons why these special transformations are useful, and we'll explore them in this section. First, let's start with the most important one: orthogonal projections also enable the decomposition of vectors in terms of a given subspace plus an orthogonal vector.

**Theorem 7.4.1** *Let $V$ be an inner product space and $P : V \to V$ be a projection. Then, $V = \ker P + \text{im} P$; that is, every vector $x \in V$ can be written as*

$$x = x_{\ker} + x_{\text{im}}, \quad x_{\ker} \in \ker P, \quad x_{\text{im}} \in \text{im} P.$$

*If $P$ is an orthogonal projection, then $x_{\text{im}} \perp x_{\ker}$.*

*Proof.* Every $x$ can be written as
$$x = (x - Px) + Px.$$

Since $P$ is idempotent, that is, $P^2 = P$, we have

$$P(\mathbf{x} - P\mathbf{x}) = P\mathbf{x} - P(P\mathbf{x})$$
$$= P\mathbf{x} - P\mathbf{x}$$
$$= \mathbf{0},$$

that is, $\mathbf{x} - P\mathbf{x} \in \ker P$. By definition, $P\mathbf{x} \in \operatorname{im} P$, so $V = \ker V + \operatorname{im} V$, which proves our main proposition.

If $P$ is an orthogonal projection, then again, by definition, $\mathbf{x}_{\operatorname{im}} \perp \mathbf{x}_{\ker}$, which is what we had to show.

In addition, orthogonal projections are self-adjoint. This might not sound like a big deal, but self-adjointness leads to several very pleasant properties.

**Theorem 7.4.2** *Let $V$ be an inner product space and $P : V \to V$ be an orthogonal projection. Then, $P$ is self-adjoint.*

*Proof.* According to the definition (7.3), all we need to show is that

$$\langle P\mathbf{x}, \mathbf{y} \rangle = \langle \mathbf{x}, P\mathbf{y} \rangle$$

holds for any $\mathbf{x}, \mathbf{y} \in V$. In the previous result, we have seen that $\mathbf{x}$ and $\mathbf{y}$ can be written as

$$\mathbf{x} = \mathbf{x}_{\ker P} + \mathbf{x}_{\operatorname{im} P}, \quad \mathbf{x}_{\ker P} \in \ker P, \quad \mathbf{x}_{\operatorname{im} P} \in \operatorname{im} P$$
$$\mathbf{y} = \mathbf{y}_{\ker P} + \mathbf{y}_{\operatorname{im} P}, \quad \mathbf{y}_{\ker P} \in \ker P, \quad \mathbf{y}_{\operatorname{im} P} \in \operatorname{im} P.$$

Since $P^2 = P$, we have

$$\langle P\mathbf{x}, \mathbf{y} \rangle = \langle P\mathbf{x}_{\ker P} + P\mathbf{x}_{\operatorname{im} P}, \mathbf{y}_{\ker P} + \mathbf{y}_{\operatorname{im} P} \rangle$$
$$= \langle \mathbf{x}_{\operatorname{im} P}, \mathbf{y}_{\ker P} + \mathbf{y}_{\operatorname{im} P} \rangle$$
$$= \underbrace{\langle \mathbf{x}_{\operatorname{im} P}, \mathbf{y}_{\ker P} \rangle}_{=0} + \langle \mathbf{x}_{\operatorname{im} P}, \mathbf{y}_{\operatorname{im} P} \rangle$$
$$= \langle \mathbf{x}_{\operatorname{im} P}, \mathbf{y}_{\operatorname{im} P} \rangle.$$

Similarly, it can be shown that $\langle \mathbf{x}, P\mathbf{y} \rangle = \langle \mathbf{x}_{\operatorname{im} P}, \mathbf{y}_{\operatorname{im} P} \rangle$. These two identities imply $\langle P\mathbf{x}, \mathbf{y} \rangle = \langle \mathbf{x}, P\mathbf{y} \rangle$, which is what we had to show.

One straightforward consequence of self-adjointness is that the kernel of orthogonal projections is the orthogonal complement of its image.

> **Theorem 7.4.3** *Let $V$ be an inner product space and $P : V \to V$ be an orthogonal projection. Then,*
>
> $$\ker P = (\operatorname{im} P)^\perp.$$

*Proof.* To prove the equality of these two sets, we need to show that *(a)* $\ker P \subseteq (\operatorname{im} P)^\perp$, and *(b)* $(\operatorname{im} P)^\perp \subseteq \ker P$.

*(a)* Let $\mathbf{x} \in \ker P$; that is, suppose that $P\mathbf{x} = \mathbf{0}$. We need to show that for any $\mathbf{y} \in \operatorname{im} P$, we have $\langle \mathbf{x}, \mathbf{y} \rangle = 0$. For this, let $\mathbf{y}_0 \in V$ such that $P\mathbf{y}_0 = \mathbf{y}$. (This is guaranteed to exist, since we took $\mathbf{y}$ from the image of $P$.) Then,

$$\begin{aligned}
\langle \mathbf{x}, \mathbf{y} \rangle &= \langle \mathbf{x}, P\mathbf{y}_0 \rangle \\
&= \langle P\mathbf{x}, \mathbf{y}_0 \rangle \\
&= \langle \mathbf{0}, \mathbf{y}_0 \rangle \\
&= 0,
\end{aligned}$$

where $P$ is self-adjoint. Thus, $\mathbf{x} \in (\operatorname{im} P)^\perp$ also holds, implying $\ker P \subseteq (\operatorname{im} P)^\perp$.

*(b)* Now, let $\mathbf{x} \in (\operatorname{im} P)^\perp$. Then, for any $\mathbf{y} \in V$, we have $\langle \mathbf{x}, P\mathbf{y} \rangle = 0$. However,

$$\langle P\mathbf{x}, \mathbf{y} \rangle = \langle \mathbf{x}, P\mathbf{y} \rangle = 0.$$

Especially, with the choice $\mathbf{y} = P\mathbf{x}$, we have $\langle P\mathbf{x}, P\mathbf{x} \rangle = 0$. Due to the positive definiteness of the inner product, this implies that $P\mathbf{x} = \mathbf{0}$, that is, $\mathbf{x} \in \ker P$.

Summing up all of the above, if $P$ is an orthogonal projection of the inner product space $V$, then

$$V = \operatorname{im} P + (\operatorname{im} P)^\perp.$$

Do you recall that when we first encountered the concept of orthogonal complements (*Definition 2.2.5*), we proved that $V = S + S^\perp$ for any finite-dimensional inner product space $V$ and its subspace $S$? We did this with the use of a special orthogonal projection. We are getting close to seeing the general pattern here.

Because the kernel of an orthogonal projection $P$ is an orthogonal complement of the image, the transformation $I - P$ is an orthogonal projection as well, with the roles of image and kernel reversed.

**Theorem 7.4.4** *Let $V$ be an inner product space and $P : V \to V$ be an orthogonal projection. Then, $I - P$ is an orthogonal projection as well, and*

$$\ker(I - P) = \operatorname{im} P, \quad \operatorname{im}(I - P) = \ker P.$$

The proof is so simple that this is left as an exercise for the reader.

One more thing to mention. If the image spaces of two orthogonal projections match, then the projections themselves are equal. This is a very strong uniqueness property, as if you think about it, this is not true for other classes of linear transformations.

**Theorem 7.4.5** *(Uniqueness of orthogonal projections)*

*Let $V$ be an inner product space and $P, Q : V \to V$ be two orthogonal projections. If $\operatorname{im} P = \operatorname{im} Q$, then $P = Q$.*

*Proof.* Because of $\ker P = (\operatorname{im} P)^{\perp}$, the equality of the image spaces also imply that $\ker P = \ker Q$.

Since $V = \ker P + \operatorname{im} P$, every $\mathbf{x} \in V$ can be decomposed as

$$\mathbf{x} = \mathbf{x}_{\ker P} + \mathbf{x}_{\operatorname{im} P}, \quad \mathbf{x}_{\ker P} \in \ker P, \quad \mathbf{x}_{\operatorname{im} P} \in \operatorname{im} P.$$

This decomposition and the equality of the kernel and image spaces give that

$$P\mathbf{x} = P\mathbf{x}_{\ker P} + P\mathbf{x}_{\operatorname{im} P} = \mathbf{x}_{\operatorname{im} P}.$$

With an identical argument, we have $Q\mathbf{x} = \mathbf{x}_{\operatorname{im} P}$, thus $P\mathbf{x} = Q\mathbf{x}$ on all vectors $\mathbf{x} \in V$. This proves $P = Q$.

In other words, given a subspace, there can be only one orthogonal projection to it. But is there any at all? Yes, and in the next section, we will see that it can be described in geometric terms.

## 7.4.2   Orthogonal projections are the optimal projections

Orthogonal projections have an extremely pleasant and mathematically useful property. In some sense, if $P : V \to V$ is an orthogonal projection, $P\mathbf{x}$ provides the optimal approximation of $\mathbf{x}$ among all vectors in $\operatorname{im} P$. To make this precise, we can state the following.

**Theorem 7.4.6** *(Construction of orthogonal transformations)*

*Let V be a finite-dimensional inner product space and $S \subseteq V$ its subspace. Then, the transformation $P : V \to V$, defined by*

$$P : \mathbf{x} \to \arg\min_{\mathbf{y} \in S} \|\mathbf{x} - \mathbf{y}\|$$

*is an orthogonal projection to S.*

In other words, since orthogonal projections to a given subspace are unique (as implied by *Theorem 7.4.5*), $P\mathbf{x}$ is the closest vector to $\mathbf{x}$ in the subspace $S$. Thus, we can denote this as $P_S$, emphasizing the uniqueness.

Besides having an explicit way to describe orthogonal projections, there is one extra benefit. Recall that previously, we showed that

$$V = \operatorname{im} P + (\operatorname{im} P)^\perp$$

holds. Since for any subspace $S$ an orthogonal projection $P_S$ exists whose image set is $S$, it also follows that $V = S + S^\perp$. Although we saw this earlier when talking about orthogonal complements (*Definition 2.2.5*), it is interesting to see a proof that doesn't require the construction of an orthonormal basis in $S$.

Interestingly, this is the point where mathematical analysis and linear algebra intersect. We don't have the tools for it yet, but using the concept of convergence, the above theorems can be generalized to infinite-dimensional spaces. Infinite-dimensional spaces are not particularly relevant to machine learning in practice, yet they provide a beautiful mathematical framework for the study of functions. Who knows, one day these advanced tools may provide a significant breakthrough in machine learning.

# 7.5 Computing eigenvalues

In the last chapter, we reached the singular value decomposition, one of the pinnacle results of linear algebra. We laid out the theoretical groundwork to get us to this point.

However, one thing is missing: computing the singular value decomposition in practice. Without this, we can't reap all the rewards this powerful tool offers. In this section, we'll develop two methods for this purpose. One offers a deep insight into the behavior of eigenvectors, but it doesn't work in practice. The other offers excellent performance, but it is hard to understand what is happening behind the formulas. Let's start with the first one, illuminating how the eigenvectors determine the effects of a linear transformation!

## 7.5.1  Power iteration for calculating the eigenvectors of real symmetric matrices

If you recall, we discovered the singular value decomposition by tracing the problem back to the spectral decomposition of symmetric matrices. In turn, we can obtain the spectral decomposition by finding an orthonormal basis from the eigenvectors of our matrix. The plan is the following: first, we define a procedure that finds an orthonormal set of eigenvectors for symmetric matrices. Then, use this to compute the singular value decomposition for arbitrary matrices.

A naive way would be to find the eigenvalues by solving the polynomial equation $\det(A - \lambda I) = 0$ for $\lambda$, then compute the corresponding eigenvectors by solving the linear equations $(A - \lambda I)\mathbf{x} = \mathbf{0}$.

However, there are problems with this approach. For an $n \times n$ matrix, the characteristic polynomial $p(\lambda) = \det(A - \lambda I)$ is a polynomial of degree $n$. Even if we could effectively evaluate $\det(A - \lambda I)$ for any lambda, there are serious issues. Unfortunately, unlike for the quadratic equation $ax^2 + bx + c = 0$, there are no formulas for finding the solutions when $n > 4$. (It is not that mathematicians were just not clever enough to find them. No such formula exists.)

How can we find an alternative approach? Once again, we use the wishful thinking approach that worked so well before. Let's pretend that we know the eigenvalues, play around with them, and see if this gives us some useful insight.

For the sake of simplicity, assume that $A$ is a small symmetric $2 \times 2$ matrix, say with eigenvalues $\lambda_1 = 4$ and $\lambda_2 = 2$. Since $A$ is symmetric, we can even find a set of corresponding eigenvectors $\mathbf{u}_1, \mathbf{u}_2$ such that $\mathbf{u}_1$ and $\mathbf{u}_2$ form an orthonormal basis. (That is, both have a unit norm and they are orthogonal to each other.) This is guaranteed by the spectral decomposition theorem (*Theorem 7.2.1*).

Thus, any $x \in \mathbb{R}^2$ can be written as $\mathbf{x} = x_1\mathbf{u}_1 + x_2\mathbf{u}_2$ for some nonzero scalars $x_1, x_2$. What happens if we apply the transformation $A$ to our vector $\mathbf{x}$? Because $\mathbf{u}_i$ is eigenvectors, we have

$$\begin{aligned}
A\mathbf{x} &= A(x_1\mathbf{u}_1 + x_2\mathbf{u}_2) \\
&= x_1 A\mathbf{u}_1 + x_2 A\mathbf{u}_2 \\
&= x_1 \lambda_1 \mathbf{u}_1 + x_2 \lambda_2 \mathbf{u}_2 \\
&= 4x_1\mathbf{u}_1 + 2x_2\mathbf{u}_2.
\end{aligned}$$

By applying $A$ one more time, we obtain

$$\begin{aligned}
A^2\mathbf{x} &= x_1 \lambda_1^2 \mathbf{u}_1 + x_2 \lambda_2^2 \mathbf{u}_2 \\
&= 4^2 x_1 \mathbf{u}_1 + 2^2 x_2 \mathbf{u}_2.
\end{aligned}$$

A pattern starts to emerge. In general, the $k$-th iteration of $A$ yields

$$A^k \mathbf{x} = x_1 \lambda_1^k \mathbf{u}_1 + x_2 \lambda_2^k \mathbf{u}_2$$

$$= 4^k x_1 \mathbf{u}_1 + 2^k x_2 \mathbf{u}_2.$$

By taking an inquisitive look at $A^k \mathbf{x}$, we can note that the contribution of $\mathbf{u}_1$ is *much* more significant than $\mathbf{u}_2$. Why? Because the coefficient $x_1 \lambda_1^k = 4^k x_1$ grows faster than $x_2 \lambda_2^k = 2^k x_2$, regardless of the value of $x_1$ and $x_2$. In technical terms, we say that $\lambda_1$ dominates $\lambda_2$.

Now, by scaling things down with $\lambda_1^k$, we can extract the eigenvector $\mathbf{u}_1$! That is,

$$\frac{A^k \mathbf{x}}{\lambda_1^k} = x_1 \mathbf{u}_1 + x_2 \left( \frac{\lambda_2}{\lambda_1} \right)^k \mathbf{u}_2$$

$$= x_1 \mathbf{u}_1 + (\text{something very small})_k.$$

If we let $k$ grow infinitely, the contribution of $\mathbf{u}_2$ to $\frac{A^k \mathbf{x}}{\lambda_1^k}$ vanishes. If you are familiar with the concept of *limits*, you could write

$$\lim_{k \to \infty} \frac{A^k \mathbf{x}}{\lambda_1^k} = x_1 \mathbf{u}_1. \tag{7.6}$$

**Remark 7.5.1 (A primer on limits)**

If you are not familiar with limits, here is a quick explanation. The identity

$$\lim_{k \to \infty} \frac{A^k \mathbf{x}}{\lambda_1^k} = x_1 \mathbf{u}_1$$

means that as $k$ grows, the quantity $\frac{A^k \mathbf{x}}{\lambda_1^k}$ gets closer and closer to $x_1 \mathbf{u}_1$, until the difference between them is infinitesimal. In practice, this means that we can approximate $x_1 \mathbf{u}_1$ by $\frac{A^k \mathbf{x}}{\lambda_1^k}$ by selecting a very large $k$.

Equation (7.6) is great news for us! All we have to do is repeatedly apply the transformation $A$ to identify the eigenvector for the dominant eigenvalue $\lambda_1$. There is one small caveat, though: we have to know the value of $\lambda_1$. We'll deal with this later, but first, let's record this milestone in the form of a theorem.

**Theorem 7.5.1** *Finding the eigenvector for the dominant eigenvalue with power iteration.*

*Let $A \in \mathbb{R}^{n \times n}$ be a real symmetric matrix. Suppose that:*

*(a) The eigenvalues of A are $\lambda_1 > \dots > \lambda_n$ (that is, $\lambda_1$ is the dominant eigenvalue).*

*(b) The corresponding eigenvectors $\mathbf{u}_1, \dots, \mathbf{u_n}$ form an orthonormal basis.*

*Let $\mathbf{x} \in \mathbb{R}^n$ be a vector such that when written as the linear combination $\mathbf{x} = \sum_{i=1}^{n} x_i \mathbf{u}_i$, the coefficient $x_1 \in \mathbb{R}$ is nonzero. Then,*

$$\lim_{k \to \infty} \frac{A^k \mathbf{x}}{\lambda_1^k} = x_1 \mathbf{u}_1. \tag{7.7}$$

Before we jump into the proof, some explanations are in order. Recall that if $A$ is symmetric, the spectral decomposition theorem (*Theorem 7.2.1*) guarantees that it can be diagonalized with a similarity transformation. In its proof (sketch), we mentioned that a symmetric matrix has:

- Real eigenvalues
- An orthonormal basis from its eigenvectors

Thus, the assumptions *(a)* and *(b)* are guaranteed, except for one caveat: the eigenvalues are not necessarily distinct. However, this rarely causes problems in practice. There are multiple reasons for this, but most importantly, matrices with repeated eigenvalues are so rare that they form a zero-probability set. (We'll learn about probability later in the book. For now, we can assume that randomly picking a matrix with repeated eigenvalues is impossible.) Thus, stumbling upon one is highly unlikely.

*Proof.* Because $\mathbf{u}_k$ is the eigenvector for the eigenvalue $\lambda_k$, we have

$$A^k \mathbf{x} = \sum_{i=1}^{n} x_i \lambda_i^k \mathbf{u}_i. \tag{7.8}$$

Thus,

$$\frac{A^k \mathbf{x}}{\lambda_1^k} = x_1 \mathbf{u}_1 + \sum_{i=2}^{n} x_i \left( \frac{\lambda_i}{\lambda_1} \right)^k \mathbf{u}_i.$$

Since $\lambda_1$ is the dominant eigenvalue, $\|\lambda_i / \lambda_1\| < 1$ for $i = 2, \dots, n$, so $(\lambda_i / \lambda_1)^k \to 0$ as $k \to \infty$. Hence,

$$\lim_{k \to \infty} \frac{A^k \mathbf{x}}{\lambda_1^k} = x_1 \mathbf{u}_1.$$

This is what we had to show.

Now, let's fix the small issue that requires us to know $\lambda_1$. Since $\lambda_1$ is the largest eigenvalue, the previous theorem shows that $A^k\mathbf{x}$ equals $x_1\lambda_1^k\mathbf{u}_1$ plus some term that is much smaller, at least compared to this dominant term. We can extract this quantity by taking the supremum norm $\|A^k\mathbf{x}\|_\infty$. (Recall that for any $\mathbf{y} = (y_1, \ldots, y_n)$, the supremum norm is defined by $\|\mathbf{y}\|_\infty = \max\{|y_1|, \ldots, |y_n|\}$. Keep in mind that the $y_i$-s are the coefficients of $\mathbf{y}$ in the original basis of our vector space, which is not necessarily our eigenvector basis $\mathbf{u}_1, \ldots, \mathbf{u}_n$.)

By factoring out $|\lambda_1|^k$ from $A^k\mathbf{x}$, we have

$$\|A^k\mathbf{x}\|_\infty = |\lambda_1|^k \left\| x_1\mathbf{u}_1 + \sum_{i=2}^n x_i \left(\frac{\lambda_i}{\lambda_1}\right)^k \mathbf{u}_i \right\|_\infty.$$

Intuitively speaking, the remainder term $\sum_{i=2}^n x_i \left(\frac{\lambda_i}{\lambda_1}\right)^k \mathbf{u}_i$ is small, thus we can approximate the norm as

$$\|A^k\mathbf{x}\|_\infty \approx |\lambda_1|^k \|x_1\mathbf{u}_1\|_\infty.$$

In other words, instead of scaling with $\lambda_1^k$, we can scale with $\|A^k\mathbf{x}\|_\infty$.

So, we are ready to describe our general eigenvector-finding procedure fully. First, we initialize a vector $\mathbf{x}_0$ randomly, then we define the recursive sequence

$$\mathbf{x}_k = \frac{A\mathbf{x}_{k-1}}{\|A\mathbf{x}_{k-1}\|_\infty}, \quad k = 1, 2, \ldots$$

Using the linearity of $A$, we can see that, in fact,

$$\mathbf{x}_k = \frac{A^k\mathbf{x}_0}{\|A^k\mathbf{x}_0\|_\infty},$$

but scaling has an additional side benefit, as we don't have to use large numbers at any computational step. With this, (7.7) implies that

$$\lim_{k\to\infty} \mathbf{x}_k = \lim_{k\to\infty} \frac{A^k\mathbf{x}_0}{\|A^k\mathbf{x}_0\|_\infty} = \mathbf{u}_1.$$

That is, we can extract the eigenvector for the dominant eigenvalue without actually knowing the eigenvalue itself.

## 7.5.2   Power iteration in practice

Let's put the power iteration method into practice! The input of our power_iteration function is a square matrix A, and we expect the output to be an eigenvector corresponding to the dominant eigenvalue.

Since this is an iterative process, we should define a condition that defines when the process should terminate. If the consecutive members of the sequence $\{\mathbf{x}_k\}_{k=1}^{\infty}$ are sufficiently close together, we arrived at a solution. That is, if, say,

$$\|\mathbf{x}_{k+1} - \mathbf{x}_k\|_2 < 1 \times 10^{-10},$$

we can stop and return the current value. However, this might never happen. For those cases, we define a cutoff point, say, $k = 100,000$, when we terminate the computation, even if there is no convergence.

To give us a bit more control, we can also manually define the initialization vector x_init.

```python
import numpy as np
def power_iteration(
    A: np.ndarray,
    n_max_steps: int = 100000,
    convergence_threshold: float = 1e-10,
    x_init: np.ndarray = None,
    normalize: bool = False
):
    """
    Performs the power iteration method to find an approximation of the dominant eigenvector
    of a square matrix.

    Parameters
    ----------
    A : np.ndarray
        A square matrix whose dominant eigenvector is to be computed.
    n_max_steps : int, optional
        The maximum number of iterations to perform. Default is 100000.
    convergence_threshold : float, optional
        The convergence threshold for the difference between successive approximations.
        Default is 1e-10.
    x_init : np.ndarray, optional
        The initial guess for the eigenvector. If None, a random vector is used.
        Default is None.
    normalize : bool, optional
        If True, the resulting vector is normalized to unit length. Default is False.

    Returns
    -------
    np.ndarray
        The approximate dominant eigenvector of the matrix `A`.
```

```
    Raises
    ------
    ValueError
        If the input matrix `A` is not square.
    """

    n, m = A.shape

    # checking the validity of the input
    if n != m:
        raise ValueError("the matrix A must be square")

    # reshaping or defining the initial vector
    if x_init is not None:
        x = x_init.reshape(-1, 1)
    else:
        x = np.random.normal(size=(n, 1))

    # performing the iteration
    for step in range(n_max_steps):
        x_transformed = A @ x      # applying the transform
        x_new = x_transformed / np.linalg.norm(x_transformed, ord=np.inf)      # scaling
        the result

        # quantifying the difference between the new and old vector
        diff = np.linalg.norm(x - x_new)
        x = x_new

        # stopping the iteration in case of convergence
        if diff < convergence_threshold:
            break

    # normalizing the result if required
    if normalize:
        return x / np.linalg.norm(x)

    return x
```

To test the method, we should use an input for which the correct output is easy to calculate by hand. Our usual recurring example

$$A = \begin{bmatrix} 2 & 1 \\ 1 & 2 \end{bmatrix}.$$

should be perfect, as we already know a lot about it.

Previously, we have seen in *Section 6.2.2* that its eigenvalues are $\lambda_1 = 3$ and $\lambda_2 = 1$, with corresponding eigenvectors $\mathbf{u}_1 = (1, 1)$ and $\mathbf{u}_2 = (-1, 1)$.

Let's see if our function correctly recovers (a scalar multiple of) $\mathbf{u}_1 = (1, 1)$!

```
A = np.array([[2, 1], [1, 2]])
u_1 = power_iteration(A, normalize=True)
u_1
```

```
array([[0.70710678],
       [0.70710678]])
```

Success! To recover the eigenvalue, we can simply apply the linear transformation and compute the proportions.

```
A @ u_1 / u_1
```

```
array([[3.],
       [3.]])
```

The result is 3, as expected.

## 7.5.3 Power iteration for the rest of the eigenvectors

Can we modify the power iteration algorithm to recover the other eigenvalues as well? In theory, yes. In practice, no. Let me elaborate!

To get a grip on how to generalize the idea, let's take another look at the equation (7.8), saying that

$$A^k \mathbf{x} = \sum_{i=1}^{n} x_i \lambda_i^k \mathbf{u}_i.$$

One of the conditions for $\frac{A^k \mathbf{x}}{\lambda_1^k}$ to converge was that $\mathbf{x}$ should have a nonzero component of the eigenvector $\mathbf{u}_1$, that is, $x_1 \neq 0$.

What if $x_1 = 0$? In that case, we have

$$A^k \mathbf{x} = x_2 \lambda_2^k \mathbf{u}_2 + \cdots + x_n \lambda_2^k \mathbf{u}_n,$$

with $x_2 \lambda_2^k \mathbf{u}_2$ becoming the dominant term.

Thus, we have

$$\frac{A^k \mathbf{x}}{\lambda_2^k} = x_2 \mathbf{u}_2 + \sum_{i=3}^{n} x_i \left( \frac{\lambda_i}{\lambda_2} \right)^k \mathbf{u}_k$$

$$= x_2 \mathbf{u}_2 + (\text{something very small})_k,$$

implying that

$$\lim_{k \to \infty} \frac{A^k \mathbf{x}}{\lambda_2^k} = x_2 \mathbf{u}_2.$$

Let's make this mathematically precise in the following theorem.

---

**Theorem 7.5.2** *(Generalized power iteration)*

*Let $A \in \mathbb{R}^{n \times n}$ be a real symmetric matrix. Suppose that:*

*(a) The eigenvalues of A are $\lambda_1 > \ldots > \lambda_n$ (that is, $\lambda_1$ is the dominant eigenvalue).*

*(b) The corresponding eigenvectors $u_1, \ldots, u_n$ form an orthonormal basis.*

*Let $\mathbf{x} \in \mathbb{R}^n$ be a vector such that, when written as a linear combination of the basis $\mathbf{u}_1, \ldots, \mathbf{u}_n$, its first nonzero component is along $\mathbf{u}_l$ for some $l = 1, 2, \ldots, n$ (that is, $\mathbf{x} = \sum_{i=l}^{n} x_i \mathbf{u}_i$).*

*Then,*

$$\lim_{k \to \infty} \frac{A^k \mathbf{x}}{\lambda_l^k} = x_l \mathbf{u}_l$$

*holds.*

---

The proof looks just like what we have seen a few times already. The question is, how can we eliminate the $\mathbf{u}_1, \ldots, \mathbf{u}_{l-1}$ components from any vector? The answer is simple: orthogonal projections (*Section 7.4*).

For the sake of simplicity, let's take a look at extracting the second dominant eigenvector with power iteration. Recall that the transformation

$$\text{proj}_{\mathbf{u}_1}(\mathbf{x}) = \langle \mathbf{x}, \mathbf{u}_1 \rangle \mathbf{u}_1$$

describes the orthogonal projection of any $\mathbf{x}$ to $\mathbf{u}$ *1*.

In concrete terms, if $\mathbf{x} = \sum_{i=1}^{n} x_i \mathbf{u}_i$, then

$$\text{proj}_{\mathbf{u}_1}(\mathbf{x}) = \text{proj}_{\mathbf{u}_1}\Big(\sum_{i=1}^{n} x_i \mathbf{u}_i\Big)$$

$$= \sum_{i=1}^{n} x_i \text{proj}_{\mathbf{u}_1}(\mathbf{u}_i)$$

$$= x_1 \mathbf{u}_1.$$

This is the exact opposite of what we are looking for! However, at this point, we can see that $I - \text{proj}_{\mathbf{u}_1}$ is going to be suitable for our purposes. This is still an orthogonal projection. Moreover, we have

$$(I - \text{proj}_{\mathbf{u}_1})\Big(\sum_{i=1}^{n} x_i \mathbf{u}_i\Big) = \sum_{i=2}^{n} x_i \mathbf{u}_i,$$

That is, $I - \text{proj}_{\mathbf{u}_1}$ eliminates the $\mathbf{u}_1$ component of $\mathbf{x}$. Thus, if we initialize the power iteration with $\mathbf{x}^* = (I - \text{proj}_{\mathbf{u}_1})(\mathbf{x})$, the sequence $\frac{A^k \mathbf{x}^*}{\|A^k \mathbf{x}^*\|_\infty}$ will converge to $\mathbf{u}_2$, the second dominant eigenvector.

How do we compute $(I - \text{proj}_{\mathbf{u}_1})(\mathbf{x})$ in practice? Recall that in the standard orthonormal basis, the matrix of $\text{proj}_{\mathbf{u}_1}$ can be written as:

$$\text{proj}_{\mathbf{u}_1} = \mathbf{u}_1 \mathbf{u}_1^T.$$

(Keep in mind that the $\mathbf{u}_i$ vectors form an orthonormal basis, so $\|\mathbf{u}_1\| = 1$.) Thus, the matrix of $I - \text{proj}_{\mathbf{u}_1}$ is:

$$I - \mathbf{u}_1 \mathbf{u}_1^T,$$

which we can easily compute.

For a general vector $\mathbf{u}$, this is how we can do this in NumPy.

```python
def get_orthogonal_complement_projection(u: np.ndarray):
    """
    Compute the projection matrix onto the orthogonal complement of the vector u.

    This function returns a projection matrix P such that for any vector v,
    P @ v is the projection of v onto the subspace orthogonal to u.

    Parameters
    ----------
    u : np.ndarray
        A 1D or 2D array representing the vector u. It will be reshaped to a column vector.
```

```
Returns
-------
np.ndarray
    The projection matrix onto the orthogonal complement of u. This matrix
    has shape (n, n), where n is the length of u.
"""

u = u.reshape(-1, 1)
n, _ = u.shape
return np.eye(n) - u @ u.T / np.linalg.norm(u, ord=2)**2
```

So, the procedure to find *all* the eigenvectors is the following.

1. Initialize a random $\mathbf{x}^{(1)}$ and use the power iteration to find $\mathbf{u}_1$.

2. Project $\mathbf{x}^{(1)}$ to the orthogonal complement of the subspace spanned by $\mathbf{u}_1$, thus obtaining

$$\mathbf{x}^{(2)} := (I - \text{proj}_{\mathbf{u}_1})(\mathbf{x}^{(1)}),$$

   which we use as the initial vector of the second round of power iteration, yielding the second dominant eigenvector $\mathbf{u}_2$.

3. Project $\mathbf{x}^{(2)}$ to the orthogonal complement of the subspace spanned by $\mathbf{u}_1$ and $\mathbf{u}_2$, thus obtaining

$$\mathbf{x}^{(3)} = (I - \text{proj}_{\mathbf{u}_2})(\mathbf{x}^{(2)})$$
$$= (I - \text{proj}_{\mathbf{u}_1, \mathbf{u}_2})(\mathbf{x}^{(1)}),$$

   which we use as the initial vector of the third round of power iteration, yielding the third dominant eigenvector $\mathbf{u}_3$.

4. Project $\mathbf{x}^{(3)}$ to...

You get the pattern. To implement this in practice, we add the find_eigenvectors function.

```
def find_eigenvectors(A: np.ndarray, x_init: np.ndarray):
    """
    Find the eigenvectors of the matrix A using the power iteration method.

    This function computes the eigenvectors of the matrix A by iteratively
    applying the power iteration method and projecting out previously found
    eigenvectors to find orthogonal eigenvectors.

    Parameters
    ----------
```

```
A : np.ndarray
    A square matrix of shape (n, n) for which eigenvectors are to be computed.

x_init : np.ndarray
    A 1D array representing the initial vector used for the power iteration.

Returns
-------
List[np.ndarray]
    A list of eigenvectors, each represented as a 1D numpy array of length n.
"""

n, _ = A.shape
eigenvectors = []

for _ in range(n):
    ev = power_iteration(A, x_init=x_init)
    proj = get_orthogonal_complement_projection(ev)
    x_init = proj @ x_init
    x_init = x_init / np.linalg.norm(x_init, ord=np.inf)
    eigenvectors.append(ev)

return eigenvectors
```

Let's test find_eigenvectors out on our old friend

$$A = \begin{bmatrix} 2 & 1 \\ 1 & 2 \end{bmatrix}!$$

```
A = np.array([[2.0, 1.0], [1.0, 2.0]])
x_init = np.random.rand(2, 1)
find_eigenvectors(A, x_init)
```

```
[array([[1.],
        [1.]]),
 array([[ 1.],
        [-1.]])]
```

The result is as we expected. (Don't be surprised that the eigenvectors are not normalized, as we haven't explicitly done so in the find_eigenvectors function.)

We are ready to actually diagonalize symmetric matrices. Recall that the diagonalizing orthogonal matrix *U* can be obtained by vertically stacking the eigenvectors one by one.

```python
def diagonalize_symmetric_matrix(A: np.ndarray, x_init: np.ndarray):
    """
    Diagonalize a symmetric matrix A using its eigenvectors.

    Parameters
    ----------
    A : np.ndarray
        A symmetric matrix of shape (n, n) to be diagonalized. The matrix should
        be square and symmetric.

    x_init : np.ndarray
        A 1D array representing the initial guess for the power iteration.

    Returns
    -------
    Tuple[np.ndarray, np.ndarray] containing:
        - U : np.ndarray
            A matrix of shape (n, n) whose columns are the normalized eigenvectors
            of A.
        - np.ndarray
            A diagonal matrix (n, n) of the eigenvalues of A, computed as U @ A @ U.T.
    """

    eigenvectors = find_eigenvectors(A, x_init)
    U = np.hstack(eigenvectors) / np.linalg.norm(np.hstack(eigenvectors), axis=0, ord=2)
    return U, U @ A @ U.T

diagonalize_symmetric_matrix(A, x_init)
```

```
(array([[ 0.70710678,  0.70710678],
        [ 0.70710678, -0.70710678]]),
 array([[ 3.00000000e+00, -3.57590301e-11],
        [-3.57589164e-11,  1.00000000e+00]]))
```

Awesome!

What's the problem? Unfortunately, power iteration is numerically unstable. For $n \times n$ matrices, where $n$ can be in the millions, this is a serious issue.

Why did we talk so much about power iteration, then? Besides being the simplest, it offers a deep insight into how linear transformations work.

The identity

$$Ax = \sum_{i=1}^{n} x_i \lambda_i \mathbf{u}_i,$$

where $\lambda_i$ and $\mathbf{u}_i$ are eigenvalue-eigenvector pairs of the symmetric matrix $A$, reflecting how eigenvectors and eigenvalues determine the behavior of the transformation.

If the power iteration is not usable in practice, how can we compute the eigenvalues? We will see this in the next section.

# 7.6  The QR algorithm

The algorithm used in practice to compute the eigenvalues is the so-called QR algorithm, proposed independently by John G. R. Francis and the Soviet mathematician Vera Kublanovskaya. This is where all of the lessons we have learned in linear algebra converge. Describing the QR algorithm is very simple, as it is the iteration of a matrix decomposition and a multiplication step.

However, understanding *why* it works is a different question. Behind the scenes, the QR algorithm combines many tools we have learned earlier. To start, let's revisit the good old Gram-Schmidt orthogonalization process.

## 7.6.1  The QR decomposition

If you recall, we encountered the Gram-Schmidt orthogonalization process (*Theorem 2.2.6*) when introducing the concept of orthogonal bases.

In essence, this algorithm takes an arbitrary basis $\mathbf{v}_1, \dots, \mathbf{v}_n$ and turns it into an orthonormal one $\mathbf{e}_1, \dots, \mathbf{e}_n$, such that $\mathbf{e}_1, \dots, \mathbf{e}_k$ spans the same subspace as $\mathbf{v}_1, \dots, \mathbf{v}_k$ for all $1 \leq k \leq n$. Since we last met this, we have gained a lot of perspective on linear algebra, so we are ready to see the bigger picture.

With the orthogonal projections defined by

$$\text{proj}_{\mathbf{e}_1, \dots, \mathbf{e}_k}(\mathbf{x}) = \sum_{i=1}^{k} \frac{\langle \mathbf{x}, \mathbf{e}_i \rangle}{\langle \mathbf{e}_i, \mathbf{e}_i \rangle} \mathbf{e}_i,$$

we can describe the Gram-Schmidt process recursively as

$$\mathbf{e}_1 = \mathbf{v}_1,$$
$$\mathbf{e}_k = \mathbf{v}_k - \text{proj}_{\mathbf{e}_1, \dots, \mathbf{e}_{k-1}}(\mathbf{v}_k),$$

where the $\mathbf{e}_k$ vectors are normalized after.

By expanding this and writing out $e_k$ explicitly, we have

$$\mathbf{e}_1 = \mathbf{v}_1$$

$$\mathbf{e}_2 = \mathbf{v}_2 - \frac{\langle \mathbf{v}_2, \mathbf{e}_1 \rangle}{\langle \mathbf{e}_1, \mathbf{e}_1 \rangle} \mathbf{e}_1$$

$$\vdots$$

$$\mathbf{e}_n = \mathbf{v}_n - \frac{\langle \mathbf{v}_n, \mathbf{e}_1 \rangle}{\langle \mathbf{e}_1, \mathbf{e}_1 \rangle} \mathbf{e}_1 - \cdots - \frac{\langle \mathbf{v}_n, \mathbf{e}_{n-1} \rangle}{\langle \mathbf{e}_{n-1}, \mathbf{e}_{n-1} \rangle} \mathbf{e}_{n-1}.$$

A pattern is starting to emerge. By arranging the $\mathbf{e}_1, \ldots, \mathbf{e}_n$ terms on one side, we obtain

$$\mathbf{v}_1 = \mathbf{e}_1$$

$$\mathbf{v}_2 = \frac{\langle \mathbf{v}_2, \mathbf{e}_1 \rangle}{\langle \mathbf{e}_1, \mathbf{e}_1 \rangle} \mathbf{e}_1 + \mathbf{e}_2$$

$$\vdots$$

$$\mathbf{v}_n = \frac{\langle \mathbf{v}_n, \mathbf{e}_1 \rangle}{\langle \mathbf{e}_1, \mathbf{e}_1 \rangle} \mathbf{e}_1 + \cdots + \frac{\langle \mathbf{v}_n, \mathbf{e}_{n-1} \rangle}{\langle \mathbf{e}_{n-1}, \mathbf{e}_{n-1} \rangle} \mathbf{e}_{n-1} + \mathbf{e}_n.$$

This is starting to resemble some kind of matrix multiplication! Recall that matrix multiplication can be viewed as taking the linear combination of columns. (Check (3.2) if you are uncertain about this.)

By horizontally concatenating the column vectors $\mathbf{v}_k$ to form the matrix $A$ and similarly defining the vector $Q$ from the $\mathbf{e}_k$-s, we obtain that

$$A = Q^* R^*$$

for some upper triangular $R$, defined by the coefficients of $\mathbf{e}_k$ in $\mathbf{v}_k$ according to the Gram-Schmidt orthogonalization. To be more precise, define

$$A = \begin{bmatrix} \mathbf{v}_1 & \cdots & \mathbf{v}_n \end{bmatrix}, \quad Q^* = \begin{bmatrix} \mathbf{e}_1 & \cdots & \mathbf{e}_n \end{bmatrix},$$

and

$$R^* = \begin{bmatrix} 1 & \frac{\langle \mathbf{v}_2, \mathbf{e}_1 \rangle}{\langle \mathbf{e}_1, \mathbf{e}_1 \rangle} & \cdots & \frac{\langle \mathbf{v}_n, \mathbf{e}_1 \rangle}{\langle \mathbf{e}_1, \mathbf{e}_1 \rangle} \\ 0 & 1 & \cdots & \frac{\langle \mathbf{v}_n, \mathbf{e}_2 \rangle}{\langle \mathbf{e}_2, \mathbf{e}_2 \rangle} \\ \vdots & \vdots & \ddots & \vdots \\ 0 & 0 & \vdots & 1 \end{bmatrix}.$$

The result $A = Q^*R^*$ is *almost* what we call the QR factorization. The columns of $Q^*$ are orthogonal (but not orthonormal), while $R^*$ is upper triangular. We can easily orthonormalize $Q^*$ by factoring out the norms columnwise, thus obtaining

$$Q = \begin{bmatrix} \frac{e_1}{\|e_1\|} & \cdots & \frac{e_n}{\|e_n\|} \end{bmatrix}, \quad R = \begin{bmatrix} \|e_1\| & \frac{\langle v_2, e_1 \rangle}{\sqrt{\langle e_1, e_1 \rangle}} & \cdots & \frac{\langle v_n, e_1 \rangle}{\sqrt{\langle e_1, e_1 \rangle}} \\ 0 & \|e_2\| & \cdots & \frac{\langle v_n, e_2 \rangle}{\sqrt{\langle e_2, e_2 \rangle}} \\ \vdots & \vdots & \ddots & \vdots \\ 0 & 0 & \vdots & \|e_n\| \end{bmatrix}.$$

It is easy to see that $A = QR$ still holds. This result is called the *QR decomposition*, and we have just proved the following theorem.

**Theorem 7.6.1** *(QR decomposition)*

*Let $A \in \mathbb{R}^{n \times n}$ be an invertible matrix. Then, there exists an orthogonal matrix $Q \in \mathbb{R}^{n \times n}$ and an upper triangular matrix $R \in \mathbb{R}^{n \times n}$ such that*

$$A = QR$$

*holds.*

As we are about to see, the QR decomposition is an extremely useful and versatile tool (like all other matrix decompositions are). Before we move forward to discuss how it can be used to compute the eigenvalues in practice, let's put what we have seen so far into code!

The QR decomposition algorithm is essentially Gram-Schmidt orthogonalization, where we explicitly memorize some coefficients and form a matrix from them. (Recall our earlier implementation in *Section 3.1.2* if you feel overwhelmed.)

```python
def projection_coeff(x: np.ndarray, to: np.ndarray):
    """
    Compute the scalar coefficient for the projection of vector x onto vector to.

    Parameters
    ----------
    x : np.ndarray
        A 1D array representing the vector onto which the projection is computed.

    to : np.ndarray
        A 1D array representing the vector onto which x is being projected.
```

```
    Returns
    -------
    float
        The scalar coefficient representing the projection of x onto to.
    """

    return np.dot(x, to)/np.dot(to, to)


from typing import List

def projection(x: np.ndarray, to: List[np.ndarray], return_coeffs: bool = True):
    """
    Computes the orthogonal projection of a vector `x` onto the subspace
    spanned by a set of vectors `to`.

    Parameters
    ----------
    x : np.ndarray
        A 1D array representing the vector to be projected onto the subspace.

    to : List[np.ndarray]
        A list of 1D arrays, each representing a vector spanning the subspace
        onto which `x` is projected.

    return_coeffs : bool, optional, default=True
        If True, the function returns the list of projection coefficients.
        If False, only the projected vector is returned.

    Returns
    -------
    Tuple[np.ndarray, List[float]] or np.ndarray
        - If `return_coeffs` is True, returns a tuple where the first element
        is the projected vector and
            the second element is a list of the projection coefficients
            for each vector in `to`.
        - If `return_coeffs` is False, returns only the projected vector.
    """

    p_x = np.zeros_like(x)
    coeffs = []

    for e in to:
        coeff = projection_coeff(x, e)
        coeffs.append(coeff)
        p_x += coeff*e

    if return_coeffs:
```

```
            return p_x, coeffs
        else:
            return p_x
```

Now we can put these together to obtain the QR factorization of an arbitrary square matrix. (Surprisingly, this works for non-square matrices as well, but we won't be concerned with this.)

```python
def QR(A: np.ndarray):
    """
    Computes the QR decomposition of matrix A using the Gram-Schmidt
    orthogonalization process.

    Parameters
    ----------
    A : np.ndarray
        A 2D array of shape (n, m) representing the matrix to be decomposed.
        The matrix A should have full column rank for a valid QR decomposition.

    Returns
    -------
    Tuple[np.ndarray, np.ndarray]
        - Q : np.ndarray
            An orthogonal matrix of shape (n, m), whose columns are orthonormal.
        - R : np.ndarray
            An upper triangular matrix of shape (m, m), representing the coefficients of the
            linear combinations of the columns of A.
    """
    n, m = A.shape

    A_columns = [A[:, i] for i in range(A.shape[1])]
    Q_columns, R_columns = [], []

    Q_columns.append(A_columns[0])
    R_columns.append([1] + (m-1)*[0])

    for i, a in enumerate(A_columns[1:]):
        p, coeffs = projection(a, Q_columns, return_coeffs=True)
        next_q = a - p
        next_r = coeffs + [1] + max(0, m - i - 2)*[0]

        Q_columns.append(next_q)
        R_columns.append(next_r)

    # assembling Q and R from its columns
    Q, R = np.array(Q_columns).T, np.array(R_columns).T
```

```
# normalizing Q's columns
Q_norms = np.linalg.norm(Q, axis=0)
Q = Q/Q_norms
R = np.diag(Q_norms) @ R
return Q, R
```

Let's try it out on a random $3 \times 3$ matrix.

```
A = np.random.rand(3, 3)
Q, R = QR(A)
```

There are three things to check: *(a)* $A = QR$, *(b)* $Q$ is an orthogonal matrix, and *(c)* $R$ is upper triangular.

```
np.allclose(A, Q @ R)
```

```
True
```

```
np.allclose(Q.T @ Q, np.eye(3))
```

```
True
```

```
np.allclose(R, np.triu(R))
```

```
True
```

Success! There is only one more question left. How does this help us in calculating the eigenvalues? Let's see that now.

## 7.6.2   Iterating the QR decomposition

Surprisingly, we can discover the eigenvalues of a matrix $A$ by a simple iterative process. First, we find the QR decomposition

$$A = Q_1 R_1,$$

and define the matrix $A_1$ by

$$A_1 = R_1 Q_1,$$

That is, we simply reverse the order of $Q$ and $R$. Then, we start it all over and find the QR decomposition of $A_1$, and so on, defining the sequence

$$A_{k-1} = Q_k R_k \quad \text{(QR decomposition)}$$
$$A_k = R_k Q_k \quad \text{(definition).}$$

(7.9)

In the long run, the diagonal elements of $A_k$ will get closer and closer to the eigenvalues of $A$. This is called the QR algorithm, which is so simple that I didn't believe it when I first saw it.

With all of our tools, we can implement the QR algorithm in a few lines.

```
def QR_algorithm(A: np.ndarray, n_iter: int = 1000):
    """
    Computes the QR algorithm to find the eigenvalues of a matrix A.

    Parameters
    ----------
    A : np.ndarray
        A square matrix of shape (n, n) for which the eigenvalues are to be computed.

    n_iter : int, optional, default=1000
        The number of iterations to run the QR algorithm.
        More iterations may lead to more accurate results,
        but the algorithm typically converges quickly.

    Returns
    -------
    np.ndarray
        A matrix that has converged, where the diagonal elements are the eigenvalues of the
        original matrix A.
        The off-diagonal elements should be close to zero.
```

```
    """

    for _ in range(n_iter):
        Q, R = QR(A)
        A = R @ Q

    return A
```

Let's test it right away.

```
A = np.array([[2.0, 1.0], [1.0, 2.0]])
QR_algorithm(A)
```

```
array([[3.00000000e+00, 2.39107046e-16],
       [0.00000000e+00, 1.00000000e+00]])
```

We are almost at the state of the art. Unfortunately, the vanilla QR algorithm has some issues, as it can fail to converge. A simple example is given by the matrix

$$A = \begin{bmatrix} 0 & 1 \\ 1 & 0 \end{bmatrix}.$$

```
A = np.array([[0.0, 1.0], [1.0, 0.0]])
QR_algorithm(A)
```

```
array([[0., 1.],
       [1., 0.]])
```

In practice, we can solve this with the introduction of shifts:

$$A_{k-1} - \alpha_k I = Q_k R_k \qquad \text{(QR decomposition)} \tag{7.10}$$

$$A_k = R_k Q_k + \alpha_k I \quad \text{(definition)}, \tag{7.11}$$

where $\alpha_k$ is some scalar. There are multiple approaches to defining the shifts themselves (Rayleigh quotient shift, Wilkinson shift, etc.), but the details lie much deeper than our study.

# 7.7 Summary

I told you that climbing the peak is not easy: so far, this was our hardest chapter yet. However, the tools we've learned are at the pinnacle of linear algebra. We started by studying two special transformations: the self-adjoint and orthogonal ones. The former ones gave the *spectral decomposition theorem*, while the latter ones gave the *singular value decomposition*.

Undoubtedly, the SVD is one of the most important results in linear algebra, stating that every rectangular matrix $A$ can be written in the form

$$A = U\Sigma V^T,$$

where $U \in \mathbb{R}^{n\times n}$, $\Sigma \in \mathbb{R}^{n\times m}$, and $V \in \mathbb{R}^{m\times m}$ are rather special: $\Sigma$ is diagonal, while $U$ and $V$ are orthogonal.

When viewing $A$ as a linear transformation, the singular value decomposition tells us that it can be written as the composition of two distance-preserving transformations (the orthogonal ones) and a simple scaling. That's quite a characterization!

Speaking of singular values and eigenvalues, how do we find them in practice? Definitely not by solving the polynomial equation

$$\det(A - \lambda I) = 0,$$

which is a computationally painful problem.

We've seen two actual methods for the task. One is the complicated, slow, unstable, but illuminating algorithm of power iteration, yielding eigenvectors for the dominant eigenvalue of a real and symmetric matrix $A$ via the limit

$$\lim_{k\to\infty} \frac{A^k \mathbf{x}_0}{\|A^k \mathbf{x}_0\|_\infty} = \mathbf{u}_1.$$

Although power iteration gives us valuable insight into the structure of such matrices, the real deal is the QR algorithm (unrelated to QR codes), originating from the vectorized version of the Gram-Schmidt algorithm. The QR algorithm is hard to intuitively understand, but despite its mystery, it provides a blazing-fast method for computing the eigenvalues in practice.

What's next? Now that we are at the peak, it's time to relax a bit and enjoy the beautiful view. The next chapter does just that.

You see, one of the most beautiful and useful topics in linear algebra is the connection between matrices and graphs. Because from the right perspective, a matrix is a graph, and we can utilize this relationship to study the structure of both. Let's see!

# 7.8 Problems

**Problem 1.** Let $\mathbf{u}_1, \ldots, \mathbf{u}_k \in \mathbb{R}^n$ be a set of linearly independent and pairwise orthogonal vectors. Show that the linear transformation

$$\mathrm{proj}_{\mathbf{u}_1, \ldots, \mathbf{u}_k}(x) = \sum_{i=1}^{k} \frac{\langle \mathbf{x}, \mathbf{u}_i \rangle}{\langle \mathbf{u}_i, \mathbf{u}_i \rangle} \mathbf{u}_i$$

is an orthogonal projection.

**Problem 2.** Let $\mathbf{u}_1, \ldots, \mathbf{u}_k \in \mathbb{R}^n$ be a set of linearly independent vectors, and define the linear transformation

$$\mathrm{fakeproj}_{\mathbf{u}_1, \ldots, \mathbf{u}_k}(\mathbf{x}) = \sum_{i=1}^{k} \frac{\langle \mathbf{x}, \mathbf{u}_i \rangle}{\langle \mathbf{u}_i, \mathbf{u}_i \rangle} \mathbf{u}_i.$$

Is this a projection? (Hint: Study the special case $k = 2$ and $\mathbb{R}^3$. You can visualize this if needed.)

**Problem 3.** Let $V$ be an inner product space and $P : V \to V$ be an orthogonal projection. Show that $I - P$ is an orthogonal projection as well, and

$$\ker(I - P) = \mathrm{im}\, P, \quad \mathrm{im}(I - P) = \ker P$$

holds.

**Problem 4.** Let $A, B \in \mathbb{R}^{n \times n}$ be two square matrix that are written in the block matrix form

$$A = \begin{bmatrix} A_{1,1} & A_{1,2} \\ A_{2,1} & A_{2,2} \end{bmatrix}, \quad B = \begin{bmatrix} B_{1,1} & B_{1,2} \\ B_{2,1} & B_{2,2} \end{bmatrix},$$

where $A_{1,1}, B_{1,1} \in \mathbb{R}^{k \times k}$, $A_{1,2}, B_{1,2} \in \mathbb{R}^{k \times l}$, $A_{2,1}, B_{2,1} \in \mathbb{R}^{l \times k}$, and $A_{2,2}, B_{2,2} \in \mathbb{R}^{l \times l}$.

Show that

$$AB = \begin{bmatrix} A_{1,1}B_{1,1} + A_{1,2}B_{2,1} & A_{1,1}B_{1,2} + A_{1,2}B_{2,2} \\ A_{2,1}B_{1,1} + A_{2,2}B_{2,1} & A_{2,1}B_{1,2} + A_{2,2}B_{2,2} \end{bmatrix}.$$

**Problem 5.** Let $A \in \mathbb{R}^{2 \times 2}$ be the square matrix defined by

$$A = \begin{bmatrix} 1 & 1 \\ 1 & 0 \end{bmatrix}.$$

*(a)* Show that the two eigenvalues of $A$ are

$$\lambda_1 = \varphi, \quad \lambda_2 = -\varphi^{-1},$$

where $\varphi = \frac{1+\sqrt{5}}{2}$ is the golden ratio.

*(b)* Show that the eigenvectors of $\lambda_1$ and $\lambda_2$ are

$$\mathbf{f}_1 = \begin{bmatrix} \varphi \\ 1 \end{bmatrix}, \quad \mathbf{f}_2 = \begin{bmatrix} -\varphi^{-1} \\ 1 \end{bmatrix},$$

and show that they are orthogonal; that is, $\langle \mathbf{f}_1, \mathbf{f}_2 \rangle = 0$.

*(c)* Let $U$ be the matrix formed by the eigenvectors of $A$, defined by

$$U = \begin{bmatrix} \varphi & -\varphi^{-1} \\ 1 & 1 \end{bmatrix}.$$

Show that

$$U^{-1} = \frac{1}{\sqrt{5}} \begin{bmatrix} 1 & \varphi^{-1} \\ -1 & \varphi \end{bmatrix}.$$

*(d)* Show that

$$A = U\Lambda U^{-1},$$

where

$$\Lambda = \begin{bmatrix} \lambda_1 & 0 \\ 0 & \lambda_2 \end{bmatrix} = \begin{bmatrix} \varphi & 0 \\ 0 & -\varphi^{-1} \end{bmatrix}.$$

*(e)* Were you wondering the purpose of all these mundane computations? Here comes the punchline. First,

show that

$$A^n = \begin{bmatrix} 1 & 1 \\ 1 & 0 \end{bmatrix}^n = \begin{bmatrix} F_{n+1} & F_n \\ F_n & F_{n-1} \end{bmatrix},$$

where $F_n$ is the $n$-th Fibonacci number, defined by the recursive sequence

$$F_0 = 0, \quad F_1 = 1, \quad F_n = F_{n-1} + F_{n-2}.$$

Consequently, show that $A = U \Lambda U^{-1}$ implies that

$$F_n = \frac{\varphi^n - (-\varphi)^{-n}}{\sqrt{5}}.$$

That's pretty cool!

# Join our community on Discord

Read this book alongside other users, Machine Learning experts, and the author himself.

Ask questions, provide solutions to other readers, chat with the author via Ask Me Anything sessions, and much more.

Scan the QR code or visit the link to join the community.

`https://packt.link/math`

# 8
# Matrices and Graphs

Now that we have gotten past the hard part (that is, the singular value decomposition and other matrix factorizations), it's time to finish our journey through linear algebra with a bang. In my teaching experience, one of students' most common concerns is the apparent disconnect between practice and theory. Among machine learning practitioners and software engineers, there's often a reluctance to touch anything that is not immediately valuable in practice.

As a mathematician, I completely get where this dread comes from. We are often taught arcane topics of no practical importance, taking valuable time away from hacking and slashing our way through data.

In this chapter, we'll look at a subject that is not immediately useful for your machine learning practice, but will pay serious dividends in the future. Considering how beautiful it is, it might be the inspiration for your next genius idea. (No promises, though.)

Let me introduce you to the single most undervalued fact of linear algebra: matrices are graphs, and graphs are matrices. Encoding matrices as graphs is a cheat code, making complex behavior simple to study.

Check out *Figure 8.1* below.

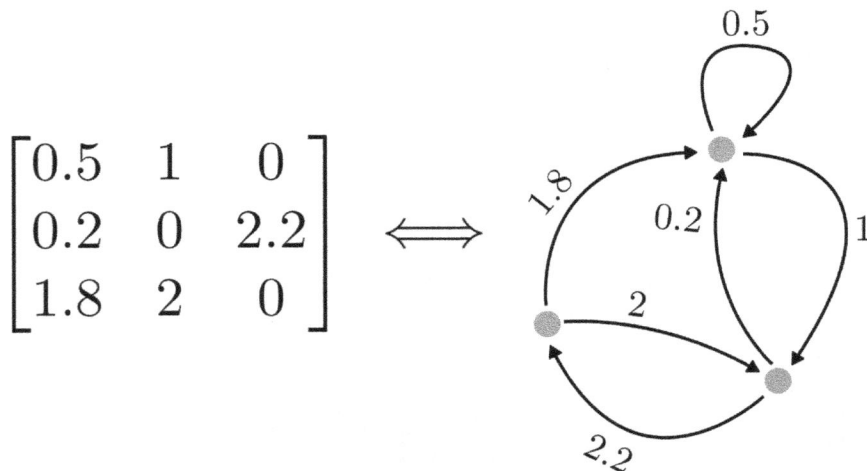

$$
\begin{bmatrix} 0.5 & 1 & 0 \\ 0.2 & 0 & 2.2 \\ 1.8 & 2 & 0 \end{bmatrix} \Longleftrightarrow
$$

Figure 8.1: A matrix and its directed graph

Can you figure out how it was constructed? Can you guess why it is useful? We'll answer these questions in the next couple of pages. Specifically, we'll see:

- What the relationship between graphs and matrices is
- How matrix multiplication can be translated to walks on the graph
- What the connectivity structure of a graph reveals about its corresponding matrix

Let's get to it!

## 8.1   The directed graph of a nonnegative matrix

If you look carefully at *Figure 8.1*, you can probably figure out how to construct a weighted graph from a matrix. Just compare each row and the outgoing edge weights for nodes.

Each row is a node, and each element represents a directed and weighted edge. Edges of zero elements are omitted. The element in the $i$-th row and $j$-th column corresponds to an edge going from $i$ to $j$. The resulting graph is called the *directed graph* (or digraph) *of the matrix*.

To unwrap the definition a bit, let's check out the previous graph of the matrix

$$
A = \begin{bmatrix} 0.5 & 1 & 0 \\ 0.2 & 0 & 2.2 \\ 1.8 & 2 & 0 \end{bmatrix} \in \mathbb{R}^{3\times3}.
$$

Here's the first row, corresponding to the edges coming out from the first node.

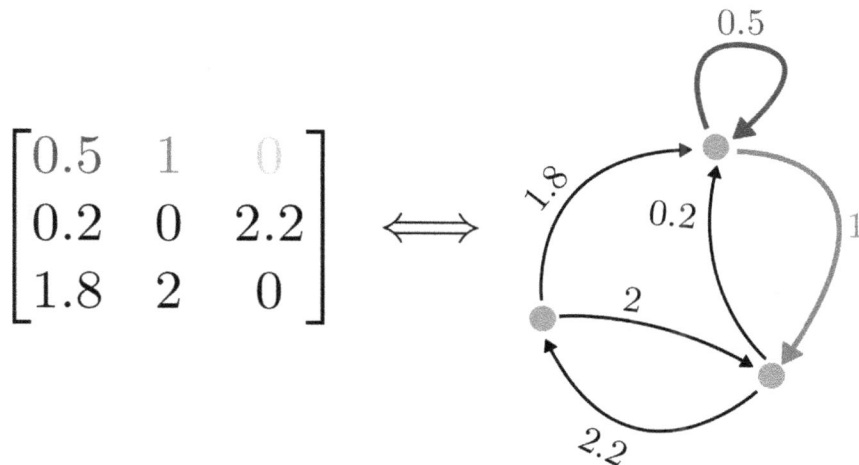

*Figure 8.2: The first row corresponds to the edges coming out from the first node*

Similarly, the first column corresponds to the edges coming into the first node.

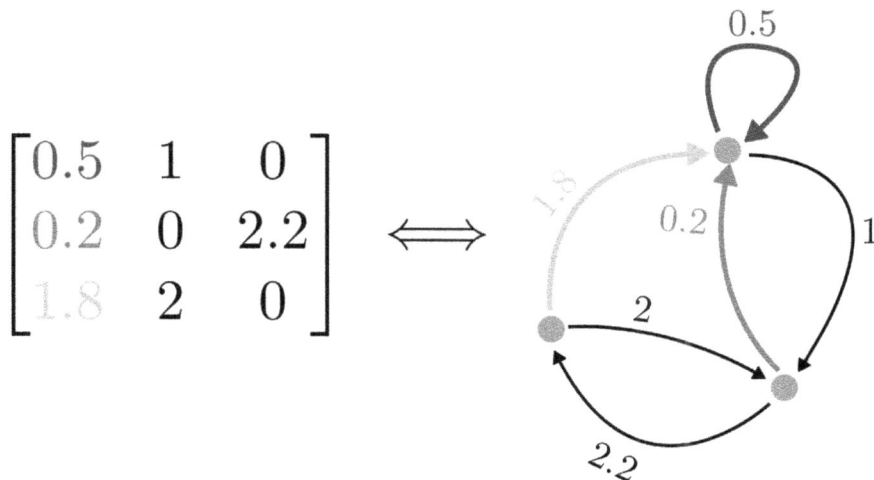

*Figure 8.3: The first column corresponds to the edges coming into the first node*

Now, we can put all of this together. *Figure 8.4* shows the full picture, with the nodes explicitly labeled.

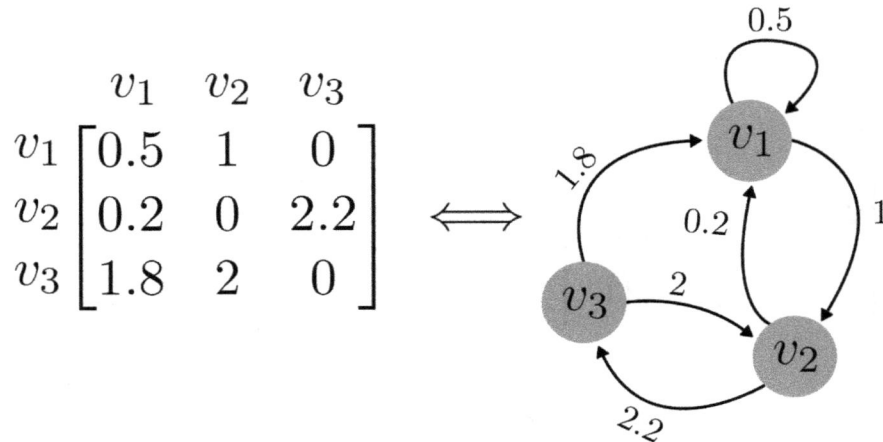

$$
\begin{array}{cc}
 & \begin{array}{ccc} v_1 & v_2 & v_3 \end{array} \\
\begin{array}{c} v_1 \\ v_2 \\ v_3 \end{array} &
\begin{bmatrix} 0.5 & 1 & 0 \\ 0.2 & 0 & 2.2 \\ 1.8 & 2 & 0 \end{bmatrix}
\end{array}
\iff
$$

Figure 8.4: Constructing the graph of a matrix

It's time to check the formal definition, which we'll split into two parts. First, let's talk about weighted and directed graphs.

**Definition 8.1.1** Let $V = \{v_1, v_2, \ldots, v_n\}$ be an arbitrary finite set. We say that $G = (V, E, w)$ is a *weighted and directed graph*, if

*(a)* $V$ represents the set of *vertices* (also called nodes).

*(b)* $E \subseteq V \times V$ represents the set of *directed edges*.

*(c)* The function $w : E \to \mathbb{R}$ represents the edge *weights*.

For example, check out *Figure 8.4*, which we can formalize as $(V, E, w)$ with $V = \{v_1, v_2, v_3\}$,

$$
\begin{aligned}
E = \{ & (v_1, v_1), (v_1, v_2), \\
       & (v_2, v_1), (v_2, v_3), \\
       & (v_3, v_1), (v_3, v_2) \},
\end{aligned}
$$

and

$$
\begin{aligned}
w(v_1, v_1) &= 0.5, & w(v_1, v_2) &= 1, \\
w(v_2, v_1) &= 0.2, & w(v_2, v_2) &= 2.2, \\
w(v_3, v_1) &= 1.8, & w(v_3, v_2) &= 2.
\end{aligned}
$$

Now, we are ready to talk about matrices.

> **Definition 8.1.2 (Irreducible and reducible matrices)**
>
> Let $A \in \mathbb{R}^{n \times n}$ be a nonnegative matrix, i.e., a matrix with only nonnegative elements. The directed weighted graph $G = (V, E, w)$ is said to be the *directed graph* (or digraph for short) *of A* if:
>
> (a) $V = \{1, 2, \dots, n\}$,
>
> (b) $(i, j) \in E$ if, and only if, $a_{i,j} > 0$,
>
> (c) $w(i, j) = a_{i,j}$.

(Sometimes, for illustrative purposes, we'll just omit the weights and assume all of them to be equal to 1.) Again, why is this useful? Because this way, we can translate algebraic questions into graph-theoretic ones. Thus, we gain access to the vast toolkit of graph theory.

## 8.2 Benefits of the graph representation

Let's talk about the concrete advantages that the graph representation offers. For one, the powers of the matrix correspond to walks in the graph. Say, for any let $A = (a_{i,j})_{i,j=1}^n \in \mathbb{R}^{n \times n}$. Its square is denoted by $A^2 = (a_{i,j}^{(2)})_{i,j=1}^n \in \mathbb{R}^{n \times n}$, where the elements $a_{i,j}^{(2)}$ are defined by

$$a_{i,j}^{(2)} = \sum_{k=1}^n a_{i,k} a_{k,j}.$$

(Note that the (2) in the superscript of $a_{i,j}^{(2)}$ is not an exponent; this is just an index indicating that $a_{i,}^{(2)}$ is the element of $A^2$.)

*Figure 8.5* shows the elements of the square matrix and its graph: all possible two-step walks are accounted for in the sum defining the elements of $A^2$.

$$A^2 = \left(a_{i,j}^{(2)}\right)_{i,j=1}^n$$
$$a_{i,j}^{(2)} = \sum_{l=1}^n a_{i,l} a_{l,j} \quad \Longleftrightarrow$$

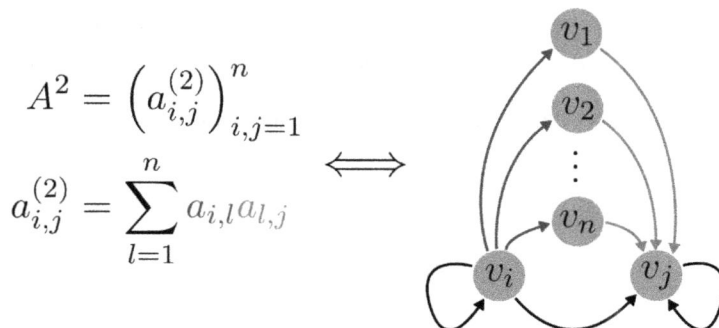

*Figure 8.5: Powers of the matrix describe walks on its directed graph*

There is much more to this connection; for instance, it gives us a deep insight into the structure of nonnegative matrices. To see how, let's talk about the concept of strongly connected components.

## 8.2.1 The connectivity of graphs

Intuitively, we can think of connectivity as the ability to reach every node from the others. To formalize this, we'll need a couple of definitions. First, the *"reach every node from the others"* part.

---

**Definition 8.2.1 (Walks on a graph)**

Let $G = (V, E, w)$ be a weighted and directed graph. The sequence $v_{k_1} v_{k_2} \ldots v_{k_l}$ is a (directed) *walk* on $G$ if $(v_{k_i}, v_{k_{i+1}}) \in E$ for all $i$.

---

(For consistency, we define walks for weighted and directed graphs, but the definition holds for simple graphs – i.e., graphs without edges and directional edges – as well. The same goes for most of the upcoming concepts.)

In general, we say that the walk $v_{k_1} v_{k_2} \ldots v_{k_l}$ starts at $v_{k_1}$ and ends at $v_{k_l}$.

The term *walk* is surprisingly descriptive, as it truly describes a walk on the directed edges, going from node to node. However, a graph-theoretic walk is a properly defined mathematical object, not just a vague intuition. Pick up a pen and a paper once again and sketch up a graph, then a couple of its walks, to understand the concept better.

What do walks have to do with connectivity? Simple: If you can reach every node from every other node, the graph is said to be connected. Since we are talking about *directed* graphs, let's add a bit of nuance to the discussion and conjure up a formal definition.

---

**Definition 8.2.2 (Strong connectivity)**

Let $G = (V, E, w)$ be a weighted and directed graph. We say that $G$ is *strongly connected* if for every $u, v \in V$, there exists a walk that starts at $u$ and ends at $v$.

---

In other words, a directed graph is strongly connected if every node can be reached from every other node. If this is not true, the graph is not strongly connected. *Figure 8.6* shows you an example of both.

*Figure 8.6* also illustrates that strong connectivity does not match the connectivity concept for simple graphs.

It's not enough to reach $u$ from $v$; you have to be able to go back as well.

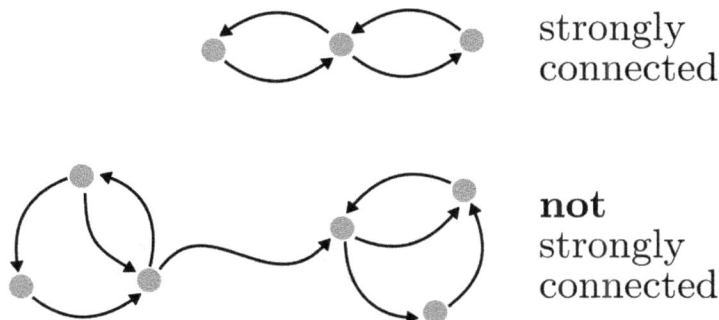

strongly
connected

**not**
strongly
connected

*Figure 8.6: Connected vs. strongly connected*

Now, let's translate what we've learned to the language of matrices. Matrices that correspond to strongly connected graphs are called irreducible. All other nonnegative matrices are called reducible. Soon, we'll see why, but first, here's the formal definition.

**Definition 8.2.3 (Irreducible and reducible matrices)**

Let $A \in \mathbb{R}^{n \times n}$ be a nonnegative matrix.

*(a) A* is called *irreducible* if its digraph is strongly connected.

*(b) A* is called *reducible* if it is not irreducible.

Let's see an example! *Figure 8.7* shows an irreducible matrix.

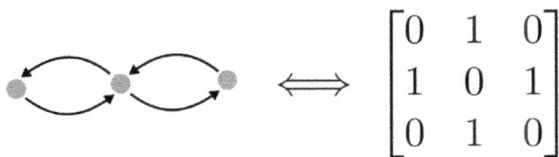

$$\Longleftrightarrow \begin{bmatrix} 0 & 1 & 0 \\ 1 & 0 & 1 \\ 0 & 1 & 0 \end{bmatrix}$$

*Figure 8.7: Strongly connected digraphs and their matrices*

Back to the general case! Even though not all digraphs are strongly connected, we can partition the nodes into strongly connected components (as *Figure 8.8* illustrates).

**1st** strongly connected component            **2nd** strongly connected component

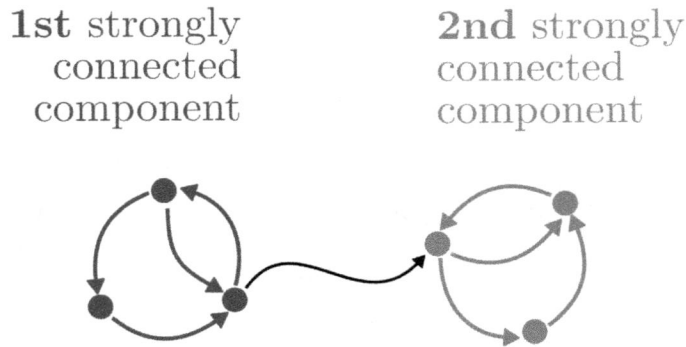

*Figure 8.8: Strongly connected components*

Let's label the nodes of this graph and construct the corresponding matrix! If you assume that the weights are simply equal to 1 and translate all the edges into rows and columns as we have learned, you'll get

$$A = \begin{bmatrix} 0 & 1 & 1 & 0 & 0 & 0 \\ 0 & 0 & 1 & 0 & 0 & 0 \\ 1 & 0 & 0 & 1 & 0 & 0 \\ 0 & 0 & 0 & 0 & 1 & 1 \\ 0 & 0 & 0 & 0 & 0 & 1 \\ 0 & 0 & 0 & 1 & 0 & 0 \end{bmatrix}.$$

That's just a big block of ones and zeroes, but you shouldn't be disappointed: there is a pattern! By dividing $A$ into blocks, the matrix of our example graph can be reduced to a simpler form:

$$A = \left[ \begin{array}{ccc|ccc} 0 & 1 & 1 & 0 & 0 & 0 \\ 0 & 0 & 1 & 0 & 0 & 0 \\ 1 & 0 & 0 & 1 & 0 & 0 \\ \hline 0 & 0 & 0 & 0 & 1 & 1 \\ 0 & 0 & 0 & 0 & 0 & 1 \\ 0 & 0 & 0 & 1 & 0 & 0 \end{array} \right] = \left[ \begin{array}{c|c} A_{1,1} & A_{1,2} \\ \hline A_{2,1} & A_{2,2} \end{array} \right],$$

where

$$A_{1,1} = \begin{bmatrix} 0 & 1 & 1 \\ 0 & 0 & 1 \\ 1 & 0 & 0 \end{bmatrix}, \quad A_{1,2} = \begin{bmatrix} 0 & 0 & 0 \\ 0 & 0 & 0 \\ 1 & 0 & 0 \end{bmatrix},$$

$$A_{2,1} = \begin{bmatrix} 0 & 0 & 0 \\ 0 & 0 & 0 \\ 0 & 0 & 0 \end{bmatrix}, \quad A_{2,2} = \begin{bmatrix} 0 & 1 & 1 \\ 0 & 0 & 1 \\ 1 & 0 & 0 \end{bmatrix}.$$

The diagonal blocks $A_{1,1}$ and $A_{2,2}$ represent graphs that are strongly connected (that is, the blocks are irreducible). Furthermore, the block below the diagonal is 0. Is this true for all nonnegative matrices?

You bet. Let's see!

## 8.3 The Frobenius normal form

In general, the block-matrix structure that we have just seen is called the Frobenius normal form. Here's the precise definition.

**Definition 8.3.1 (Frobenius normal form)**

Let $A \in \mathbb{R}^{n \times n}$ be a nonnegative matrix. $A$ is said to be in *Frobenius normal form* if it can be written in the block matrix form

$$A = \begin{bmatrix} A_1 & A_{1,2} & \dots & A_{1,k} \\ 0 & A_2 & \dots & A_{2,k} \\ \vdots & \vdots & \ddots & \vdots \\ 0 & 0 & \dots & A_k \end{bmatrix},$$

where $A_1, \dots, A_k$ are irreducible matrices.

Let's reverse the question: can we transform an arbitrary nonnegative matrix into the Frobenius normal form? Yes, and with the help of directed graphs, this is much easier to show than purely using algebra. Here is the famous theorem in full form.

**Theorem 8.3.1** *(The existence of the Frobenius normal form)*

*Let $A \in \mathbb{R}^{n \times n}$ be a nonnegative matrix. There exists a permutation matrix $P \in \mathbb{R}^{n \times n}$ such that $P^T A P$ is in Frobenius normal form.*

Rigorously spelling out the proof of *Theorem 8.3.1* is quite complicated. However, the ideas behind the proof are simple to show. Thus, we'll take the less rigorous, more fun route.

So, why is the Frobenius normal form a big deal and what on Earth is a permutation matrix? Let's dive into it.

## 8.3.1  Permutation matrices

Mathematics is often done from concrete to abstract. That's why we are often start with special cases: what happens if we multiply a 2 x 2 matrix by

$$P_{1,2} = \begin{bmatrix} 0 & 1 \\ 1 & 0 \end{bmatrix},$$

a simple zero-one matrix? With a quick calculation, we can verify that

$$\begin{bmatrix} 0 & 1 \\ 1 & 0 \end{bmatrix} \begin{bmatrix} a & b \\ c & d \end{bmatrix} = \begin{bmatrix} c & d \\ a & b \end{bmatrix},$$

$$\begin{bmatrix} a & b \\ c & d \end{bmatrix} \begin{bmatrix} 0 & 1 \\ 1 & 0 \end{bmatrix} = \begin{bmatrix} b & a \\ d & c \end{bmatrix},$$

that is,

1. it switches the rows when multiplied from the left,
2. and it switches the columns when multiplied from the right.

Multiplying by P from both the left and right compounds the effects: it switches rows and columns, as

$$\begin{bmatrix} 0 & 1 \\ 1 & 0 \end{bmatrix} \begin{bmatrix} a & b \\ c & d \end{bmatrix} \begin{bmatrix} 0 & 1 \\ 1 & 0 \end{bmatrix} = \begin{bmatrix} d & c \\ b & a \end{bmatrix}$$

shows. (By the way, this is a similarity transformation, as our special zero-one matrix is its own inverse. This is not an accident; more about it later.)

Why are we looking at this? Because behind the scenes, this transformation doesn't change the underlying graph structure, just relabels its nodes!

You can easily verify this by hand, but *Figure 8.9* illustrates this as well.

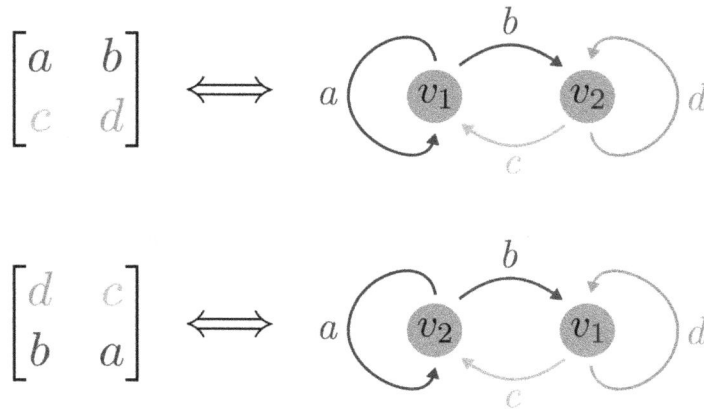

*Figure 8.9: Relabeling the nodes*

A similar phenomenon is true in the general $n \times n$ case. Here, we define the so-called transposition matrices by switching the $i$-th and $j$-th rows of the identity matrix, for example:

$$P_{1,2} = \begin{bmatrix} 0 & 1 & 0 & 0 & 0 \\ 1 & 0 & 0 & 0 & 0 \\ 0 & 0 & 1 & 0 & 0 \\ 0 & 0 & 0 & 1 & 0 \\ 0 & 0 & 0 & 0 & 1 \end{bmatrix} \in \mathbb{R}^{5 \times 5}, \quad P_{3,5} = \begin{bmatrix} 1 & 0 & 0 & 0 & 0 \\ 0 & 1 & 0 & 0 & 0 \\ 0 & 0 & 0 & 0 & 1 \\ 0 & 0 & 0 & 1 & 0 \\ 0 & 0 & 1 & 0 & 0 \end{bmatrix} \in \mathbb{R}^{5 \times 5}.$$

The two most important properties of the permutation matrices are $P_{i,j}^T = P_{i,j}$ and $P_{i,j}^T P_{i,j} = I$. That is, their inverse is their transpose.

Multiplication with a transposition matrix has the same effect: it switches rows from the left and columns from the right. To be precise,

1. $P_{i,j}A$ switches the $i$-th and $j$-th *rows* of $A$,
2. and $AP_{i,j}$ switches the $i$-th and $j$-th *columns* of $A$.

Most importantly, the similarity transformation

$$P_{i,j}AP_{i,j}$$

relabels the $i$-th and $j$-th nodes of $A$'s digraph, leaving the graph structure invariant.

Now, about the aforementioned permutation matrices. A permutation matrix is simply a product of transposition matrices:

$$P = P_{i_1,i_2} P_{i_3,i_4} \dots P_{i_{2k-1},i_{2k}}.$$

Permutation matrices inherit some properties from their building blocks. Most importantly,

1. their inverse is their transpose,
2. and a similarity transformation with them is just a relabeling of nodes that leave the graph structure invariant.

To see this latter one, consider that

$$P^T A P = (P_{i_1,i_2} P_{i_3,i_4} \dots P_{i_{2k-1},i_{2k}})^T A (P_{i_1,i_2} P_{i_3,i_4} \dots P_{i_{2k-1},i_{2k}})$$
$$= (P_{i_{2k-1},i_{2k}} \dots (P_{i_1,i_2} A P_{i_1,i_2}) \dots P_{i_{2k-1},i_{2k}}),$$

succesively relabeling the nodes. (Recall that transposing a matrix product switches up the order, and transposition matrices are their own transposes.) Conversely, every node relabeling is equivalent to a similarity transformation with a well-constructed permutation matrix.

Why are we talking about this? Because the proper labeling of nodes is key to the Frobenius normal form.

## 8.3.2   Directed graphs and their strongly connected components

Now, let's talk about graphs. We'll see how every digraph decomposes into strongly connected components. Let's see a concrete example:

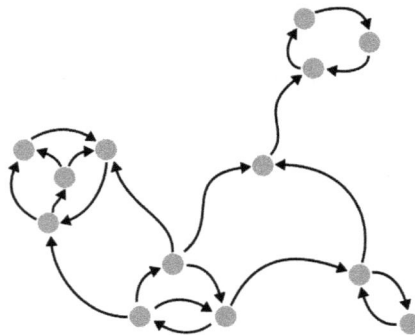

*Figure 8.10: A directed graph (that is complex enough for us to study)*

This'll be our textbook example. How many nodes can be reached from a given node? Not necessarily all. Say, for the point highlighted in *Figure 8.11*, only a portion of the graph is accessible.

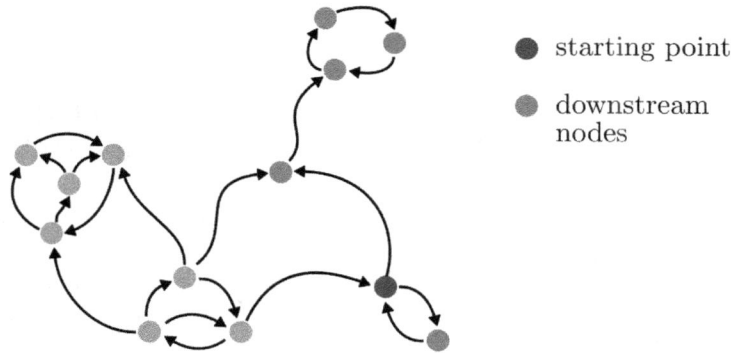

*Figure 8.11: Downstream nodes for a single starting point*

However, the set of mutually reachable nodes is much smaller: *Figure 8.12* shows that in our example, it consists of only two points.

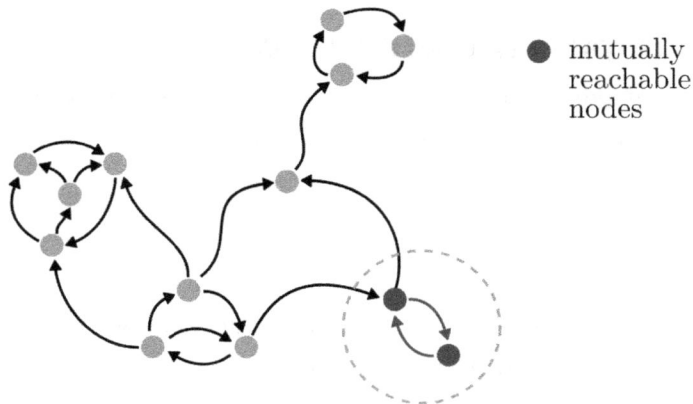

*Figure 8.12: Mutually reachable nodes*

Algebraically speaking, "*a* and *b* are mutually reachable from each other" is a special relation that partitions the set of nodes into disjoint subsets such that

1. two nodes from the same subset are mutually reachable from each other,
2. and two nodes from different subsets are not mutually reachable.

The subsets of this partition are called the strongly connected components, and we can always decompose a directed graph in this way.

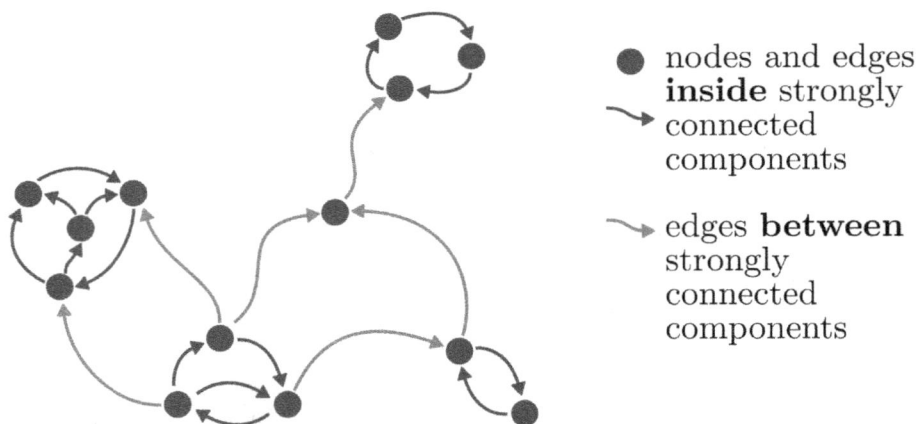

nodes and edges **inside** strongly connected components

edges **between** strongly connected components

*Figure 8.13: Strongly connected components of our example graph*

Now, let's connect everything together (not in a graph way but, you know, in a wholesome mathematical one)!

## 8.3.3 Putting graphs and permutation matrices together

We are two steps away from proving that every nonnegative square matrix can be transformed into the Frobenius normal form with a permutation matrix. Here is the plan.

1. Construct the graph for our nonnegative matrix.
2. Find the strongly connected components.
3. Relabel the nodes in a clever way.

And that's it! Why? Because, as we have seen, relabeling is the same as a similarity transform with a permutation matrix. There's just one tiny snag: what is the clever way? I'll show you.

First, we "skeletonize" the graph: merge the components together, as well as any edges between them.

Consider each component as a black box: we don't care what's inside, only about their external connections.

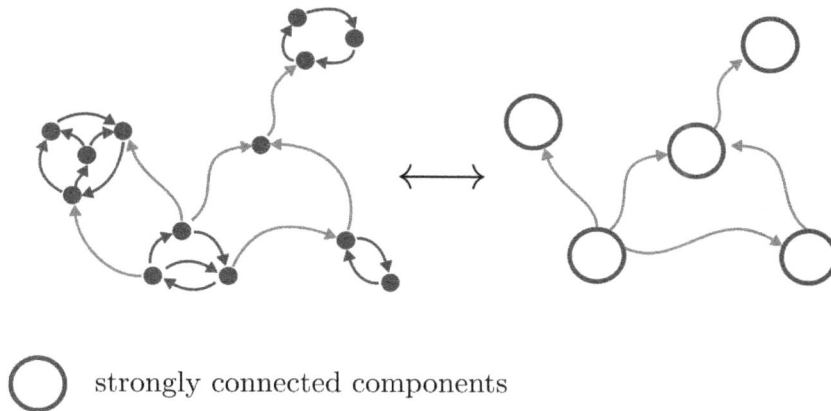

strongly connected components

*Figure 8.14: Strongly connected components*

In this skeleton, we can find components that cannot be entered from other components. These will be our starting points, the zeroth-class components. In our example, we only have one.

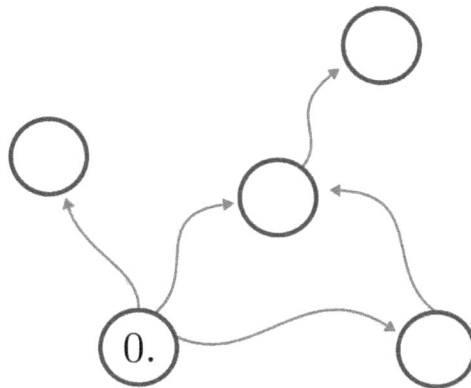

*Figure 8.15: Finding the "zeroth" component*

Now, things get a bit tricky. We number each component by the longest path from the farthest zero-class component from which it can be reached.

This is hard to even read, let alone understand. *Figure 8.16* illustrates the process.

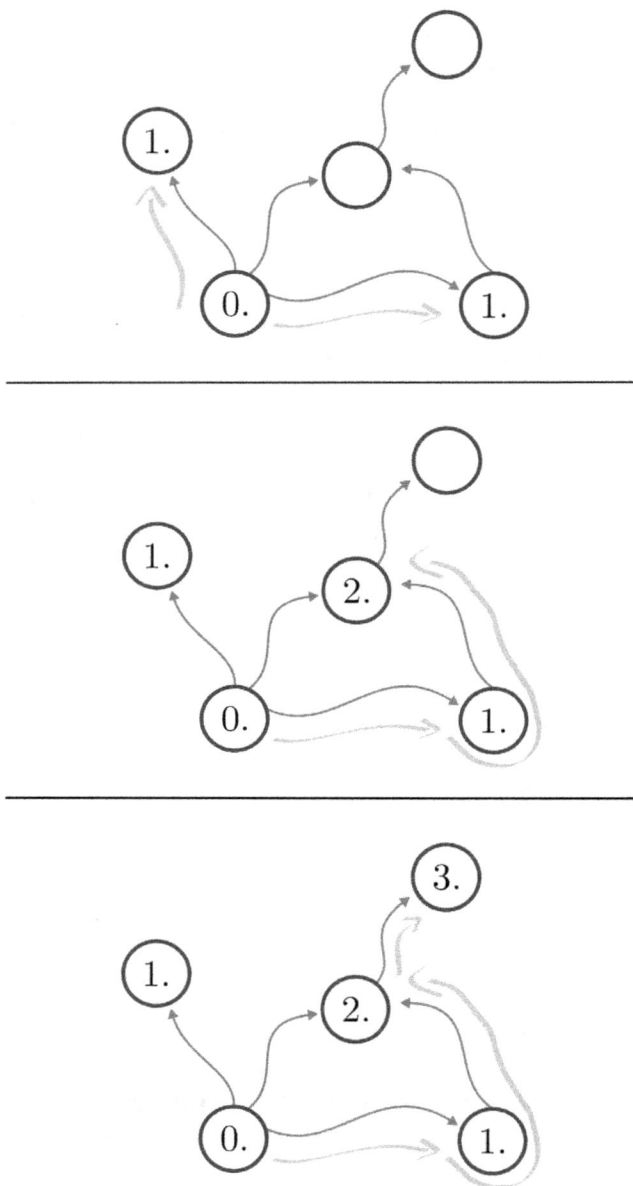

*Figure 8.16: Numbering the components*

The gist is that if you can reach an $m$-th class from an $n$-th class, then $n < m$. In the end, we have something like *Figure 8.17*.

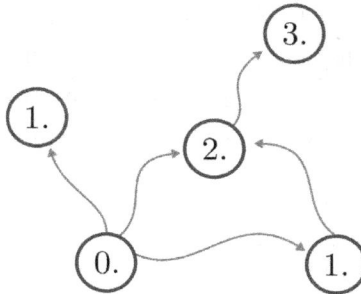

*Figure 8.17: Numbered components*

This defines an ordering on the components (a partial ordering, if you would like to be precise).

Now, we label the nodes inside such that

1. higher-order classes come first,
2. and consecutive indices are labeling nodes from the same component if possible.

This is how it goes.

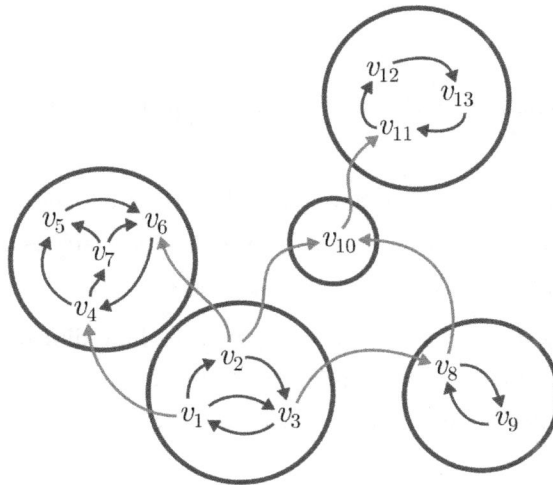

*Figure 8.18: Labeling the nodes*

Here is the matrix in this particular example, with zeros and ones for simplicity:

$$
\begin{bmatrix}
0 & 1 & 1 & 1 & 0 & 0 & 0 & 0 & 0 & 0 & 0 & 0 & 0 \\
0 & 0 & 1 & 0 & 0 & 1 & 0 & 0 & 0 & 1 & 0 & 0 & 0 \\
1 & 0 & 0 & 0 & 0 & 0 & 0 & 1 & 0 & 0 & 0 & 0 & 0 \\
0 & 0 & 0 & 0 & 1 & 0 & 1 & 0 & 0 & 0 & 0 & 0 & 0 \\
0 & 0 & 0 & 0 & 0 & 1 & 0 & 0 & 0 & 0 & 0 & 0 & 0 \\
0 & 0 & 0 & 1 & 0 & 0 & 0 & 0 & 0 & 0 & 0 & 0 & 0 \\
0 & 0 & 0 & 0 & 1 & 1 & 0 & 0 & 0 & 0 & 0 & 0 & 0 \\
0 & 0 & 0 & 0 & 0 & 0 & 0 & 0 & 1 & 1 & 0 & 0 & 0 \\
0 & 0 & 0 & 0 & 0 & 0 & 0 & 1 & 0 & 0 & 0 & 0 & 0 \\
0 & 0 & 0 & 0 & 0 & 0 & 0 & 0 & 0 & 0 & 1 & 0 & 0 \\
0 & 0 & 0 & 0 & 0 & 0 & 0 & 0 & 0 & 0 & 0 & 1 & 0 \\
0 & 0 & 0 & 0 & 0 & 0 & 0 & 0 & 0 & 0 & 0 & 0 & 1 \\
0 & 0 & 0 & 0 & 0 & 0 & 0 & 0 & 0 & 0 & 1 & 0 & 0 \\
\end{bmatrix}
$$

With that, the ideas behind the proof of *Theorem 8.3.1* are clear! Now, we also finally understand why irreducible matrices are called *irreducible*: as they describe strongly connected graphs, they cannot be further decomposed into smaller blocks in a meaningful way.

## 8.4 Summary

With the study of the connections between linear algebra and graph theory, our journey through linear algebra is over.

In this and the previous seven chapters, we have learned that vectors and matrices are not merely data structures that store observations and measurements. Vectors and matrices possess a rich and beautiful geometric structure, describing data *and* their transformations at the same time!

First, we learned that vectors live in so-called *vector spaces*, the high-dimensional generalizations of the three-dimensional space we are living in (which might be 26-dimensional, according to some string theorists, but let's stick to the Earth for now). We can measure lengths and distances via norms, most often defined by

$$
\|\mathbf{x}\| = \sum_{i=1}^{n} x_i^2, \quad \mathbf{x} \in \mathbb{R}^n,
$$

or measure angles (among others) via inner products, most often defined by

$$\langle \mathbf{x}, \mathbf{y} \rangle = \sum_{i=1}^{n} x_i y_i$$

$$= \cos(\alpha) \|\mathbf{x}\| \|\mathbf{y}\|.$$

From a mathematical perspective, matrices originate from the linear transformation of vector spaces, i.e., functions of the form $f : U \to V$, satisfying the linearity relation $f(a\mathbf{x} + b\mathbf{y}) = af(\mathbf{x}) + bf(\mathbf{y})$. Matrices arise from the algebraic representation of linear transformations by expressing them in the form

$$f(\mathbf{x}) = A_f \mathbf{x} = \begin{bmatrix} a_{1,1} & a_{1,2} & \dots & a_{1,m} \\ a_{2,1} & a_{2,2} & \dots & a_{2,m} \\ \vdots & \vdots & \ddots & \vdots \\ a_{n,1} & a_{n,2} & \dots & a_{n,m} \end{bmatrix} \begin{bmatrix} x_1 \\ x_2 \\ \vdots \\ x_n \end{bmatrix} = \begin{bmatrix} \sum_{i=1}^{n} a_{1,i} x_i \\ \sum_{i=1}^{n} a_{2,i} x_i \\ \vdots \\ \sum_{i=1}^{n} a_{n,i} x_i \end{bmatrix},$$

allowing us to reason about data transformations from a geometric perspective. This is an extremely powerful tool in machine learning. Think about it: $A\mathbf{x}$ can be a regression model, a layer in a neural network, or various other machine learning building blocks. Ultimately, this is why we want to study vector spaces: the data lives there, and data transformations are described by matrices.

However, building a model doesn't stop at linear algebra. To capture more complex patterns, we need *nonlinearities*. For instance, consider the famous Sigmoid function, defined by

$$\sigma(x) = \frac{1}{1 + e^{-x}}.$$

The transformation defined by $\sigma(A\mathbf{x})$ (where $\sigma$ is applied elementwise) is a simple logistic regression model, allowing us to perform binary classification on our multidimensional feature space. Iterating on this idea, the expression

$$N(\mathbf{x}) = \sigma(B\sigma(A\mathbf{x}))$$

defines a two-layer neural network.

So, the next part of our journey is into the domain of *calculus*, where we'll learn what functions really are, how we build predictive models from them, and how we fit these models by tuning the parameters with gradient descent.

Let's go!

# 8.5  Problems

**Problem 1.** Let $G = (V, E)$ be a directed graph and let $u, v \in V$ be two of its nodes. Show that if there exists a walk from $u$ to $v$, then there exists a walk without repeated edges and repeated vertices.

**Problem 2.** Let $G = (V, E)$ be a strongly connected directed graph. Show that $|E| \geq |V|$, where $|S|$ denotes the number of elements in the set $S$. (In other words, show that in order to be strongly connected, $G$ must have at least as many edges as nodes.)

**Problem 3.** Let $A \in \mathbb{R}^{n \times n}$ be an irreducible matrix. Is $A^2$ also reducible? (If yes, prove it. If no, show a counterexample.)

**Problem 4.** Let $A \in \mathbb{R}^{4 \times 4}$ be the matrix defined by

$$A = \begin{bmatrix} 0 & 0 & 1 & 0 \\ 0 & 0 & 0 & 1 \\ 1 & 0 & 0 & 0 \\ 0 & 1 & 0 & 0 \end{bmatrix}.$$

Find the permutation matrix $P$ that transforms $A$ to a Frobenius normal form!

# Join our community on Discord

Read this book alongside other users, Machine Learning experts, and the author himself.

Ask questions, provide solutions to other readers, chat with the author via Ask Me Anything sessions, and much more.

Scan the QR code or visit the link to join the community.

https://packt.link/math

# References

[1] VanderPlas, J. (2016). *Python Data Science Handbook*. O'Reilly Media, Inc.

[2] Strang, G. (2005). *Linear Algebra and Its Applications* (4th ed.). Brooks Cole.

[3] Lax, P. D. (2007). *Linear Algebra and Its Applications*. Wiley.

[4] Trefethen, L. N., & Bau III, D. (1997). *Numerical Linear Algebra* (1st ed.). Philadelphia: SIAM. ISBN 978-0-89871-361-9.

[5] Axler, S. (2024). *Linear Algebra Done Right* (4th ed.). Springer.

# PART II
# CALCULUS

This part comprises the following chapters:

# 9
# Functions

*"Mathematicians are like Frenchmen: whatever you say to them they translate into their own language and forthwith it is something entirely different."*

— *Johann Wolfgang von Goethe*

It's time we tackle the next big subject of machine-learning-math: functions and calculus. What do we have to do with functions?

As we've seen before, a predictive model is nothing but a multivariate parametric function. Linear regression? $A\mathbf{x}$. Logistic regression? $\sigma(A\mathbf{x})$. Neural networks? A sequence of $\sigma(A\mathbf{x})$-s, and a multitude of other layers.

But it's not just the description of models, it's fitting them to the data. Say, for a simple linear regression model, "training" is minimizing the mean squared error; that is, finding

$$\mathrm{argmin}_{a,b \in \mathbb{R}} \frac{1}{n} \sum_{i=1}^{n} (a x_i + b - y_i)^2,$$

where $x_i$ is the training data, and $y_i$ is the ground truth. This is done via *differentiation*, one of humanity's most essential inventions. (Whether we can call mathematical discoveries *inventions* is a matter of constant debate.)

For a function $f(x)$, its derivative at $x_0$ is defined by

$$f'(x_0) = \lim_{x \to x_0} \frac{f(x) - f(x_0)}{x - x_0},$$

describing the rate of change in terms of $x$. There's a lot to unravel: what is that strange $\lim_{x \to x_0}$ symbol? What do the ratio $\frac{f(x) - f(x_0)}{x - x_0}$ represent? We'll get to all of it in due time. The gist is, differentiation is the driving force of gradient descent, the number one algorithm for training neural networks.

But it does not end here either. Differentiation has an equally important counterpart: *integration*. The formula

$$\int_a^b f(x)dx$$

is called the *integral* of $f(x)$, describing the signed area under its graph. Integration is not as prevalent in practice, but it's an essential tool for theory. For instance, half of probability theory is built around it. Say, the famous expected value is defined in terms of an integral; as are several notable probability distributions.

In the next couple of chapters, we'll dive deep into *calculus*, the theory of univariate functions. Why? Because even though machine learning is multivariate, the ideas we develop here will serve as the foundations for all the math to come. Let's get to work!

# 9.1   Functions in theory

Everyone has an intuitive understanding of what functions are. At one point or another, all of us have encountered this concept. For most of us, a function is a curve drawn with a continuous line onto a representation of the Cartesian coordinate system.

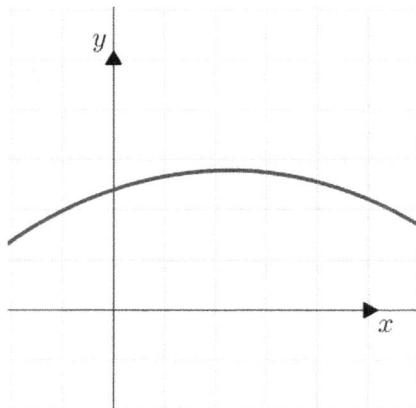

*Figure 9.1: Definitely looks like a function*

However, in mathematics, intuitions can often lead us to false conclusions. Often, there is a difference between *what something is* and *how you think about it* – what your mental models are. To give an example from a real-life scenario in machine learning, consider the following piece of code:

```python
import numpy as np


def cross_entropy_loss(X, y):
    """
    Args:
        X: numpy.ndarray of shape (n_batch, n_dim).
            Contains the predictions in form of a probability distribution.
        y: numpy.ndarray of shape (n_batch, 1).
            Contains class labels for each data point in X.

    Returns:
        loss: numpy.float.
            Cross entropy loss of the predictions.
    """

    exp_x = np.exp(X)
    probs = exp_x / np.sum(exp_x, axis=1, keepdims=True)
    log_probs = - np.log([probs[i, y[i]] for i in range(len(probs))])
    loss = np.mean(log_probs)

    return loss
```

Suppose that you wrote this function, and it is in your codebase somewhere. Depending on our needs, we might think of it as *cross-entropy loss*, but in reality, this is a 579-character-long string in the Python language, eventually processed by an interpreter. However, when working with it, we often use a mental model that compacts this information into easily usable chunks, like the three words *cross-entropy loss*. When we reason about high-level processes like training a neural network, abstractions such as this allow us to move further and step bigger.

But sometimes, things don't go our way. When this function throws an error and crashes the computations, *cross-entropy loss* will not cut it. Then, it is time to unravel the definition and put everything under a magnifying glass. What could have hindered your thinking before is now essential.

These principles are also true for theory, not just for practice. Mathematics is a balancing act between logical precision and a clear understanding, two often contradicting objectives.

Let's go back to our starting point: functions in a mathematical sense. One possible mental model, as mentioned, is *a curve drawn with a continuous line*. It allows us to reason about functions visually and intuitively answer some questions. However, this particular mental model can go very wrong.

To give an example, does *Figure 9.2* depict a function?

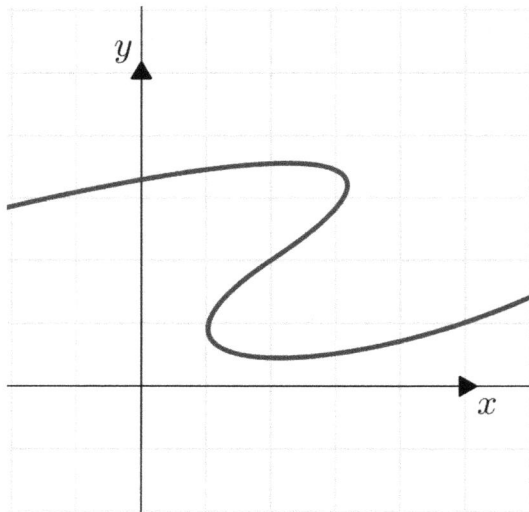

*Figure 9.2: Is this a function?*

Even though this curve is drawn with a continuous line, this is *not* a function, as there are values with multiple images, which cannot happen. To avoid confusion later, we have to build the foundations of our discussion if we were to talk about mathematical objects. In this chapter, our goal is to establish a basic dictionary to properly understand the objects we are working with in machine learning.

## 9.1.1 The mathematical definition of a function

Let's dive straight into the deep water and see the exact *mathematical definition* of functions! (Don't worry if you don't understand it for the first read. I'll explain everything in detail. This is the usual experience when encountering a definition for the first time.)

**Definition 9.1.1 (Functions)**

Let $X$ and $Y$ be two sets. The subset $f \subseteq X \times Y$ is a *function* if for every $x \in X$, there is at most one $y \in Y$ such that $(x, y) \in f$.

(The set $X \times Y$ denotes the *Cartesian product* of $X$ and $Y$. If you are not familiar with the concept, check out *Appendix C*.)

For simplicity, we introduce the notation

$$f : X \to Y,$$

which is short for $f$ *is a function from $X$ to $Y$*.

Note that $X$ and $Y$ can be any set. In the examples we encounter, these are usually the set of real numbers or vectors, but there is no such restriction.

To visualize the definition, we can draw two sets and arrows pointing from elements of $X$ to elements of $Y$. Each element $(x, y) \in f$ represents an arrow, pointing from $x$ to $y$.

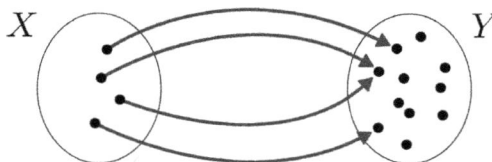

*Figure 9.3: A function, as arrows between two sets*

The only criteria is that there can be at most one arrow starting from any $x \in X$. This is why *Figure 9.2* is *not* a function.

Defining a function as a subset is mathematically precise but very low level. To be more useful, we can introduce an abstraction by defining functions with *formulas*, such as

$$f : \mathbb{R} \to \mathbb{R}, \quad x \mapsto x^2,$$

or simply $f(x) = x^2$ in short. This is how most of us think about functions when working with them.

Now that we are familiar with the definition, we should get to know some of the most basic structural properties of functions.

## 9.1.2   Domain and image

We saw that, in essence, functions are arrows between sets. At this point, we don't know anything useful about them. When is a function invertible? How can we find their minima and maxima? Why should we even care? Probably you have a bunch of questions here. Slowly but surely, we will cover all of these.

The first steps in our journey are concerned with the *sets* from which arrows start and point. There are two important sets in a function's life: its *domain* and *image*.

**Definition 9.1.2  (Domain and image of functions)**

Let $f : X \to Y$ be a function. The sets

$$\mathrm{dom} f := \{x \in X : \text{there is an } y \in Y \text{ such that } f(x) = y\} \subseteq X$$

and

$$\operatorname{im} f := \{y \in Y : \text{there is an } x \in X \text{ such that } f(x) = y\} \subseteq Y$$

are respectively called the *domain* and *image* of $f$.

In other words, the domain is the subset of $X$ where arrows start; the image is the subset of $Y$ where arrows point.

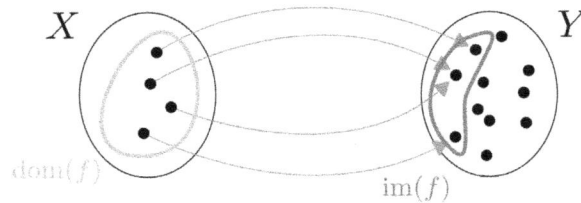

*Figure 9.4: The domain and image of a function*

Why is this important? For one, these are directly related to the *invertibility* of a function. If you consider the "points and arrows" mental representation, inverting a function is as simple as flipping the direction of the arrows. When can we do it? In some cases, doing this might not even result in a function, as *Figure 9.5* shows.

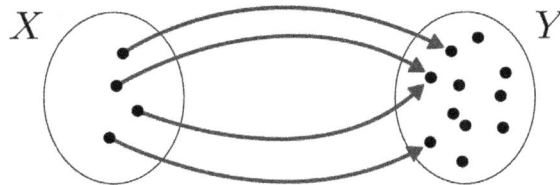

*Figure 9.5: This function is not invertible, as reversing the arrows doesn't give a well-defined function*

To put the study of functions on top of solid theoretical foundations, we introduce the concept of *injective*, *surjective* and *bijective* functions.

**Definition 9.1.3 (Surjective, injective, and bijective functions)**

Let $f : X \to Y$ be an arbitrary function.

*(a) f* is *injective* if for every $y \in Y$ there is at most one $x \in X$ such that $f(x) = y$. (We often say that an injective $f$ is a *one-to-one* function.)

(b) $f$ is *surjective* if for every $y \in Y$ there is an $x \in X$ such that $f(x) = y$. (We often say that a surjective $f$ maps $X$ *onto* $Y$.)

(c) $f$ is *bijective* if it is injective and surjective.

In terms of arrows, injectivity means that every element of the image has at most one arrow pointing to it, while surjectivity is that every element indeed has at least one arrow. When both are satisfied, we have a bijective function, one that can be inverted properly. When the inverse $f^{-1}$ exists, it is unique. Both $f^{-1} \circ f$ and $f \circ f^{-1}$ equal to the identity function in their respective domains.

Let's see some concrete examples! For instance,

$$f : \mathbb{R} \to \mathbb{R}, \quad f(x) = x^2$$

is not injective nor surjective. (Ponder on this a bit if you don't understand it right away. It helps if you draw a figure.)

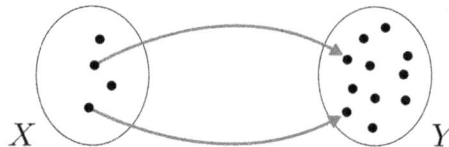

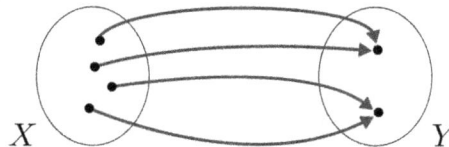

*Figure 9.6: Injective, surjective, and bijective functions*

On the contrary,

$$g : \mathbb{R} \to \mathbb{R}, \quad g(x) = x^3$$

is both, so it is bijective and invertible with inverse $g^{-1}(x) = x^{1/3}$.

Invertible functions behave nicely, and from a certain perspective, they are much better to work with.

## 9.1.3    Operations with functions

Functions, just like numbers, have operations defined on them. Two numbers can be multiplied and added together, but can you do the same with functions? Without any difficulty, they can be added together and multiplied with a scalar as

$$(f + g)(x) := f(x) + g(x),$$
$$(cf)(x) := cf(x),$$

where $c$ is some scalar.

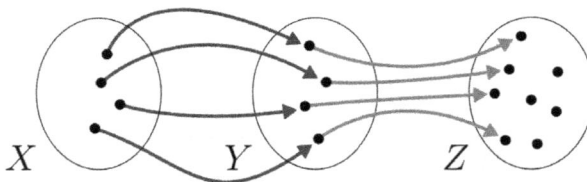

*Figure 9.7: Composing functions*

Another essential operation is *composition*. Let's consider the famous logistic regression for a minute! The estimator itself is defined by

$$f(x) = \sigma(ax + b),$$

where

$$\sigma(x) = \frac{1}{1 + e^{-x}}$$

is the sigmoid function. The estimator $f(x)$ is the *composition* of two functions: $l(x) = ax + b$ and the sigmoid function, so

$$f(x) = \sigma\big(l(x)\big).$$

As $\sigma$ maps $\mathbb{R}$ onto $[0,1]$, we can think about $f(x)$ as the probability that $x$ belongs to the positive class.

*Figure 9.7* shows the points-and-arrows illustration of function composition.

To give one more example, a neural network with several hidden layers is just the composition of a bunch of functions. The output of each layer is fed into the next one, which is exactly how composition is defined.

In general, if $f : Y \to Z$ and $g : X \to Y$ are two functions, then their composition is formally defined by

$$f \circ g : X \to Z, \quad x \mapsto f(g(x)).$$

Note that in $(f \circ g)(x) = f(g(x))$, the function $g$ is applied first. The order of application is often a point of confusion, so keep this in mind.

---

**Remark 9.1.1 (Function addition as composition)**

Given the functions

$$f : X \to \mathbb{R}, \quad g : X \to \mathbb{R},$$

we can define their sum by

$$(f + g)(x) := f(x) + g(x).$$

Believe it or not, this is yet another form of function composition. Why? Define the function add by

$$\text{add} : X \times X \to \mathbb{R}, \quad \text{add}(x_1, x_2) = x_1 + x_2.$$

Now we can write addition as

$$(f + g)(x) = \text{add}\big(f(x), g(x)\big).$$

---

Composition is an *extremely* powerful tool. In fact, so powerful that given a small set of cleverly defined building blocks, "almost every function" can be obtained as the composition of these blocks. (I put "almost every function" in quotes because if we want to stay mathematically precise here, long detours are needed. To keep ourselves focused, let's allow ourselves to be a little hand-wavy here.)

## 9.1.4   Mental models of functions

So far, we have seen that functions are defined as arrows drawn between elements of two sets. This, although being mathematically rigorous, does not give us useful mental models to reason about them. As you'll surely see by the end of our journey, in mathematics, the key is often to find the right way to look at things.

Regarding functions, one of the most common and useful mental models is their *graph*.

If $f : \mathbb{R} \to \mathbb{R}$ is a function mapping a real number to a real number, we can *visualize* it using its graph, defined by

$$\operatorname{graph}(f) := \{(x, f(x)) : x \in \mathbb{R}\}.$$

This set of points can be drawn in the two-dimensional plane. For instance, in the case of the famous *rectified linear unit* (ReLU)

$$\operatorname{ReLU}(x) = \begin{cases} 0 & \text{if } x < 0 \\ x & \text{if } 0 \le x, \end{cases}$$

the graph looks like this.

```python
import numpy as np
import matplotlib.pyplot as plt

x = np.linspace(-10, 10, 400)
relu = np.maximum(0, x)

with plt.style.context("seaborn-v0_8-white"):
    plt.figure()
    plt.plot(x, relu, label="ReLU(x)", color="blue")
    plt.axhline(0, color='black', linewidth=0.8, linestyle="--")
    plt.title("ReLU", fontsize=14)
    plt.xlabel("x", fontsize=12)
    plt.ylabel("ReLU(x)", fontsize=12)
    plt.grid(alpha=0.3)
    plt.show()
```

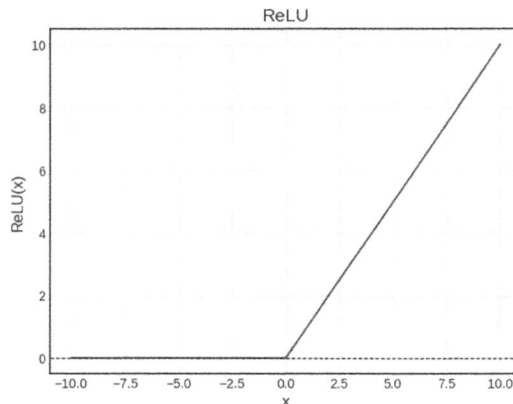

*Figure 9.8: Graph of the ReLU function*

Although identifying functions with their graphs can be useful, it is not generalizable for more complex mappings. Visualizing it is challenging if the function's domain and image are not the set of real numbers.

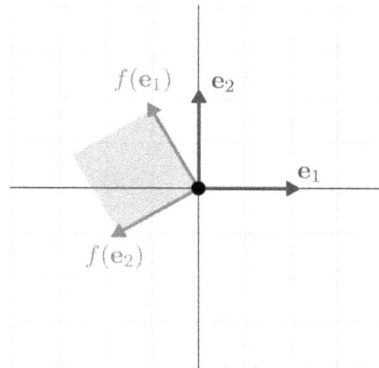

*Figure 9.9: Functions as a transformation of the space. Here, the vectors of the space are rotated around the origin*

When dealing with neural networks, probably the best way to think about functions (that is, layers in this context) is as *transformations of the underlying feature space.* A simple example is a rotation in the two-dimensional Euclidean plane, as *Figure 9.9* shows.

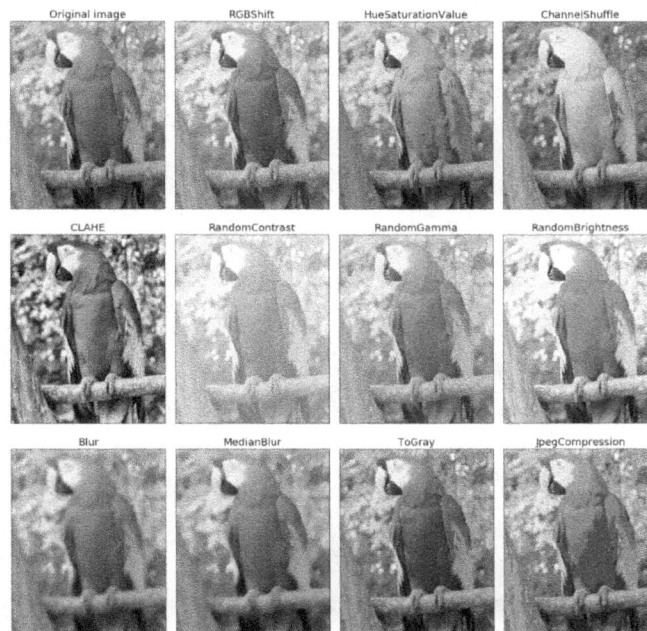

*Figure 9.10: Examplar applications of image transformations available in Albumenations*

Image transformations provide a set of more complex examples. You rarely think about image blur as a transformation between spaces, but this is the case. After all, an image is just a huge vector in some vector space.

Image operations as transformations, as done by the Albumentations (`https://albumentations.ai/`) library. Source of the image: Albumentations: Fast and Flexible Image Augmentations by Alexander Buslaev, Vladimir I. Iglovikov, Eugene Khvedchenya, Alex Parinov, Mikhail Druzhinin and Alexandr A. Kalinin (`https://www.mdpi.com/2078-2489/11/2/125`).

In essence, a neural network is simply a stack of transformations, each taking its input from the output of the previous one. As you'll see, what makes them special is that the transformations are not hand-engineered but *learned* from the data.

# 9.2 Functions in practice

In our study of functions, we started from arrows between sets and ended up with mental models such as formulas and graphs. For pure mathematical purposes, these models are perfectly enough to conduct thorough investigations. However, once we leave the realm of theory and start putting things into practice, we must think about how functions are represented in programming languages.

In Python, functions are defined using a straightforward syntax. For instance, this is how the $square(x) = x^2$ function can be implemented.

```python
def square(x):
    return x**2
```

The result is an object of type `function`. (In Python, everything is an object.)

```python
type(square)
```

```
function
```

Functions are called using the `()` operator.

```python
square(12)
```

```
144
```

Python is well-known for its simplicity, and functions are no exception. However, this doesn't mean that they are limited in features – quite the contrary: you can achieve a lot with the clever use of functions.

## 9.2.1 Operations on functions

There are three operations that we want to do on functions: composition, addition, multiplication. The easiest way is to call the functions themselves and fall back to the operations defined for the number types. To see an example, let's implement the $cube(x) = x^3$ function and add/multiply/compose it with square.

```python
def cube(x):
    return x**3
x = 2

square(x) + cube(x)    # addition
```

```
12
```

```python
square(x)*cube(x)      # multiplication
```

```
32
```

```python
square(cube(x))        # composition
```

```
64
```

However, there is a major problem. If you take another look at the function operations, you can notice that they *take* functions and *return* functions. For instance, the composition is defined by

$$\text{compose} : f, g \mapsto \underbrace{\left(x \mapsto f(g(x))\right)}_{\text{the composed function}},$$

with a function as a result. We did no such thing by simply passing the return value to the outer function. There is no function object to represent the composition.

In Python, functions are first-class objects, meaning that we can pass them to other functions and return them from functions. (This is an absolutely fantastic feature, but if this is the first time you've encountered this, it might take some time to get used to.)

Thus, we can implement the compose function above by using the first-class function feature.

```
def compose(f, g):

    def composition(*args, **kwargs):
        return f(g(*args, **kwargs))

    return composition
square_cube_composition = compose(square, cube)

square_cube_composition(2)
```

64

Addition and multiplication can be done just like this. (They are even assigned as an exercise problem *Section 9.4*.)

## 9.2.2   Functions as callable objects

The standard way of doing function definitions is not a good fit for an application that is essential for us: parametrized functions. Think about the case of linear functions of the form $ax + b$, where $a$ and $b$ are parameters. On the first try, we can do something like this.

```
def linear(x, a, b):
    return a*x + b
```

Passing the parameters as arguments seems to work, but there are serious underlying issues. For instance, functions can have a lot of parameters. Even if we compact parameters into multidimensional arrays, we might need to deal with dozens of such arrays. Passing them around manually is error-prone, and we usually have to work with multiple functions. For example, neural networks are composed of several layers. Each layer is a parameterized function, and their composition yields a predictive model.

We can solve this issue by applying the classical object-oriented principle of *encapsulation*, implementing functions as callable objects. In Python, we can do this by implementing the magic __call__ method for the class.

```
class Linear:
    def __init__(self, a, b):
        self.a = a
        self.b = b
```

```
    def __call__(self, x):
        return self.a*x + self.b
f = Linear(2, -1) # this represents the function f(x) = 2*x - 1
f(2.1)
```

```
3.2
```

This way, we can store, access, and modify the parameters using attributes.

```
f.a, f.b
```

```
(2, -1)
```

Since there can be a lot of parameters, we should implement a method that collects them together in a dictionary.

```
class Linear:
    def __init__(self, a, b):
        self.a = a
        self.b = b

    def __call__(self, x):
        return self.a*x + self.b

    def parameters(self):
        return {"a": self.a, "b": self.b}


f = Linear(2, -1)
f.parameters()
```

```
{'a': 2, 'b': -1}
```

Interactivity is one of the most useful features of Python. In practice, we frequently find ourselves working in the REPL, inspecting objects and calling functions by hand. We often add a concise string representation for our classes for these situations.

By default, printing a `Linear` instance results in a cryptic message.

```
f
```

```
<__main__.Linear at 0x7c9fbef31190>
```

This is not very useful. Besides the class name and its location in the memory, we haven't received any information. We can change this by implementing the __repr__ method responsible for returning the string representation for our object.

```
class Linear:
    def __init__(self, a, b):
        self.a = a
        self.b = b

    def __call__(self, x):
        return self.a*x + self.b

    def __repr__(self):
        return f"Linear(a={self.a}, b={self.b})"

    def parameters(self):
        return {"a": self.a, "b": self.b}
f = Linear(2, -1)
f
```

```
Linear(a=2, b=-1)
```

This looks much better! Adding a pretty string representation seems like a small thing, but this can go a long way when doing machine learning engineering in the trenches.

## 9.2.3 Function base class

The `Linear` class that we have just seen is only the tip of the iceberg. There are hundreds of function families that are used in machine learning. We'll implement many of them eventually, and to keep the interfaces consistent, we are going to add a base class from which all others will be inherited.

```
class Function:
    def __init__(self):
        pass
```

```
    def __call__(self, *args, **kwargs):
        pass

    def parameters(self):
        return dict()
```

With this, we can implement functions and function families in the following way.

```
import numpy as np


class Sigmoid(Function):            # the parent class is explicitly declared
    def __call__(self, x):
        return 1/(1 + np.exp(-x))

sigmoid = Sigmoid()
sigmoid(2)
```

```
np.float64(0.8807970779778823)
```

Even though we haven't implemented the `parameters` method for the `Sigmoid` class, it is inherited from the base class.

```
sigmoid.parameters()
```

```
{}
```

For now, let's keep the base class as simple as possible. During the course of this book, we'll progressively enhance the `Function` base class to cover all the methods a neural network and its layers need. (For instance, gradients.)

## 9.2.4 Composition in the object-oriented way

Recall how we did function composition (in *Section 9.2.1*) when working with plain Python functions? Syntactically, that can work with our `Function` class as well, although there is a huge issue: the return value is not a `Function` type.

```
composed = compose(Linear(2, -1), Sigmoid())
composed(2)
```

```
np.float64(0.7615941559557646)
```

```
isinstance(composed, Function)
```

```
False
```

This kind of composition doesn't inherit the interface we need.

```
composed.parameters()
```

```
---------------------------------------------------------------------
AttributeError                            Traceback (most recent call last)
Cell In[33], line 1
----> 1 composed.parameters()

AttributeError: 'function' object has no attribute 'parameters'
```

To fix the issue, we implement function composition as a child of the Function base class. Recall that composition is a *function*, taking two functions as input and returning one:

$$\text{compose} : f, g \mapsto \underbrace{\left( x \mapsto f(g(x)) \right)}_{\text{the composed function}} .$$

Keeping this in mind, this is how we can do composition.

```python
class Composition(Function):
    def __init__(self, *functions):
        self.functions = functions

    def __call__(self, x):

        for f in reversed(self.functions):
            x = f(x)

        return x
```

(Note that due to how composition is defined, we iterate through the list of functions in the reverse order. That's because $(f \circ g)(x) = f(g(x))$, we apply $g$ first.)

```
composed = Composition(Linear(2, -1), Sigmoid())
composed(2)
```

```
np.float64(0.7615941559557646)
```

This way, we get to keep the `Function` interface.

```
composed.parameters()
```

```
{}
```

```
isinstance(composed, Function)
```

```
True
```

# 9.3 Summary

Now that we've learned what functions really are, I bet your world is a bit shaken. Functions as graphs drawn with continuous lines, sure. Maybe even expressions like $f(x) = x^2 + 1$. But functions as dots and arrows?

Surprisingly, the dot-and-arrow representation is the closest to the true definition. Graphs and expressions come after that. This is what we've learned in this chapter, and by now, I feel like a magician. I've shown you mathematical objects and revealed that, deep inside, they are not what you think. We did this with vectors, matrices, and now, with functions.

Along with the *"what's behind the curtain?"* tricks, putting theory into practice also became an established pattern for us. So, we've used object-oriented Python to get a taste of what a function is like.

Next, we'll turn our lenses to an even higher level of magnification. For us, the most important functions map numbers to numbers. But what's a number? Even in programming languages, we have several different number types, like `int`-s, `float`-s, `double`-s, and so on. These are deeply rooted in math, and familiarity with the structure of numbers is a must for every developer and engineer.

See you in the next chapter!

# 9.4  Problems

**Problem 1.** Which of these functions are injective, surjective, or bijective? Find the inverse of bijective functions.

*(a)* $f : \mathbb{R} \to (0, \infty), x \mapsto e^x$

*(b)* $g : \mathbb{R} \to [0, \infty), x \mapsto x^2$

*(c)* $h : [0, \infty) \to [0, \infty), x \mapsto x^2$

*(d)* $\sin : \mathbb{R} \to [0, 1], x \mapsto \sin(x)$

*(e)* $\tan : \mathbb{R} \to \mathbb{R}, x \mapsto \tan(x)$,

**Problem 2.** Find a function $f : \mathbb{R} \to \mathbb{R}$ such that $(f \circ f)(x) = -x$.

**Problem 3.** Can any real function $g : \mathbb{R} \to \mathbb{R}$ be obtained as $g = f \circ f$ for some $f : \mathbb{R} \to \mathbb{R}$?

**Problem 4.** Following the example of the composition in *Section 9.2.1*, implement

- the add function, taking $f$ and $g$, returning $f + g$,
- the mul function, taking $f$ and $g$, returning $fg$,
- and the div function, taking $f$ and $g$, returning $f/g$.

# Join our community on Discord

Read this book alongside other users, Machine Learning experts, and the author himself.

Ask questions, provide solutions to other readers, chat with the author via Ask Me Anything sessions, and much more.

Scan the QR code or visit the link to join the community.

`https://packt.link/math`

# 10

# Numbers, Sequences, and Series

*"It's like asking why is Ludwig van Beethoven's Ninth Symphony beautiful. If you don't see why, someone can't tell you. I know numbers are beautiful. If they aren't beautiful, nothing is."*

— *Paul Erdős*

When I was about to take my first mathematical analysis course at the university, coming straight from high school, I wondered why we would spend several lectures on real numbers. At the time, I was confident in my knowledge and thought that I knew what numbers were. This was my first painful encounter with the Dunning–Kruger effect: the less you know, the more confident you are. Suffice to say, after a few classes, I was left confused about numbers, taking a while to finally understand them.

If you look at numbers under a magnifying glass, they become extremely complex. In this chapter, we are going to make sense of them. To look ahead and keep machine learning in our sights, consider that gradient descent (you know, the optimization algorithm that is used *everywhere*) is not possible for functions that are not differentiable. In turn, a function $f$ is differentiable at $x$ if the limit

$$\lim_{x \to y} \frac{f(x) - f(y)}{x - y}$$

exists. To understand limits, we must understand real numbers first.

Another good reason to dig deep into the patterns and structures of numbers: they are beautiful (as said above by Paul Erdős, one of the greatest mathematicians ever). There is a particular joy to understanding seemingly familiar things on a deep level. Even though you might not use this knowledge every day, it teaches you perspective about the objects you encounter during your work.

So, let's get started!

# 10.1  Numbers

There are five famous classes of numbers that one has to know in order to become adept in mathematics:

- natural numbers, denoted by $\mathbb{N}$,
- integers, denoted by $\mathbb{Z}$,
- rational numbers, denoted by $\mathbb{Q}$,
- real numbers, denoted by $\mathbb{R}$,
- and finally, complex numbers, denoted by $\mathbb{C}$.

These classes increase in order, that is,

$$\mathbb{N} \subseteq \mathbb{Z} \subseteq \mathbb{Q} \subseteq \mathbb{R} \subseteq \mathbb{C}.$$

In this section, we are going to concern ourselves with the first four. (Complex numbers will get their own chapter.)

## 10.1.1  Natural numbers and integers

*Natural numbers* are simply defined as

$$\mathbb{N} := \{1, 2, 3, \dots\}.$$

Sometimes 0 is included; sometimes it is not. Believe it or not, after a few thousand years, mathematicians still cannot decide whether or not 0 is a natural number. This problem might sound comical, but trust me, I have seen senior professors almost get into a fistfight upon debating this issue. For some people, this is a religious question.

I don't particularly care, and neither should you. I propose using the more common and practical definition, which is the one without 0. When we really need to talk about the natural numbers AND 0, I will use the notation $\mathbb{N}_0 = \{0, 1, 2, \dots\}$.

The cardinality of the set of natural numbers is countably infinite. In fact, countability is defined as $|\mathbb{N}|$. (If you are not familiar with the concept of *cardinality*, check out *Appendix C*.)

To be able to express negative and zero quantities, we extend natural numbers to obtain the set of *integers*, defined by

$$\mathbb{Z} = \{\dots, -2, -1, 0, 1, 2, \dots\}.$$

So far, so good. Integers are also countable: one can enumerate all of its elements by

$$0, 1, -1, 2, -2, 3, -3, \dots.$$

One significant advantage of integers over natural numbers is that they contain the additive inverse for each element. In plain English, if $n \in \mathbb{Z}$, then so does $-n \in \mathbb{Z}$. This makes it possible to define all kinds of algebraic structures over the integers, giving us mathematical tools to reason about phenomena modeled by them.

Note that if $n, m \in \mathbb{Z}$, then $n + m \in \mathbb{Z}$. In mathematical terminology, we say that $\mathbb{Z}$ is *closed to addition*.

To summarize, $\mathbb{Z}$ is

- closed to addition,
- and every element has an additive inverse.

These two properties will guide us on how to go from natural numbers to real numbers. Each extension is constructed so that these two properties hold, but for different operations.

## 10.1.2  Rational numbers

So, we obtained $\mathbb{Z}$ from $\mathbb{N}$ by extending it with 0 and the additive inverses for each element. What about the multiplicative inverses? This idea leads us to the concept of *rational numbers*, numbers that can be written as a ratio of two integers. It is defined by

$$\mathbb{Q} = \left\{ \frac{p}{q} \ : \ p, q \in \mathbb{Z}, q \neq 0 \right\},$$

It is both closed to multiplication and every element (except 0) has a multiplicative inverse. This is not just a *l'art pour l'art* mathematical construction: rational numbers model quantities that are all around us. "7.9 seconds" to "100.0 km/h". 78.4 kilograms to carry. 0.5 pizza to eat. You get it.

It might be surprising, but $\mathbb{Q}$ is also countable.

One easy way to prove this is to notice that it can be obtained as the countable union of countable sets:

$$\mathbb{Q} = \bigcup_{p \in \mathbb{Z}} \left\{ \frac{p}{q} : q \in \mathbb{Z} \setminus \{0\} \right\}.$$

(If you are not familiar with the basic set operations like union and setminus, check out *Appendix C.*)

Since the union of countable sets is countable, $\mathbb{Q}$ is countable as well. Another (and perhaps more visual) way to see this is to simply enumerate them in the sequence, illustrated by *Figure 10.1.*

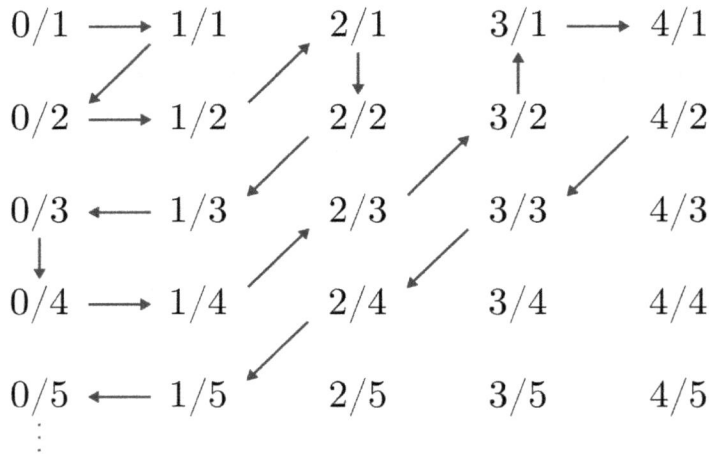

*Figure 10.1: Enumeration of rational numbers, where the arrows indicate the ordering*

Rational numbers can be written in decimal form, like $\frac{1}{2} = 0.5$, for example. In general, the following is true.

**Theorem 10.1.1** *Any rational number $x$ can be represented as a:*

*(a) finite decimal*

$$x = x_0 \ldots x_k . x_{k+1} \ldots x_n, \quad x_i \in \{0, 1, 2, \ldots, 9\}$$

*(b) or a repeating decimal*

$$x = x_0 \ldots x_k . x_{k+1} \ldots \dot{x}_n \ldots \dot{x}_m, \quad x_i \in \{0, 1, 2, \ldots, 9\},$$

*where the decimals between the two dots repeat infinitely. (This can be just a single digit as well.)*

Note that the decimal representation is not unique: for example, 1.0 and $0.\dot{9}$ are equal.

The above theorem fully characterizes rational numbers. But what about numbers with an infinite decimal form that does not repeat?

Like the famous mathematical constant $\pi$ describing the half circumference of the unit circle, that is,

$$\pi = 3.14159265358979323846264338327950288419716939937510...,$$

with no repeating patterns. These are called *irrational* numbers, and together with rationals, they make up the *real numbers*.

## 10.1.3   Real numbers

The simplest way to imagine real numbers is a line, where each point represents a number.

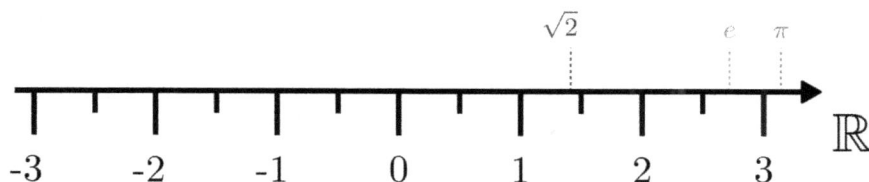

*Figure 10.2: The real number line*

If we temporarily let mathematical correctness slide, we can say that

$$\mathbb{R} = \text{finite decimals} \cup \text{infinite repeating decimals} \cup \text{infinite nonrepeating decimals}.$$

Real numbers are also the first we have encountered in our journey that are not countable, and we will prove this! Its proof is so beautiful that it belongs in The Book, a collection of the most elegant and beautiful mathematical proofs.

**Theorem 10.1.2** $\mathbb{R}$ *is not countable.*

*Proof.* To show that $\mathbb{R}$ is not countable, let's take an indirect approach: assume that it is countable and deduce a contradiction. This method is called an *indirect proof*, a top-tier tool in a mathematician's toolkit.

Since $[0, 1) \subseteq \mathbb{R}$, it is enough to show that $[0, 1)$ is not countable. If it is countable, we can enumerate it:

$$[0, 1) = \{a_1, a_2, ...\}.$$

We can write out the decimal forms of these:

$$a_1 = 0.a_{11}a_{12}a_{13}\ldots,$$
$$a_2 = 0.a_{21}a_{22}a_{23}\ldots,$$
$$a_3 = 0.a_{31}a_{32}a_{33}\ldots.$$
$$\vdots$$

Let's focus on the diagonal! By changing the digits there, we define

$$\hat{a}_{nn} := \begin{cases} 5, & \text{if } a_{nn} \neq 5, \\ 1, & \text{if } a_{nn} = 5. \end{cases}$$

Can the number

$$\hat{a} := 0.\hat{a}_{11}\hat{a}_{22}\hat{a}_{33}\ldots$$

be found in the sequence $\{a_1, a_2, \ldots\}$? No, because the $i$-th decimal of $a_i$ and $\hat{a}$ must be different for all $i \in \mathbb{N}$! We have constructed $\hat{a}$ by changing the i-th decimal of $a_i$.

To summarize, our assumption that $[0, 1)$ can be enumerated leads to a contradiction because we have found an element that cannot possibly be in our enumeration. So, $[0, 1)$ is not countable, hence $\mathbb{R}$ is not countable as well. This is what we needed to show!

The method of proof that you have seen above is called Cantor's diagonal argument. This is a beautiful and powerful idea, and although we won't encounter it anymore, it is the key to proving several difficult theorems. (Like Gödel's famous incompleteness theorems, essentially stating that an axiomatic system is either inexpressive or inconsistent. These threw a huge monkey wrench into the machinery of mathematics at the beginning of the 20th century.)

Notice that real numbers break the pattern we observed previously. Integers were constructed by extending the natural numbers with additive inverses and closing them to addition; rationals were obtained the same way, except doing it for multiplication. As we shall see later, real numbers follow a similar process: we obtain them from rationals by closing them to *limits*.

But what are limits? Let's see – by studying sequences, the objects that provide the context of limits!

# 10.2 Sequences

Sequences lie at the very heart of mathematics. Sequences and their limits describe long-term behavior, like the (occasional) convergence of gradient descent to a local optimum. By definition, a sequence is an enumeration of mathematical objects.

The elements of a sequence can be any mathematical object, like sets, functions, or Hilbert spaces. (Whatever those might be.) For us, sequences are composed of numbers. We formally denote them as

$$\{a_n\}_{n=1}^{\infty}, \quad a_n \in \mathbb{R}.$$

For simplicity, the subscripts and the superscripts are often omitted, so don't panic if you see $\{a_n\}$, as it is just an abbreviation. (Or $a_n$. Mathematicians love abbreviations.) If all elements of the sequence belong to a set $A$, we often write $\{a_n\} \subseteq A$.

Sequences can be bidirectional as well. Those are denoted as $\{a_n\}_{n=-\infty}^{\infty}$. We don't need them for now, but they will frequently come up in the context of probability distributions later.

Sometimes we don't need an entire sequence, just a *subsequence*. We will not do anything special with them just yet, but here is the formal definition.

**Definition 10.2.1 (Subsequences)**

Let $a_n{}_{n=1}^{\infty}$, and let $n_k{}_{k=1}^{\infty} \subseteq \mathbb{N}$ be a strictly increasing sequence of natural numbers. Then, the sequence $\{a_{n_k}\}_{k=1}^{\infty}$ is a subsequence of $\{a_n\}$.

Think of it as throwing elements away from a sequence.

## 10.2.1 Convergence

One of the most important aspects of sequences is their asymptotic behavior, or in other words, what they do in the long term. A particular property we often look for is *convergence*. In plain English, the sequence $\{a_n\}$ converges to $a$ if no matter how small of an interval $(a - \varepsilon, a + \varepsilon)$ we define (where $\varepsilon$ can be *really* small), eventually all of the elements of $\{a_n\}$ fall into it.

The following is the mathematically precise definition of convergence.

> **Definition 10.2.2 (Convergence of sequences)**
>
> The sequence $\{a_n\} \subseteq \mathbb{R}$ is said to *converge* to some $a \in \mathbb{R}$ if for every $\varepsilon > 0$, there is a cutoff index $n_0 \in \mathbb{N}$ such that
>
> $$|a_n - a| < \varepsilon$$
>
> holds for all indices $n > n_0$. The value $a$ is said to be the limit of $\{a_n\}$, and we write
>
> $$\lim_{n \to \infty} a_n = a$$
>
> or
>
> $$a_n \to a \quad (n \to \infty).$$

Note that the cutoff index $n_0$ depends on $\varepsilon$. We could write $n_0(\varepsilon)$ to emphasize this dependency, but we rarely do so. To avoid referencing and naming the cutoff index $n_0$ all the time, we often simply say that a given property *"holds for all n large enough."* (Did I mention that mathematicians love abbreviations?)

In plain English, the definition means that no matter how small of an interval you enclose $a$ in, all members of the sequence will eventually fall into it.

Although mathematically extremely precise and correct, this definition doesn't give us a lot of tools to show if a sequence is convergent or not. First, we have to conjure up the limit $a$ and then construct the cutoff indexes. For example, consider $a_n := \frac{1}{n}$.

To make our job easier, we can plot this to visualize the situation.

```python
import numpy as np
import matplotlib.pyplot as plt


with plt.style.context("seaborn-v0_8"):
    plt.figure(figsize=(8, 5))
    plt.scatter(range(1, 21), [1/n for n in range(1, 21)])
```

```
plt.xticks(range(1, 21, 2))
plt.title("the 1/n sequence")
plt.show()
```

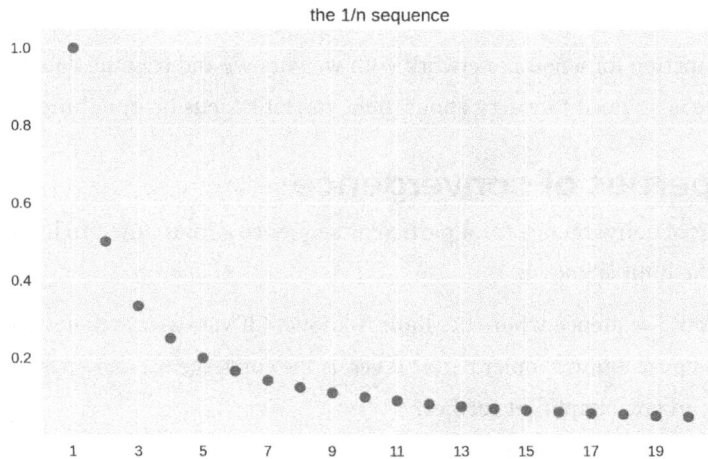

*Figure 10.3: The $1/n$ sequence*

Here, we can explicitly construct the cutoff index $n_0$ for every $\varepsilon$. Since we want to have

$$\frac{1}{n} < \varepsilon,$$

we can reorganize the inequality to obtain

$$\frac{1}{\varepsilon} < n.$$

So,

$$n_0 := \left\lfloor \frac{1}{\varepsilon} \right\rfloor + 1$$

will do the job.

We had it easy in this example, but this is pretty much as far as we can go with the definition. For example, how do you show the convergence of

$$a_n := \left( \frac{1}{n} + \frac{1}{n+1} + \cdots + \frac{1}{2n} \right)^{-1}$$

with the definition only? You don't.

There are more advanced tools for this, as we shall see. (By the way, $\lim_{n\to\infty} a_n = \frac{1}{\ln 2}$). For sequences that are defined recursively and there is no analytic formula available, like

$$\{L(\vec{w}_n, \vec{x}, \vec{y})\}_{n=1}^\infty,$$

where $L$ is the loss function for a neural network with weights $\vec{w}_n$ and training data $(\vec{x}, \vec{y})$, we have even more complications. There is no need to worry about them yet; let's focus on one thing at a time.

## 10.2.2   Properties of convergence

In essence, the study of convergence for a particular sequence comes down to breaking it into simpler and simpler parts until the limit is known.

1. Is this a "famous" sequence where the limit is known? If yes, we are done. If not, go to the next step.
2. Can you decompose it into simpler parts? If yes, is the convergence known for them? If the convergence is unknown, can you simplify it further?

We can do this because convergence has some particularly nice properties, as summarized in the theorem below.

**Theorem 10.2.1** *(Properties of convergence)*

*Let $\{a_n\}$ and $\{b_n\}$ be two convergent sequences with*

$$\lim_{n\to\infty} a_n = a \quad and \quad \lim_{n\to\infty} b_n = b.$$

*Then:*

*(a)*

$$\lim_{n\to\infty}(a_n + b_n) = a + b,$$

*(b)*

$$\lim_{n\to\infty} ca_n = ca \quad for\ all\ c \in \mathbb{R},$$

*(c)*

$$\lim_{n\to\infty} a_n b_n = ab,$$

*(d) and if $a_n \neq 0$ and $a \neq 0$, then*

$$\lim_{n\to\infty} \frac{1}{a_n} = \frac{1}{a}.$$

The properties *(a)* and *(b)* together are called *linearity of convergence*.

As we shall see later, the *continuity* of functions also provides a great tool to study convergence properties of a sequence. In fact, continuity is nothing more than the interchangeability of limits and functions:

$$\lim_{n\to\infty} f(x_n) = f\Big( \lim_{n\to\infty} x_n \Big).$$

One essential property of convergent sequences is that under certain circumstances, they preserve inequalities. This is true of function limits as well, so it is important for us.

**Theorem 10.2.2** *(The transfer principle)*

*Let $\{a_n\}_{n=1}^{\infty}$ be a convergent sequence. If $a_n \geq \alpha$ holds for all $n \in \mathbb{N}$, where $\alpha \in \mathbb{R}$ is some lower bound, then $\lim_{n\to\infty} a_n \geq \alpha$.*

*Proof.* We are going to do this indirectly. If $\lim_{n\to\infty} a_n < \alpha$, then by the definition of convergence, $|a_n - a| < \frac{|a-\alpha|}{2}$ for all large $n$. This means that those $a_n$-s are actually below $\alpha$, contradicting our assumptions.

This proof is straightforward to understand if you draw a figure and visualize what happens, so I encourage you to do so. The identical result is true if we replace $\geq$ with $\leq$ in the above, with the wording of the proof staying the same.

Note that if $a_n > \alpha$ for all $n$, $\lim_{n\to\infty} a_n > \alpha$ is not guaranteed! The best example to show this is $a_n := 1/n$, which converges to 0, although all of its terms are positive.

As a corollary, we obtain a tool that will be very useful for showing the convergence of particular sequences.

**Corollary 10.2.1** *(The squeeze principle)*

*Let $\{a_n\}_{n=1}^{\infty}, \{b_n\}_{n=1}^{\infty},$ and $\{c_n\}_{n=1}^{\infty}$ be three sequences such that $a_n \leq b_n \leq c_n$ for all large enough n. If*

$$\lim_{n\to\infty} a_n = \lim_{n\to\infty} c_n = \alpha,$$

*then*

$$\lim_{n\to\infty} b_n = \alpha.$$

In other words, squeezing $\{b_n\}$ between two convergent sequences that have the same limit implies convergence of $b_n$ to the joint limit.

## 10.2.3 Famous convergent sequences

Because convergence behaves nicely with respect to certain operations (*Section 10.2.2*), we study sequences by decomposing them into building blocks. Let's see the most important ones that will be useful for us later!

**Example 1.** For any $x \geq 0$,

$$\lim_{n \to \infty} x^n = \begin{cases} 0 & \text{if } 0 \leq x < 1, \\ 1 & \text{if } x = 1, \\ \infty & \text{if } x > 1. \end{cases} \tag{10.1}$$

If you think about it for a minute, this is easy to see. The $x = 0$ and $x = 1$ cases are trivial. Regarding the others, because taking the logarithm turns exponentiation into multiplication, we have $\log x^n = n \log x$. So,

$$\lim_{n \to \infty} n \log x = \begin{cases} -\infty & \text{if } 0 < x < 1, \\ \infty & \text{if } x > 1. \end{cases}$$

Since the logarithm is increasing and invertible, (10.1) follows.

**Example 2.** For any $x \geq 0$,

$$\lim_{n \to \infty} x^{1/n} = \begin{cases} 1 & \text{if } x > 0, \\ 0 & \text{if } x = 0. \end{cases} \tag{10.2}$$

Similarly to the previous example, this can be shown with the use of logarithms.

**Example 3.** Let's consider the sequences

$$a_n = \log n, \quad b_n = n, \quad c_n = 2^n, \quad d_n = n!.$$

Can you order them according to the speed of growth? This is quite important throughout computer science, as these could represent time complexities. It's almost folklore that logarithmic time complexity beats linear, which beats exponential, which beats factorial. In other terms,

$$\lim_{n \to \infty} \frac{\log n}{n} = \lim_{n \to \infty} \frac{n}{2^n} = \lim_{n \to \infty} \frac{2^n}{n!} = 0.$$

**Example 4.** We have just learned that $n$ grows faster than $\log n$. From this, it follows that

$$\lim_{n \to \infty} \sqrt[n]{n} = 1,$$

a surprising result (at least, it surprised me as a young student)! To show this, we apply the good old logarithm trick:

$$\begin{aligned} \lim_{n \to \infty} \sqrt[n]{n} &= \lim_{n \to \infty} n^{\frac{1}{n}} \\ &= \lim_{n \to \infty} e^{\log n^{\frac{1}{n}}} \\ &= \lim_{n \to \infty} e^{\frac{\log n}{n}} \\ &= e^0 = 1. \end{aligned}$$

## 10.2.4 The role of convergence in machine learning

Convergence is everywhere. You've just come across this concept for the first time, so you might not realize its importance just yet. However, it is central to mathematics and machine learning.

Just to look ahead and give a few examples, *differentiation* is defined by a limit:

$$f'(x) := \lim_{y \to x} \frac{f(x) - f(y)}{x - y}.$$

Regarding derivatives, *integrals* (the "inverse" of differentiation) are limits of convergent sequences. For instance,

$$\int_0^1 x^2 \, dx = \lim_{n \to \infty} \sum_{k=1}^{n} \frac{k^2}{n^3}.$$

Because integrals are limits, so is every quantity calculated with integration, such as *expected values*, like for the standard normal distribution,

$$\mathbb{E}[\mathcal{N}(0, 1)] = \int_{-\infty}^{\infty} x \frac{1}{\sqrt{2\pi}} e^{-x^2/2} \, dx.$$

Convergence is also central to probability and statistics.

There are two famous theorems: the *law of large numbers*, stating that

$$\lim_{n \to \infty} \frac{1}{n} \sum_{k=1}^{n} X_k = \mu$$

holds; and the *central limit theorem*, which says

$$\sqrt{n}\left(\frac{X_1 + \cdots + X_n}{n} - \mu\right) \to \mathcal{N}(\mu, \sigma^2) \quad (n \to \infty)$$

in distribution, for independent and identically distributed random variables $X_1, X_2, \ldots$ with finite expected value $\mathbb{E}[X_i] = \mu$ and variance $\text{var}(X_i) = \sigma^2$. They are both very important in machine learning and neural networks; for instance, the law of large numbers is one of the fundamental ideas behind stochastic gradient descent.

Even the gradient descent optimization process is a recursively defined sequence of model weights, converging toward an optimum where the model best fits the data.

We will talk about all of these in detail. So, even if you don't understand these right now, don't worry. It'll become clear soon. Before finishing up with sequences, we shall discuss what happens when a sequence is *not* convergent.

## 10.2.5   Divergent sequences

We have talked about how convergent sequences are everywhere, and they are at the core of mathematics and machine learning. However, not all sequences are convergent.

Think about the following example:

$$a_n := \sin(n).$$

When plotted, this is how it looks.

```python
with plt.style.context("seaborn-v0_8"):
    plt.figure(figsize=(8, 5))
    plt.scatter(range(1, 21), [np.sin(n) for n in range(1, 21)])
    plt.xticks(range(1, 21, 2))
    plt.title("the sin(n) sequence")
    plt.show()
```

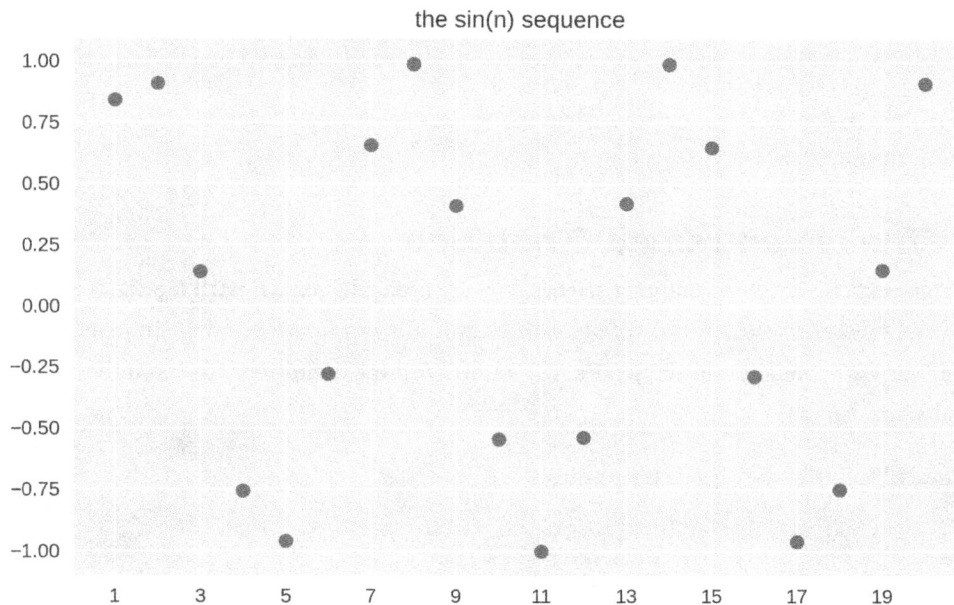

*Figure 10.4: The sin(n) sequence*

Although it is hard to prove, this sequence does not converge. Its value is constantly oscillating in $[-1, 1]$. We call non-convergent sequences *divergent*. Among these, there is a special kind of divergence: approaching infinity.

---

**Definition 10.2.3 ($\infty$-divergence)**

The sequence $\{a_n\}$ is said to be $\infty$-divergent if for every arbitrarily large number $x$, there is a cutoff index $n_0$ such that

$$a_n > x$$

holds for all indices $n > n_0$.

---

We denote $\infty$-divergence by writing $x_n \to \infty$. Analogously, $(-\infty)$-divergent sequences can be defined.

An obvious example is $\{n\}$ or $\{n \log n\}$. These are all across computer science as well: the runtime of algorithms given the number of steps or the size of the input is $\infty$-divergent.

When you see something like $a_n = O(n)$, it means that there is a constant $c$ such that

$$0 \leq a_n \leq cn$$

holds for all large enough $n$.

## 10.2.6 The big and small O notation

If you have some experience with computer science, you are probably familiar with the big O small O notation. There, it is used to express the runtime of algorithms, but it is not limited to that. In general, it is used to compare the long-term behavior of sequences. Let's start with the definitions first, and then I'll explain the intuition and some use cases.

> **Definition 10.2.4 (Big and small O notation)**
>
> Let $\{a_n\}_{n=1}^{\infty}$ and $\{b_n\}_{n=1}^{\infty}$ be two arbitrary sequences. We say that:
>
> (a) $b_n = O(a_n)$, if there is a constant $C > 0$ such that $|b_n| \leq Ca_n$ for all sufficiently large $n$
>
> (b) $b_n = o(a_n)$, if for every $\varepsilon > 0$, there exists a cutoff index $N \in \mathbb{N}$ such that $|b_n| \leq \varepsilon a_n$ for every $n > N$.

In plain English, "$b_n$ is big O of $a_n$" means that $b_n$ grows roughly at the same rate as $a_n$, while "$b_n$ is small O of $a_n$" means that $b_n$ is an order of magnitude smaller than $a_n$.

So, when we say that the runtime of an algorithm is $O(n)$ steps where $n$ is the input size, we mean that the algorithm will finish in $Cn$ steps. Often, we don't care about the constant multiplier since it doesn't mean an order of magnitude difference in the long run.

## 10.2.7 Real numbers are sequences

Now that we have familiarized ourselves with the concept of convergent sequences, we shall take another look at rational and real numbers. When extending the classes of numbers going from $\mathbb{N}$ to $\mathbb{R}$, we pick an operation, close the set with respect to it, and add inverse elements to that operation.

Extending $\mathbb{N}$ with additive inverses $-n$ for all $n \in \mathbb{N}$ yields $\mathbb{Z}$. Extending $\mathbb{Z}$ with multiplicative inverses $1/n$ for all $n$ and closing it for multiplication yields $\mathbb{Q}$. The pattern is seemingly different in the case of $\mathbb{R}$, but this is not the case. After understanding what convergence is, we have the tools to see why.

Consider the following sequence:

$$a_n := \left(1 + \frac{1}{n}\right)^n, \quad n = 1, 2, \ldots.$$

Since rational numbers are closed to addition and multiplication, we see that $a_n$ is rational. However,

$$\lim_{n \to \infty} \left( 1 + \frac{1}{n} \right)^n = e,$$

which is the famous Euler constant, is *not* rational.

Thus, we have found what is missing: $\mathbb{Q}$ is not closed to taking limits. So, we can obtain the set of real numbers by closing $\mathbb{Q}$ to taking limits.

The fact that every irrational number can be approximated with rational numbers as close as possible is often under-appreciated. Think about this: can you represent all real numbers with a computer? Nope. This follows from a simple cardinality argument: the number of possible floats is finite, but there are uncountably many real numbers. However, certain numbers (like $\pi$ or $e$) are essential in engineering calculations and simulations. Without approximations, working with irrational numbers would be unfeasible.

Speaking of $e$: how would you approximate it for computational purposes? In theory, it is enough to take a large enough $n$ and use the value $(1 + 1/n)^n$. In practice, there are several potential problems: the convergence might be slow, and taking a large power of a quantity so close to 1 can be numerically unstable.

However, there is a solution: the form

$$e = \lim_{N \to \infty} \sum_{n=0}^{N} \frac{1}{n!}$$

solves both of these problems! From a numerical point of view, addition is much better than multiplication. Moreover, as $n!$ grows extremely rapidly, the term $\frac{1}{n!}$ becomes negligible even for small $n$-s. Thus, the convergence is fast. Check this out.

```python
from math import factorial

x = range(1, 21)

e_def = [(1 + 1/n)**n for n in x]
e_sum = [np.sum([1/factorial(k) for k in range(n)]) for n in x]
with plt.style.context("seaborn-v0_8"):
    plt.figure(figsize=(8, 5))
    plt.scatter(x, e_def, label="(1 + 1/n) ** n")
    plt.scatter(x, e_sum, label="sum approximation")
    plt.xticks(range(1, 21, 2))
    plt.title("Approximating the value of e")
    plt.legend()
    plt.show()
```

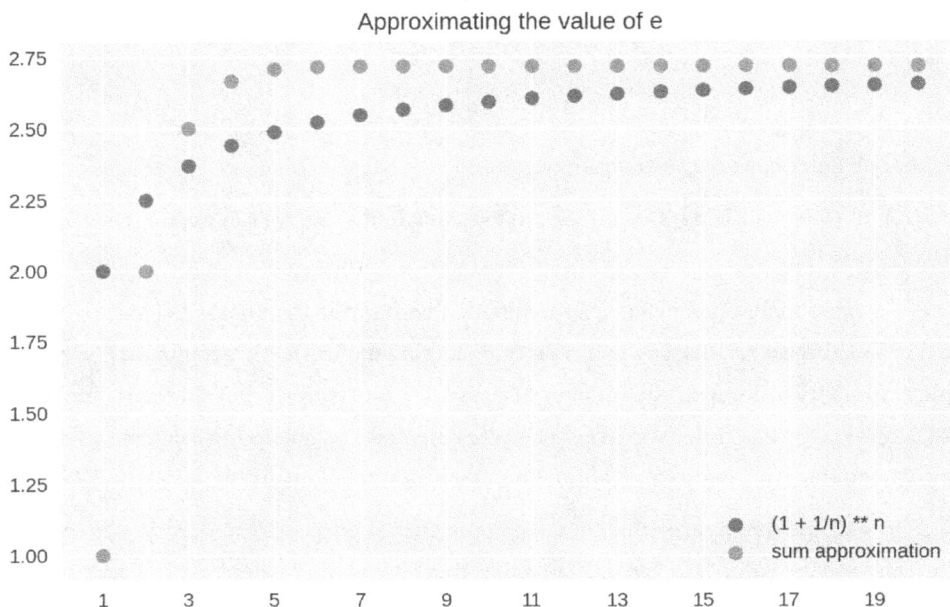

Figure 10.5: *Approximating e: Definition vs. series*

Expressions of form $\sum_{n=0}^{\infty} a_n$ are called *series*, and they are one of the most important mathematical objects, both in theory and practice. Let's see what they are and how can we work with them!

# 10.3 Series

There is a great pizza place in my hometown that I used to visit quite a lot. There, each pizza is packed in a box that contains a pizza coupon. Ten pizza coupons can be exchanged for a free pizza. This begs the question: how much pizza do you get with a single purchase?

Immediately, you'll receive a pizza. You'll also get 1/10-th of a pizza in the form of a coupon, making the "value" of the purchase at least $1 + \frac{1}{10}$ pizzas.

However, upon exchanging ten coupons, you get another one. Thus, a coupon represents $\frac{1}{10} + \frac{1}{100}$ pizzas. Continuing with this logic, we'll obtain that the true value of each purchase is

$$\sum_{n=0}^{\infty} \frac{1}{10^n} = 1 + \frac{1}{10} + \frac{1}{100} + \dots.$$

What is the value of this number? To find out, we'll take a look at *infinite series*, a main pillar of mathematics. Infinite series (or *series* for short) are sums of the form

$$\sum_{n=1}^{\infty} a_n.$$

Adding infinitely many terms together might seem like a trivial matter, but I assure you, this is far from the truth. For instance, consider the sum

$$\sum_{n=0}^{\infty} (-1)^n = 1 - 1 + 1 - 1 + \dots.$$

On one hand,

$$\sum_{n=0}^{\infty} (-1)^n = (1 - 1) + (1 - 1) + \dots$$
$$= 0 + 0 + \dots$$
$$= 0,$$

but on the other,

$$\sum_{n=0}^{\infty} (-1)^n = 1 + (-1 + 1) + (-1 + 1) + \dots$$
$$= 1 + 0 + 0 + \dots$$
$$= 1.$$

Which one is it? Zero or one? It is neither. We'll see why in this section.

# 10.3.1  Convergent and divergent series

The natural way to make sense of the infinite series

$$\sum_{n=1}^{\infty} a_n$$

is by taking the limit of the so-called partial sums

$$S_N = \sum_{n=1}^{N} a_n.$$

This is formalized by the following definition.

---

**Definition 10.3.1  (Convergent and divergent series)**

Let $\{a_n\}_{n=1}^{\infty}$ be an arbitrary real sequence. The infinite series $\sum_{n=1}^{\infty} a_n$ is defined by

$$\sum_{n=1}^{\infty} a_n := \lim_{N\to\infty} \sum_{n=1}^{N} a_n.$$

If the above limit exists, we say that $\sum_{n=1}^{\infty} a_n$ is *convergent*. Otherwise, it is *divergent*.

---

Sounds simple enough. Let's see some examples!

**Example 1.** The geometric series, given by

$$\sum_{n=0}^{\infty} q^n = \frac{1}{1-q}, \quad q \in (-1, 1).$$

As $S_N + q^{N+1} = S_{N+1} = 1 + qS_N$, it follows that $S_N = \frac{1-q^{N+1}}{1-q}$. Thus,

$$\sum_{n=0}^{\infty} q^n = \lim_{N\to\infty} \frac{1-q^{N+1}}{1-q} = \begin{cases} \text{undefined} & \text{if } q \leq -1, \\ \frac{1}{1-q} & \text{if } -1 < q < 1, \\ \infty & \text{if } 1 \leq q. \end{cases}$$

This is where the famous formula $\sum_{n=1}^{\infty} \frac{1}{2^n} = 1$ comes from. (*Figure 10.6* illustrates this fact as well.)

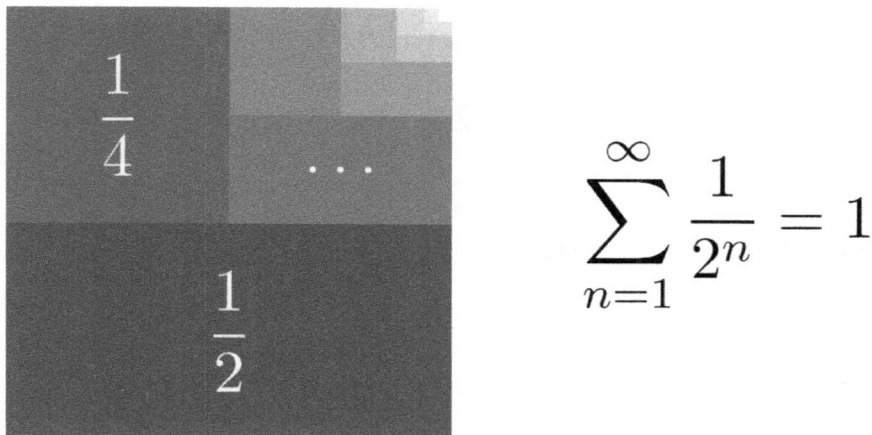

$$\sum_{n=1}^{\infty} \frac{1}{2^n} = 1$$

*Figure 10.6: The visual proof of the convergence of a geometric series for $q = 1/2$*

The geometric series is also the one that appears in our introductory pizza coupon example. Now, we can see that the value of a single purchase is

$$\sum_{n=0}^{\infty} \frac{1}{10^n} = \frac{1}{1 - \frac{1}{10}} = \frac{10}{9}$$

pizzas.

**Example 2.** The harmonic series, given by

$$\sum_{n=1}^{\infty} \frac{1}{n} = \infty.$$

Why is the harmonic series divergent? To see why, first notice that

$$\sum_{n=2^k}^{2^{k+1}-1} \frac{1}{n} \geq \sum_{n=2^k}^{2^{k+1}-1} \frac{1}{2^{k+1}}$$
$$= \frac{2^k}{2^{k+1}}$$
$$= \frac{1}{2}.$$

Thus, by grouping the terms appropriately, we obtain that

$$\sum_{n=1}^{\infty} \frac{1}{n} = \sum_{k=0}^{\infty} \left( \sum_{n=2^k}^{2^{k+1}-1} \frac{1}{n} \right)$$

$$\geq \sum_{k=0}^{\infty} \frac{1}{2}$$

$$= \infty.$$

If you are having trouble imagining this, here is the plot.

```
xs = range(1, 41)
an = [1/n for n in xs]
ys = np.cumsum(an)

with plt.style.context("seaborn-v0_8"):
    plt.figure(figsize=(8, 5))
    plt.scatter(xs, ys)
    plt.xticks(range(1, 41, 5))
    plt.title("the harmonic series")
    plt.show()
```

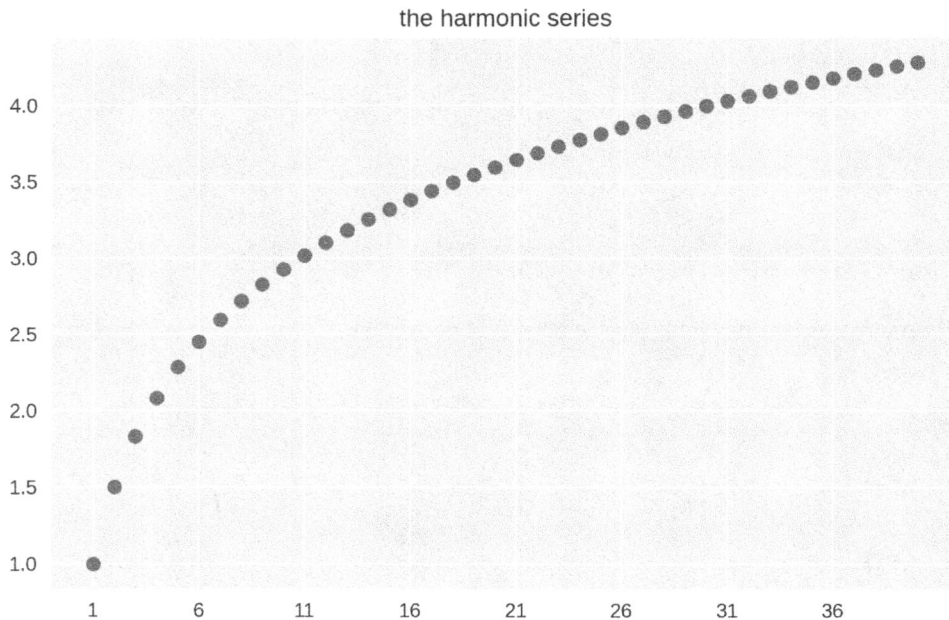

*Figure 10.7: The harmonic series*

According to the Euler-Maclaurin formula,

$$\sum_{n=1}^{N} \frac{1}{n} \approx \log N + \gamma,$$

where $\gamma \approx 0.5772156649\ldots$ is the famous Euler–Mascheroni constant. Check it out. (When the base is omitted, log denotes the natural logarithm, also often denoted by ln.)

```python
with plt.style.context("seaborn-v0_8"):
    plt.figure(figsize=(8, 5))
    plt.plot(xs, np.log(xs) + np.euler_gamma, c="r", linewidth=5, zorder=1,
    label="log(x + y)")
    plt.scatter(xs[::2], ys[::2], label="harmonic series")
    plt.xticks(range(1, 41, 5))
    plt.title("the harmonic series")
    plt.legend()
    plt.show()
```

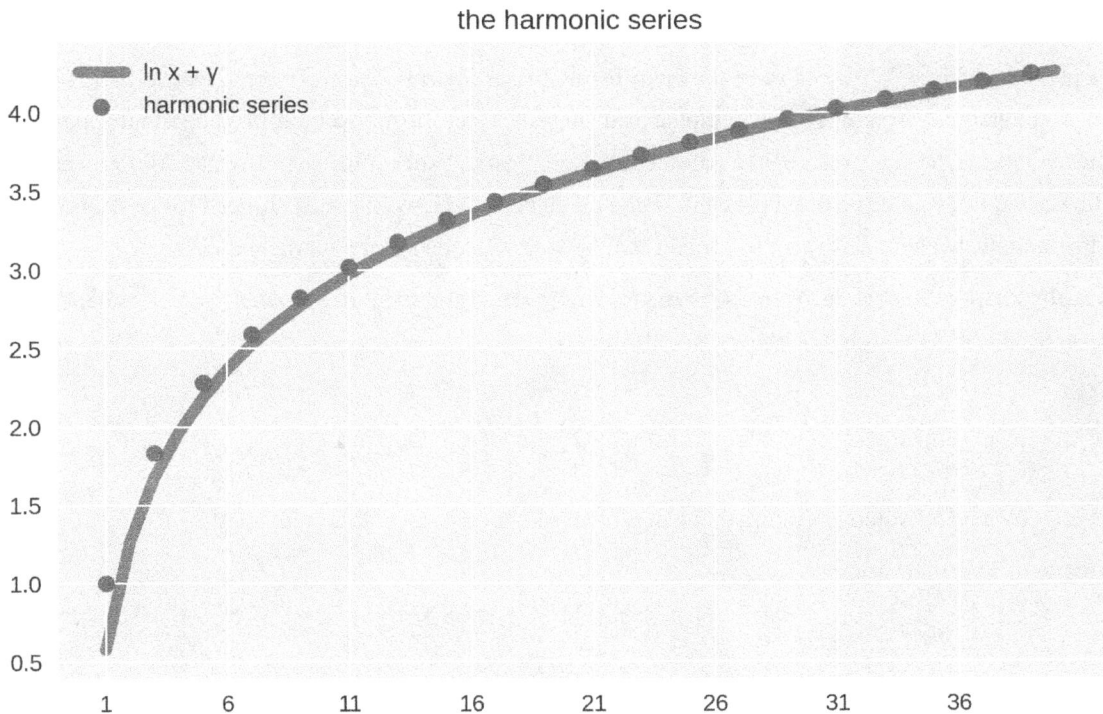

*Figure 10.8: The harmonic series and the $\log(x + \gamma)$ function*

**Example 3.** The alternating harmonic series, given by

$$\sum_{n=1}^{\infty} (-1)^{n+1} \frac{1}{n} = \log 2.$$

Surprisingly, the alternating harmonic series is convergent, with the sum $\log 2$.

**Example 4.** The Basel problem. Believe it or not, solving polynomial equations and evaluating infinite sums were some kind of sport for mathematicians of previous eras. One of the most famous ones was the Basel problem, concerning the infinite sum of inverse squares. In 1735, the legendary Euler showed that

$$\sum_{n=1}^{\infty} \frac{1}{n^2} = \frac{\pi^2}{6},$$

which is a stunning result. What on Earth is $\pi$ doing there? The constant $\pi$ is defined as the half circumference of a circle of radius 1, and seeing it pop up in the infinite sum of inverse squares is puzzling to say the least. (There is an explanation: the numbers $1/n^2$ are the (scaled) Fourier coefficients of the function $f(x) = x$, and the sum of the Fourier coefficients always evaluates to the integral of the function over the interval $[-\pi, \pi]$. However, this method is beyond the scope of this book.)

On a personal note, $\sum_{n=1}^{\infty} \frac{1}{n^2} = \frac{\pi^2}{6}$ was the favorite identity of my calculus professor, a teacher who I loved so much that I ended up specializing in mathematical analysis. Once, he hit his head in an accident. Upon arrival of the ambulance, he was asked by the paramedic to talk, just to gauge if his cognitive capabilities were intact. So, he started to explain the Basel problem to the medical staff. He was almost shipped to the trauma center, but fortunately, his wife was there to explain that he is normal; he's just a mathematician.

**Example 5.** Apéry's constant. We've seen the harmonic series and the Basel problem, so what about the sum of inverse cubes? Although it is known that the series

$$\sum_{n=1}^{\infty} \frac{1}{n^3}$$

is convergent and its value is irrational, we don't have a closed-form solution just yet!

**Example 6.** The alternating series

$$\sum_{n=0}^{\infty} (-1)^n.$$

This is the first example of a divergent series: as

$$\sum_{n=0}^{N}(-1)^n = \begin{cases} 1 & \text{if } n \text{ is even} \\ 0 & \text{if } n \text{ is odd,} \end{cases}$$

the limit $\lim_{N\to\infty}\sum_{n=0}^{N}(-1)^n$ doesn't exist. This is the resolution of the seemingly paradoxical result from the introduction: as $\sum_{n=0}^{\infty}(-1)^n$ is divergent, associativity breaks down.

**Example 7.** The famous Euler number, defined by $e = \lim_{n\to\infty}\left(1 + \frac{1}{n}\right)^n$, is also given by the infinite series $e = \sum_{n=0}^{\infty}\frac{1}{n!}$.

Currently, this is beyond our scope, but later, we'll see why this is true. (Spoiler alert: This is a so-called Taylor series, which we'll talk about in *Chapter 12*.)

This leads to the following question: is there a method that tells if a series is convergent or not? Finding a closed formula for the partial sums is not always possible. I urge you to try finding one for $\sum_{n=0}^{N}1/n!$, if you don't believe me.

## 10.3.2 Properties of series

Let's take a look at some of the most important properties of series. First, it is immediately clear that the general term of a convergent series should converge to 0 as well.

**Proposition 10.3.1** *Let $\sum n = 1^{\infty}a_n$ be a convergent series. Then $\lim_{N\to\infty}a_N = 0$.*

*Proof.* As $a_N = \sum_{n=1}^{N}a_n - \sum_{n=1}^{N-1}a_n$, we have

$$\lim_{N\to\infty}a_N = \lim_{N\to\infty}\left(\sum_{n=1}^{N}a_n - \sum_{n=1}^{N-1}a_n\right)$$

$$= \lim_{N\to\infty}\sum_{n=1}^{N}a_n - \lim_{N\to\infty}\sum_{n=1}^{N-1}a_n$$

$$= \sum_{n=1}^{\infty}a_n - \sum_{n=1}^{\infty}a_n$$

$$= 0,$$

which is what we had to show.

*Proposition 10.3.1* can be used to quickly gauge the convergence of certain series. For instance, $\sum_{n=1}^{\infty} \frac{n+1}{n}$ cannot be convergent, as $\lim_{n\to\infty} \frac{n+1}{n} = 1$.

On the other hand, **note that the reverse of *Proposition 10.3.1* is not true**: there are divergent series with a general term converging to 0. One immediate example is the harmonic series $\sum_{n=1}^{\infty} \frac{1}{n}$.

Just like sequences, convergent series behave with respect to addition and scalar multiplication, as we expect.

> **Theorem 10.3.1** *(Linearity of convergent series)*
>
> *Let $\sum n = 1^{\infty} a_n$ and $\sum n = 1^{\infty} b_n$ be two convergent series, and let $\alpha, \beta \in \mathbb{R}$ be two arbitrary real numbers. Then,*
>
> $$\alpha \sum_{n=1}^{\infty} a_n + \beta \sum_{n=1}^{\infty} b_n = \sum_{n=1}^{\infty} (\alpha a_n + \beta b_n).$$

The proof is a direct consequence of *Theorem 10.2.1*. Feel free to work it out by hand for practice!

Note that *Theorem 10.3.1* does not apply to divergent series. For instance,

$$\sum_{n=0}^{\infty} \left[ (-1)^n + (-1)^{n+1} \right] = 0,$$

However,

$$\sum_{n=0}^{\infty} (-1)^n \quad \text{and} \quad \sum_{n=0}^{\infty} (-1)^{n+1}$$

are both divergent.

The product of a series is slightly more *convoluted*. We'll deal with this later in the chapter, I promise.

## 10.3.3   Conditional and absolute convergence

Let's turn the weirdness up a notch. Recall the alternating harmonic series $\sum_{n=1}^{\infty} (-1)^{n+1} \frac{1}{n}$, whose sum is log 2. What happens if we rearrange its terms?

Instead of alternating between one odd and one even term, let's do one odd and two even. (Odd and even with respect to their indices.)

This is illustrated by *Figure 10.9*.

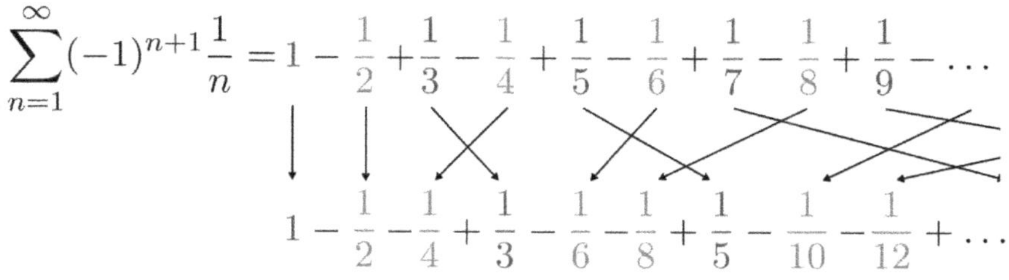

$$\sum_{n=1}^{\infty}(-1)^{n+1}\frac{1}{n} = 1 - \frac{1}{2} + \frac{1}{3} - \frac{1}{4} + \frac{1}{5} - \frac{1}{6} + \frac{1}{7} - \frac{1}{8} + \frac{1}{9} - \cdots$$

$$1 - \frac{1}{2} - \frac{1}{4} + \frac{1}{3} - \frac{1}{6} - \frac{1}{8} + \frac{1}{5} - \frac{1}{10} - \frac{1}{12} + \cdots$$

*Figure 10.9: A rearrangement of the alternating harmonic series*

The change is small, but carries a profound impact. As it turns out, by simply rearranging the terms, we change the value of the sum! Check out *Figure 10.10* to see why.

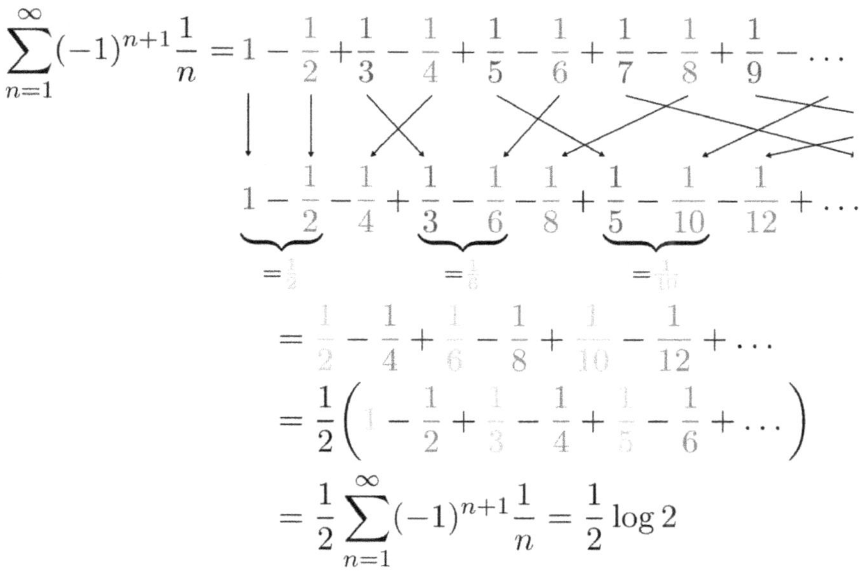

$$\sum_{n=1}^{\infty}(-1)^{n+1}\frac{1}{n} = 1 - \frac{1}{2} + \frac{1}{3} - \frac{1}{4} + \frac{1}{5} - \frac{1}{6} + \frac{1}{7} - \frac{1}{8} + \frac{1}{9} - \cdots$$

$$1 - \underbrace{\frac{1}{2} - \frac{1}{4}}_{=\frac{1}{2}} + \underbrace{\frac{1}{3} - \frac{1}{6}}_{=\frac{1}{6}} - \frac{1}{8} + \underbrace{\frac{1}{5} - \frac{1}{10}}_{=\frac{1}{10}} - \frac{1}{12} + \cdots$$

$$= \frac{1}{2} - \frac{1}{4} + \frac{1}{6} - \frac{1}{8} + \frac{1}{10} - \frac{1}{12} + \cdots$$

$$= \frac{1}{2}\left(1 - \frac{1}{2} + \frac{1}{3} - \frac{1}{4} + \frac{1}{5} - \frac{1}{6} + \cdots\right)$$

$$= \frac{1}{2}\sum_{n=1}^{\infty}(-1)^{n+1}\frac{1}{n} = \frac{1}{2}\log 2$$

*Figure 10.10: A rearrangement of the alternating harmonic series, explained*

This happens because rearrangements are not valid operations in the world of series. However, the notion of convergence can be refined in a way that'll enable us to differentiate between rearrangeable and non-rearrangeable series.

Enter *absolute* and *conditional* convergence.

> **Definition 10.3.2 (Absolute and conditional convergence)**
>
> Let $\sum_{n=1}^{\infty} a_n$ be an infinite series.
>
> (a) If $\sum_{n=1}^{\infty} |a_n|$ is convergent, then $\sum_{n=1}^{\infty} a_n$ is called *absolutely convergent*.
>
> (b) If $\sum_{n=1}^{\infty} a_n$ is convergent but not absolutely convergent, then it is called *conditionally convergent*.

For example, the geometric series $\sum_{n=0}^{\infty} \left(-\frac{1}{2}\right)^n$ is absolutely convergent, while the alternating harmonic series $\sum_{n=1}^{\infty} (-1)^n \frac{1}{n}$ is only conditionally convergent.

If a series is absolutely convergent, then it is convergent as well. (We'll skip the proof, as the minute technical details are not that important for us.)

## 10.3.4   Revisiting rearrangements

Being absolutely or conditionally convergent has a profound impact on the behavior of the series. One prime example is its rearrangeability. As it turns out, absolutely convergent series can be rearranged, while conditionally convergent series go crazy.

Mathematically speaking, rearrangements can be formalized via permutations of the index set.

> **Definition 10.3.3 (Permutations)**
>
> Let $A$ be an arbitrary set. The mapping $\sigma : A \to A$ is called a *permutation* of $A$ if it is bijective.

It seems quite abstract, but permutations are easy to grasp. Think about the index set as the increasing sequence

$$1, 2, 3, 4, 5, 6, \dots.$$

(It may include 0, or might even start from a larger number.) The simple permutation

$$\sigma(n) = \begin{cases} 2k - 1 & \text{if } n = 2k, \\ 2k & \text{if } n = 2k - 1 \end{cases}$$

swaps the neighboring even and odd numbers, turning the index set into

$$\sigma(1) = 2, \sigma(2) = 1, \sigma(3) = 4, \sigma(4) = 3, \dots.$$

For a general series $\sum_{n=1}^{\infty} a_n$, its rearrangement given by the permutation $\sigma$ is $\sum_{n=1}^{\infty} a_{\sigma(n)}$.

So, can you rearrange a convergent series? If it is absolutely convergent, without a doubt.

---

**Theorem 10.3.2** *Let $\sum_{n=1}^{\infty} a_n$ be an absolutely convergent series and $\sigma : \mathbb{N} \to \mathbb{N}$ be an arbitrary permutation. Then, the rearrangement $\sum_{n=1}^{\infty} a_{\sigma(n)}$ is convergent as well and*

$$\sum_{n=1}^{\infty} a_{\sigma(n)} = \sum_{n=1}^{\infty} a_n.$$

---

We are not going to prove this here, but the essence is, convergent series absolutely behave as we expect: we can change the order of terms without affecting the result.

What about conditionally convergent series? We have seen that rearranging can change the value, but the situation is much more interesting. Meet the Riemann rearrangement theorem.

---

**Theorem 10.3.3** *(The Riemann rearrangement theorem)*

*Let $\sum_{n=1}^{\infty} a_n$ be a conditionally convergent series and let $c \in \mathbb{R}$ be an arbitrary real number. Then, there exists a permutation $\sigma : \mathbb{N} \to \mathbb{N}$ such that*

$$\sum_{n=1}^{\infty} a_{\sigma(n)} = c.$$

---

This is quite wild. The Riemann rearrangement theorem states that if you give me an arbitrary real number, I can conjure a rearrangement that'll change the value to the number you gave me.

We won't prove this either, but here is an intuitive explanation. Suppose that you want to rearrange the series to change its sum to 10. You start putting the positive terms in descending order to the front, and when the partial sum overshoots 10, you continue with the negative terms. When the partial sum undershoots 10, you turn to positive values once more, and you keep repeating this until infinity. The property of conditional convergence guarantees that this method works.

Now that we are familiar with the subtle differences between absolute and conditional convergence, let's turn to one of the burning questions we should have asked already: how do we know if a series is convergent or not?

## 10.3.5 Convergence tests for series

Showing if a series is convergent or not is a hard task. Finding a closed form for the partial sums is often impossible, even in the seemingly simplest cases, such as $\sum_{n=1}^{\infty} \frac{1}{n^2}$.

What to do, then? The simplest way is to *compare* the series with another one that we are familiar with.

---

**Theorem 10.3.4** *(The direct comparison test)*

Let $\sum_{n=1}^{\infty} a_n$ be an arbitrary series.

(a) If $\sum_{n=1}^{\infty} b_n$ is absolutely convergent and $|a_n| \leq |b_n|$ for all $n$ after a certain cutoff, $\sum_{n=1}^{\infty} a_n$ is absolutely convergent.

(a) If $\sum_{n=1}^{\infty} b_n = \infty$ and $|b_n| \leq |a_n|$ for all $n$ after a certain cutoff, then $\sum_{n=1}^{\infty} a_n = \infty$.

---

Here is a simple use case for *Theorem 10.3.4*: the series

$$\sum_{n=1}^{\infty} \frac{1}{n^{\alpha}}.$$

For $\alpha > 2$, since $\frac{1}{n^{\alpha}} < \frac{1}{n^2}$, the series

$$\sum_{n=1}^{\infty} \frac{1}{n^{\alpha}}$$

is *convergent* via comparison to

$$\sum_{n=1}^{\infty} \frac{1}{n^2}.$$

On the other hand, for $\alpha < 1$, as $\frac{1}{n^{\alpha}} > \frac{1}{n}$, $\sum_{n=1}^{\infty} \frac{1}{n^{\alpha}}$ is *divergent* via comparison to the harmonic series.

Although the direct comparison test is powerful, it has a significant downside: you have to conjure up a series for comparison. This is not always simple. Thus, we need other tests to show convergence. We'll talk about two of them: the root test and the alternating series test.

---

**Theorem 10.3.5** *(The root test)*

Let $\sum_{n=1}^{\infty} a_n$ be an arbitrary series. If there exists a positive integer $N$ such that

$$\sqrt[n]{|a_n|} < 1$$

for all $n \geq N$, then $\sum_{n=1}^{\infty} a_n$ is absolutely convergent.

---

The most important use case for the root test is showing the convergence of $\sum_{n=0}^{\infty} \frac{1}{n!}$ (which sums up to the Euler constant $e$, but we have no way to evaluate the sum yet). We'll have one soon, coming from the most unexpected place. But I don't want to spoil the fun yet.

About that sum. As $\lim_{n \to \infty} \sqrt[n]{n!} = \infty$ (which, trust me, it really is), $\lim_{n \to \infty} \frac{1}{\sqrt[n]{n!}} = 0$. Thus, the conditions of *Theorem 10.3.5* are satisfied, hence $\sum_{n=1}^{\infty} \frac{1}{n!}$ is convergent.

The disadvantage of the root test is that studying the behavior of $\sqrt[n]{a_n}$ can be surprisingly hard. We have to be true calculus ninjas to handle all the bounds and limits the root test will throw at us. So, we'll take a look at one more convergence test.

---

**Theorem 10.3.6** *(The alternating test, a.k.a. the Leibniz criterion)*

*Let $\{a_n\}_{n=0}^{\infty}$ be a sequence whose terms are either all nonnegative or all negative. If*

*(a) $|a_n|$ decreases monotonically*

*(b) and $\lim_{n\to\infty} a_n = 0$,*

*then $\sum_{n=0}^{\infty}(-1)^n a_n$ is conditionally convergent.*

---

It's that simple. The alternating test is the easiest to apply, but in turn, it only implies conditional convergence. With this, we can finally see that the alternating harmonic series $\sum_{n=1}^{\infty}(-1)^{n+1}\frac{1}{n}$ is conditionally convergent.

## 10.3.6 The Cauchy product of series

To close this chapter, let's talk about the *product* of two series. We've seen that the linear combination of convergent series behaves nicely, but what about the product? This is slightly more complicated than addition. Here is the exact result.

---

**Theorem 10.3.7** *(Mertens' theorem)*

*Let $\sum_{n=0}^{\infty} a_n$ and $\sum_{n=0}^{\infty} b_n$ be two convergent series, and assume that at least one of them converges absolutely. Then, the series*

$$\sum_{n=0}^{\infty}\left(\sum_{k=0}^{n} a_k b_{n-k}\right) \tag{10.3}$$

*is convergent and*

$$\left(\sum_{k=0}^{\infty} a_k\right)\left(\sum_{n=0}^{\infty} b_n\right) = \sum_{n=0}^{\infty}\left(\sum_{k=0}^{n} a_k b_{n-k}\right).$$

---

The series (10.3) is called the *Cauchy product* of $\sum_{n=0}^{\infty} a_n$ and $\sum_{n=0}^{\infty} b_n$. Note that in the sum $\sum_{k=0}^{n} a_k b_{n-k}$, the indices of each term $a_k b_{n-k}$ sum up to $n$.

Instead of an exact proof, here is an intuitive explanation. Let's unpack what is going on when we take the product of sums.

As the product is calculated term by term, we have

$$\left(\sum_{k=0}^{\infty} a_k\right)\left(\sum_{n=0}^{\infty} b_n\right) = \sum_{k=0}^{\infty}\left(a_k \sum_{n=0}^{\infty} b_n\right)$$

$$= a_0(b_0 + b_1 + b_2 + \dots)$$
$$+ a_1(b_0 + b_1 + b_2 + \dots)$$
$$+ a_2(b_0 + b_1 + b_2 + \dots)$$
$$+ \dots$$
$$= a_0 b_0 + a_0 b_1 + a_0 b_2 + \dots$$
$$+ a_1 b_0 + a_1 b_1 + a_1 b_2 + \dots$$
$$+ a_2 b_0 + a_2 b_1 + a_2 b_2 + \dots$$
$$+ \dots.$$

After spelling out the product term by term, the terms $a_k b_l$ are arranged in a table. Upon taking the product of $\sum_{n=0}^{\infty} a_n$ and $\sum_{n=0}^{\infty} b_n$, we simply take the terms in this table and sum them up, row by row.

However, we can sum the terms diagonally. By taking a look at the table

$$a_0 b_0 + a_0 b_1 + a_0 b_2 + \dots$$
$$a_1 b_0 + a_1 b_1 + a_1 b_2 + \dots$$
$$a_2 b_2 + a_2 b_1 + a_2 b_2 + \dots$$

you can notice that the diagonals sum up to $\sum_{k=0}^{n} a_k b_{n-k}$. Thus, by taking all of them into account, we obtain

$$\left(\sum_{k=0}^{\infty} a_k\right)\left(\sum_{n=0}^{\infty} b_n\right) = \sum_{n=0}^{\infty}\left(\sum_{k=0}^{n} a_k b_{n-k}\right).$$

Of course, this is not an exact proof, as we've used rearrangements without showing absolute convergence.

## 10.4  Summary

Until this chapter, our mathematical study was quite close to machine learning. Vectors, matrices, functions: they are all there at the ground zero of theory and practice.

This time, however, we've gone far below the surface. We rarely work directly with sequences in practice, but despite appearances, they are all over the place, providing a solid theoretical foundation for everything that is quantitative.

So, what did we learn? What numbers are, for one. Going from natural numbers to real numbers is nothing short of a revelation, allowing us to see the evolution of the concept of a *number*. But deep down, sequences hold the concept of numbers together. And whenever we talk about sequences, limits and convergence enter the picture.

To summarize, gradient descent is about the limit of the sequence

$$x_{n+1} = x_n - hf'(x_n),$$

converging to a local minima of $f$ if the stars are aligned. In the following chapters, our main goal is to understand $x_{n+1} = x_n - hf'(x_n)$. What is $f'(x)$, and why does $x_n$ converge to a local minimum? To see the full picture, we need to study differentiation and integration, or in other words, *calculus*.

However, to get there, there's one more step to take. It's time to move beyond sequences and study the concept of limits in the context of functions!

## 10.5 Problems

**Problem 1.** Which sequences are convergent, $\infty$-divergent, or neither?

(a) $a_n = 2^{-n^2}$

(b) $b_n = \sqrt{2^n}$

(c) $c_n = \frac{n}{n+1}$

(d) $d_n = \sin(1/n)$

(e) $e_n = e^{n(-1)^n}$

**Problem 2.** Calculate the following limits.

(a) $\lim_{n \to \infty} \frac{5n^2+2}{3n^2-12}$

(b) $\lim_{n \to \infty} \ln \left( \frac{5n^2+2}{3n^2-12} \right)$

(c) $\lim_{n \to \infty} \frac{n}{\sqrt{1+2n^2}}$

(d) $\lim_{n \to \infty} \frac{1}{n} \sin(n)$

(e) $\lim_{n \to \infty} \left( 1 + \frac{2}{n} \right)^n$

**Problem 3.** Which series are absolute convergent, conditionally convergent, or divergent? (Use the direct comparison test, the root test, and the alternating series test.)

(a) $\sum_{n=0}^{\infty} e^{-n^2}$

(b) $\sum_{n=0}^{\infty} (-1)^n \sin\left(\frac{1}{n}\right)$

(c) $\sum_{n=0}^{\infty} \frac{n}{2^n}$

(d) $\sum_{n=1}^{\infty} \frac{n+1}{n^2}$

(e) $\sum_{n=0}^{\infty} \frac{1}{2^n+n}$

# Join our community on Discord

Read this book alongside other users, Machine Learning experts, and the author himself.

Ask questions, provide solutions to other readers, chat with the author via Ask Me Anything sessions, and much more.

Scan the QR code or visit the link to join the community.

`https://packt.link/math`

# 11

# Topology, Limits, and Continuity

In the previous chapter, we learned all (that's relevant to us) about numbers, sequences, and series. These are the foundational objects of calculus: numbers define sequences, sequences define limits, and limits define almost every quantity that interests us. However, there's a snag. Let's look ahead and take a look at the definition of the derivative:

$$f'(y) = \lim_{x \to y} \frac{f(x) - f(y)}{x - y}$$

If you're feeling a sense of déjà vu, don't be surprised. We looked at this exact formula in the introduction of the previous chapter as well, and we are much closer to understanding it. We have learned about limits, but there seems to be an issue: limits were defined in terms of sequences, and the expression $\lim_{x \to y}$ whatever$(x)$ does not seem to be it.

What is it, then? This is what we'll learn in this chapter, starting with the topology of real numbers. Let's go!

## 11.1   Topology

According to the Cambridge English Dictionary, the word "topology" means

> *the way the parts of something are organized or connected.*

From a mathematical perspective, topology studies the *local* properties of structures and spaces.

In machine learning, we are often interested in *global* properties like minima and maxima but only have *local* tools to search for them. One example is the derivative of functions. Derivatives describe the slope of the tangent plane, and as *Figure 11.1* illustrates, this doesn't change if the function is modified away from the point where the derivative is taken.

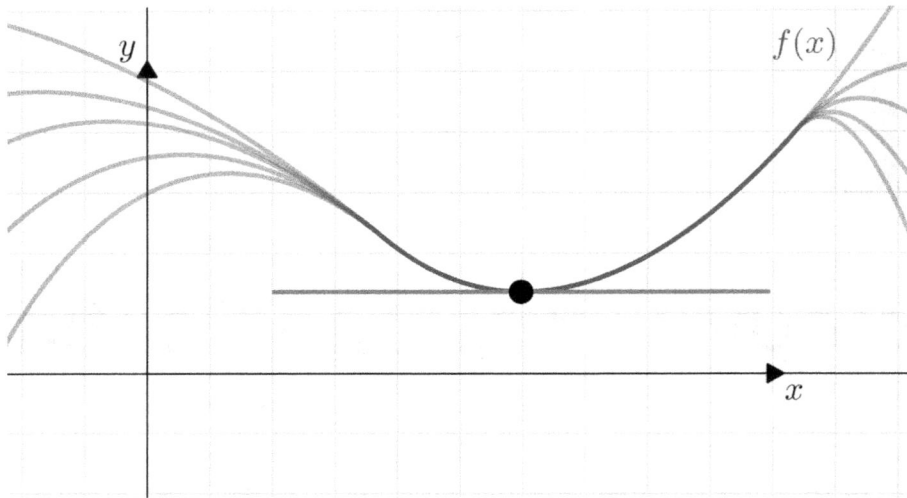

*Figure 11.1: The derivatives of two functions are equal if the functions are equal in any small interval around the point*

In mathematics, local properties are handled in terms of *sequences* and *neighborhoods*. We learned about sequences in the last chapter, and now we will tackle the subject of neighborhoods.

We are going to focus on three fundamental aspects:

- open and closed sets,
- the behavior of sequences within sets,
- and their smallest and largest elements, upper and lower bounds.

Our main goal with mathematical analysis is to understand *gradient descent*, a fundamental tool for training models. To do that, we need to understand *limits*. For that, we need understand *sequences* and *real numbers*, leading us deep into the rabbit hole where we are now.

Think of it as learning the Python language versus learning TensorFlow or PyTorch. Since we want to do machine learning, we ultimately want to learn a high-level framework. However, if we lack the understanding of the basic keywords in Python, like `import` or `def`, we are not ready to learn and productively use advanced tools. Sequences, open and closed sets, limits, and others are the fundamental building blocks of mathematical analysis, the language of optimization.

## 11.1.1   Open and closed sets

Let's start our discussion with *open* and *closed* sets! (In this chapter, when we refer to something as a *subset* or *set*, it is implicitly assumed to be within $\mathbb{R}$.)

---

**Definition 11.1.1 (Open and closed sets)**

Let $A \subseteq \mathbb{R}$ be a subset of the real numbers.

*(a)* $A$ is open if for every $x \in A$, there is a $\varepsilon > 0$ such that $(x - \varepsilon, x + \varepsilon) \subseteq A$.

*(b)* $A$ is closed if its complement $\mathbb{R} \setminus A$ is open.

---

Before we start analyzing the properties of open and closed sets, here are some key examples for building up useful mental models.

**Example 1.** Intervals of the form $(a, b) = \{x \in \mathbb{R} : a < x < b\}$ are open. This can be easily seen by picking any $x \in (a, b)$ and letting $\varepsilon = \min\{|x - a|/2, |x - b|/2\}$. Essentially, we take the distance from the closest endpoint and cut that in half. Any point that is closer to $x$ than the half-distance of the closest endpoint will also be in $(a, b)$.

**Example 2.** Intervals of the form $[a, b] = \{x \in \mathbb{R} : a \leq x \leq b\}$ are closed. Indeed, its complement is $\mathbb{R} \setminus [a, b] = (-\infty, a) \cup (b, \infty)$. Using the reasoning above, it is easy to see that $(-\infty, a) \cup (b, \infty)$ is open.

**Example 3.** Intervals of the form $(a, b] = \{x \in \mathbb{R} : a < x \leq b\}$ are neither open nor closed. To see that it is not open, observe that no interval containing $b$ is fully within $(a, b]$, since $b$ is an endpoint. For similar reasons, its complement $\mathbb{R} \setminus (a, b] = (-\infty, a] \cup (b, \infty)$ is not open.

An important takeaway from the last example is that if a set is *not closed*, it doesn't mean that it is *open* and vice versa.

We can rephrase the definition of openness by introducing the concept of *neighborhoods*. The neighborhoods of a given point $x$ are the open intervals $(a, b)$ containing $x$. With this terminology, any set $A$ is open if, for any $x \in A$, there exists a neighborhood of $x$ that is fully contained within $A$. From this aspect, openness means that there is still "room to move" from any point.

The most fundamental property of open and closed sets is their behavior under union and intersection. As this holds for *any* collection of open and closed sets, we need to introduce a new notation here. Recall that if we have the sets $A_1, \ldots, A_n$, their union can be abbreviated as

$$\cup_{k=1}^{n} A_k = A_1 \cup \cdots \cup A_n.$$

This is denoted similarly for the intersection. If we have countable sets, we can even write

$$\bigcup_{k=1}^{\infty} A_k,$$

but what can we do if we have an uncountable collection of sets? Say, $A_c = [0, c]$ for all $c \in (0, \infty)$. In this case, we use the expression

$$\bigcup_{c>0} A_c.$$

For a general set $\Gamma$, we talk about the collection of sets defined by $\{A_\gamma\}_{\gamma \in \Gamma}$, and the union/intersection

$$\bigcup_{\gamma \in \Gamma} A_\gamma, \quad \bigcap_{\gamma \in \Gamma} A_\gamma.$$

Now, let's see our first theorem!

---

**Theorem 11.1.1** *Let $\{A_\gamma\}_{\gamma \in \Gamma}$ be an arbitrary collection of sets.*

*(a) If each $A_\gamma$ is open, then*

$$\bigcup_{\gamma \in \Gamma} A_\gamma$$

*is also open.*

*(b) If each $A_\gamma$ is closed, then*

$$\bigcap_{\gamma \in \Gamma} A_\gamma$$

*is also closed.*

---

*Proof. (a)* Suppose that $A_\gamma$, $\gamma \in \Gamma$ are open sets and let $x \in \cup_{\gamma \in \Gamma} A_\gamma$. Because $x$ is in the union, there is some $\gamma_0 \in \Gamma$ such that $x \in A_{\gamma_0}$. Because $A_{\gamma_0}$ is open, there is a small neighborhood $(a, b)$ of $x$ such that $(a, b) \subseteq A_{\gamma_0}$. Because of this, $(a, b) \subseteq \cup_{\gamma \in \Gamma} A_\gamma$, which is what we had to show.

*(b)* Now let $A_\gamma$, $\gamma \in \Gamma$ be closed sets. In this case, De Morgan's laws (*Theorem C.2.2*) imply that $\mathbb{R} \setminus (\cap_{\gamma \in \Gamma} A_\gamma) = \cup_{\gamma \in \Gamma} (\mathbb{R} \setminus A_\gamma)$. Since each $A_\gamma$ is closed, $\mathbb{R} \setminus A_\gamma$ is open. As we have previously seen, the union of open sets is open.

The closedness and openness of a set influence its behavior regarding set sequences. The first fundamental result regarding this is Cantor's axiom.

**Theorem 11.1.2** *(Cantor's axiom)*

*Let $I_n = [a_n, b_n] \subseteq \mathbb{R}$ be a sequence of intervals such that $I_{n+1} \subseteq I_n$ for every $n \in \mathbb{N}$. Then, the intersection $\cap_{n=1}^{\infty} I_n$ is nonempty.*

This seemingly simple proposition is a profound property of real numbers, one that ultimately follows from their mathematical construction. Cantor's axiom is not true, for instance, if we talk about subsets of $\mathbb{Q}$ instead of $\mathbb{R}$. Think about a sequence of rational numbers $a_n \to \pi$ that approximates $\pi$ from below, and another sequence $b_n \to \pi$ that approximates $\pi$ from above, that is,

$$a_n < \pi < b_n, \quad a_n, b_n \in \mathbb{Q}.$$

The intersection of the intervals $[a_n, b_n]$ only contains $\pi$, which is *not* rational. Thus, in the space of rational numbers, $\cap_{n=1}^{\infty}[a_n, b_n] = \varnothing$, therefore Cantor's axiom doesn't hold there.

There is an old proverb about losing the war because of a nail in a horseshoe. It goes something like this:

> *For want of a nail the shoe was lost.*
> *For want of a shoe the horse was lost.*
> *For want of a horse the rider was lost.*
> *For want of a rider the message was lost.*
> *For want of a message the battle was lost.*
> *For want of a battle the kingdom was lost.*
> *And all for the want of a horseshoe nail.*

Think about Cantor's axiom as the nail in the horseshoe. Without it, we can't talk about taking limits of sequences. Without limits, there are no gradients. Without gradients, there is no gradient descent, and consequently, we can't fit machine learning models.

## 11.1.2 Distance and topology

Originally, we defined open sets in terms of small open intervals like $(x - \varepsilon, x + \varepsilon)$. We called a set open if you could squeeze in such a small interval for each of its points. By taking a step of abstraction, we can rephrase the definition in terms of norms (*Definition 2.1.1*).

From this viewpoint, an interval $(x - \varepsilon, x + \varepsilon)$ is the same as a one-dimensional open *ball*. Given a normed space $V$ with the norm $|\cdot|$, the ball of radius $r > 0$ centered at $\mathbf{x}$ is defined by

$$B(\mathbf{x}, r) := \{\mathbf{y} \in V : \|\mathbf{x} - \mathbf{y}\| < r\}.$$

Equivalently, a ball of radius $r$ is the set of points with distance less than $r$ from the center point. In the Euclidean spaces $\mathbb{R}^n$, with the norm $\|\mathbf{x}\| = \sqrt{x_1^2 + \cdots + x_n^2}$, this matches our intuitive understanding. This is illustrated in *Figure 11.2*.

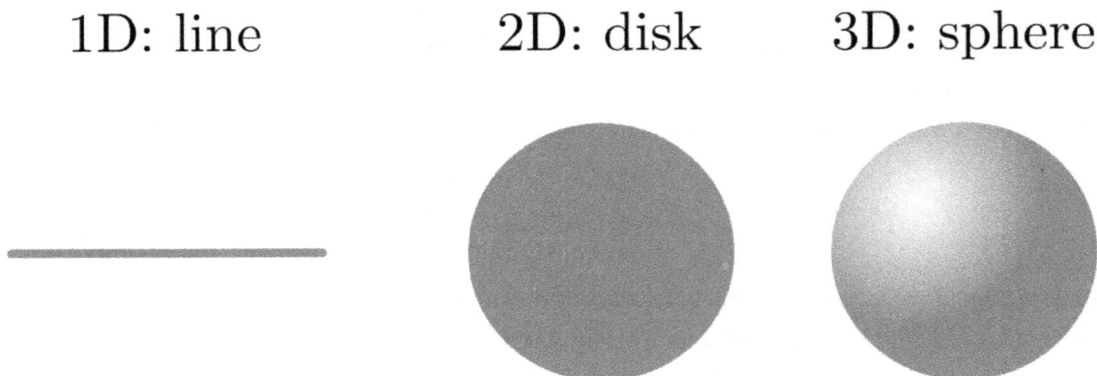

# 1D: line      2D: disk    3D: sphere

*Figure 11.2: Balls, from one to three dimensions*

However, in one dimension, the Euclidean norm simplifies to $\|x\| = |x|$. Thus, we have

$$B(x, r) = \{y \in \mathbb{R} : |x - y| < r\}$$
$$= (x - r, x + r).$$

We don't often think about the interval $(x - \varepsilon, x + \varepsilon)$ as the one-dimensional ball $B(x, \varepsilon)$. However, making this connection will make it easy to later extend the topology of $\mathbb{R}$ to $\mathbb{R}^n$, which is where we want to work eventually.

With norms and balls, we can rephrase the definition of open sets in the following way.

**Definition 11.1.2 (Open sets, second take)**

Let $A \subseteq \mathbb{R}$ be a subset of the real numbers.

*A is open* if for every $x \in A$, there is a one-dimensional ball $B(x, \varepsilon)$ of radius $\varepsilon$ centered around $x$ such that $B(x, \varepsilon) \subseteq A$.

Thus, in a sense, open sets are determined by open balls.

### 11.1.3   Sets and sequences

Closed sets can be characterized in terms of their sequences. The following theorem shows an equivalent definition of closed sets, giving us a helpful way of thinking about them.

---

**Theorem 11.1.3   *(Characterization of closed sets with sequences)***

*Let $A \subseteq \mathbb{R}$ be a set. The following are equivalent.*

*(a) $A$ is closed.*

*(b) If $\{a_n\}_{n=1}^{\infty} \subseteq A$ is a convergent sequence, then $\lim_{n \to \infty} a_n \in A$.*

---

*Proof.* To prove that the two statements are equivalent, we have to show two things: that *(a)* implies *(b)* and that *(b)* implies *(a)*. Don't worry if this proof seems too complicated when you read it the first time. If you don't understand it right away, I suggest thinking about $A$ as a closed interval and drawing a figure. You can also skip it since I will refer back to this fact every time we need it later.

**First, let's see that *(a)* implies *(b)*.** Thus, suppose that $A$ is closed and $\{a_n\}_{n=1}^{\infty} \subseteq A$ is a convergent sequence, $a := \lim_{n \to \infty} a_n$. We have to show that $a \in A$, and we are going to do this by contradiction. The plan is the following: assume that $a \notin A$ and deduce that $\{a_n\}$ must eventually leave $A$.

Indeed, suppose that $a \in \mathbb{R} \setminus A$. Because $A$ is closed, $\mathbb{R} \setminus A$ is open, so there is a small neighborhood $(a - \varepsilon, a + \varepsilon) \subseteq \mathbb{R} \setminus A$. In plain English, this means that we can *separate* $a$ from $A$. This contradicts the fact that $\{a_n\} \subseteq A$ and $a_n \to a$, because according to the definition of convergence (*Definition 10.2.2*), eventually all members of the sequence have to fall into $(a - \varepsilon, a + \varepsilon)$. This is a contradiction, so $a \in A$.

**Second, we will show that *(b)* implies *(a)*,** that is, if the limit of every convergent sequence of $A$ is also in $A$, then the set is closed. Our goal is to show that $\mathbb{R} \setminus A$ is open. More precisely, if $x \in \mathbb{R} \setminus A$, we want to find a small neighborhood $(x - \varepsilon, x + \varepsilon)$ that is disjointed from $A$. Again, we can show this via contradiction.

Suppose that no matter how small $\varepsilon > 0$ is, we can find an $a \in A \cap (x - \varepsilon, x + \varepsilon)$. Thus, we can define a sequence $\{a_n\}_{n=1}^{\infty}$ such that $a_n \in A \cap (x - 1/n, x + 1/n)$. Due to the construction $\lim_{n \to \infty} a_n = a$, and as $A$ is closed to taking limits according to the premise *(b)*, this would imply that $a \in A$, which is a contradiction. This is what we had to show.

This result also explains the origins of the terminology *closed*. A closed set is such because it is *closed to limits*.

### 11.1.4   Bounded sets

From a (very) high-level view, machine learning can be described as an optimization problem. For inputs $\mathbf{x}$ and predictions $\mathbf{y}$, we are looking at a parametrized family of functions $f(\mathbf{x}, \mathbf{w})$, where our parameters are condensed in the variable $\mathbf{w}$.

Given a set of samples and observations, our goal is to find the minimum of the set

$$\left\{ \mathrm{Loss}(f(\mathbf{x}, \mathbf{w}), \mathbf{y}) \ : \ \mathbf{w} \in \mathbb{R}^{\text{a very large number}} \right\} \subseteq \mathbb{R}, \tag{11.1}$$

and the parameter configuration $\mathbf{w}$ where the optimum is attained. To make sure that our foundations are not missing this building block, we are going to take some time to study this.

---

**Definition 11.1.3 (Bounded sets)**

Let $A \subseteq \mathbb{R}$.

*(a) A* is *bounded from below* if there is an $m \in \mathbb{R}$ such that $m < x$ for all $x \in A$. The number $m$ is called a lower bound.

*(b) A* is *bounded from above* if there is an $M \in \mathbb{R}$ such that $x < M$ for all $x \in A$. Similar to before, $M$ is called an upper bound.

*(c) A* is *bounded* if it is bounded from above and below.

---

Being bounded means that we can include the set in a large interval $[m, M]$. For optimization, there are a few essential quantities that are related to bounds: minimal and maximal elements, smallest upper bounds, and largest lower bounds.

Let's start with formalizing the concept of the smallest and largest elements within a set.

---

**Definition 11.1.4 (Minimal and maximal elements)**

Let $A \subseteq \mathbb{R}$.

*(a)* $m \in A$ is called the *minimal element* of $A$ if for every $a \in A$, $m \leq a$ holds. The minimal element is denoted by $m = \min A$.

*(b)* $M \in A$ is called the *maximal element* of $A$ if for every $a \in A$, $a \leq M$ holds. The maximal element is denoted by $M = \max A$.

---

As usual, let's see some examples.

**Example 1.** The set $A = [0, 1] \cup [2, 3]$ has both minimal and maximal elements. Its minimal element is 0, and its maximal element is 3.

**Example 2.** The set $A = \{\frac{1}{n} \ : \ n \in \mathbb{N}\}$ has no minimal element. 0 is the largest lower bound, but since it is not in the set, it is not a minimal element.

The key takeaway is that minimal and maximal elements do not necessarily exist. To avoid the inconvenience, we need related quantities that always exist.

These will be the *infimum* and *supremum*.

---

**Definition 11.1.5  (Infimum and supremum)**

Let $A \subseteq \mathbb{R}$.

*(a)* inf $A \in \mathbb{R}$ is the largest lower bound of $A$ if $m \leq$ inf $A$ holds for every lower bound $m$. This is called the *infimum*.

*(b)* sup $A \in \mathbb{R}$ is the smallest upper bound of $A$ if $M \geq$ sup $A$ holds for every upper bound $M$. This is called the *supremum*.

---

We won't go into great detail here, but the infimum and supremum always exist. However, it is essential to note that there is a sequence $\{a_n\}_{n=1}^{\infty} \subseteq A$ such that $a_n \to$ inf $A$. (This is true for the supremum as well.)

# 11.1.5   Compact sets

With the concept of infimum and supremum, we can formalize the optimization problem for machine learning described by (11.1) as

$$\inf \left\{ \text{Loss}(f(\mathbf{x}, \mathbf{w}), \mathbf{y}) \ : \ \mathbf{w} \in \mathbb{R}^n \right\},$$

where this number represents the smallest possible value of the loss function and $w$ is the parameter of our model.

However, there is a significant issue with the $\mathbf{w} \in \mathbb{R}^n$ part. First, our parameter space is high-dimensional. In practice, $n$ can be in the millions. Besides that, we are looking at an unbounded parameter space, where such an optimum might not even exist. Finally, is there even a parameter $\mathbf{w}$ where the infimum is attained? After all, this is what we are primarily interested in.

We can restrict the parameter space to a closed and bounded set to fix these issues. These sets are so prevalent that they have their own name: *compact sets*.

---

**Definition 11.1.6  (Compact sets)**

The set $A \subseteq \mathbb{R}$ is *compact* if it is bounded and closed.

---

We love compact sets. Even though their definition seems straightforward, these two properties have profound consequences regarding optimization. At this point, we are not ready to talk about this in detail, but we can find minima or maxima in practice because continuous functions behave nicely on compact sets.

There is a key result about compact sets that will constantly resurface during our studies of functions: the Bolzano-Weierstrass theorem.

**Theorem 11.1.4** *(Bolzano-Weierstrass theorem)*

*In a compact set, every sequence has a convergent subsequence.*

*Proof.* Let $A \subseteq \mathbb{R}$ be a compact set, and $\{a_n\}_{n=1}^{\infty}$ be an arbitrary sequence in $A$.

Because $A$ is compact, there exists an interval $I_1 := [m, M]$ that contains $A$ in its entirety. By cutting this interval in half, we obtain $[m, (m+M)/2]$ and $[(m+M)/2, M]$. At least one of these will contain infinitely many points from $\{a_n\}$; let that be $I_2$. Repeating this process will yield a sequence of closed intervals $I_1 \supseteq I_2 \supseteq I_3 \dots$. The length of $I_k$ is $(M-m)/2^{k-1}$, so eventually these will get really small.

Due to the construction of these intervals, we can also define a subsequence $\{a_{n_k}\}_{k=1}^{\infty}$ by selecting $a_{n_k}$ such that $a_{n_k} \in I_k$.

According to Cantor's axiom (*Theorem 11.1.2*), $\cap_{k=1}^{\infty} I_k$ is nonempty, so let $a \in \cap_{k=1}^{\infty} I_k$. Because both $a_{n_k}$ and $a$ are elements of $I_k$, we have

$$|a - a_{n_k}| \leq \frac{M-m}{2^{k-1}}, \quad k \in \mathbb{N}.$$

This shows that $\lim_{k \to \infty} a_{n_k} = a$, which is what we had to show.

The technique we used here is called lion catching. How does a mathematician catch a lion in the desert? By cutting the desert in half. The lion will be located in one half or the other. This section can be cut in half repeatedly until the area becomes smaller. Thus, the lion will be trapped there eventually.

Now that we understand the topological structure of real numbers and the relation to convergent sequences, we can move forward and finally talk about the limit of functions!

# 11.2 Limits

Recall that in *Section 10.2* about sequences, we defined limits of convergent sequences. Intuitively, limits capture the notion that eventually, all elements get as close to the limit as we wish. This concept can be extended to functions as well.

---

**Definition 11.2.1 (Limits of functions)**

Let $f : \mathbb{R} \to \mathbb{R}$ be an arbitrary function. We say that

$$\lim_{x \to x_0} f(x) = a$$

if for every sequence $x_n \to x_0$, where $x_n$ does not equal $x_0$ for all $n$,

$$\lim_{n \to \infty} f(x_n) = a$$

holds.

---

Right off the bat, there are two essential things to note.

1. The limit of a function is defined in terms of limits of sequences. If all possible sequences of the form $\{f(x_n)\}$ with $x_n \to x_0$ have the same limit, then $\lim_{x \to x_0} f(x)$ is defined as the common limit.
2. With $\infty$-divergent sequences, limits at $\pm\infty$ are defined.

Here is a figure that helps visualize the process.

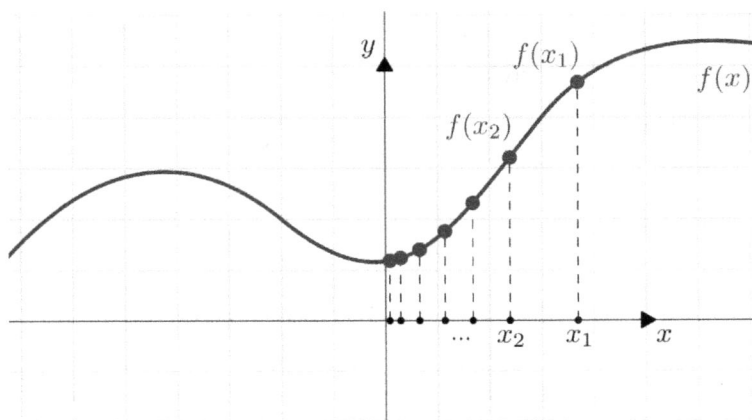

*Figure 11.3: Illustration of limits: as $x_n$ gets closer to 0, $f(x_n)$ gets closer to $\lim_{n \to \infty} f(x_n)$*

To further illustrate the concept of limits for functions, let's see some examples.

**Example 1.** Define the function

$$f(x) = \begin{cases} 1 & \text{if } x = 0, \\ 0 & \text{otherwise,} \end{cases}$$

illustrated by *Figure 11.4*. In other words, $f(x)$ is 0 everywhere except at 0, where it is 1.

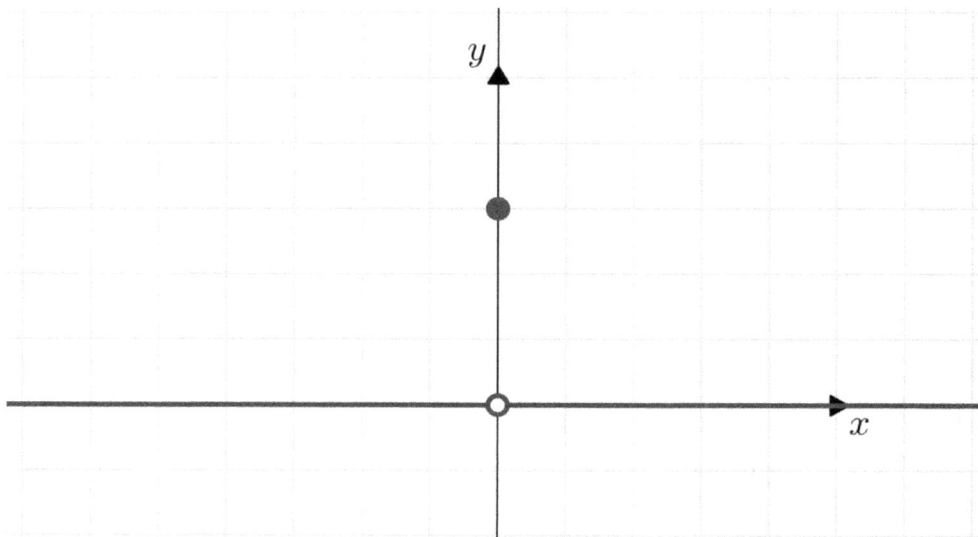

*Figure 11.4: Plot of $f(x)$*

Does $\lim_{x\to 0} f(x)$ exist? Yes. Because for any sequence $x_n \to 0$ that is not 0, the limit $\lim_{n\to\infty} f(x_n) = 0$. On the other hand, note that $\lim_{x\to 0} f(x) \neq f(0)$.

**Example 2.** For our second example, define

$$D(x) = \begin{cases} 1 & \text{if } x \in \mathbb{Q}, \\ 0 & \text{otherwise.} \end{cases} \tag{11.2}$$

This is the (in)famous Dirichlet function, which is hard to imagine and impossible to plot: its value is 1 at rationals and 0 at irrationals. Not surprisingly, $\lim_{x\to x_0} D(x)$ does not exist for all $x_0$, because rational and irrational numbers are "dense": every number $x_0$ can be obtained as a limit of rationals and as a limit of irrationals.

Since limits of functions are defined as the common limit of sequences, many of its properties are inherited from sequences. How sequences behave under operations (*Theorem 10.2.1*) determines how function limits behave.

**Theorem 11.2.1 (Operations and limits)**

*Let f and g be two functions.*

*(a)*

$$\lim_{x \to x_0} f(x) + g(x) = \lim_{x \to x_0} f(x) + \lim_{x \to x_0} g(x),$$

*(b)*

$$\lim_{x \to x_0} cf(x) = c \lim_{x \to x_0} f(x) \quad \text{for all } c \in \mathbb{R},$$

*(c)*

$$\lim_{x \to x_0} f(x)g(x) = \lim_{x \to x_0} f(x) \lim_{x \to x_0} g(x),$$

*(d) If $f(x) \neq 0$ in some small interval $(x_0 - \varepsilon, x_0 + \varepsilon)$ around $x_0$ and $\lim_{x \to x_0} f(x) \neq 0$, then*

$$\lim_{x \to x_0} \frac{1}{f(x)} = \frac{1}{\lim_{x \to x_0} f(x)}.$$

*Proof.* This follows directly from *Theorem 10.2.1*.

Similarly as we have seen for convergent sequences, *(a)* and *(b)* above are referred to as the *linearity* of limits.

Remember how the big and small O notation (*Definition 10.2.4*) expressed asymptotic properties of sequences? We have a similar tool for functions as well.

**Definition 11.2.2 (Big and small O notation for functions)**

Let $f : \mathbb{R} \to \mathbb{R}$ and $g : \mathbb{R} \to \mathbb{R}$ be two arbitrary functions. We say that

*(a)* $g(x) = O(f(x))$ as $x \to a$ if there is a constant $C$ such that for some $\delta > 0$, we have $|g(x)| \leq Cf(x)$ for all $x \in (a - \delta, a) \cup (a, a + \delta)$. Similarly, $g(x) = O(f(x))$ as $x \to \infty$ if there is a constant $C$ and a cutoff number $N$ such that $|g(x)| \leq Cf(x)$ holds for all $x > N$.

*(b)* $g(x) = o(f(x))$ as $x \to a$ if for any $\varepsilon > 0$, there is a $\delta > 0$ such that we have $|g(x)| \leq \varepsilon f(x)$ for all $x \in (a - \delta, a) \cup (a, a + \delta)$. Similarly, $g(x) = o(f(x))$ as $x \to \infty$ if for any $\varepsilon > 0$, there is a cutoff number $N$ such that $|g(x)| \leq \varepsilon f(x)$ for all $x > N$.

## 11.2.1  Equivalent definitions of limits

If you have a sharp eye (and some experience in mathematics), you might have already posed the question: won't showing convergence of $\{f(x_n)\}$ for *all* sequences $x_n \to x_0$ be difficult?

Indeed, it is often not the most convenient way to think about function limits. Another equivalent definition expresses limits in terms of smaller and smaller neighborhoods around the point in question.

---

**Theorem 11.2.2** *(Limits as error terms)*

*Let $f : \mathbb{R} \to \mathbb{R}$ be an arbitrary function and $x_0 \in \mathbb{R}$. Then, the following are equivalent.*

*(a)* $\lim_{x \to x_0} f(x) = a$.

*(b) For every $\varepsilon > 0$, there exists a small neighborhood $(x_0 - \delta, x_0 + \delta)$ around $x_0$ such that for every $x \in (x_0 - \delta, x_0) \cup (x_0, x_0 + \delta)$,*

$$|f(x) - a| < \varepsilon$$

*holds.*

---

*Proof.* $(a) \implies (b)$. We are going to do this indirectly, so we assume that $(a)$ holds and $(b)$ is not true. The negation of $(b)$ states that there is a $\varepsilon > 0$ such that for every $\delta > 0$, there is an $x \in (x_0 - \delta, x_0) \cup (x_0, x_0 + \delta)$ such that $|f(x) - a| > \varepsilon$. (If you don't see why this is the negation, check out the *Appendix A* about logic.)

Now we define a sequence that will contradict $(a)$. If we select $\delta = 1/n$, we can let $x_n$ be the one in $(x_0 - \delta, x_0) \cup (x_0, x_0 + \delta)$ such that $|f(x_n) - a| > \varepsilon$, as guaranteed by our assumption that $(b)$ is false. Due to its construction, $\{f(x_n)\}_{n=1}^{\infty}$ does not converge to $a$. This contradicts $(a)$, which completes our indirect proof.

$(b) \implies (a)$. Let $\{x_n\}_{n=1}^{\infty}$ be an arbitrary sequence that converges to $x_0$. If $n$ is large enough (that is, larger than some cutoff index $N$), $x_n$ will fall into $(x_0 - \delta, x_0 + \delta)$, where $\delta > 0$ is an arbitrary small constant. Since $(b)$ says that $|f(x_n) - a| < \varepsilon$ here for all such $n$, we have $\lim_{n \to \infty} f(x_n) = a$ by the definition of convergence *(Definition 10.2.2)*.

In plain English, this theorem says that $f(x)$ gets arbitrarily close to $\lim_{x \to x_0} f(x)$ if $x$ is close enough to $x_0$. Definitions similar to *(b)* are called epsilon-delta definitions.

There is yet another equivalent definition that, although it might seem trivial, is a useful mental model when thinking about limits.

**Theorem 11.2.3** *Let $f : \mathbb{R} \to \mathbb{R}$ be an arbitrary function and $x_0 \in \mathbb{R}$. Then, the following are equivalent.*

*(a)* $\lim_{x \to x_0} f(x) = a$.

*(b) There exists a function* error$(x)$ *such that* $\lim_{x \to x_0}$ error$(x) = 0$ *and*

$$f(x) = a + \text{error}(x).$$

*Proof.* (a) $\implies$ (b). Due to how limits behave with respect to operations (*Theorem 11.2.1*), it is easy to see that

$$\text{error}(x) := f(x) - a$$

satisfies the requirements.

(b) $\implies$ (a). Again, this is trivial because of the linearity of the limit operation:

$$\lim_{x \to x_0} f(x) = \lim_{x \to x_0} \big( a + \text{error}(x) \big) = a.$$

This is what we had to show.

Often, we don't need to know the exact limits of a function; it is enough to know that the limit is above or below a specific bound.

To give a specific example, we will look slightly ahead and talk about differentiation. I'll explain everything in detail in the next chapter, but the derivative of a function $f$ at the point $x_0$ is defined as the limit

$$\lim_{x \to x_0} \frac{f(x) - f(x_0)}{x - x_0}.$$

If the function is increasing, we have

$$\frac{f(x) - f(x_0)}{x - x_0} \geq 0,$$

which, given the things we are about to see, implies that the derivative is positive.

Without any further ado, let's see the result!

> **Theorem 11.2.4** *(The transfer principle for functions)*
>
> Let $f : \mathbb{R} \to \mathbb{R}$ be an arbitrary function. If $f(x) \geq \alpha$ for all $x \in (a - \delta, a) \cup (a, a + \delta)$ and some $\delta >$ constant and $\alpha \in \mathbb{R}$ lower bound, then $\lim_{x \to a} f(x) \geq \alpha$ if the limit exists.

*Proof.* Due to the definition of function limits, this is the immediate consequence of the transfer principle for convergent sequences (*Theorem 10.2.2*).

There are a couple of special limits that come up all the time in calculations. These are the building blocks for calculating more complicated limits, as they can often be reduced to a form for which the limit is known.

We won't include the proofs here, as they are not that useful for our purposes (which is understanding how machine learning algorithms work).

**Theorem 11.2.5** *(a)*

$$\lim_{x \to 0} \frac{\sin x}{x} = 1. \tag{11.3}$$

*(b)*

$$\lim_{x \to 0} x \log x = 0. \tag{11.4}$$

*(c)*

$$\lim_{x \to \infty} x^k e^{-ax} = 0, \quad a \in (0, \infty), \quad k = 0, 1, 2, \dots \tag{11.5}$$

With the extension of limits from sequences to functions, we saw that if the limit exists, it is not necessarily equal to the function's value at the given point. However, when it does, the function is much easier to handle. This is called *continuity*, and this is what we'll discuss next.

# 11.3  Continuity

If I asked you to conjure up a random function from your mind, I am almost certain that you would come up with one that is both continuous and differentiable. (unless you have weird tastes, as many mathematicians do).

However, the vast majority of functions are neither. In terms of cardinality, if you count all real functions $f : \mathbb{R} \mapsto \mathbb{R}$, it turns out that there are $2^c$ of them in total, but the subset of continuous ones have cardinality $c$. It is hard to imagine such quantities: $c$ and $2^c$ are both infinite, but, well, $2^c$ is *more* infinite. Yeah, I know. Set theory is weird. (Recall that $c$ denotes the cardinality of the set of real numbers. If you would like a refresher on the topic, check out the set theory appendix *Appendix C.*)

Overall, as we shall see, continuity and differentiability allow us to do meaningful work with functions. For instance, the usual gradient descent-based optimization for neural networks doesn't work if the loss function and the layers are not differentiable. That alone would throw a huge wrench into the cogs of machine learning since this is used all the time in the deep learning part of the field.

This section explores how these concepts work together and ultimately enable us to train neural networks. So, let's dive straight into the deep water and precisely define the notion of continuity!

> **Definition 11.3.1 (Continuity)**
>
> Let $f : \mathbb{R} \to \mathbb{R}$ be an arbitrary function. We say that $f$ is continuous at $a$ if
>
> $$\lim_{x \to a} f(x) = f(a)$$
>
> holds.

In other words, continuity means that if $x$ is close to $y$, then $f(x)$ will also be close to $f(y)$. This is how most of our mental models work. This is also what we want from many machine learning models. For example, if $f$ is a model that takes images and decides if they feature a cat or not, we would expect that after changing a few pixels on $x$, the prediction would stay the same. (However, this is definitely not the case in general, which is exploited by certain adversarial attacks.)

We can rephrase the above definition by unpacking the limits. If you think it through a bit, it is easy to see that continuity of $f$ at $a$ is equivalent to

$$\lim_{n \to \infty} f(a_n) = f\left( \lim_{n \to \infty} a_n \right)$$

for all convergent sequences $a_n \to a$. In other words, the limit and function application are interchangeable.

We are going to use this very frequently.

As usual, we'll see some examples first. We'll revisit the ones we saw *Section 11.2*.

**Example 1.** Let's revisit

$$f(x) = \begin{cases} 1 & \text{if } x = 0, \\ 0 & \text{otherwise.} \end{cases}$$

While $f(x)$ is not continuous at 0 since $\lim_{x \to 0} f(x) = 0 \neq f(0)$, as we have seen before, $f(x)$ is continuous everywhere else (since it is constant 0).

Note that even though the function is not continuous at 0, the limit *does* exist!

**Example 2.** What about the Dirichlet function $D(x)$? (See (11.2) for the definition.) Since the limit doesn't even exist, this is a nowhere near a continuous function.

**Example 3.** Define

$$f(x) = \begin{cases} x & \text{if } x \in \mathbb{Q}, \\ -x & \text{otherwise.} \end{cases}$$

Surprisingly, $f(x)$ is continuous at 0, but nowhere else. As you can see, (almost) nothing is off the table with continuity. Functions, in general, can be wild objects, and without certain regularity conditions, optimizing them is extremely hard. In essence, this chapter aims to understand when and how we can optimize functions that we used to do when training a machine learning model.

One final example!

**Example 4.** We call a function an *elementary function* if it can be obtained by taking a finite sum, product, and combination of

- constant functions,
- power functions $x, x^2, x^3, \ldots$,
- $n$-th root functions $x^{1/2}, x^{1/3}, x^{1/4}, \ldots$,
- exponential functions $a^x$,
- logarithms $\log_a x$,
- trigonometric and inverse trigonometric functions $\sin x, \cos x, \arcsin x, \arccos x$,
- hyperbolic and inverse hyperbolic functions $\sinh x, \cosh x, \sinh^{-1} x, \cosh^{-1} x$.

For instance,

$$f(x) = \sin(x^2 + e^x) - \frac{1 - 3x + 5x^4}{2x - \sqrt{x}}$$

is an elementary function. Elementary functions are continuous wherever they are defined. This is going to be extremely useful for us since showing the continuity of a complicated function like $f(x)$ is hard with the definition alone. This way, if it is elementary, we know it is continuous. This will also be true for multivariate functions (like a neural network).

## 11.3.1 Properties of continuous functions

A typical pattern in mathematics, as you have seen when discussing the properties of convergent sequences in *Theorem 10.2.1*, is to study certain properties on basic building blocks first, then show how it behaves with respect to operations.

We are going to follow a similar pattern to the previous example regarding the continuity of elementary functions.

**Theorem 11.3.1** *Let $f$ and $g$ be two functions.*

*(a) If $f$ and $g$ are continuous at $a$, then $f + g$ and $fg$ are also continuous at $a$.*

*(b) If $f$ and $g$ are continuous at $a$ and $g(a) \neq 0$, then $f/g$ is also continuous at $a$.*

*(c) If $g$ is continuous at $a$ and $f$ is continuous at $g(a)$, then $f \circ g$ is also continuous at $a$.*

*Proof.* (a) and (b) follow directly from the properties of limits (*Theorem 11.2.1*).

To see (c), we simply let $\{a_n\}_{n=1}^{\infty}$ be a sequence that converges to $a$. Then, assuming that $f$ is continuous at $g(a)$ and $g$ is continuous at $a$, we have

$$\lim_{n\to\infty} f(g(a_n)) = f\big(\lim_{n\to\infty} g(a_n)\big) = f\big(g\big(\lim_{n\to\infty} a_n\big)\big) = f(g(a)),$$

which shows the continuity of $f \circ g$ at $a$.

So far, we have only defined continuity at a single point. In general, a function $f : \mathbb{R} \to \mathbb{R}$ is continuous on the set $A$ simply if it is continuous at its every point.

We have arrived at the point that partly explains why we love continuous functions and compact sets.

The reason is simple: functions that are continuous on compact sets are bounded there and attain their optima.

**Theorem 11.3.2** *Let $f$ be continuous on a compact set $K$. There exists $\alpha, \beta \in K$ such that $f(\alpha) \le f(x) \le f(\beta)$ holds for all $x \in K$.*

Let $\{\alpha_n\}_{n=1}^{\infty} \subseteq K$ be a sequence such that $f(\alpha_n) \to \inf\{f(x) : x \in K\}$. (This is guaranteed to exist, as follows from the properties of infimum and supremum.)

Now, according to the Bolzano-Weierstrass theorem (*Theorem 11.1.4*), $\{\alpha_n\}_{n=1}^{\infty}$ has a convergent subsequence $\{\alpha_{n_k}\}_{k=1}^{\infty}$ with $\lim_{k\to\infty} \alpha_{n_k} = \alpha$. Since $K$ is compact, $\alpha \in K$. (Recall that a compact set is closed, and closed sets contain the limits of their convergent sequences, as *Theorem 11.1.3* implies.)

Because $f$ is continuous, we have

$$f(\alpha) = f\left( \lim_{k\to\infty} \alpha_{n_k} \right) = \lim_{k\to\infty} f(\alpha_{n_k}) = \inf\{f(x) : x \in K\},$$

which is what we had to show. An identical argument shows the existence of a $\beta \in K$ such that $f(x) \le f(\beta)$ for all $x \in K$.

This statement is not true for sets that are not closed and bounded. For example, $f(x) = \frac{1}{x}$ is continuous on $(0, 1]$, but has no upper bound.

## 11.4   Summary

I admit, the fine details of topology, limits, and continuity can feel complex and abstract. However, in my experience, taking the hardest path is the most rewarding, especially when learning technical topics. To sum up, we've learned

- what topology is,
- what it has to do with sequences and limits,
- how to take limits of functions,
- and finally, what a continuous function is.

Now that we are familiar with all the above, we are ready to tackle a subject at the heart of machine learning: differentiation. We'll look at how to analyze functions and what makes a function "behave nicely."

If you think through what machine learning is really about, you'll find that it is quite straightforward from a bird's eye view. In essence, all we want to do is 1. Design parameterized functions to explain the relationships between data and observations and 2. Find the parameters that best fit our data.

To find models that are expressive enough yet easy to work with, we need to restrict ourselves to functions that satisfy certain properties. The two most important are *continuity* and *differentiability*. Now that we have seen what continuity is, we can move on to study differentiable functions.

In the following chapters, we will exclusively deal with univariate real functions. This is to introduce concepts without adding many layers of complexity at once. In later chapters, we will slowly turn toward multivariate functions, and by the time we get to machine learning, we will have mastered their use.

## 11.5   Problems

**Problem 1.** Which of the following subsets of $\mathbb{R}$ are open, closed, or neither?

*(a)* $\mathbb{Z}$

*(b)* $\mathbb{Q}$

*(c)* $\cap_{n=1}^{\infty}(-\frac{1}{n}, \frac{1}{n})$

*(d)* $\cup_{n=1}^{\infty}[0, 1 - \frac{1}{n}]$

**Problem 2.** Let $A \subseteq \mathbb{R}$ be an arbitrary set. Show that there exists a sequence $\{a_n\}_{n=1}^{\infty} \subseteq A$ such that

$$\lim_{n \to \infty} a_n = \sup A.$$

(An identical statement is true for inf $A$ as well, which can be shown in the same way.)

**Problem 3.** Let $D(x)$ be the Dirichlet function, defined by

$$D(x) = \begin{cases} 1 & \text{if } x \in \mathbb{Q}, \\ 0 & \text{otherwise.} \end{cases}$$

Give a mathematically rigorous proof that the limit $\lim_{x \to x_0} D(x)$ does not exist for any $x_0 \in \mathbb{R}$.

**Problem 4.** Let $X$ be an arbitrary set, and let $\tau \subseteq P(X)$ be a collection of its subsets. (Recall that $P(A)$ denotes all subsets of $A$.) The structure $(X, \tau)$ is called a *topological space*, if

1. $\varnothing \in \tau$ and $X \in \tau$,

2. For any collection of sets in $\tau$, the union is also in $\tau$,

3. For any finite collection of sets in $\tau$, the intersection is also in $\tau$.

The sets in $\tau$ are called *open sets*. Show that the following are topological spaces.

*(a)* $X$ is any set, $\tau = \{\varnothing, X\}$.

*(b)* $X$ is any set, $\tau = P(X)$.

*(c)* $(\mathbb{N}, \tau)$, where $\tau = \{S \subseteq \mathbb{N} : 0 \notin \tau\}$.

**Problem 5.** Let $f : \mathbb{R} \to \mathbb{R}$ be an arbitrary function. Show that $f$ is continuous at $x_0$ if and only if for every $\varepsilon > 0$ there exists a $\delta > 0$ such that for any $x \in (x_0 - \delta, x_0 + \delta)$, $|f(x) - f(x_0)| < \varepsilon$.

The above is an equivalent definition of continuity, called the *epsilon-delta* definition.

**Problem 6.** Let $f : \mathbb{R} \to \mathbb{R}$ be an arbitrary function. Show that $f$ is continuous everywhere if and only if the set

$$f(X) = \{f(x) : x \in X\}$$

is open for every open set $X$!

# Join our community on Discord

Read this book alongside other users, Machine Learning experts, and the author himself.

Ask questions, provide solutions to other readers, chat with the author via Ask Me Anything sessions, and much more.

Scan the QR code or visit the link to join the community.

https://packt.link/math

# 12

# Differentiation

*I turn with terror and horror from this lamentable scourge of continuous functions with no derivatives.*

*— Charles Hermite*

In the history of science, a few milestones are as significant as inventing the wheel. Even among these, differentiation is a highlight. With the invention of calculus, Newton created mechanics as we know it. Differentiation is all over science and engineering, and as it turns out, it's a key component of machine learning as well.

Why? Because of optimization! As it turns out, the extremal points of a function can be characterized in terms of their derivative, and these extremal points can be iteratively found via gradient descent.

In this chapter, we'll learn what differentiation is, what its origins are, and most importantly, how to use it in practice. Let's go!

# 12.1   Differentiation in theory

Instead of jumping straight into the mathematical definition, let's start our discussion with a straightforward example: a point-like object moving along a straight line. Its movement is fully described by the time-distance plot (*Figure 12.1*), which shows its distance from the starting point at a given time.

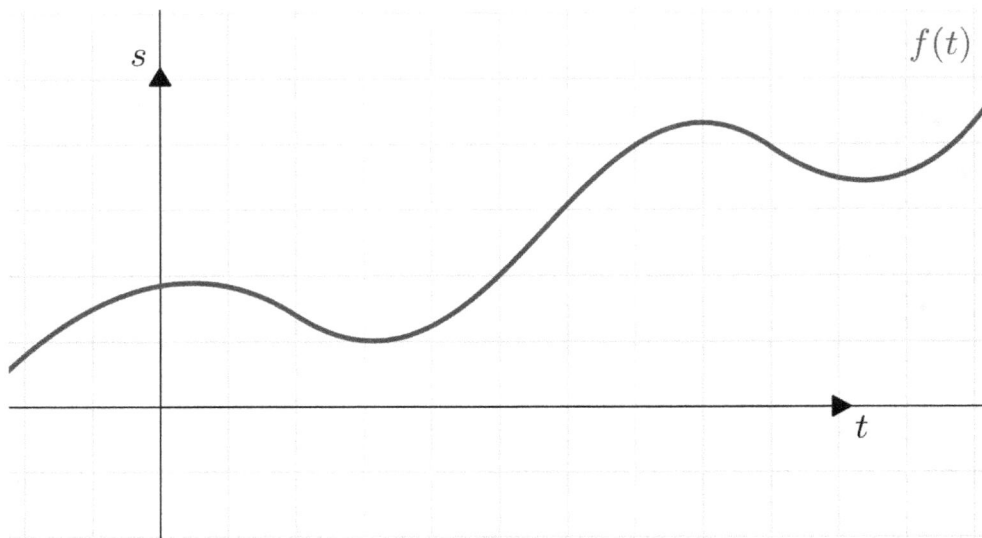

*Figure 12.1: Time-distance plot of our moving object*

Our goal is to calculate the object's velocity at a given time. In high school, we learned that

$$\text{average velocity} = \frac{\text{distance}}{\text{time}}.$$

To put this into a quantitative form, if $f(t)$ denotes the time-distance function, and $t_1 < t_2$ are two arbitrary points in time, then

$$\text{average velocity between } t_1 \text{ and } t_2 = \frac{f(t_2) - f(t_1)}{t_2 - t_1}.$$

Expressions like $\frac{f(t_2)-f(t_1)}{t_2-t_1}$ are called *differential quotients*. Note that if the object moves backwards, the average velocity is negative. (As opposed to *speed*, which is always positive. Velocity is speed and direction.)

The average velocity has a simple geometric interpretation: if you replace the object's motion with a constant velocity motion moving with that average, you'll end up at exactly the same place. In graphical terms, this is equivalent of connecting $(t_1, f(t_1))$ and $(t_2, f(t_2))$ with a single line.

The average velocity is just the slope of this line. This is visualized by *Figure 12.2.*

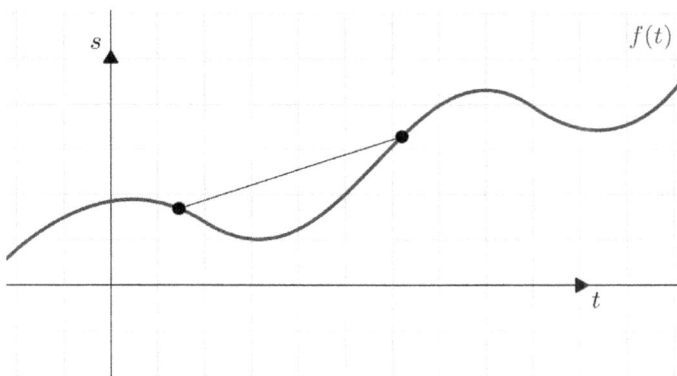

*Figure 12.2: Average velocity between $t_1$ and $t_2$*

Given this, we can calculate the exact velocity at a single time point $t$, which we'll denote with $v(t)$. The idea is simple: the average speed in the small time-interval between $t$ and $t + \Delta t$ should get closer and closer to $v(t)$ if $\Delta t$ is small enough. ($\Delta t$ can be negative as well.)

So,

$$v(t) = \lim_{\Delta t \to 0} \frac{f(t + \Delta t) - f(t)}{\Delta t}, \tag{12.1}$$

if the above limit exists. *Figure 12.3* illustrates the limit defined by (12.1). There, we can see that as $\Delta t$ gets closer and closer to 0, the slope of the line connecting $(t, f(t))$ to $(t + \Delta t, f(t + \Delta t))$ gets closer and closer to the slope of the tangent.

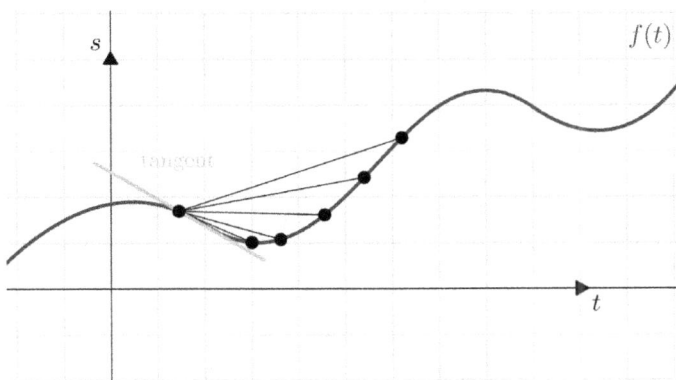

*Figure 12.3: Approximating the speed at t*

Following our geometric intuition, we see that $v(t)$ is simply the slope of the tangent line of $f$ at $t$. Keeping this in mind, we are ready to introduce the formal definition.

**Definition 12.1.1 (Differentiability)**

Let $f : \mathbb{R} \to \mathbb{R}$ be an arbitrary function. We say that $f$ is *differentiable* at $x_0 \in \mathbb{R}$ if the limit

$$\frac{df}{dx}(x_0) = \lim_{x \to x_0} \frac{f(x) - f(x_0)}{x - x_0}$$

exists. If so, this is called the *derivative* of $f$ at $x_0$.

In other words, if $f$ describes a time-distance function of a moving object, then the derivative is simply its velocity.

Similar to continuity, differentiability is a local property. However, we'll be more interested in functions that are differentiable (almost) everywhere. In those cases, the derivative is a function, often denoted with $f'(x)$.

Sometimes it is confusing that $x$ can denote the variable of $f$ and the exact point where the derivative is taken. Here is a quick glossary of terms to clarify the difference between derivative and derivative function.

- $\frac{df}{dx}(x_0)$: derivative of $f$ with respect to the variable $x$ at the point $x_0$. This is a *scalar*, also denoted with $f'(x_0)$.
- $\frac{df}{dx}$: derivative function of $f$ with respect to the variable $x$. This is a *function*, also denoted with $f'$.

**Remark 12.1.1 (Variables in the limit)**

Don't let the change in notation from $t$ and $t + \Delta t$ to $x_0$ and $x$ confuse you; the limit that defines the derivative means exactly the same as before:

$$\lim_{h \to 0} \frac{f(x_0 + h) - f(x_0)}{h} = \lim_{x \to x_0} \frac{f(x) - f(x_0)}{x - x_0}.$$

Also note that

$$\lim_{x \to x_0} \frac{f(x) - f(x_0)}{x - x_0} = \lim_{x \to x_0} \frac{f(x_0) - f(x)}{x_0 - x}.$$

On occasion, we might even use $x$ and $y$ instead of $x_0$ and $x$, writing

$$f'(x) = \lim_{y \to x} \frac{f(x) - f(y)}{x - y}.$$

We'll use whichever is more convenient.

Let's see some examples!

**Example 1.** $f(x) = x$. For any $x$, we have

$$\lim_{y \to x} \frac{f(x) - f(y)}{x - y} = \lim_{y \to x} \frac{x - y}{x - y} = 1.$$

Thus, $f(x) = x$ is differentiable everywhere and its derivative is the constant function $f'(x) = 1$.

**Example 2.** $f(x) = x^2$. Here, we have

$$\begin{aligned}
\lim_{y \to x} \frac{f(x) - f(y)}{x - y} &= \lim_{y \to x} \frac{x^2 - y^2}{x - y} \\
&= \lim_{y \to x} \frac{(x - y)(x + y)}{x - y} \\
&= \lim_{y \to x} x + y \\
&= 2x.
\end{aligned}$$

So, $f(x) = x^2$ is differentiable everywhere and $f'(x) = 2x$. Later, when talking about elementary functions, we'll see the general case $f(x) = x^k$.

**Example 3.** $f(x) = |x|$ at $x = 0$. For this, we have

$$\lim_{y \to 0} \frac{f(0) - f(y)}{0 - y} = \lim_{y \to 0} \frac{|y|}{y}.$$

Since

$$\frac{|y|}{y} = \begin{cases} 1 & \text{if } y > 0, \\ -1 & \text{if } y < 0, \end{cases}$$

this limit *does not exist*, as it is illustrated on *Figure 12.4*. This is our first example of a non-differentiable function. However, $|x|$ is differentiable everywhere else.

It is worth drawing a picture here to enhance our understanding of differentiability. Recall that the value of the derivative at a given point equals the slope of the tangent line to the function's graph.

Since $|x|$ has a sharp corner at 0, the tangent line is not well-defined.

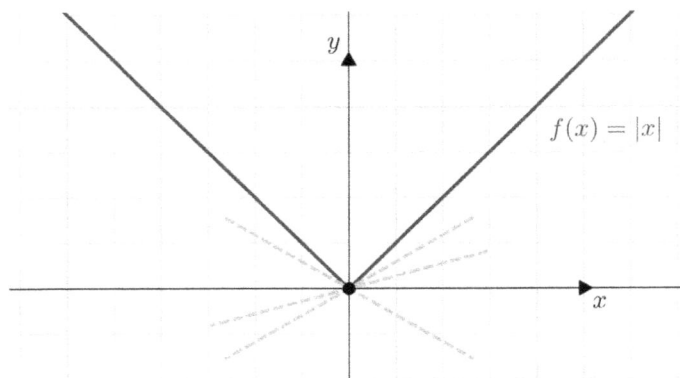

*Figure 12.4: Tangent planes of $f(x) = |x|$ at 0*

Differentiability means no sharp corners in the graph, so differentiable functions are often called *smooth*. This is one reason we prefer differentiable functions: the rate of change is tractable.

Next, we'll see an equivalent definition of differentiability, involving local approximation with a linear function. From this perspective, differentiability means manageable behavior: no wrinkles, corners, or sharp changes in value.

## 12.1.1   Equivalent forms of differentiation

To *really* understand derivatives and differentiation, we are going to take a look at it from another point of view: local linear approximations.

Approximation is a very natural idea in mathematics. Say, have you ever thought about what happens when you punch sin(2.18) into a calculator? We cannot express the function sin with finitely many additions and multiplications, so we have to *approximate* it. In practice, we use functions of the form

$$p(x) = p_0 + p_1 x + \cdots + p_n x^n,$$

which, can be evaluated easily. These are called *polynomials*, and they are just a finite combination of additions and multiplications.

Can we just replace functions with polynomials to make computations easier? (Even at the cost of perfect precision.)

It turns out that we can, and differentiation is one way to do so. In essence, the derivative desribes the best *local approximation with a linear function*.

The following theorem makes this clear.

---

**Theorem 12.1.1** *(Differentiation as a local linear approximation)*

Let $f : \mathbb{R} \to \mathbb{R}$ *be an arbitrary function. The following are equivalent.*

*(a) $f$ is differentiable at $x_0$.*

*(b) there is an $\alpha \in \mathbb{R}$ such that*

$$f(x) = f(x_0) + \alpha(x - x_0) + o(|x - x_0|) \quad as\ x \to x_0. \tag{12.2}$$

---

Recall that the small O notation (see *Definition 11.2.2*) means that the function is an order of magnitude smaller around $x_0$ than the function $|x - x_0|$.

If exists, the $\alpha$ in the above theorem is the derivative $f'(x_0)$. In other words, $f(x)$ can be locally written as

$$f(x_0) + f'(x_0)(x - x_0) + o(|x - x_0|). \tag{12.3}$$

*Proof.* To show the equivalence of two statements, we have to prove that differentiation implies the desired property and vice versa. Although this might seem complicated, it is straightforward and entirely depends on how functions can be written as their limit plus an error term (*Theorem 11.2.2*).

*(a)* $\implies$ *(b).* The existence of the limit

$$\lim_{x \to x_0} \frac{f(x) - f(x_0)}{x - x_0} = f'(x_0)$$

implies that we can write the slope of the approximating tangent in the form

$$\frac{f(x) - f(x_0)}{x - x_0} = f'(x_0) + \text{error}(x),$$

where $\lim_{x \to x_0} \text{error}(x) = 0$.

With some simple algebra, we obtain

$$f(x) = f(x_0) + f'(x_0)(x - x_0) + \text{error}(x)(x - x_0).$$

Since the error term tends to zero as $x$ goes to $x_0$, $\text{error}(x)(x - x_0) = o(|x - x_0|)$, which is what we wanted to show.

*(b)* $\implies$ *(a)*. Now, repeat what we did in the previous part, just in reverse order. We can rewrite

$$f(x) = f(x_0) + \alpha(x - x_0) + o(|x - x_0|)$$

in the form

$$\frac{f(x) - f(x_0)}{x - x_0} = \alpha + o(1),$$

which, according to what we have used before, implies that

$$\lim_{x \to x_0} \frac{f(x) - f(x_0)}{x - x_0} = \alpha.$$

So, $f$ is differentiable at $x_0$ and its derivative is $f'(x_0) = \alpha$.

One huge advantage of this form is that it will be easily generalized to multivariate functions. Even though we are far from it, we can get a glimpse. Multivariate functions map *vectors* to scalars, so the ratio

$$\frac{f(\mathbf{x}) - f(\mathbf{x_0})}{\mathbf{x} - \mathbf{x_0}}, \quad \mathbf{x}, \mathbf{x_0} \in \mathbb{R}^n$$

is not even defined. (Since we can't divide with a vector.) However, the expression

$$f(\mathbf{x}) = f(\mathbf{x_0}) + \nabla f(\mathbf{x_0})^T (\mathbf{x} - \mathbf{x_0}) + o(\|\mathbf{x} - \mathbf{x_0}\|)$$

makes perfect sense, since $\nabla f(\mathbf{x_0})^T (\mathbf{x} - \mathbf{x_0})$ is a scalar. Here, $\nabla f(\mathbf{x_0})$ denotes the *gradient* of $f$, that is, the multivariable version of derivatives. $\nabla f(\mathbf{x_0})$ is an n-dimensional vector. Don't worry if you are not familiar with this notation, we'll cover everything in due time. The take-home message is that this alternative definition will be more convenient for us in the future.

Another advantage of the locally-best-approximation mindset is that *Theorem 12.1.1* can be generalized to higher derivatives.

Check out the following theorem.

**Theorem 12.1.2** *(Taylor's theorem)*

Let $f : \mathbb{R} \to \mathbb{R}$ *function that is n times differentiable at* $x_0$. *Then*

$$f(x) = \sum_{k=0}^{n} \frac{f^{(k)}(x_0)}{k!}(x - x_0)^k + o(|x - x_0|^n)$$

*holds, where* $f^{(k)}(x_0)$ *denotes the k-th derivative of f at* $x_0$.

(Note that the zeroth derivative $f^{(0)}$ equals to $f$.)

In other words, *Theorem 12.1.2* says that if $f$ is differentiable enough times, it can be written as a polynomial plus a small error term. For infinitely differentiable functions, Taylor's theorem gives rise to the famous Taylor expansion.

**Definition 12.1.2 (Taylor expansion)**

Let $f : \mathbb{R} \to \mathbb{R}$ be a function that is differentiable at $x_0$ infinitely many times. The series defined by

$$f(x) \sim \sum_{k=0}^{\infty} \frac{f^{(k)}(x_0)}{k!}(x - x_0)^k$$

is called the *Taylor expansion of f around* $x_0$.

To give you an example, as $(e^x)' = e^x$, the Taylor expansion of $e^x$ around 0 is

$$e^x = \sum_{k=0}^{\infty} \frac{1}{k!}x^k.$$

Note that the equality sign is not an accident: the Taylor expansion of $e^x$ equals $e^x$! (This is not always the case.) Now we see why $e = \sum_{k=0}^{\infty} \frac{1}{k!}$, as we hinted earlier when discussing sequences and series.

In other words,

$$e^x \approx \sum_{k=0}^{n} \frac{1}{k!} x^k,$$

meaning that on any interval $[-\alpha, \alpha]$ and for any arbitrarily small $\varepsilon > 0$,

$$\left| e^x - \sum_{k=0}^{n} \frac{1}{k!} x^k \right| < \varepsilon$$

holds if $n$ is large enough. In practice, this polynomial is evaluated to approximate the value of $e^x$.

## 12.1.2 Differentiation and continuity

As the following theorem states, differentiation is a more strict condition than continuity.

**Theorem 12.1.3** *Differentiable functions are continuous.*

*If $f : \mathbb{R} \to \mathbb{R}$ is differentiable at $a$, it is also continuous there.*

*Proof.* We'll use *Theorem 12.1.1* to prove the result. With the general form (12.2), we have

$$\lim_{x \to x_0} f(x) = \lim_{x \to x_0} \left( f(x_0) + f'(x_0)(x - x_0) + o(|x - x_0|) \right) = f(x_0),$$

which shows the continuity of $f$ at $x_0$.

Note that the previous theorem is not true the other way around: a function can be continuous, but not differentiable. (As the example $f(x) = |x|$ at $x = 0$ shows.)

This can be taken to the extremes: there are functions that are continuous everywhere but differentiable nowhere. One of the first examples was provided by Weierstrass (from the Bolzano-Weierstrass theorem). The function itself is defined by the infinite sum

$$W(x) = \sum_{n=0}^{\infty} a^n \cos(b^n \pi x),$$

where $a \in (0, 1)$, $b$ is a positive odd integer, and $ab > 1 + 3\pi/2$.

I agree, this definition feels totally random, and you are probably wondering: how did the author come up with it? To get a grip on this function, imagine this as the superposition of cosine waves with smaller and smaller amplitude but higher and higher frequency. Remember that differentiation implies "no sharp corners"?

This definition puts a sharp corner at every point on the real line.

Its graph is a fractal curve with self-similarity, as illustrated below.

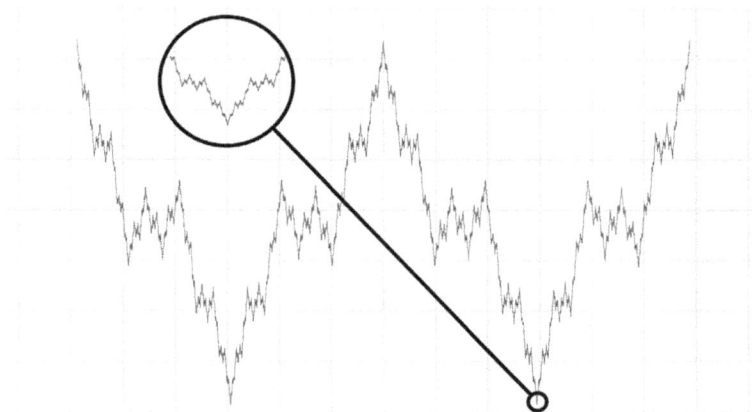

*Figure 12.5: Graph of the Weierstrass function. Source:* https://en.wikipedia.org/wiki/Weierstrass_function

Examples such as this inspired the opening quote of the section:

> *I turn with terror and horror from this lamentable scourge of continuous functions with no derivatives.*
> — *Charles Hermite*

19th-century mathematicians certainly did not think much about nondifferentiable functions. However, there are much more of them than differentiable ones. We won't go into the details, but amongst all continuous functions, the set of ones that are differentiable at at least one point is *meagre*. Meagre is a proper technical term for sets, and although we don't need to know what it means exactly, its name implies that it is extremely small.

Now that we understand what the derivative is, it's time to put theory into practice. How do we compute derivatives, and how do we work with them in machine learning? We'll see in the next section.

## 12.2 Differentiation in practice

During our first encounter with differentiation, we saw that computing derivatives by the definition

$$f'(x_0) = \lim_{x \to x_0} \frac{f(x_0) - f(x)}{x_0 - x}$$

can be really hard in practice if we encounter convoluted functions such as $f(x) = \cos(x) \sin(e^x)$. Similar to convergent sequences and limits, using the definition of differentiation won't get us far—the complexity piles on fast. So, we have to find ways to decompose the complexity into its fundamental building blocks.

## 12.2.1 Rules of differentiation

First, we'll look at the simplest of operations: scalar multiplication, addition, multiplication, and division.

**Theorem 12.2.1** *(Rules of differentiation)*

*Let* $f : \mathbb{R} \to \mathbb{R}$ *and* $g : \mathbb{R} \to \mathbb{R}$ *be two arbitrary functions and let* $x \in \mathbb{R}$. *Suppose that both* $f$ *and* $g$ *is differentiable at* $x$. *Then*

*(a)* $(cf)'(x) = cf'(x)$ *for all* $c \in \mathbb{R}$,

*(b)* $(f + g)'(x) = f'(x) + g'(x)$,

*(c)* $(fg)'(x) = f'(x)g(x) + f(x)g'(x)$ *(the product rule)*,

*(d)* $\left(\frac{f}{g}\right)'(x) = \frac{f'(x)g(x)-f(x)g'(x)}{g(x)^2}$ *if* $g(x) \neq 0$ *(the quotient rule)*.

*Proof.* *(a)* and *(b)* is a direct consequence of the *Theorem 11.2.1*.

To show *(c)*, we have to do a bit of algebra:

$$
\begin{aligned}
\lim_{y\to x} \frac{f(x)g(x) - f(y)g(y)}{x - y} &= \lim_{x\to y} \frac{f(x)g(x) - f(y)g(x) + f(y)g(x) - f(y)g(y)}{x - y} \\
&= \lim_{y\to x} \frac{f(x)g(x) - f(y)g(x)}{x - y} + \lim_{x\to y} \frac{f(y)g(x) - f(y)g(y)}{x - y} \\
&= \lim_{y\to x} \left[\frac{f(x) - f(y)}{x - y} g(x)\right] + f(y) \lim_{x\to y} \frac{g(x) - g(y)}{x - y} \\
&= f'(x)g(x) + f(x)g'(x),
\end{aligned}
$$

from which *(c)* follows.

For *(d)*, we are going to start with the special case of $(1/g)'$. We have

$$
\begin{aligned}
\lim_{y\to x} \frac{\frac{1}{g(x)} - \frac{1}{g(y)}}{x - y} &= \lim_{y\to x} \frac{1}{g(x)g(y)} \frac{g(y) - g(x)}{x - y} \\
&= -\frac{g'(x)}{g(x)^2},
\end{aligned}
$$

from which the general case follows by applying *(c)* to $f$ and $1/g$.

There is one operation which we haven't covered in the previous theorem: function composition. In the study of neural networks, composition plays an essential role. Each layer can be thought of as a function, which are composed together to form the entire network.

---

**Theorem 12.2.2** *(Chain rule/Leibniz rule)*

*Let $f : \mathbb{R} \mapsto \mathbb{R}$ and $g : \mathbb{R} \mapsto \mathbb{R}$ be two arbitrary functions and let $x \in \mathbb{R}$. Suppose that $g$ is differentiable at $x$ and $f$ is differentiable at $g(x)$. Then*

$$(f \circ g)'(x) = f'(g(x))g'(x)$$

*holds.*

---

*Proof.* First, we rewrite the differential quotient into the following form:

$$\lim_{y \to x} \frac{f(g(x)) - f(g(y))}{x - y} = \lim_{y \to x} \frac{f(g(x)) - f(g(y))}{g(x) - g(y)} \frac{g(x) - g(y)}{x - y}$$

$$= \lim_{y \to x} \frac{f(g(x)) - f(g(y))}{g(x) - g(y)} \lim_{x \to y} \frac{g(x) - g(y)}{x - y}.$$

Because $g$ is differentiable at $x$, it is also continuous there, so $\lim_{y \to x} g(y) = g(x)$. So, the first term can be rewritten as

$$\lim_{y \to x} \frac{f(g(x)) - f(g(y))}{g(x) - g(y)} = \lim_{y \to g(x)} \frac{f(y) - f(g(x))}{y - g(x)} = f'(g(x)).$$

Since $g$ is differentiable at $x$, the second term is $g'(x)$. Thus, we have

$$\lim_{y \to x} \frac{f(g(x)) - f(g(y))}{x - y} = f'(g(x))g'(x),$$

which is what we had to show.

As neural networks are just huge composed functions, their derivative is calculated with the repeated application of the chain rule. (Although the derivatives of its layers are vectors and matrices since they are multivariable functions.)

## 12.2.2 Derivatives of elementary functions

Following the already familiar pattern, now we calculate the derivatives for the most important class: elementary functions. There are a few that we will encounter all the time, like in the mean squared error, cross-entropy, Kullback-Leibler divergence, etc.

> **Theorem 12.2.3** *(Derivatives of elementary functions)*
>
> *(a)* $(x^0)' = 0$ *and* $(x^n)' = nx^{n-1}$, *where* $n \in \mathbb{Z} \setminus 0$,
>
> *(b)* $(\sin x)' = \cos x$ *and* $(\cos x)' = -\sin x$,
>
> *(c)* $(e^x)' = e^x$,
>
> *(d)* $(\log x)' = \frac{1}{x}$.

You don't necessarily have to know how to prove these. I'll include the proof of *(a)*, but feel free to skip it, especially if this is your first encounter with calculus. What you have to remember, though, are the derivatives themselves. (However, I'll refer back to this part when necessary.)

*Proof.* *(a)* It is easy to see that for $n = 0$, the derivative $(x^0)' = 0$. The case $n = 1$ is also simple: calculating the differential quotient shows that $(x)' = 1$. For the case $n \geq 2$, we are going to employ a small trick. Writing out the differential quotient for $f(x) = x^n$, we obtain

$$\frac{x^n - y^n}{x - y},$$

which we want to simplify. If you don't have a lot of experience in math, it might seem like magic, but $x^n - y^n$ can be written as

$$x^n - y^n = (x - y)(x^{n-1} + x^{n-2}y + \cdots + xy^{n-2} + y^{n-1})$$
$$= (x - y)\sum_{k=0}^{n-1} x^{n-1-k}y^k.$$

This can be seen easily by calculating the product:

$$(x - y)\sum_{k=0}^{n-1} x^{n-1-k}y^k. = x^n + \left[x^{n-1}y + \cdots + xy^{n-1}\right]$$
$$- \left[x^{n-1}y + \cdots + xy^{n-1}\right] - y^n$$
$$= x^n - y^n.$$

Thus, we have

$$\lim_{y \to x} \frac{x^n - y^n}{x - y} = \lim_{x \to y} \frac{(x - y) \sum_{k=0}^{n-1} x^{n-1-k} y^k}{x - y}$$

$$= \lim_{y \to x} \sum_{k=0}^{n-1} x^{n-1-k} y^k$$

$$= n x^{n-1}.$$

So, $(x^n)' = n x^{n-1}$. With this and the rules of differentiation, we can calculate the derivative of any polynomial $p(x) = \sum_{k=0}^{n} p_k x^k$ as

$$p'(x) = \sum_{k=1}^{n} k p_k x^{k-1}.$$

The case $n < 0$ follows from $x^{-n} = 1/x^n$ using the rules of differentiation.

With these rules under our belt, we can calculate the derivatives for some of the most famous activation functions.

The most classical one, the *sigmoid* function is defined by

$$\sigma(x) = \frac{1}{1 + e^{-x}}.$$

Since it is an elementary function, it is differentiable everywhere. To calculate its derivative, we can use the quotient rule:

$$\sigma'(x) = \left( \frac{1}{1 + e^{-x}} \right)' = \frac{e^{-x}}{(1 + e^{-x})^2}$$

$$= \frac{1}{1 + e^{-x}} \frac{e^{-x}}{1 + e^{-x}} \tag{12.4}$$

$$= \sigma(x)(1 - \sigma(x)).$$

Now that we have the sigmoid and its derivative, let's plot them together!

```python
def sigmoid(x):
    return 1/(1 + np.exp(-x))

def sigmoid_prime(x):
    return sigmoid(x) - sigmoid(x)**2
```

```python
import numpy as np
import matplotlib.pyplot as plt

xs = np.linspace(-10, 10, 1000)

with plt.style.context("seaborn-v0_8"):
    plt.title("Sigmoid and its derivative")
    plt.plot(xs, [sigmoid(x) for x in xs], label="Sigmoid")
    plt.plot(xs, [sigmoid_prime(x) for x in xs], label="Sigmoid prime")
    plt.legend()
    plt.tight_layout()
    plt.show()
```

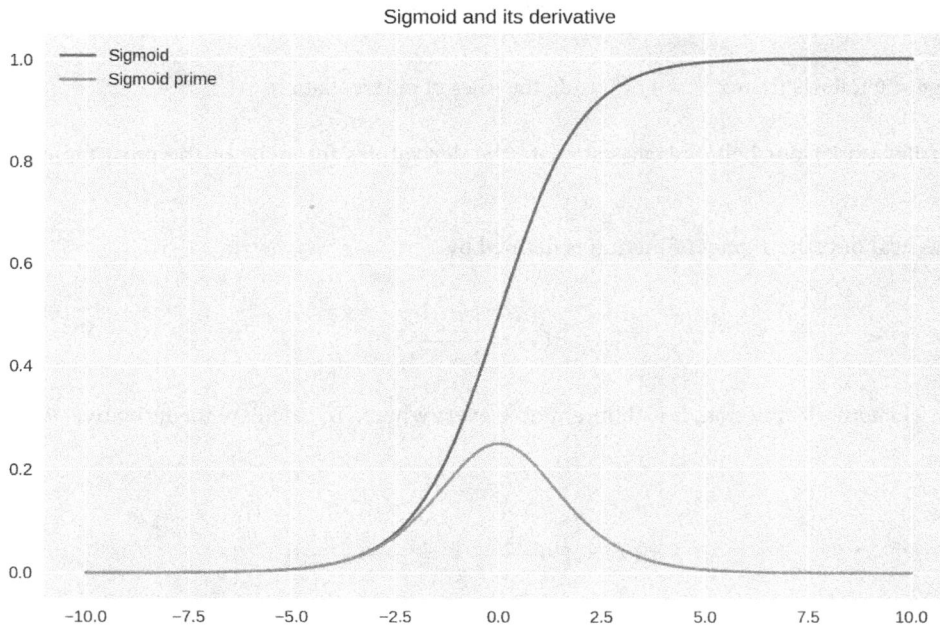

*Figure 12.6: Sigmoid and its derivative*

Another popular activation function is the *ReLU*, defined by

$$\text{ReLU}(x) = \begin{cases} x & \text{if } x > 0, \\ 0 & \text{otherwise.} \end{cases}$$

Let's plot its graph first!

```python
def relu(x):
    if x > 0:
        return x
    else:
        return 0


xs = np.linspace(-5, 5, 1000)

with plt.style.context("seaborn-v0_8"):
    plt.title("Graph of the ReLU function")
    plt.plot(xs, [relu(x) for x in xs], label="ReLU")
    plt.legend()
    plt.tight_layout()
    plt.show()
```

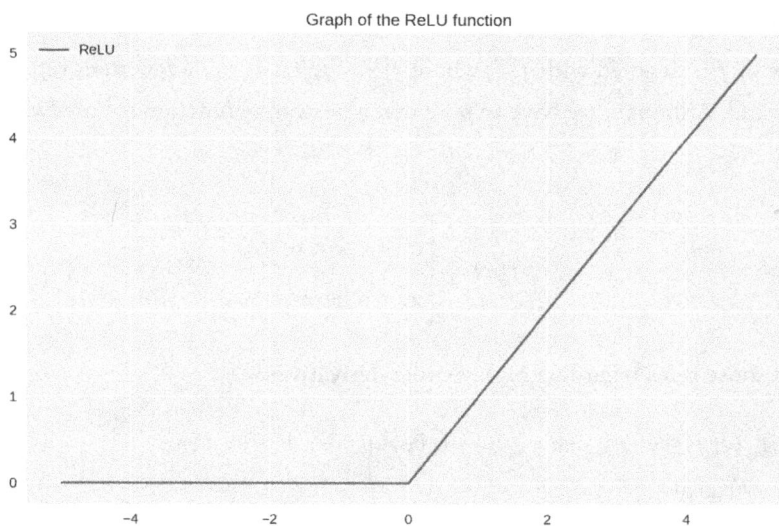

*Figure 12.7: Graph of the ReLU function*

By looking at it, we can suspect that it is not differentiable at 0. Indeed, since

$$\frac{\text{ReLU}(x) - \text{ReLU}(0)}{x} = \begin{cases} 1 & \text{if } x > 0, \\ 0 & \text{if } x < 0, \end{cases}$$

the limit of the differential quotient doesn't exist.

However, besides 0, it is differentiable and

$$\text{ReLU}'(x) = \begin{cases} 1 & \text{if } x > 0, \\ 0 & \text{if } x < 0. \end{cases}$$

Even though ReLU is not differentiable at 0, this is not a problem in practice. When performing backpropagation, it is extremely unlikely that $\text{ReLU}'(x)$ will receive 0 as its input. Even if this is the case, the derivative can be artificially extended to zero by defining it as 0.

## 12.2.3   Higher-order derivatives

One last thing to do before we move on is to talk about higher-order derivatives. Because derivatives are *functions*, it is a completely natural idea to calculate the *derivative of derivatives*. As we will see when studying the basics of optimization in *Chapter 14*, the second derivatives contain quite a lot of essential information regarding minima and maxima.

The $n$-th derivative of $f$ is denoted with $f^{(n)}$, where $f^{(0)} = f$. There are a few rules regarding them that are worth keeping in mind. Although, we have to note that a derivative function is not always differentiable, as the example

$$f(x) = \begin{cases} 0 & \text{if } x < 0, \\ x^2 & \text{otherwise} \end{cases}$$

shows. Now, about those rules regarding higher-order derivatives.

> **Theorem 12.2.4**  *Let $f : \mathbb{R} \to \mathbb{R}$ and $g : \mathbb{R} \to \mathbb{R}$ be two arbitrary functions.*
>
> *(a) $(f + g)^{(n)} = f^{(n)} + g^{(n)}$*
>
> *(b) $(fg)^{(n)} = \sum_{k=0}^{n} \binom{n}{k} f^{(n-k)} g^{(k)}$*

*Proof. (a)* trivially follows from the linearity of differentiation.

Regarding *(b)*, we are going to use proof by induction. For $n = 1$, the statement simply says that $(fg)' = f'g + fg'$, as we have seen before.

Now, we assume that it is true for $n$ and deduce the $n + 1$ case. For this, we have

$$(fg)^{(n+1)} = \left( (fg)^{(n)} \right)' = \sum_{k=0}^{n} \binom{n}{k} \left( f^{(n-k)} g^{(k)} \right)'$$

$$= \sum_{k=0}^{n} \binom{n}{k} \left[ f^{(n-k+1)} g^{(k)} + f^{(n-k)} g^{(k+1)} \right]$$

$$= \sum_{k=0}^{n} \binom{n}{k} f^{(n-k+1)} g^{(k)} + \sum_{k=0}^{n} \binom{n}{k} f^{(n-k)} g^{(k+1)}$$

$$= \binom{n}{0} f^{(n+1)} g + \left[ \sum_{k=1}^{n} \binom{n}{k} f^{(n+1-k)} g^{(k)} \right] + \left[ \sum_{k=0}^{n-1} \binom{n}{k} f^{(n-k)} g^{(k+1)} \right] + \binom{n}{n} f g^{(n+1)}.$$

First, we note that $\binom{n}{0} = \binom{n+1}{0} = 1$ and $\binom{n}{n} = \binom{n+1}{n+1} = 1$. Second, the recursive relation for binomial coefficients says that

$$\binom{n+1}{k} = \binom{n}{k} + \binom{n}{k-1}.$$

With a simple reindexing, we have

$$\sum_{k=0}^{n-1} \binom{n}{k} f^{(n-k)} g^{(k+1)} = \sum_{k=1}^{n} \binom{n}{k-1} f^{(n+1-k)} g^{(k)},$$

so we can join the two sums together and obtain

$$(fg)^{(n+1)} = \binom{n+1}{0} f^{(n+1)} g + \sum_{k=1}^{n} \left[ \binom{n}{k} + \binom{n}{k-1} \right] f^{(n+1-k)} g^{(k)} + \binom{n+1}{n+1} f g^{(n+1)}$$

$$= \sum_{k=0}^{n+1} \binom{n+1}{k} f^{(n+1-k)} g^{(k)},$$

which is what we had to show.

## 12.2.4  Extending the Function base class

Now that we have several tools under our belt to calculate derivatives, it's time to think about implementations. Since we have our own Function base class (*Section 9.2.3*), a natural idea is to implement the derivative as a method. This is a simple solution that is in line with object-oriented principles as well, so we should go for it!

```
class Function:
    def __init__(self):
        pass

    def __call__(self, *args, **kwargs):
        pass

    # new interface element for
    # computing the derivative
    def prime(self):
        pass

    def parameters(self):
        return dict()
```

To see a concrete example, let's revisit the sigmoid function, whose derivative is given by (12.4):

$$\sigma'(x) = \sigma(x)(1 - \sigma(x)).$$

```
class Sigmoid(Function):
    def __call__(self, x):
        return 1/(1 + np.exp(-x))

    def prime(self, x):
        return self(x) - self(x)**2
```

Simple implementation, powerful functionality. Now that we have the derivatives covered, let's move towards calculating the derivative of more complex functions!

## 12.2.5  The derivative of compositions

At this point, I have probably emphasized the importance of function compositions and the chain rule (*Theorem 12.2.2*) dozens of times. We have finally reached a point when we are ready to implement a simple neural network and compute its derivative! (Of course, our methods will be far more refined in the end, but still, this is a milestone.)

How can we calculate the derivative for a composition of *n* functions?

To see the pattern, let's map out the first few cases. For $n = 2$, we have the good old chain rule

$$\left(f_2(f_1(x))\right)' = f_2'(f_1(x)) \cdot f_1'(x).$$

For $n = 3$, we have

$$\left(f_3(f_2(f_1(x)))\right)' = f_3'(f_2(f_1(x))) \cdot f_2'(f_1(x)) \cdot f_1'(x).$$

Among the multitude of parentheses, we can notice a pattern. First, we should calculate the value of the composed function $f_3 \circ f_2 \circ f_1$ at $x$ while storing the intermediate results, then pass these to the appropriate derivatives and take the product of the result.

```python
class Composition(Function):
    def __init__(self, *functions):
        self.functions = functions

    def __call__(self, x):
        for f in self.functions:
            x = f(x)

        return x

    def prime(self, x):
        forward_pass = [x]

        for f in self.functions:
            try:
                x = f(x)
                forward_pass.append(x)
            except ValueError as e:
                print(f"Error in function {f}: {e}")
                return np.nan

        forward_pass.pop()     # removing the last element, as we won't need it

        derivative = np.prod([f.prime(x) for f, x in zip(self.functions, forward_pass)])

        return derivative
```

To see if our implementation works, we should test it on a simple test case, say for

$$f_1(x) = 2x,$$
$$f_2(x) = 3x,$$
$$f_3(x) = 4x.$$

The derivative of the composition $(f_3 \circ f_2 \circ f_1)(x) = 24x$ should be constant 24.

```python
class Linear(Function):
    def __init__(self, a, b):
        self.a = a
        self.b = b

    def __call__(self, x):
        return self.a*x + self.b

    def prime(self, x):
        return self.a

    def parameters(self):
        return {"a": self.a, "b": self.b}


f = Composition(Linear(2, 0), Linear(3, 0), Linear(4, 0))

xs = np.linspace(-10, 10, 1000)
ys = [f.prime(x) for x in xs]

with plt.style.context("seaborn-v0_8"):
    plt.title("The derivative of f(x) = 24x")
    plt.plot(xs, ys, label="f prime")
    plt.legend()
    plt.tight_layout()
    plt.show()
```

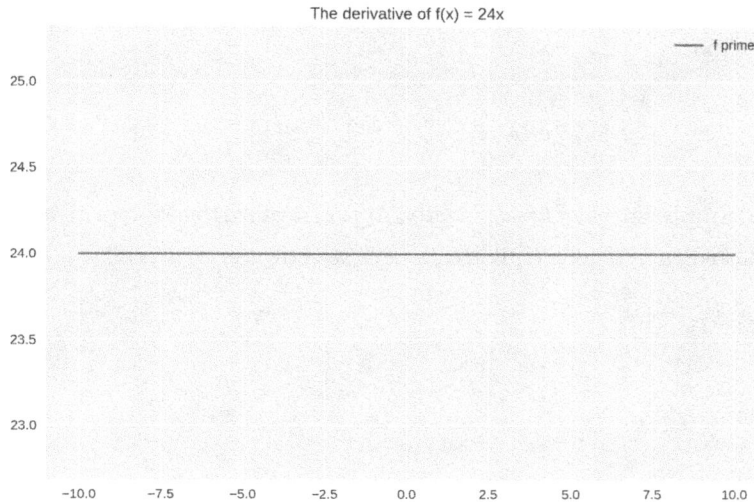

*Figure 12.8: The derivative of $f(x) = 24x$*

Success! Even though we're only dealing with single-variable functions for now, our `Composition` is going to be the skeleton for neural networks.

## 12.2.6 Numerical differentiation

So far, we have seen that in the cases when at least some formula is available for the function in question, we can apply the rules of differentiation (see *Theorem 12.2.1*) to obtain the derivative.

However, in practice, this is often not the case. For instance, think about the case when the function represents a recorded audio signal. If we can't compute the derivative exactly, a natural idea is to *approximate* it, that is, provide an estimate that is sufficiently close to the real value.

For the sake of example, suppose that we don't know the exact formula of our function to be differentiated, which is secretly the good old sine function.

```
def f(x):
    return np.sin(x)
```

Recall that, by definition, the derivative is given by

$$f'(x) = \lim_{h \to 0} \frac{f(x+h) - f(x)}{h}.$$

Since we can't take limits inside a computer (as computers can't deal with infinity), the second best thing to do is to approximate this by

$$\Delta_h f(x) = \frac{f(x + h) - f(x)}{h},$$

where $h > 0$ is an arbitrarily small but fixed quantity. $\Delta_h f(x)$ is called the *forward difference quotient*. In theory, $\Delta_h f(x) \approx f'(x)$ holds when $h$ is sufficiently small. Let's see how they perform!

```python
def delta(f, h, x):
    return (f(x + h) - f(x))/h

def f_prime(x):
    return np.cos(x)


hs = [3.0, 1.0, 0.1]
xs = np.linspace(-5, 5, 100)
f_prime_ys = [f_prime(x) for x in xs]

with plt.style.context("seaborn-v0_8"):
    _colormap = plt.cm.hot_r
    plt.figure(figsize=(10, 5))
    plt.title("Approximating the derivative with finite differences")

    true_color = _colormap(0.99)   # Get a fixed color for the true derivative
    for i, h in enumerate(hs):
        ys = [delta(f, h, x) for x in xs]
        blend_ratio = 1 - (len(hs) - i) / len(hs)   # Progressively blend closer to the
        true color
        approx_color = _colormap(blend_ratio)
        plt.plot(xs, ys, label=f"h = {h}", color=approx_color)

    plt.plot(xs, f_prime_ys, label="the true derivative", color=true_color, linewidth=2)
    plt.legend()
    plt.tight_layout()
    plt.show()
```

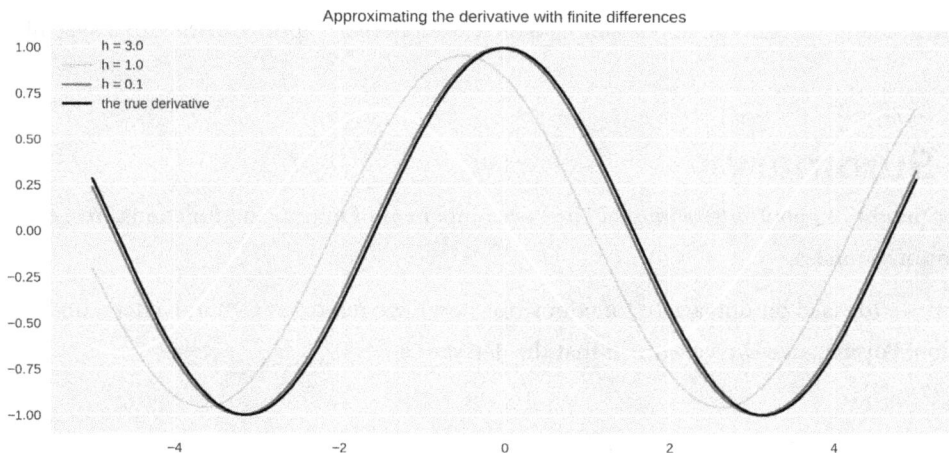

*Figure 12.9: Approximating the derivative with finite differences*

Although the $\Delta_h f(x)$ functions *seem* to be close $f'(x)$, when $h$ is small, there is a plethora of potential issues. For one, $\Delta_h f(x) = \frac{f(x+h)-f(x)}{h}$ only approximates the derivative from the right of $x$, as $h > 0$. To solve this, one might use the *backward difference quotient*

$$\nabla_h f(x) = \frac{f(x) - f(x - h)}{h},$$

but that seem to have the same problems. The crux of the issue is that if $f$ is differentiable at some $x$, then

$$\frac{f(x + h) - f(x)}{h} \approx \frac{f(x) - f(x - h)}{h},$$

but only if $h$ is very small, and the "good enough" choice for $h$ can vary from point to point.

A middle ground is provided by the so-called symmetric difference quotients, defined by

$$\delta_h f(x) = \frac{f(x + h) - f(x - h)}{2h}, \quad h \in (0, \infty),$$

which is the average of forward and backward differences: $\delta_h f(x) = \frac{\Delta_h f(x) + \nabla_h f(x)}{2}$. These three approximators are called *finite differences*. With their repeated application, we can approximate higher-order derivatives as well.

Even though symmetric differences are provably better, the approximation errors can still be significantly amplified in the long run.

All things considered, we are not going to use finite differences for machine learning in practice. However, as we'll see, the gradient descent method is simply a forward difference approximation of a special differential equation.

## 12.3  Summary

This chapter taught us about *differentiation*, the key component of optimizing functions. Yes, even functions with millions of variables.

Even though we focused on univariate functions (for now), we managed to build a deep understanding of differentiation. For instance, we've learned that the derivative

$$f'(x) = \lim_{y \to x} \frac{f(x) - f(y)}{x - y}$$

describes the slope of the tangent line drawn to the graph of $f$ at $x$, which describes the velocity if $f$ is the trajectory of a one-dimensional motion. From the perspective of physics, the derivative describes the rate of change.

However, from the perspective of mathematics, differentiation offers much more than the rate of change: we've seen that a differentiable function can be written in the form

$$f(x) = f(x_0) + f'(x_0)(x - x_0) + o(|x - x_0|)$$

around some $x_0 \in \mathbb{R}$. In other words, locally speaking, a differentiable function is a linear part plus a small error term. Unlike the limit-of-quotients definition, this will generalize for multiple variables without an issue. Moreover, we can apply a similar idea to obtain the so-called Taylor expansion

$$f(x) = \sum_{k=0}^{n} \frac{f^{(k)}(x_0)}{k!}(x - x_0)^k + o(|x - x_0|^n),$$

allowing us to approximate transcendental functions like $\log x, \sin x, \cos x, e^x$ with polynomials.

In addition to the theory, we also learned about computing derivatives in practice.

This is done by either 1) decomposing complex functions into their building blocks, then calculating the derivative using the rules

$$(cf)'(x) = cf'(x),$$
$$(f + g)'(x) = f'(x) + g'(x),$$
$$(fg)'(x) = f'(x)g(x) + f(x)g'(x),$$
$$(f \circ g)'(x) = f'(g(x))g'(x),$$

or 2) approximating the derivative with finite differences like

$$\Delta_h f(x) = \frac{f(x + h) - f(x)}{h},$$

where $h > 0$ is a small constant. The former method is the foundation of backpropagation, while the latter lies at the heart of gradient descent. These are the methods that truly enable training huge neural networks.

Now that we understand differentiation, it's time to talk about its counterpart: integration. Let's go!

## 12.4  Problems

**Problem 1.** Calculate the derivative of the $\tanh(x)$ function defined by

$$\tanh(x) = \frac{e^x - e^{-x}}{e^x + e^{-x}}.$$

**Problem 2.** Define the function

$$f(x) = \begin{cases} 0 & \text{if } x < 0, \\ x^2 & \text{otherwise.} \end{cases}$$

Find the derivative of $f(x)$. Is $f'(x)$ differentiable everywhere?

**Problem 3.** Define the function

$$f(x) = \begin{cases} x^2 & \text{if } x \in \mathbb{Q}, \\ -x^2 & \text{otherwise.} \end{cases}$$

Show that

*(a) f is differentiable at 0, and $f'(0) = 0$, (b) and f is nowhere else differentiable.*

**Problem 4.** Calculate the derivatives of the following functions.

*(a) $f(x) = e^{-\frac{x^2}{2}}$*

*(b) $f(x) = x^2 e^{\sin x}$*

*(c) $f(x) = \sin(\cos x^2)$*

*(d) $f(x) = \frac{x^2+1}{x^2-1}$*

**Problem 5.** Find the Taylor expansion of the following functions around 0.

*(a) $f(x) = \sin x$*

*(b) $f(x) = \cos x$*

*(c) $f(x) = \log x$*

*(d) $f(x) = e^{\{-x}2\}$*

**Problem 6.** Find the Taylor expansion of the function

$$f(x) = \begin{cases} e^{-\frac{1}{x^2}} & \text{if } x \neq 0, \\ 0 & \text{if } x = 0. \end{cases}$$

around 0. Is the Taylor expansion of $f$ equal to $f$?

# Join our community on Discord

Read this book alongside other users, Machine Learning experts, and the author himself.

Ask questions, provide solutions to other readers, chat with the author via Ask Me Anything sessions, and much more.

Scan the QR code or visit the link to join the community.

`https://packt.link/math`

# 13

# Optimization

If someone gave you a function defined by some tractable formula, how would you find its minima and maxima? Take a moment and conjure up some ideas before moving on.

The first idea that comes to mind for most people is to evaluate the function for all possible values and simply find the optimum. This method immediately breaks down due to multiple reasons. We can only perform finite evaluations, so this would be impossible. Even if we cleverly define a discrete search grid and evaluate only there, this method takes an unreasonable amount of time.

Another idea is to use some kind of inequality to provide an ad hoc upper or lower bound, then see if this bound can be attained. Sadly, this is nearly impossible for more complicated functions, like losses for neural networks.

However, derivatives provide an extremely useful way to optimize functions. In this chapter, we will study the relationship between derivatives and optimal points, and algorithms on how to find them. Let's go!

# 13.1   Minima, maxima, and derivatives

Intuitively, the notion of minima and maxima is simple. Take a look at *Figure 13.1* below.

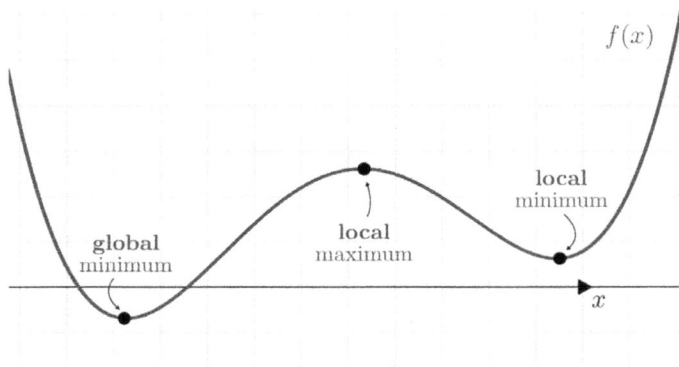

*Figure 13.1: Local and global optima*

Peaks of hills are the maxima, and the bottoms of valleys are the minima. Minima and maxima are collectively called extremal or optimal points. As our example demonstrates, we have to distinguish between local and global optima. The graph has two valleys, and although both have a bottom, one of them is lower than the other.

The really interesting part is finding these, as we'll see next. Let's consider our example above to demonstrate how derivatives are connected to local minima and maxima.

If we use our geometric intuition, we see that the tangents are horizontal at the peaks of the hills and the bottoms of the valleys. Intuitively, if the tangent line is not horizontal, then there's an elevation or decline in the graph. This is illustrated by *Figure 13.2*.

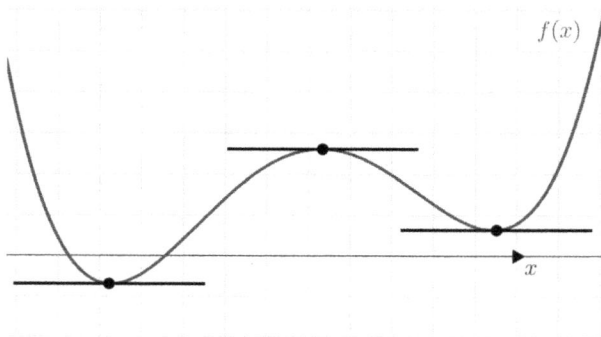

*Figure 13.2: Tangents at local and global optima*

In terms of derivatives, since they describe the slope of the tangent, it means that the derivative should be 0 there.

If we think about the function as the description of a motion along the real line, derivatives say that the motion stops there and changes direction. It slows down first, stops, then immediately starts in the opposite direction. For instance, in the local maxima case, the function increases up until that point, where it starts decreasing.

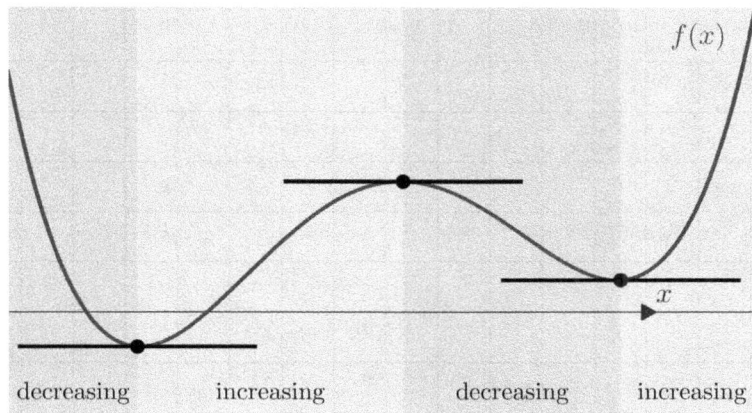

*Figure 13.3: The flow of the function*

Again, we can describe this monotonicity behavior in terms of derivatives. Notice that when the function increases, the derivative is positive (the object in motion has a positive speed). On the other hand, decreasing parts have a negative derivative.

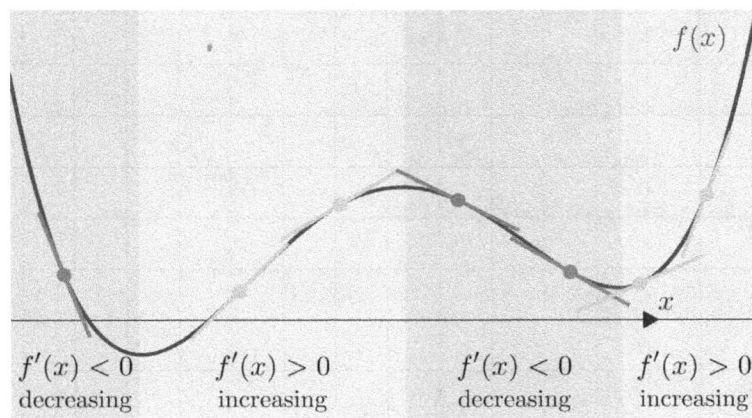

*Figure 13.4: The sign of the derivatives*

We can go ahead and put these intuitions into a mathematical form. First, we'll start with the definitions' monotonicity, and their relation to the derivative. Then, we'll connect all the dots and see how this comes together to characterize the optima.

**Definition 13.1.1 (Locally increasing and decreasing functions)**

Let $f : \mathbb{R} \to \mathbb{R}$ be an arbitrary function and let $a \in \mathbb{R}$. We say that

*(a) $f$ is locally increasing at $a$ if there is a neighborhood $(a - \delta, a + \delta)$ such that*

$$f(a) \begin{cases} \geq f(x), & \text{if } x \in (a - \delta, a), \\ \leq f(x), & \text{if } x \in (a, a + \delta), \end{cases}$$

*(b) and $f$ is strictly locally increasing at $a$ if there is a neighborhood $(a - \delta, a + \delta)$ such that*

$$f(a) \begin{cases} > f(x), & \text{if } x \in (a - \delta, a), \\ < f(x), & \text{if } x \in (a, a + \delta). \end{cases}$$

The locally decreasing and strictly locally decreasing properties are defined similarly, with the inequalities reversed.

For differentiable functions, the behavior of the derivative describes their local behavior in terms of monotonicity.

**Theorem 13.1.1** *Let $f : \mathbb{R} \to \mathbb{R}$ be an arbitrary function that is differentiable at some $a \in \mathbb{R}$.*

*(a) If $f'(a) \geq 0$, then $f$ is locally increasing at $a$.*

*(b) If $f'(a) > 0$, then $f$ is strictly locally increasing at $a$.*

*(c) If $f'(a) \leq 0$, then $f$ is locally decreasing at $a$.*

*(d) If $f'(a) < 0$, then $f$ is strictly locally decreasing at $a$.*

*Proof.* We will only show *(a)*, since the rest of the proofs go the same way. Due to how limits are defined (*Definition 11.2.1*),

$$\lim_{x \to a} \frac{f(x) - f(a)}{x - a} = f'(a) \geq 0$$

means that once $x$ gets close enough to $a$, that is, $x$ is from a small neighborhood $(a - \delta, a + \delta)$,

$$\frac{f(x) - f(a)}{x - a} \geq 0, \quad x \in (a - \delta, a + \delta)$$

holds. If $x > a$, then because the differential quotient is nonnegative, $f(x) \geq f(a)$ must hold. Similarly, for $x < a$, the nonnegativity of the differential quotient implies that $f(x) \leq f(a)$.

The proofs for *(b)*, *(c)*, and *(d)* are almost identical, with the obvious changes in the inequalities.

The propositions related to *not strict* monotonicity are true the other way around as well.

**Theorem 13.1.2** *Let $f : \mathbb{R} \to \mathbb{R}$ be an arbitrary function that is differentiable at some $a \in \mathbb{R}$.*

*(a) If $f$ is locally increasing at $a$, then $f'(a) \geq 0$.*

*(b) If $f$ is locally decreasing at $a$, then $f'(a) \leq 0$.*

*Proof.* Similar to before, we will only show the proof of *(a)*, since *(b)* can be done in the same way. If $f$ is locally increasing at $a$, then the differential quotient is positive:

$$\frac{f(x) - f(a)}{x - a} \geq 0.$$

Using the transfer principle of limits (*Theorem 11.2.4*), we obtain

$$f'(a) = \lim_{x \to a} \frac{f(x) - f(a)}{x - a} \geq 0,$$

which is what we had to prove.

After all this setup, we are ready to study local optima. What can the derivative tell us about them? Let's see!

## 13.1.1 Local minima and maxima

As we have seen in the introduction, the tangent at the extremal points is horizontal. Now it is time to put this introduction into a mathematically correct form.

**Definition 13.1.2 (Local minima and maxima)**

Let $f : \mathbb{R} \to \mathbb{R}$ be an arbitrary function and let $a \in \mathbb{R}$.

*(a) $a$ is a local minimum*, if there is a neighborhood $(a - \delta, a + \delta)$ such that for every $x \in (a - \delta, a + \delta)$, $f(a) \leq f(x)$ holds.

(b) $a$ is a *strict local minimum*, if there is a neighborhood $(a - \delta, a + \delta)$ such that for every $x \in (a - \delta, a + \delta)$, $f(a) < f(x)$ holds.

(c) $a$ is a *local maximum*, if there is a neighborhood $(a - \delta, a + \delta)$ such that for every $x \in (a - \delta, a + \delta)$, $f(x) \leq f(a)$ holds.

(d) $a$ is a *strict local maximum*, if there is a neighborhood $(a - \delta, a + \delta)$ such that for every $x \in (a - \delta, a + \delta)$, $f(x) < f(a)$ holds.

Extremal points have their global versions as well. The sad truth is, even though we always want global optima, we only have the tools to find local ones.

**Definition 13.1.3 (Global minima and maxima)**

Let $f : \mathbb{R} \to \mathbb{R}$ be an arbitrary function and let $a \in \mathbb{R}$.

(a) $a$ is a *global minimum* if $f(a) \leq f(x)$ holds for every $x \in \mathbb{R}$.

(b) $a$ is a *global maximum* if $f(x) \leq f(a)$ holds for every $x \in \mathbb{R}$.

Note that a global optimum is also a local optimum, but not the other way around.

**Theorem 13.1.3** *Let $f : \mathbb{R} \to \mathbb{R}$ be an arbitrary function that is differentiable at some $a \in \mathbb{R}$. If $f$ has a local minima or maxima at $a$, then $f'(a) = 0$.*

*Proof.* According to *Theorem 13.1.1*, if $f'(a) \neq 0$, then it is either strictly increasing or decreasing locally. Since this contradicts our assumption that $a$ is a local optimum, the theorem is proven.

(In case you are interested, this was the principle of contraposition (*Theorem B.5.4*) in action. From the negation of the conclusion, we have shown the negation of the premises.)

It is very important to emphasize that the theorem is not true the other way around. For instance, the function $f(x) = x^3$ is strictly increasing everywhere, yet $f'(0) = 0$.

In general, we call this behavior *inflection*. So, $f(x) = x^3$ is said to have an inflection point at 0. Inflection means a change in behavior, which reflects the switch in its derivative from decreasing to increasing in this case. (The multidimensional analogue of inflection is called a "saddle," as we shall see later.)

So, we are not at our end goal yet, as the other half of the promised characterization is missing.

The derivative is zero at the local extremal points, but can we come up with a criterion that implies the existence of minima or maxima?

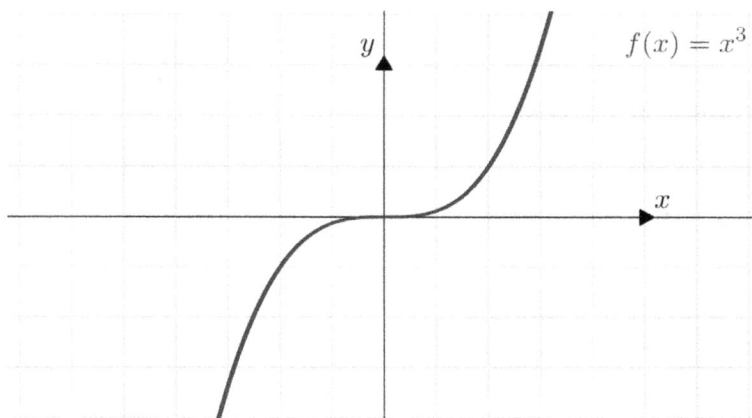

*Figure 13.5: Graph of $f(x) = x^3$ as a counterexample to show that $f'(0) = 0$ doesn't imply local optimum*

With the utilization of second derivatives, this is possible.

## 13.1.2 Characterization of optima with higher order derivatives

Let's take a second look at our example, considering the local behavior of $f'$ this time, not just its sign. In *Figure 13.6*, the derivative is plotted along with our function.

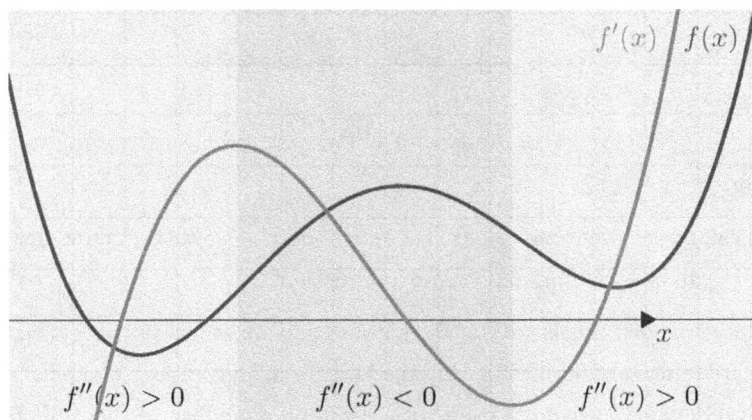

*Figure 13.6: The function and its derivative*

The pattern seems simple: an increasing derivative implies a local minimum, a decreasing one means a local maximum. This aligns with our intuition about derivative as speed: local maximum means that the object is going in a positive direction, then stops and starts reversing.

We can make this mathematically precise with the following theorem.

---

**Theorem 13.1.4** *(The second derivative test)*

Let $f : \mathbb{R} \to \mathbb{R}$ *be an arbitrary function that is twice differentiable at some* $a \in \mathbb{R}$.

*(a) If* $f'(a) = 0$ *and* $f''(a) > 0$, *then* $a$ *is a local minimum.*

*(b) If* $f'(a) = 0$ *and* $f''(a) < 0$, *then* $a$ *is a local maximum.*

---

*Proof.* Once again, we will only prove *(a)*, since the proof of *(b)* is almost identical.

First, as we saw when discussing the relation between derivatives and monotonicity (*Theorem 13.1.1*), $f''(a) > 0$ implies that $f'$ is strictly locally increasing at $a$. Since $f'(a) = 0$, this means that

$$f'(x) \begin{cases} \leq 0 & \text{if } x \in (a - \delta, a] \\ \geq 0 & \text{if } x \in [a, a + \delta) \end{cases}$$

for some $\delta > 0$. Because of *Theorem 13.1.1*, $f$ is locally decreasing in $(a - \delta, a]$ and locally increasing in $[a, a + \delta)$. This can only happen if $a$ is a local minimum.

In summary, the method of finding the extrema of a function $f$ is the following.

1. Solve $f'(x) = 0$. Its solutions $\{x_1, \dots, x_n\}$ — called *critical points* — are the candidates that *can be* extremal points. (But not necessarily all of them are.)

2. Check the sign of $f''(x_i)$ for all solutions $x_i$. If $f''(x_i) > 0$, it is a local minimum. If $f''(x_i) < 0$, it is a local maximum.

If $f''(x_i) = 0$, we still can't draw any conclusions. The functions $x^4$, $-x^2$, and $x^3$ show that critical points with zero second derivatives can be local minima, maxima, or neither.

Even though we have a "recipe," this is still far from enough for practical purposes. Not counting that our functions of interest are multivariable, calculating the derivative and solving $f'(x) = 0$ is not tractable. For loss functions of neural networks, we don't even bother writing out a formula because, for a composition of hundreds of functions, it can be unreasonably complex.

## 13.1.3   Mean value theorems

In some cases, we can extract a lot of information about the derivatives without explicitly calculating them. These results are extremely useful in cases where we don't have an explicit formula for the function or the formula might be too huge. (Like in the case of neural networks.) In the following, we'll get to meet the famous *mean value theorems*, connecting the function's behavior at the endpoints and inside an interval.

First, we start with a special case that states that the function attains the same value at the end of some interval $[a, b]$, then its derivative is zero somewhere inside the interval.

**Theorem 13.1.5  *(Rolle's mean value theorem)***

*Let $f : \mathbb{R} \to \mathbb{R}$ be a differentiable function and suppose that $f(a) = f(b)$ for some $a \neq b$. Then there exists a $\xi \in (a, b)$ such that $f'(\xi) = 0$.*

*Proof.* If you are a visual person, take a look at *Figure 13.7*. This is what we need to show.

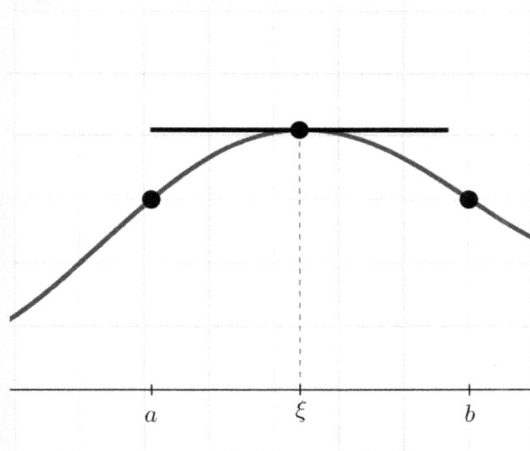

*Figure 13.7: Rolle's theorem*

To be mathematically precise, there are two cases. First, if $f$ is constant on $[a, b]$, then its derivative is zero on the entire interval.

If $f$ is not constant, then it attains some value $c$ inside $(a, b)$ that is not equal to $f(a) = f(b)$. For simplicity, suppose that $c > f(a)$. (The argument that follows goes through in the $c < f(a)$ case with some obvious changes.) Since $f$ is continuous, it attains its maximum there at a point $\xi \in [a, b]$. (See *Theorem 11.3.2*.) According to what we have just seen regarding the relation of local maxima and the derivative (*Theorem 13.1.3*), $f'(\xi) = 0$, which is what we had to show.

Rolle's theorem is an important stepping stone towards Lagrange's mean value theorem, which we will show in the following.

**Theorem 13.1.6** *(Lagrange's mean value theorem)*

*Let $f : \mathbb{R} \to \mathbb{R}$ be a differentiable function and $[a, b]$ an interval for some $a \neq b$. Then there exists a $\xi \in (a, b)$ such that*

$$f'(\xi) = \frac{f(b) - f(a)}{b - a}$$

*holds.*

*Proof.* Again, let's start with a visualization to get a grip on the theorem. *Figure 13.8* shows what we need to show.

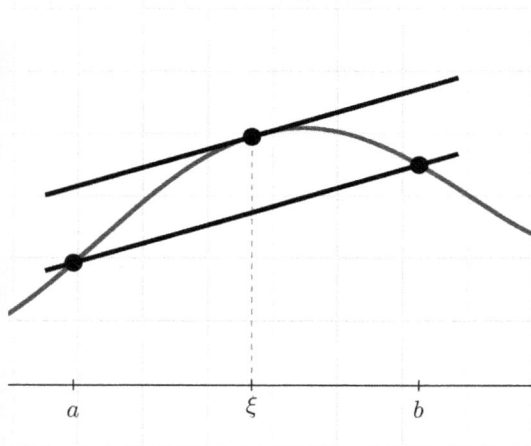

Figure 13.8: Lagrange's mean value theorem

Recall that $\frac{f(b)-f(a)}{b-a}$ is the slope of the line going through $(a, f(a))$ and $(b, f(b))$. This line is described by the function

$$\frac{f(b) - f(a)}{b - a}(x - a) + f(a),$$

as given by the point-slope equation of a line. Using this, we introduce the function

$$g(x) := f(x) - \left( \frac{f(b) - f(a)}{b - a}(x - a) + f(a) \right).$$

We can apply Rolle's theorem to $g(x)$, since $g(a) = g(b) = 0$. Thus, for some $\xi \in (a, b)$, we have

$$g'(\xi) = 0 = f'(\xi) - \frac{f(b) - f(a)}{b - a},$$

implying $f'(\xi) = \frac{f(b) - f(a)}{b - a}$, which is what we had to show.

Why are mean value theorems so important? In mathematics, they serve as a cornerstone in several results. To give you one example, think about integration. (Perhaps you are familiar with this concept already. Don't worry if not, we are going to study it in detail later.) Integration is essentially the inverse of differentiation: if $F'(x) = f(x)$, then

$$\int_a^b f(x)dx = F(b) - F(a),$$

which will be a simple consequence of Lagrange's mean value theorem.

# 13.2   The basics of gradient descent

We need to solve two computational problems to train neural networks:

- computing the derivative of the loss $L(w)$,
- and finding its minima using the derivative.

Finding the minima by solving $\frac{d}{dw}L(w) = 0$ is not going to work in practice. There are several problems. First, as we have seen, not all solutions are minimal points: there are maximal and inflection points as well. Second, solving this equation is not feasible except in the simplest cases, like for linear regression with the mean squared error. Training a neural network is *not* a simple case.

Fortunately for us, machine learning practitioners, there is a solution: gradient descent! The famous gradient descent provides a way to tackle the complexity of finding the exact solution, enabling us to do machine learning on a large scale. Let's see how it's done!

## 13.2.1   Derivatives, revisited

When we first explored the concept of the derivative in *Chapter 12*, we saw its many faces. We have learned that the derivative can be thought of as

- speed (when the function describes a time-distance graph of a moving object),
- the slope of the tangent line of a function,
- and the best linear approximator at a given point.

To understand how gradient descent works, we'll see yet another interpretation: derivatives as *vectors*. For any differentiable function $f(x)$, the derivative $f'(x)$ can be thought of as a one-dimensional vector. If $f'(x)$ is positive, it points to the right. If it is negative, it points to the left. We can visualize this by drawing a horizontal vector to every point of $f(x)$-s graph, where the length represents $|f'(x)|$ and the direction represents the sign. This is illustrated by *Figure 13.9*.

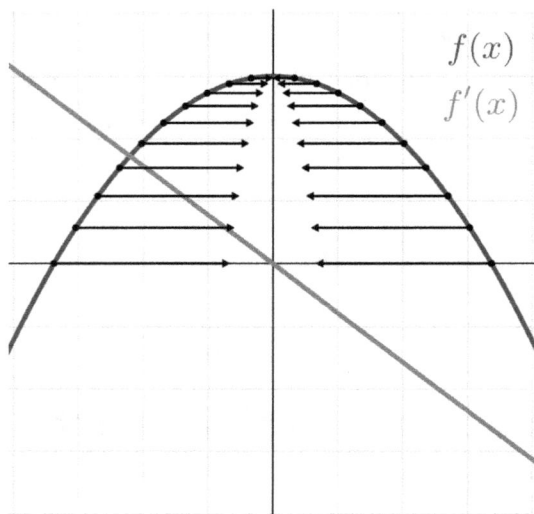

*Figure 13.9: The derivative as a vector*

Do you recall how monotonicity is characterized by the sign of the derivative? (As *Theorem 13.1.1* states.) Negative derivative means a decreasing function, and positive means an increasing function. In other words, this implies that the derivative, as a vector, *points towards the direction of the increase.*

Imagine yourself as a hiker on the $x$-$y$ plane, where $y$ signifies the height. How would you climb a mountain ahead of you? By taking a step towards the direction of increase; that is, following the derivative. If you are not there yet, you can still take another (perhaps smaller) step in the right direction, over and over again until you arrive. If you are right at the top, the derivative is zero, so you won't move anywhere.

This process is illustrated by *Figure 13.10.*

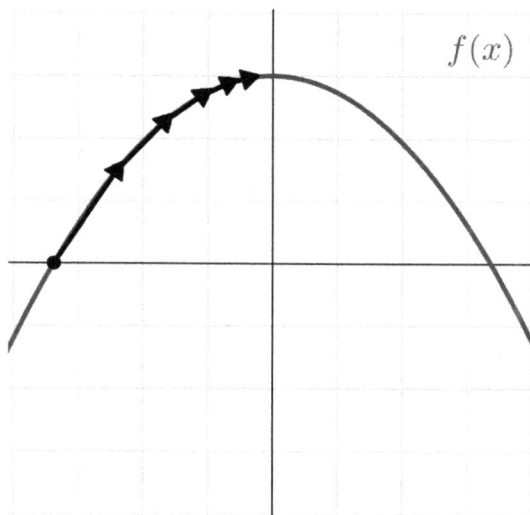

*Figure 13.10: Climbing a mountain, one step at a time*

What you have seen here is gradient *ascent* in action. Now that we understand the main idea, we are ready to tackle the mathematical details.

## 13.2.2   The gradient descent algorithm

Let $f : \mathbb{R} \to \mathbb{R}$ be a differentiable function which we want to *maximize*, that is, find

$$x_{\text{max}} = \text{argmax}_{x \in \mathbb{R}} f(x).$$

Based on our intuition, the process is quite simple. First, we conjure up an arbitrary starting point $x_0$, then define the sequence

$$x_{n+1} := x_n + h f'(x_n), \tag{13.1}$$

where $h \in (0, \infty)$ is a parameter of our gradient descent algorithm, called the *learning rate*.

In English, the formula $x_n + hf'(x_n)$ describes taking a small step from $x_n$ towards the direction of the increase, with step size $hf'(x_n)$. (Recall that the sign of the derivative shows the direction of the increase.)

If things go our way, the sequence $x_n$ converges to a local maximum of $f$. However, things do not always go our way. We'll discuss this when talking about the issues of gradient descent.

But what about finding *minima*? In machine learning, we are trying to minimize loss functions. There is a simple trick: the minima of $f(x)$ is the maxima of $-f(x)$. So, since $\left(-f\right)' = -f'$, the definition of the approximating sequence $x_n$ changes to

$$x_{n+1} := x_n - hf'(x_n).$$

This is gradient descent in a nutshell.

## 13.2.3   Implementing gradient descent

At this point, we have all the knowledge to implement the gradient descent algorithm. We'll use the previously introduced Function base class; here it is again so you don't have to look up the class definition.

```python
class Function:
    def __init__(self):
        pass

    def __call__(self, *args, **kwargs):
        pass

    def prime(self):
        pass

    def parameters(self):
        return dict()
```

As usual, I encourage you to try implementing your version of gradient descent before looking at mine. Coding is one of the most effective ways to learn, even in the age of AI – *especially* in the age of AI.

```python
def gradient_descent(
    f: Function,
    x_init: float,                # the initial guess
    learning_rate: float = 0.1,   # the learning rate
    n_iter: int = 1000,           # number of steps
    return_all: bool = False      # if true, returns all intermediate values
):
```

```
    xs = [x_init]    # we store the intermediate results for visualization

    for n in range(n_iter):
        x = xs[-1]
        grad = f.prime(x)
        x_next = x - learning_rate*grad
        xs.append(x_next)

    if return_all:
        return xs
    else:
        return x
```

Let's test the gradient descent out on a simple example, say $f(x) = x^2$! If all goes according to plan, the algorithm should find the minimum $x = 0$ in no time.

```
class Square(Function):
    def __call__(self, x):
        return x**2

    def prime(self, x):
        return 2*x

f = Square()

gradient_descent(f, x_init=5.0)
```

```
7.688949513507002e-97
```

The result is as expected: our `gradient_descent` function successfully finds the minimum.

To visualize what happens, we can plot the process in its entirety. As we'll reuse the same plot, here's a general function that does this job.

```
import numpy as np
import matplotlib.pyplot as plt


def plot_gradient_descent(f, xs: list, x_min: float, x_max: float, label: str = "f(x)"):
    ys = [f(x) for x in xs]

    grid = np.linspace(x_min, x_max, 1000)
    fs = [f(x) for x in grid]
```

```python
with plt.style.context("seaborn-v0_8-whitegrid"):
    plt.figure(figsize=(8, 8))
    plt.plot(grid, fs, label=label, c="b", lw=2.0)
    plt.plot(xs, ys, label="gradient descent", c="r", lw=4.0)
    plt.scatter(xs, ys, c="r", s=100.0)
    plt.legend()
    plt.show()

xs = gradient_descent(f, x_init=5.0, n_iter=25, learning_rate=0.2, return_all=True)
plot_gradient_descent(f, xs, x_min=-5, x_max=5, label="x²")
```

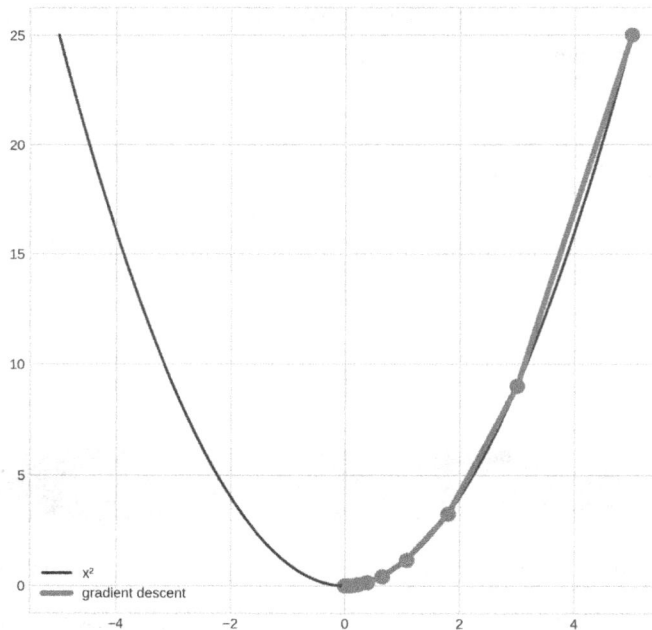

Figure 13.11: Finding the minima of $f(x) = x^2$ by gradient descent

So, is it all happiness and sunshine? No, but that's fine. Let's see what can go wrong, and how we can fix it.

## 13.2.4 Drawbacks and caveats

Even though the idea behind gradient descent is sound, there are several issues. During our journey in machine learning, we'll see most of these issues fixed by variants of the algorithm, but it is worth looking at the potential problems of the base version at this point.

First, the base gradient descent can get infinitely stuck at a local minima.

To illustrate this, let's take a look at the $f(x) = \cos(x) + \frac{1}{2}x$ function, which has no global minima, only local ones.

```python
class CosPlusSquare(Function):
    def __call__(self, x):
        return np.sin(x) + 0.5*x

    def prime(self, x):
        return np.cos(x) + 0.5

f = CosPlusSquare()
xs = gradient_descent(f, x_init=7.5, n_iter=20, learning_rate=0.2, return_all=True)
plot_gradient_descent(f, xs, -10, 10, label="sin(x) + 0.5x")
```

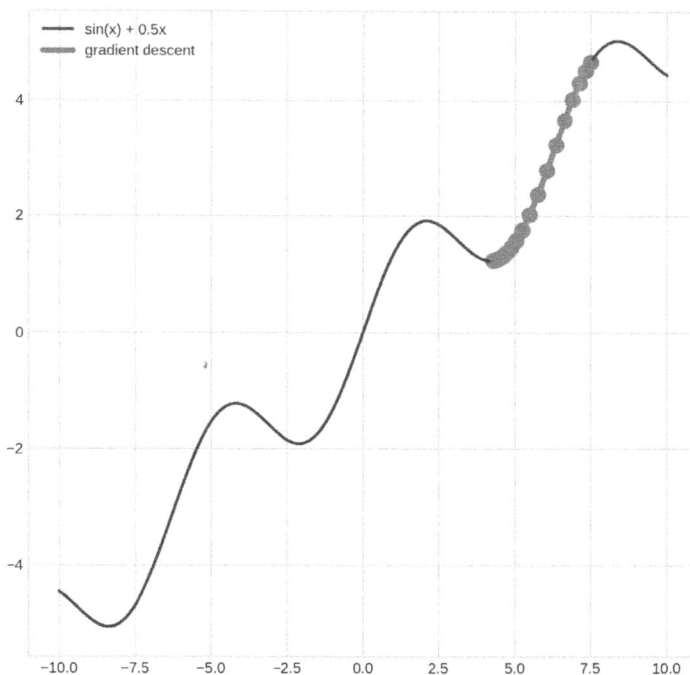

*Figure 13.12: Running the gradient descent on $f(x) = \sin(x) + \frac{1}{2}x$*

Note that if the initial point $x_0$ is selected poorly, the algorithm is much less effective. In other words, sensitivity to the initial conditions is another weakness. It might not seem that much of an issue in the simple one-variable case that we have just seen. However, it is a significant headache in the million-dimensional parameter spaces that we encounter when training neural networks.

The starting point is not the only parameter of the algorithm; it depends on the learning rate $h$ as well. There are several potential mistakes here: a too large learning rate results in the algorithm bouncing all around the space, never finding an optimum. On the other hand, a too small one results in an extremely slow convergence.

In the case of $f(x) = x^2$, starting the gradient descent from $x_0 = 1.0$ with a learning rate of $h = 1.05$, the algorithm diverges, with $x_n$ oscillating at a larger and larger amplitude.

```
f = Square()

xs = gradient_descent(f, x_init=1.0, n_iter=20, learning_rate=1.05, return_all=True)
plot_gradient_descent(f, xs, -8, 8, label="x²")
```

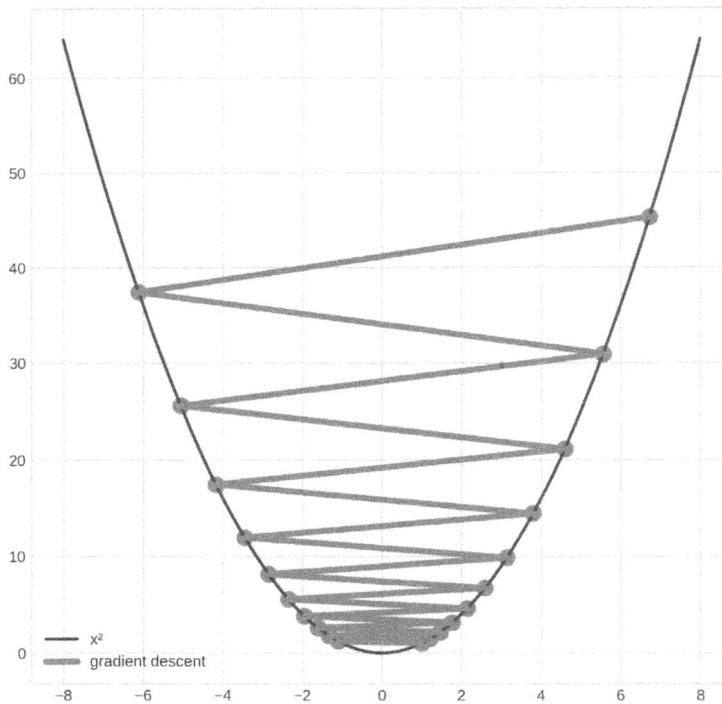

*Figure 13.13: Gradient descent, as it overshoots the optimum because of the large learning rate*

Can you come up with some solution ideas to these problems? No need to work anything out, just take a few minutes to brainstorm and make a mental note about what comes to mind. In the later chapters, we'll see several proposed solutions for all of these problems, but putting some time into this is a very useful exercise.

However, if you have an eye for detail, you might ask: does gradient descent always converge to a local optimum? Why does it work so well in practice? Let's take a look.

# 13.3   Why does gradient descent work?

*Young man, in mathematics you don't understand things. You just get used to them. — John von Neumann*

In the practice of machine learning, we use gradient descent so much that we get used to it. We hardly ever question *why* it works.

What's usually told is the mountain-climbing analogue: to find the peak (or the bottom) of a bumpy terrain, one has to look at the direction of the steepest ascent (or descent), and take a step in that direction. This direction is desribed by the gradient, and the iterative process of finding local extrema by following the gradient is called gradient ascent/descent. (Ascent for finding peaks, descent for finding valleys.)

However, this is not a mathematically precise explanation. There are several questions left unanswered, and based on our mountain-climbing intuition, it's not even clear if the algorithm works.

Without a precise understanding of gradient descent, we are practically flying blind. In this section, our goal is to look behind gradient descent and reveal the magic behind it.

Understanding the "whys" of gradient descent starts with one of the most beautiful areas of mathematics: differential equations.

## 13.3.1   Differential equations 101

What is a differential equation? Equations play an essential role in mathematics; this is common wisdom, but there is a deep truth behind it. Quite often, equations arise from modeling systems such as interactions in a biochemical network, economic processes, and thousands more. For instance, modelling the metabolic processes in organisms yields linear equations of the form

$$Ax = b, \quad A \in \mathbb{R}^{n \times n}, \quad x, b \in \mathbb{R}^n$$

where the vectors $x$ and $b$ represent the concentration of molecules (where $x$ is the unknown), and the matrix $A$ represents the interactions between them. Linear equations are easy to solve, and we understand quite a lot about them.

However, the equations we have seen so far are unfit to model dynamical systems, as they lack a time component. To describe, for example, the trajectory of a space station orbiting around Earth, we have to describe our models in terms of *functions and their derivatives*.

For instance, the trajectory of a swinging pendulum can be described by the equation

$$x''(t) + \frac{g}{L} \sin x(t) = 0, \tag{13.2}$$

where

- $x(t)$ describes the angle of the pendulum from the vertical,
- $L$ is the length of the (massless) rod that our object of mass $m$ hangs on,
- and $g$ is the gravitational acceleration constant $\approx 9.81m/s^2$.

According to the original interpretation of differentiation, if $x(t)$ describes the movement of the pendulum at time $t$, then $x'(t)$ and $x''(t)$ describe the velocity and the acceleration of it, where the differentiation is taken with respect to the time $t$.

(In fact, the differential equation (13.2) is a direct consequence of Newton's second law of motion.)

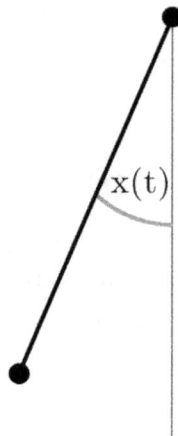

*Figure 13.14: A swinging pendulum*

Equations involving functions and their derivatives, such as (13.2), are called *ordinary differential equations*, or ODEs for short. Without any overexaggeration, their study has been the main motivating force of mathematics since the 17th century. Trust me when I say this: differential equations are one of the most beautiful objects in mathematics. As we are about to see, the gradient descent algorithm is, in fact, an approximate solution of differential equations.

The first part of this section will serve as a quickstart to differential equations. I am mostly going to follow the fantastic *Nonlinear Dynamics and Chaos* book by Steven Strogatz. If you ever feel the desire to dig deep into dynamical systems, I wholeheartedly recommend this book to you. (This is one of my favorite math books ever – it reads like a novel. The quality and clarity of its exposition serves as a continuous inspiration for my writing.)

## 13.3.2   The (slightly more) general form of ODEs

Let's dive straight into the deep waters and start with an example to get a grip on differential equations. Quite possibly, the simplest example is the equation

$$x'(t) = x(t),$$

where the differentiation is taken with respect to the time variable $t$. If, for example, $x(t)$ is the size of a bacterial colony, the equation $x'(t) = x(t)$ describes its population dynamics if the growth is unlimited. Think about $x'(t)$ as the rate at which the population grows: if there are no limitations in space and nutrients, every bacterial cell can freely replicate whenever possible. Thus, since every cell can freely divide, the speed of growth matches the colony's size.

In plain English, the solutions of the equation $x'(t) = x(t)$ are functions whose derivatives are themselves. After a bit of thinking, we can come up with a family of solutions: $x(t) = ce^t$, where $c \in \mathbb{R}$ is an arbitrary constant. (Recall that $e^t$ is an elementary function, and we have seen that its derivative is itself in *Theorem 12.2.3*.)

If you are a visual person, some of the solutions are plotted on *Figure 13.15*.

There are two key takeaways here: differential equations describe dynamical processes that change in time, and they can have multiple solutions. Each solution is determined by two factors: the equation $x'(t) = x(t)$, and an initial condition $x(0) = x^*$. If we specify $x(0) = x^*$, then the value of $c$ is given by

$$x(0) = ce^0 = c = x^*.$$

Thus, ODEs have a bundle of solutions, each one determined by the initial condition.

So, it's time to discuss differential equations in more general terms!

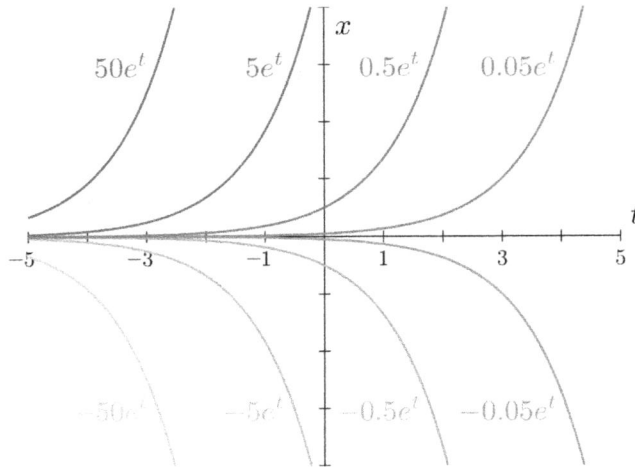

*Figure 13.15: Some solutions of the exponential growth equation*

**Definition 13.3.1 (Ordinary differential equations in one dimension)**

Let $f : \mathbb{R} \to \mathbb{R}$ be a differentiable function. The equation

$$x'(t) = f(x(t)) \tag{13.3}$$

is called a *first-order homogeneous ordinary differential equation.*

When it is clear, the dependence on $t$ is often omitted, so we only write $x' = f(x)$. (Some resources denote the time derivative by $\dot{x}$, a notation that can be originated from Newton. We will not use this, though it is good to know.)

The term "first-order homogeneous ordinary differential equation" doesn't exactly roll off the tongue, and it is overloaded with heavy terminology. So, let's unpack what is going on here.

The *differential equation* part is clear: it is a functional equation that involves derivatives. Since the time $t$ is the only variable, the differential equation is *ordinary*. (As opposed to differential equations involving multivariable functions and partial derivatives, but more on those later.) As only the first derivative is present, the equation becomes *first-order*. Second-order would involve second derivatives, and so on. Finally, since the right-hand side $f(x)$ doesn't explicitly depend on the time variable $t$, the equation is *homogeneous in time*. Homogeneity means that the rules governing our dynamical system don't change over time.

Don't let the $f(x(t))$ part scare you! For instance, in our example $x'(t) = x(t)$, the role of $f$ is cast to the identity function $f(x) = x$. In general, $f(x)$ establishes a relation between the quantity $x(t)$ (which can be position, density, etc) and its derivative, that is, its rate of change.

As we have seen, we think in terms of differential equations and initial conditions that pinpoint solutions among a bundle of functions. Let's put this into a proper mathematical definition!

---

**Definition 13.3.2 (Initial value problems)**

Let $x' = f(x)$ be a first-order homogeneous ordinary differential equation and let $x_0 \in \mathbb{R}$ be an arbitrary value. The system

$$\begin{cases} x' & = f(x) \\ x(t_0) & = x_0 \end{cases}$$

is called an *initial value problem*. If a function $x(t)$ satisfies both conditions, it is said to be a *solution* to the initial value problem.

---

Most often, we select $t_0$ to be 0. After all, we have the freedom to select the origin of the time as we want.

Unfortunately, things are not as simple as they seem. In general, differential equations and initial value problems are tough to solve. Except for a few simple ones, we cannot find exact solutions. (And when I say we, I include every person on the planet.) In these cases, there are two things that we can do: either we construct approximate solutions via numeric methods or turn to qualitative methods that study the behavior of the solutions without actually finding them.

We'll talk about both, but let's turn to the qualitative methods first. As we'll see, looking from a geometric perspective gives us a deep insight into how differential equations work.

## 13.3.3 A geometric interpretation of differential equations

When finding analytic solutions is not feasible, we look for a *qualitative* understanding of the solutions, focusing on the local and long-term behavior instead of formulas.

Imagine that, given a differential equation

$$x'(t) = f(x(t)),$$

you are interested in a particular solution that assumes the value $x^*$ at time $t_0$.

For instance, you could be studying the dynamics of a bacterial colony and want to provide a predictive model to fit your latest measurement $x(t_0) = x^*$. In the short term, where will your solutions go?

We can immediately notice that if $x(t_0) = x^*$ and $f(x) = 0$, then the constant function $x(t) = x$ is a solution! These are called *equilibrium solutions*, and they are extremely important. So, let's make a formal definition!

**Definition 13.3.3 (Equilibrium solutions)**

Let

$$x'(t) = f(x(t)) \tag{13.4}$$

be a first-order homogeneous ODE, and let $x^* \in \mathbb{R}$ be an arbitrary point. If $f(x^*) = 0$, then $x^*$ is called an *equilibrium point* of the equation $x' = f(x)$.

For equilibrium points, the constant function $x(t) = x^*$ is a solution of (13.4). This is called an *equilibrium solution*.

Think about our recurring example, the simplest ODE $x'(t) = x(t)$. As mentioned, we can interpret this equation as a model of unrestricted population growth under ideal conditions. In that case, $f(x) = x$, and this is zero only for $x = 0$. Therefore, the constant $x(t) = 0$ function is a solution. This makes perfect sense: if a population has zero individuals, no change is going to happen in its size. In other words, the system is in equilibrium.

This is like a pendulum that stopped moving and reached its resting point at the bottom. However, pendulums have two equilibria: one at the top and one at the bottom. (Let's suppose that the mass is held by a massless rod. Otherwise, it would collapse.) At the bottom, you can push the hanging mass all you want and it'll return to rest. However, at the top, any small push would disrupt the equilibrium state, to which it would never return.

To shed light on this phenomenon, let's look at another example: the famous *logistic equation*

$$x'(t) = x(t)(1 - x(t)). \tag{13.5}$$

From a population dynamics perspective, if our favorite equation $x'(t) = x(t)$ describes the unrestricted growth of a bacterial colony, the logistic equation models the population growth under a resource constraint. If we assume that 1 is the total capacity of our population, the growth becomes more difficult as the size approaches this limit. Thus, the population's rate of change $x'(t)$ can be modelled as $x(t)(1 - x(t))$, where the term $1 - x(t)$ slows down the process as the colony nears the sustain capacity.

We can write the logistic equation in the general form (13.3) by casting the role $f(x) = x(1 - x)$. Do you recall *Theorem 13.1.1* about the relation of derivatives and monotonicity? Translated to the differential equation $x' = f(x)$, this reveals the *flow* of our solutions! To be specific,

$$x(t) \text{ is } \begin{cases} \text{increasing} & \text{if } f(x) > 0, \\ \text{decreasing} & \text{if } f(x) < 0, \\ \text{constant} & \text{if } f(x) = 0. \end{cases}$$

We can visualize this in the so-called *phase portrait*.

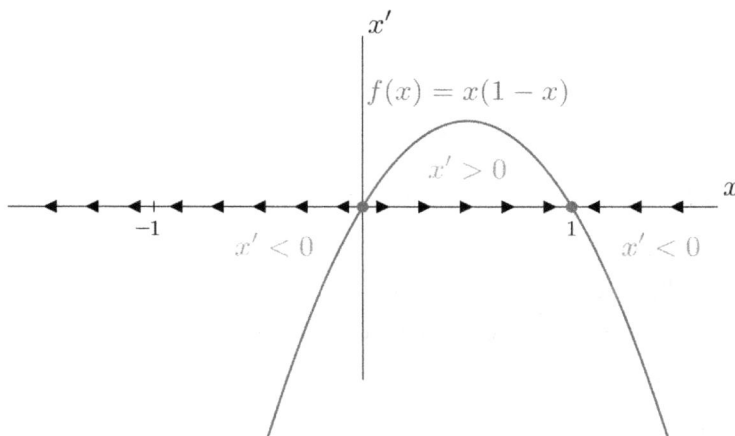

Figure 13.16: *The flow of solutions for* $x' = x(1 - x)$, *visualized on the phase portrait. (The arrows represent the direction where the solutions for given initial values are headed.)*

Thus, the monotonicity describes long-term behavior:

$$\lim_{t \to \infty} x(t) = \begin{cases} 1 & \text{if } x'(0) > 0, \\ 0 & \text{if } x'(0) = 0, \\ -\infty & \text{if } x'(0) < 0. \end{cases} \qquad (13.6)$$

With a little bit of calculation (whose details are not essential for us), we can obtain that we can write the solutions as noindent

$$x(t) = \frac{1}{1 + ce^{-t}},$$

where $c \in \mathbb{R}$ is an arbitrary constant. For $c = 1$, this is the famous sigmoid function.

You can check by hand that these are indeed solutions. We can even plot them, as shown in *Figure 13.17* below.

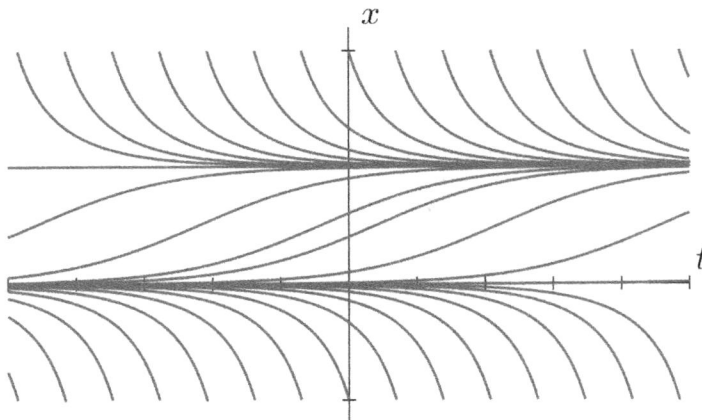

*Figure 13.17: Solutions of the logistic differential equation $x' = x(1 - x)$*

As we can see in *Figure 13.17*, the monotonicity of the solutions are as we predicted in (13.6).

We can characterize the equilibria based on the long-term behavior of nearby solutions. (In the case of our logistic equation, the equilibria are 0 and 1.) This can be connected to the *local* behavior of $f$: if it decreases around the equilibrium $x^*$, it attracts the nearby solutions. On the other hand, if $f$ increases around $x^*$, then the nearby solutions are repelled.

This gives rise to the concept of *stable* and *unstable* equilibria.

**Definition 13.3.4 (Stable and unstable equilibria)**

Let $x' = f(x)$ be a first-order homogeneous ordinary differential equation, and suppose that $f$ is differentiable. Moreover, let $x^*$ be an equilibrium of the equation.

$x^*$ is called a *stable* equilibrium if there is a neighborhood $(x^* - \varepsilon, x^* + \varepsilon)$ around $x^*$ such that for all $x_0 \in (x^* - \varepsilon, x^* + \varepsilon)$, the solution of the initial value problem

$$\begin{cases} x' & = f(x) \\ x(0) & = x_0 \end{cases}$$

converges towards $x^*$. (That is, $\lim_{t \to \infty} x(t) = x^*$ holds.)

If $x^*$ is not stable, it is called *unstable*.

In the case of the logistic ODE $x' = x(1 - x)$, $x^* = 1$ is a stable and $x^* = 0$ is an unstable equilibrium. This makes sense given its population dynamics interpretation: the equilibrium $x^* = 1$ means that the population is at maximum capacity. If the size is slightly above or below the capacity 1, some specimens die due to starvation, or the colony reaches its constraints. On the other hand, no matter how small the population is, it won't ever go extinct in this ideal model.

Recall how the derivatives characterize the monotonicity of differentiable functions by *Theorem 13.1.1*? With this, we have a simple tool that can help us decide whether a given equilibrium is stable or not.

> **Theorem 13.3.1** *Let $x' = f(x)$ be a first-order homogeneous ordinary differential equation, and suppose that $f$ is differentiable. Moreover, let $x^*$ be an equilibrium point of the equation.*
>
> *If $f'(x) < 0$, then $x$ is a stable equilibrium.*

The concept of stable equilibrium is fundamental, even in the most general cases. At this point, it's time to take a few steps backward and remind ourselves why we are here: to understand gradient descent. If stable equilibria remind you of a local minimum which a gradient descent process converges towards, it is not an accident. We are ready to see what's behind the scenes.

## 13.3.4　A continuous version of gradient ascent

Now, let's talk about *maximizing* a function $F : \mathbb{R} \to \mathbb{R}$. Suppose that $F$ is twice differentiable, and we denote its derivative by $F' = f$. Luckily, the local maxima of $F$ can be found with the help of its second derivative (*Theorem 13.1.4*) by looking for $x^*$ where $f(x) = 0$ and $f'(x) < 0$.

Does this look familiar? If $f(x) = 0$ indeed holds, then $x(t) = x$ is an equilibrium solution; and since $f'(x^*) < 0$, it attracts the nearby solutions as well. This means that if $x_0$ is drawn from the basin of attraction and $x(t)$ is the solution of the initial value problem

$$\begin{cases} x' & = f(x) \\ x(0) & = x_0, \end{cases} \tag{13.7}$$

then $\lim_{t \to \infty} x(t) = x^*$. In other words, the solution converges towards $x^*$, a local maxima of $F$! This is gradient ascent in a continuous version.

We are happy, but there is an issue. We've talked about how hard solving differential equations are. For a general $F$, we have no prospects to actually find the solutions. Fortunately, we can *approximate* them.

## 13.3.5   Gradient ascent as a discretized differential equation

When studying differentiation in practice, we have seen that derivatives can be approximated numerically by the forward difference

$$x'(t) \approx \frac{x(t+h) - x(t)}{h},$$

where $h > 0$ is a small step size. If $x(t)$ is indeed the solution for the initial value problem (13.7), we are in luck! Using forward differences, we can take a small step from 0 and approximate $x(h)$ by substituting the forward difference into the differential equation. To be precise, we have

$$\frac{x(h) - x(0)}{h} \approx f(x(0)),$$

from which

$$x(h) \approx x(0) + hf(x(0))$$

follows. By defining $x_0$ and $x_1$ by

$$x_0 := x(0),$$
$$x_1 := x_0 + hf(x_0),$$

we have $x_1 \approx x(h)$. If this looks like the first step of the gradient ascent (13.1) to you, you are on the right track. Using the forward difference once again, this time from the point $x(h)$, we obtain

$$x(2h) \approx x(h) + hf(x(h))$$
$$\approx x_1 + hf(x_1),$$

thus by defining $x_2 := x_1 + hf(x_1)$, we have $x_2 \approx x(2h)$. Notice that in $x_2$, two kinds of approximation errors are accumulated: first the forward difference, then the approximation error of the previous step.

This motivates us to define the recursive sequence

$$x_0 := x(0),$$
$$x_{n+1} := x_n + hf(x_n),$$

(13.8)

which approximates $x(nh)$ with $x_n$, as this is implied by the very definition. This recursive sequence is the gradient ascent itself, and the small step $h$ is the learning rate!

Check (13.1) if you don't believe me. (13.8) is called the Euler method.

Without going into the details, if $h$ is small enough and $f$ "behaves properly," the Euler method will converge to the equilibrium solution $x^*$. (Whatever proper behavior might mean.)

We only have one more step: to turn everything into gradient *descent* instead of ascent. This is extremely simple, as gradient descent is just applying gradient ascent to $-f$. Think about it: minimizing a function $f$ is the same as maximizing its negative $-f$. And with that, we are done! The famous gradient descent is a consequence of dynamical systems converging towards their stable equilibria, and this is beautiful.

## 13.3.6   Gradient ascent in action

To see gradient ascent (that is, the Euler method) in action, we should go back to our good old example: the logistic equation (13.5). So, suppose that we want to find the local maxima of the function

$$F(x) = \frac{1}{2}x^2 - \frac{1}{3}x^3,$$

plotted in *Figure 13.18*.

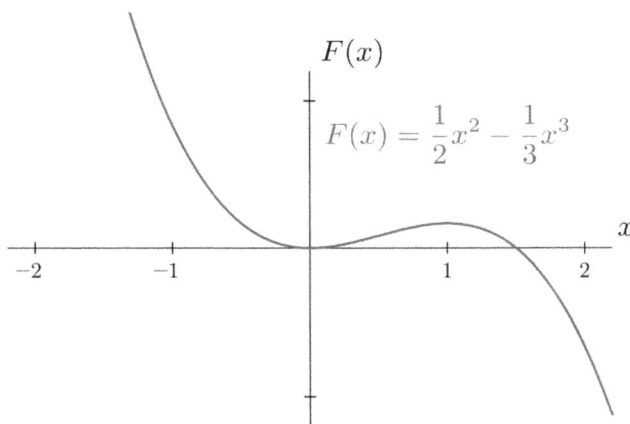

*Figure 13.18: The graph of $F(x) = \frac{1}{2}x^2 - \frac{1}{3}x^3$*

First, we can use what we learned and find the maxima using the derivative $f(x) = F'(x) = x(1-x)$, concluding that there is a local maximum at $x^* = 1$. (Don't just take my word for it, check out *Theorem 13.1.4* and work it out!)

Since $f(x) = F'(x) = 0$ and $f'(x) < 0$, the point $x$ is a stable equilibrium of the logistic equation

$$x' = x(1 - x).$$

Thus, if the initial value $x(0) = x_0$ is sufficiently close to $x^* = 1$, the solution $x(t)$ of the initial value problem

$$\begin{cases} x' & = x(1-x), \\ x(0) & = x_0, \end{cases}$$

then $\lim_{t \to \infty} x(t) = x^*$. (In fact, we can select any initial value $x_0$ from the infinite interval $(0, \infty)$, and the convergence will hold). Upon discretization via the Euler method, we obtain the recursive sequence

$$x_0 = x(0),$$

$$x_{n+1} = x_n + hx_n(1 - x_n).$$

This process is visualized by *Figure 13.19*.

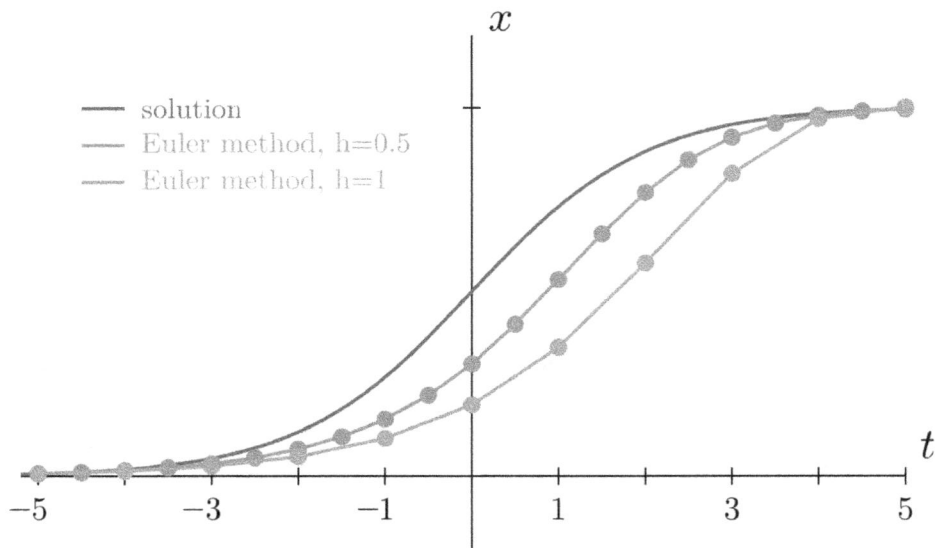

*Figure 13.19: Solving $x' = x(1-x)$ via the Euler method. (For visualization purposes, the initial value was set at $t_0 = -5$.)*

We can even take the discrete solution provided by the Euler method, and plot it on the $x - F(x)$ plane.

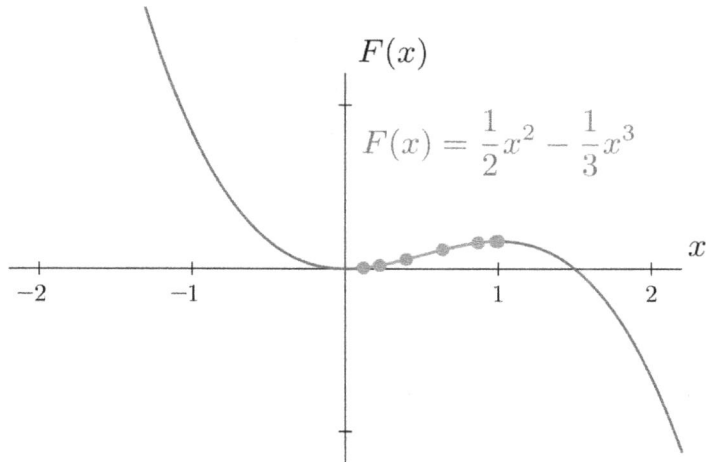

*Figure 13.20: Mapping the Euler method on the x, F(x) plane*

If you check *Figure 13.20* out, you can see that this is the gradient ascent for $F$! If you consider that $F' = f$ and consider that the solution given by the Euler method is

$$x_0 := x(0),$$
$$x_{n+1} := x_n + hF'(x_n),$$

(13.9)

you can notice that (13.9) is exactly how we defined gradient ascent.

## 13.4 Summary

Finally, we've done it. Until this chapter, we haven't been that close to machine learning, but now, we are right at the heart of it. Gradient descent is the number one algorithm to train neural networks. Yes, even state-of-the-art ones.

It all starts with calculus. To reach the heights of gradient descent, we studied the relations between monotonicity, local extrema, and the derivative. The pattern is simple: if $f'(a) > 0$, then $f$ is *increasing*, but if $f'(a) < 0$, then $f$ is *decreasing* around $a$. Speaking in terms of physics, if the speed is positive, the object is moving away, but if the speed is negative, the object is coming closer.

Based on this observation, we can deduce necessary and sufficient conditions to find local minima and maxima: if $f'(a) = 0$

- and if $f''(a) > 0$, then $a$ is a *local minimum*,

- but if $f'(a) = 0$ and $f''(a) < 0$, then $a$ is a *local maximum*.

So, finding the local extrema should be as simple as solving $f'(a) = 0$, right? In theory, no, because the case $f''(a) = 0$ is undetermined. In practice, still no, because even finding $f'$ is hard for complex $f$-s, let alone solving $f'(x) = 0$.

However, there's a way. We can take an iterative approach with gradient descent: if the learning rate $h$ and starting point $x_0$ are selected appropriately, the recursive sequence defined by

$$x_{n+1} = x_n - hf'(x_n)$$

converges to a local minimum.

As always, when one problem is solved, a dozen others are created. For example, the gradient descent can fail to converge or get stuck in the local minimum instead of finding the global one. But that's the least of our problems. The real issue is that we have to optimize functions of multiple variables in practice, often in the range of billions of parameters.

Before we move on to the study of multivariable calculus, there's one more topic to go. I've hinted at *integration*, the mysterious "inverse differentiation" a couple of times. It's time to see what it is and why it is indispensable to study advanced mathematics.

# 13.5  Problems

**Problem 1.** Find the local minima and maxima of $f(x) = \sin x$.

**Problem 2.** Use the second derivative test to find the local minima and maxima of $f(x) = 2x^3 + 5x^2 + 4x + 6$.

**Problem 3.** Let $f : \mathbb{R} \to \mathbb{R}$ be a differentiable function. The recursive sequence defined by

$$\delta_{n+1} = \alpha\delta_n - hf'(x_n),$$
$$x_{n+1} = x_n + \delta_n,$$

where $\delta_0 = 0$ and $x_0$ is arbitrary, is called *gradient descent with momentum*. Implement it!

# Join our community on Discord

Read this book alongside other users, Machine Learning experts, and the author himself.

Ask questions, provide solutions to other readers, chat with the author via Ask Me Anything sessions, and much more.

Scan the QR code or visit the link to join the community.

`https://packt.link/math`

# 14

# Integration

When we first encountered the concept of *derivatives* in *Chapter 12*, we introduced it through an example from physics. As Newton created it, the derivative describes the velocity of a moving object as calculated from its time-distance graph. In other words, the velocity can be *derived* from the time-distance information.

Can the distance be reconstructed given the velocity? In a sense, this is the inverse of differentiation.

Questions such as these are hard to answer if we only look at the most general case, so let's consider a special one. Suppose that our object is moving with a constant velocity $v(t) = v_0 \frac{m}{s}$, for a duration of $T$ seconds. With some elementary logic, we can conclude that the total distance traveled is $v_0 T$ meters.

When taking a look at the *time-velocity* plot, we can immediately see that the distance is *the area under the time-velocity* function graph $v(t) = v_0$.

The graph of $v(t)$ describes a rectangle with width $v_0$ and length $T$, hence its area is indeed $v_0 T$.

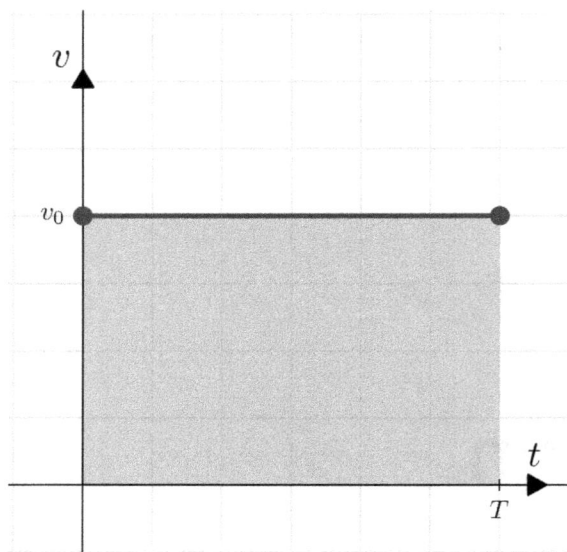

*Figure 14.1: Time-velocity plot of an object, moving with constant velocity*

Does the area under $v(t)$ equal the distance traveled in the general case? For instance, what happens when the time-velocity plot looks something like this?

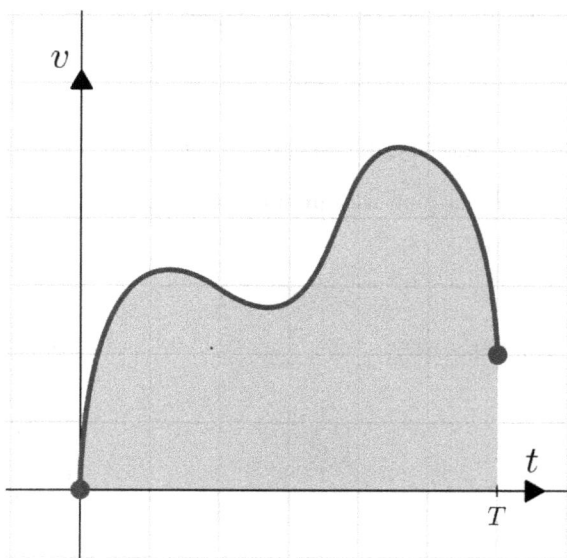

*Figure 14.2: Time-velocity plot of an object, moving with a changing velocity*

The speed is not constant here. In this case, we can do a simple trick: partition the time interval $[0, T]$ into smaller ones and approximate the object's motion as a constant-speed motion on each of these intervals.

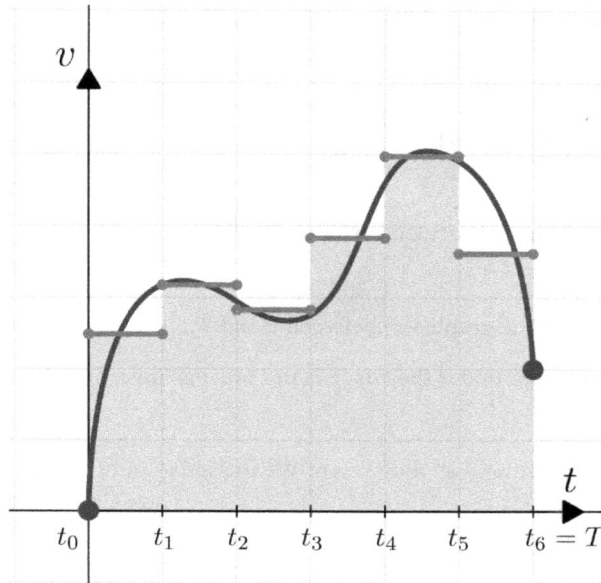

*Figure 14.3: Approximating with a constant velocity motion*

If the time intervals $[t_i, t_{i+1}]$ are sufficiently granular, the distance traveled will roughly match a constant velocity motion with the average velocity at $[t_i, t_{i+1}]$. That is, if we introduce the notation

$$v_i := \text{average velocity during the time interval } [t_{i-1}, t_i], \quad i = 1, 2, \dots, n,$$

we should have

$$\sum_{i=1}^{n} v_i(t_i - t_{i-1}) \approx \text{total distance traveled during } [0, T].$$

Let's think about this whole process as approximating the function $v(t)$ with a stepwise constant function $v_{\text{approx}}(t)$. From this angle, we have

$$\sum_{i=1}^{n} v_i(t_i - t_{i-1}) = \text{area under } v_{\text{approx}}(t),$$

and

$$\text{area under } v_{\text{approx}}(t) \approx \text{area under } v(t).$$

(Very) loosely speaking, if the granularity of the time intervals $[t_i, t_{i+1}]$ gets infinitesimally small, the approximations turn into equality. Thus,

$$\text{total distance traveled during } [0, T] = \text{area under } v(t) \text{ in } [0, T].$$

There are two key points that we need to remember: if $s(t)$ is the *distance* traveled and $v(t)$ is the velocity, then

- $v(t)$ is the derivative $s'(t)$,
- and $s(T)$ is the area under the graph $v(t)$ between $0$ and $T$.

In other words, calculating the area under the curve is the same as inverting differentiation. This process is called *integration*.

Unfortunately, things are not as simple as they seem. We missed a lot of mathematical detail in the above discussion. For one, does the sum

$$\sum_{i=1}^{n} v_i(t_i - t_{i-1})$$

converge if the partition of $[0, T]$ gets more granular? Does the limit depend on the partitions? Can we even define the area under the "graph" for all functions? Like the Dirichlet function, defined by

$$D(x) = \begin{cases} 1 & \text{if } x \in \mathbb{Q}, \\ 0 & \text{otherwise.} \end{cases} \tag{14.1}$$

How do we calculate limits of $\sum_{i=1}^{n} v_i(t_i - t_{i-1})$ in practice? In addition, what does all of this have to do with machine learning?

Fasten your seatbelts! Here comes the rigorous study of integration, clearing up all of these questions.

## 14.1   Integration in theory

Let's build a solid theoretical foundation for the intuitive explanation! Let $f : [a, b] \to \mathbb{R}$ be an arbitrary *bounded* function, and our goal is to calculate the signed area under the graph. (Note that the signed area is negative if the graph goes below the $x$ axis. In the time-speed graph example above, this is equivalent to moving backward, thus decreasing the distance traveled from the starting point.)

Let $a = x_0 < x_1 < ... < x_n = b$ an arbitrary partition of the interval $[a, b]$.

For notational convenience, we'll denote this partition as $X = \{x_0, \ldots, x_n\}$ as well. The *granularity* (or *mesh*) of $X$ is defined by

$$|X| := \max_{i=1,\ldots,n} |x_i - x_{i-1}|,$$

which is the length of the biggest gap in $X$. Note that the partition is not necessarily uniform, so $|x_i - x_{i-1}|$ is not constant.

We are going to use an argument similar to the squeeze principle (*Corollary 10.2.1*) to make the approximation idea rigorous. (You know, the one where we replaced the speed of a moving object with a piecewise constant one.) Instead of using the averages of $f(x)$ on each interval $[x_{i-1}, x_i]$, we are going to provide an upper and lower estimation by using

$$m_i := \inf_{x \in [x_{i-1}, x_i]} f(x)$$

and

$$M_i := \sup_{x \in [x_{i-1}, x_i]} f(x).$$

Mathematically speaking, the infimum and the supremum are much easier to work with than the average. Now we can approximate $f(x)$ with a piecewise constant function from both above and below. This is visualized by *Figure 14.4*.

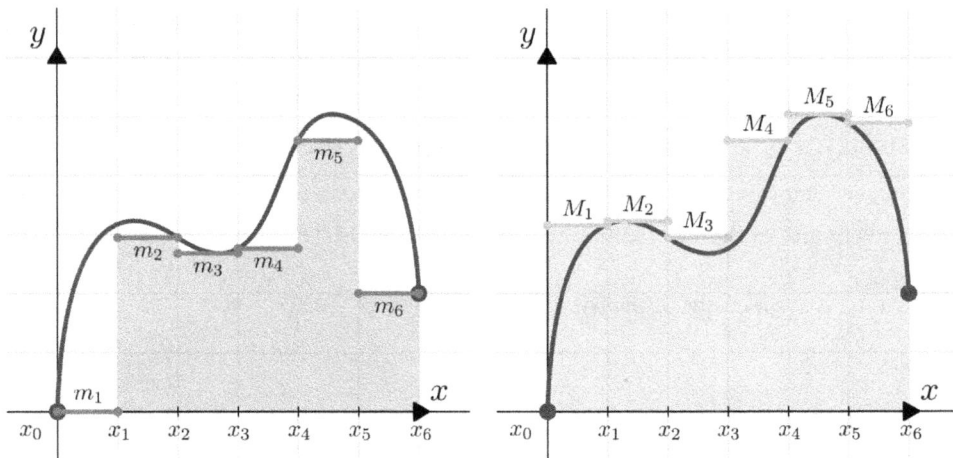

Figure 14.4: Estimating the area under the curve of $f$ using the partition $X$

Our plan is to squeeze the area between the lower and upper sums

$$L[f, X] := \sum_{i=1}^{n} m_i(x_i - x_{i-1}) \tag{14.2}$$

and

$$U[f, X] := \sum_{i=1}^{n} M_i(x_i - x_{i-1}), \tag{14.3}$$

then study if these two match. (As usual, the dependence on $f$ and $X$ will be omitted if it is clear from the context.)

It is clear from the construction that

$$L[f, X] \leq \text{area under the graph} \leq U[f, X]$$

holds for any partition $X$.

As the granularity of the partition $X$ goes to zero, hopefully, both $L[f, X]$ and $U[f, X]$ will converge to the same number. Intuitively, this common limit should be the "area under the function graph," but currently, our notion of the area is not general enough to make such bold statements. For instance, how would you define the "area" under the Dirichlet function, defined by (14.1)? As we shall see soon, integration will generalize our heuristic notion of area. To get to that point, we have a lot to do. First, we'll take a closer look at the *partitions*.

## 14.1.1 Partitions and their refinements

We need to introduce some basic facts about refining partitions to construct mathematically correct arguments regarding the convergence of the approximating sums $L[f, X]$ and $U[f, X]$.

**Definition 14.1.1 (Refinement of partitions)**

Let $X = \{x_0, \ldots, x_n\}$ and $Y = \{y_0, \ldots, y_m\}$ be two partitions of $[a, b]$. We say that $Y$ is a *refinement* of $X$ if $X \subseteq Y$.

We can visualize this easily.

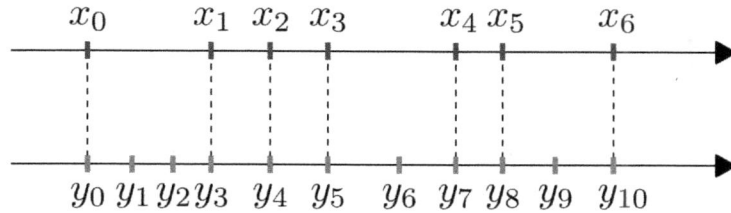

*Figure 14.5: The partition Y, as a refinement of X*

Refinements are vital for understanding why integration works. One of the core reasons is the following result.

---

**Definition 14.1.2 (Monotonicity of upper and lower sums)**

Let $f : [a, b] \rightarrow \mathbb{R}$ be a bounded function and $X$ and $Y$ be two partitions of $[a, b]$. Suppose that $Y$ is a refinement of $X$. Then

$$L[f, X] \leq L[f, Y] \tag{14.4}$$

and

$$U[f, Y] \leq U[f, X]. \tag{14.5}$$

---

*Proof.* We are going to show $L[f, X] \leq L[f, Y]$, as (14.5) follows from a similar argument. Suppose that $x_{i-1} \leq y_j \leq \cdots \leq y_l \leq x_i$. Mathematically speaking, we have

$$\inf_{x\in[x_{i-1},x_i]} f(x) \leq \inf_{x\in[y_{k-1},y_k]} f(x), \quad k = j + 1, \dots, l.$$

Since $x_i - x_{i-1} = \sum_{k=j+1}^{l} y_k - y_{k-1}$, the above implies that

$$\begin{aligned}
\inf_{x\in[x_{i-1},x_i]} f(x)(x_i - x_{i-1}) &= \sum_{k=j+1}^{l} \inf_{x\in[x_{i-1},x_i]} f(x)(y_k - y_{k-1}) \\
&\leq \sum_{k=j+1}^{l} \inf_{x\in[y_{k-1},y_k]} f(x)(y_k - y_{k-1}).
\end{aligned} \tag{14.6}$$

Don't worry if these mathematical formalisms make this hard to follow. Just take a look at *Figure 14.6* below, which summarizes all that we have done so far.

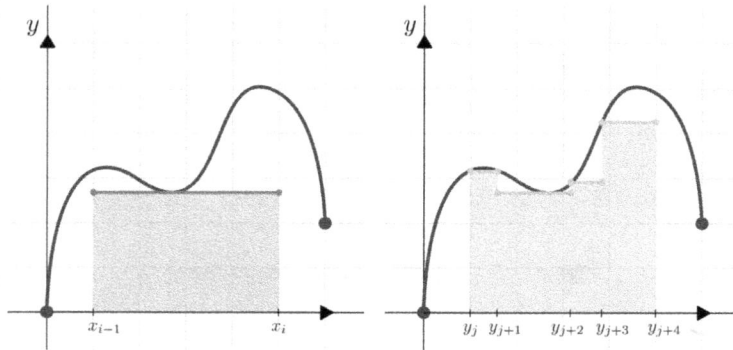

*Figure 14.6: Refinement of lower sums*

Since $L[f, X]$ and $L[f, Y]$ are composed from parts like in (14.6), summing over $i$ in the above immediately yields $L[f, X] \leq L[f, Y]$.

We are almost there. There is one thing left for us to show: that for any two partitions, the lower sum is always smaller than the upper sum. Hence, the squeeze principle (*Corollary 10.2.1*) could be applied to show that the lower and upper sums converge to the same limit in some instances.

For that, we need a simple but fundamental fact about partitions.

**Proposition 14.1.1** *Let $X$ and $Y$ be two partitions of $[a, b]$. Then there exists a partition $Z$ that is a refinement of both $X$ and $Y$.*

*Proof.* It is easy to see that $Z = X \cup Y$ satisfies our requirements.

The above $Z$ is called a *mutual refinement* of $X$ and $Y$. We can show a fundamental relation between the upper and lower sums with this idea.

**Proposition 14.1.2** *Let $f : [a, b] \to \mathbb{R}$ be a bounded real function and let $X$ and $Y$ be two partitions of the interval $[a, b]$. Then*

$$L[f, X] \leq U[f, Y]$$

*holds.*

*Proof.* Let $Z$ be a mutual refinement of $X$ and $Y$, as guaranteed by the previous result. Then, (14.4) and (14.5) imply that

$$L[f, X] \leq L[f, Z] \leq U[f, Z] \leq U[f, Y],$$

which is what we wanted to show.

## 14.1.2 The Riemann integral

Let's denote the set of all partitions on $[a, b]$ by $\mathcal{F}[a, b]$:

$$\mathcal{F}[a, b] = \{X \ : \ X \text{ is a partition of } [a, b]\}.$$

Now we are ready to define the integral of the function as the single value that separates upper and lower sums.

**Definition 14.1.3 (Riemann-integrability)**

Let $f : [a, b] \to \mathbb{R}$ be a bounded function. We say that $f$ is *Riemann-integrable* (or just integrable) on $[a, b]$ if

$$\sup_{X \in \mathcal{F}[a,b]} L[f, X] = \inf_{X \in \mathcal{F}[a,b]} U[f, X].$$

This value is called the Riemann integral (or just the integral) of $f$ over $[a, b]$, denoted by

$$\int_a^b f(x)dx.$$

The function $f$ in $\int_a^b f(x)dx$ is called the *integrand*. How do we calculate the integral itself? The hard way is to define a sequence of partitions $X_n$ and show that

$$\lim_{n \to \infty} L[f, X_n] = \lim_{n \to \infty} U[f, X_n],$$

so this number is necessarily $\int_a^b f(x)dx$. We'll see the easy way soon, but let's see an example demonstrating this process.

Let's calculate $\int_0^1 x^2 dx$!

The simplest is to use the uniform partition $X_n = i/n_{i=0}^n$, obtaining

$$L[x^2, X_n] = \sum_{i=1}^{n} \left(\frac{i-1}{n}\right)^2 \frac{1}{n}$$

$$= \frac{1}{n^3} \sum_{i=1}^{n} (i-1)^2.$$

Since $\sum_{k=1}^{n} k^2 = \frac{n(n+1)(2n+1)}{6}$ (as it can be shown by induction), it is easy to see that

$$\lim_{n \to \infty} L[x^2, X_n] = \frac{1}{3}.$$

With a similar argument, you can check that $\lim_{n \to \infty} U[x^2, X_n] = \frac{1}{3}$ as well, thus, $\int_0^1 x^2 dx$ exists and

$$\int_0^1 x^2 dx = \frac{1}{3}.$$

Although this method works for simple cases such as $f(x) = x^2$, it breaks down for more complex functions, as calculating limits of upper and lower sums can be difficult. In addition, selecting the right partition is also a challenge. For instance, can you calculate $\int_0^\pi \sin(x) dx$ by the definition?

Because we are lazy (just like any good mathematician), we want to find a general method to calculate integrals. Lower and upper sums are needed to make the notion of an integral mathematically precise. Combined with the squeeze principle (*Corollary 10.2.1*), they are used to provide a definition.

However, other tools become available once we know that a function is integrable. Such as the general approximating sum, as we are about to see next.

**Theorem 14.1.1** *Let $f : \mathbb{R} \to \mathbb{R}$ be an arbitrary bounded function, and let $X_n = \{x_{0,n}, \dots, x_{n,n}\}$ be a sequence of partitions on $[a, b]$ such that $|X_n| \to 0$. Then $f$ is integrable if and only if the limit*

$$\lim_{n \to \infty} \sum_{i=1}^{n} f(\xi_i)(x_{i,n} - x_{i-1,n})$$

*exists and in this case,*

$$\lim_{n \to \infty} \sum_{i=1}^{n} f(\xi_i)(x_{i,n} - x_{i-1,n}) = \int_a^b f(x) dx$$

*holds, where $\xi_i \in [x_{i-1,n}, x_{i,n}]$.*

(Note that $|X_n| \to 0$ means that the length of the largest subinterval of $X_n$ gets extremely small. In other words, the resolution of $X_n$ gets extremely large.)

We will not prove the above theorem, as the proof is technical and doesn't provide any valuable insight. However, the point is clear: local infima and suprema in lower and upper sums can be replaced with any local value.

For simplicity, we'll denote this sum by

$$S[f, X, \xi_X] = \sum_{i=1}^{n} f(\xi_i)(x_i - x_{i-1}) \tag{14.7}$$

for any $X = \{x_0, \dots, x_n\}$ and $\xi_X = \{\xi_1, \dots, \xi_n\}$ with $\xi_i \in [x_{i-1}, x_i]$.

## 14.1.3   Integration as the inverse of differentiation

Now that we understand the mathematical definition of the integral, it is time to find some tools that enable its use in practice. The most important result is the Newton-Leibniz formula, named after Isaac Newton and Gottfried Wilhelm Leibniz, the inventors of calculus. (Fun fact: these men discovered calculus independently and were mortal enemies throughout their lives.)

**Theorem 14.1.2** *(The fundamental theorem of calculus, a.k.a. the Newton-Leibniz formula)*

*Let $f : \mathbb{R} \to \mathbb{R}$ a function that is integrable on $[a, b]$ and suppose that there is an $F : \mathbb{R} \to \mathbb{R}$ such that $F'(x) = f(x)$. Then*

$$\int_a^b f(x)dx = F(b) - F(a) \tag{14.8}$$

*holds.*

In other words, by defining $x \mapsto F(a) + \int_a^x f(x)dx$, we can effectively reconstruct a function from its derivative.

*Proof.* Let $a = x_0 < x_1 < \cdots < x_n = b$ be an arbitrary partition of $[a, b]$. According to Lagrange's mean value theorem (*Theorem 13.1.6*), there exists a $\xi_i \in (x_{i-1}, x_i)$ for all $i = 1, \dots, n$ such that

$$F(x_i) - F(x_{i-1}) = F'(\xi_i)(x_i - x_{i-1})$$
$$= f(\xi_i)(x_i - x_{i-1}).$$

Thus, we can sum these numbers up, eliminating all but the first and last elements:

$$\sum_{i=1}^{n} f(\xi_i)(x_i - x_{i-1}) = \sum_{i=1}^{n} F(x_i) - F(x_{i-1})$$

$$= F(b) - F(a).$$

On the other hand, due to the properties of lower and upper sums, we have

$$L[f, X] \leq \sum_{i=1}^{n} f(\xi_i)(x_i - x_{i-1}) \leq U[f, X]$$

Since $f$ is integrable, the squeeze principle (*Corollary 10.2.1*) and *Theorem 14.1.1* imply that

$$\int_a^b f(x)dx = \sum_{i=1}^{n} f(\xi_i)(x_i - x_{i-1}) = F(b) - F(a)$$

must hold. This is what we had to show.

**Remark 14.1.1 (Increments of functions)**

For simplicity, the increments of a function $F$ on the interval $[a, b]$ is also denoted by

$$\left[F(x)\right]_{x=a}^{x=b} := F(b) - F(a).$$

Thus, according to the fundamental theorem of calculus (*Theorem 14.1.2*),

$$\int_a^b f(x)dx = \left[F(x)\right]_{x=a}^{x=b}$$

holds if $F'(x) = f(x)$.

Note that integration is insensitive towards changing the values of $f(x)$ at countably many points. To be more precise, suppose that $f : \mathbb{R} \to \mathbb{R}$ is a function that is integrable on $[-1, 1]$. Let's change its value at a single point and define

$$f^*(x) = \begin{cases} f(0) + 1 & \text{if } x = 0 \\ f(x) & \text{otherwise.} \end{cases}$$

If a partition is given by $-1 = x_0 < \ldots < x_{k-1} \le 0 < x_k < \ldots < x_n = 1$, then

$$\left|L[f, X] - L[f^*, X]\right| = \left|\underbrace{\inf_{x \in [x_{k-1}, x_k]} f(x) - \inf_{x \in [x_{k-1}, x_k]} f^*(x)}_{=:m_k}\right|(x_k - x_{k-1})$$

and

$$\left|U[f, X] - U[f^*, X]\right| = \left|\underbrace{\sup_{x \in [x_{k-1}, x_k]} f(x) - \sup_{x \in [x_{k-1}, x_k]} f^*(x)}_{=:M_k}\right|(x_k - x_{k-1})$$

holds. We can select the partition such that $x_k - x_{k-1} < \varepsilon$ for some arbitrary $\varepsilon > 0$, thus,

$$\left|L[f, X] - L[f^*, X]\right| \quad \text{and} \quad \left|U[f, X] - U[f^*, X]\right|$$

can be made as small as needed. This implies that

$$\int_a^b f(x)dx = \int_a^b f^*(x)dx.$$

Hence, saying that integration is the inverse of differentiation is mathematically a bit imprecise. Given a differentiable function $F(x)$, its derivative is unique, but there are infinitely many functions whose integral $F(a) + \int_a^x g(y)dy$ reconstructs $F$.

The fundamental theorem of calculus allows us to formulate Lagrange's mean value theorem (*Theorem 13.1.6*) in terms of integrals.

---

**Theorem 14.1.3** *(The mean value theorem for definite integrals)*

Let $f : \mathbb{R} \to \mathbb{R}$ *a function that is continuous on* $[a, b]$. *Then there exists an* $\xi \in [a, b]$ *such that*

$$\int_a^b f(x)dx = (b - a)f(\xi).$$

---

*Proof.* According to the fundamental theorem of calculus (*Theorem 14.1.2*), the function

$$F(t) = \int_a^t f(x)dx$$

is differentiable on $[a, b]$ and $F'(t) = f(t)$.

Thus, Lagrange's mean value theorem (*Theorem 13.1.6*) gives that

$$\frac{F(b) - F(a)}{b - a} = f(\xi)$$

for some $\xi \in (a, b)$, from which

$$\int_a^b f(x)dx = F(b) - F(a)$$

$$= (b - a)f(\xi)$$

follows. This is what we had to show.

After all this theory, you might ask: what does integration have to do with machine learning? Without being mathematically rigorous, here is a (very) brief overview of what's to come.

First, you can think about integration as a continuous generalization of the arithmetic mean. As you can see, for equidistant partitions, an approximating sum

$$S[f, X, \xi] = \frac{1}{n} \sum_{i=1}^{n} f(\xi_i)$$

is exactly the average of $f(\xi_1), \ldots, f(\xi_n)$. In machine learning, averages are frequently used to express various quantities, like the mean-squared error. Think about it: loss functions are often averages of certain individual losses. On a fine enough scale, averages become integrals.

Along with linear algebra and calculus, the central pillar of machine learning is *probability theory and statistics*, which gives us a way to model the world based on our observations. Probability and statistics are the logic of science and decision-making. There, integration is used to express probabilities, expected value, information, and much more. Without a rigorous theory of integration, we cannot build probabilistic models beyond a certain point.

## 14.2  Integration in practice

Even though we understand what an integral is, we are far from computing them in practice. As opposed to differentiation, analytically evaluating integrals can be *really* difficult and sometimes downright impossible. The formula (14.8) suggests that the key is to find the function whose derivative is the integrand, called the antiderivative or primitive function. This is harder than you think. Nevertheless, there are several tools for this, and we are going to devote this section to studying the most important ones.

Often, the key is finding the antiderivative, so we introduce the notation

$$F(x) = \int f(x)dx,$$

for the functions where $F' = f$. (Sometimes we abbreviate this to $F = \int f \, dx$.) Note that since $(F +$ some constant$)' = F'$, the antiderivative $\int f(x)dx$ is not uniquely determined. However, this is not an issue for us, as the Newton-Leibniz formula states that

$$\int_a^b f(x)dx = F(b) - F(a).$$

Thus, any additional constants would be eliminated.

With this under our belt, we are ready to dig deep into evaluating integrals in practice.

## 14.2.1   Integrals and operations

As we have seen this several times (for instance, when discussing the rules of differentiation in *Theorem 12.2.1*), the relations of an operation with addition, multiplication, and possibly others are extremely useful for gaining insight and developing practical tools.

This is the same for integration as well. Similar to before, the linearity of the integral is our main tool to evaluate it.

---

**Theorem 14.2.1   *(Linearity of the Riemann integral)***

*Let $f, g : \mathbb{R} \to \mathbb{R}$ be two functions that are integrable on $[a, b]$. Then*

*(a)* $\int_a^b \big(f(x) + g(x)\big)dx = \int_a^b f(x)dx + \int_a^b g(x)dx,$

*(b)* $\int_a^b cf(x)dx = c \int_a^b f(x)dx.$

---

*Proof.* (a) If $f$ and $g$ are integrable, then for any $\varepsilon > 0$, there are partitions $X_f, X_g$ such that

$$\int_a^b f(x)dx - \varepsilon \le L[f, X_f] \le U[f, X_f] \le \int_a^b f(x)dx + \varepsilon$$

and

$$\int_a^b g(x)dx - \varepsilon \le L[g, X_g] \le U[g, X_g] \le \int_a^b g(x)dx + \varepsilon,$$

where the lower and upper sums are defined by (14.2) and (14.3). So, for the mutual refinement $X = X_f \cup X_g$, we have

$$L[f, X_f] \leq L[f, X],$$
$$L[g, X_g] \leq L[g, X],$$
$$U[f, X] \leq U[f, X_f],$$
$$U[g, X] \leq U[g, X_g]$$

due to the *Proposition 14.1.2.* Thus,

$$\int_a^b f(x)dx + \int_a^b g(x)dx - 2\varepsilon \leq L[f, X_f] + L[g, X_g]$$
$$\leq L[f, X] + L[g, X]$$
$$\leq S[f, X, \xi_X] + S[g, X, \xi_X]$$
$$\leq U[f, X] + U[g, X]$$
$$\leq U[f, X_f] + U[g, X_g]$$
$$\leq \int_a^b f(x)dx + \int_a^b g(x)dx + 2\varepsilon,$$

where $S$ is defined by (14.7). From this definition, it can also be seen that

$$S[f + g, X, \xi_X] = S[f, X, \xi_X] + S[g, X, \xi_X].$$

Thus,

$$\left| \int_a^b f(x)dx + \int_a^b g(x)dx - S[f + g, X, \xi_X] \right| \leq 2\varepsilon,$$

implying that

$$\lim_{|X| \to 0} S[f + g, X, \xi_X] = \int_a^b f(x)dx + \int_a^b g(x)dx$$
$$= \int_a^b \big(f(x) + g(x)\big)dx.$$

*Theorem 14.1.1* regarding the approximating sum $S$ implies that $f + g$ is integrable on $[a, b]$ and

$$\int_a^b \big(f(x) + g(x)\big)\,dx = \int_a^b f(x)\,dx + \int_a^b g(x)\,dx.$$

*(b)* This follows from the fact that $S[cf, X, \xi_X] = cS[f, X, \xi_X]$.

## 14.2.2   Integration by parts

As we have learned when studying the rules of differentiation (*Theorem 12.2.1*), for an arbitrary $f$ and $g$, we have

$$(fg)' = f'g + fg'.$$

Applying this logic to the antiderivatives,

$$fg = \int (f'g + fg')\,dx$$

holds. Rearranging the equation a bit, we obtain the formula of *integration by parts*:

$$\int f'g\,dx = fg - \int fg'\,dx. \tag{14.9}$$

This is summed up in the following theorem.

**Theorem 14.2.2  *(Integration by parts)***

*Let $f, g : \mathbb{R} \to \mathbb{R}$ be two functions. If both are differentiable on the interval $[a, b]$, then*

$$\int_a^b f'(x)g(x)\,dx = \big[f(x)g(x)\big]_{x=a}^{x=b} - \int_a^b f(x)g'(x)\,dx$$

*holds.*

How is this useful for us? Consider a situation where finding the antiderivative of $f$ and the derivative of $g$ is easy, but the antiderivative of the product $fg$ is hard. For example, can you quickly calculate the following?

$$\int x \log x\,dx$$

Applying (14.9) with the roles $f'(x) = x$ and $g(x) = \log x$ immediately yields

$$\int x \log x = \frac{1}{2}x^2 \log x - \int \frac{1}{2}x dx$$
$$= \frac{1}{2}x^2 \log |x| - \frac{1}{4}x^2 + C,$$

where $C \in \mathbb{R}$ is an arbitrary constant.

## 14.2.3   Integration by substitution

As the integration by parts formula is the "opposite" of the differentiation rule for products, there is an analogue for the chain formula as well. Recall that for two differentiable functions, we had

$$(f \circ g)'(x) = f'(g(x))g'(x).$$

Translating this to the language of integrals, we obtain the following result.

**Theorem 14.2.3** *(Integration by substitution)*

Let $f, g : \mathbb{R} \to \mathbb{R}$ be integrable functions on $[a, b]$. Suppose that $f$ is continuous and $g$ is differentiable. Then

$$\int_a^b f(g(y))g'(y)dy = \int_{g(a)}^{g(b)} f(x)dx$$

holds.

This is called *integration by substitution*. To give you an example of its use, consider

$$\int x \sin(x^2)dx.$$

With the roles $y(x) = x^2$, we have

$$\int x \sin(x^2)dx = \frac{1}{2}\int \sin y dy$$
$$= -\frac{1}{2}\cos y + C$$
$$= -\frac{1}{2}\cos(x^2) + C,$$

where $C \in \mathbb{R}$ is an arbitrary constant.

Integration by parts and substitution are our main weapons for calculating integrals on paper. Most of the integrals one might encounter can be solved with the creative (and possibly iterated) application of these two rules. The recipe is simple: find the antiderivative, then use the Newton-Leibniz formula (*Theorem 14.1.2*) to compute the value of the integral.

However, there is a serious issue: antiderivatives can be extremely hard to find, maybe even impossible. This makes integrals difficult to compute symbolically. For instance, consider

$$\int e^{-x^2} dx,$$

where the function $e^{-x^2}$ describes the well-known Gaussian bell curve. As surprising as it is, $\int e^{-x^2} dx$ cannot be described with a closed formula! (That is, one that uses a finite number of operations and only elementary functions.) It's not that mathematicians were not clever enough to find a closed formula for the antiderivative; this doesn't exist.

Thus, computing integrals is much simpler to do numerically. This is in stark contrast with differentiation, which is easy to do symbolically but hard numerically.

## 14.2.4 Numerical integration

Instead of using symbolic computation to get the exact value of an integral, we will resolve to approximation once again. Previously, *Theorem 14.1.1* showed us that an integral is the limit of the Riemann-sums:

$$\int_a^b f(x)dx = \lim_{n \to \infty} \sum_{i=1}^n f(\xi_i)(x_{i,n} - x_{i-1,n}), \tag{14.10}$$

where $X_n = \{x_{0,n}, \dots, x_{n,n}\}$ is a partition of $[a, b]$ and $\xi_i \in [x_{i-1,n}, x_{i,n}]$ are arbitrary intermediate values.

In other words, if $n$ is large enough, the sum $\sum_{i=1}^n f(\xi_i)(x_{i,n} - x_{i-1,n})$ is close to $\int_a^b f(x)dx$. There are two crucial issues: first, how to select the partition and the intermediate values; second, how fast is the convergence?

If we want to make (14.10) useful, we have to devise a concrete method that prescribes the $x_i$-s, $\xi_i$-s, and tells us how large of an $n$ we should select. This is an extremely rich subject that has been the focus of studies ever since the introduction of integration. So, there is a lot to talk about here. To keep things simple, let's just focus on the essentials.

The most straightforward method is to select a uniform partition, then approximate the area under the function curve with a sequence of trapezoids.

That is, let $X = \{a, a + \frac{b-a}{n}, a + 2\frac{b-a}{n}, \dots, b\}$ be an equidistant partition, which we'll use to estimate the integral via calculating the areas determined by the trapezoids given by the partition and the graph, as illustrated in *Figure 14.7*.

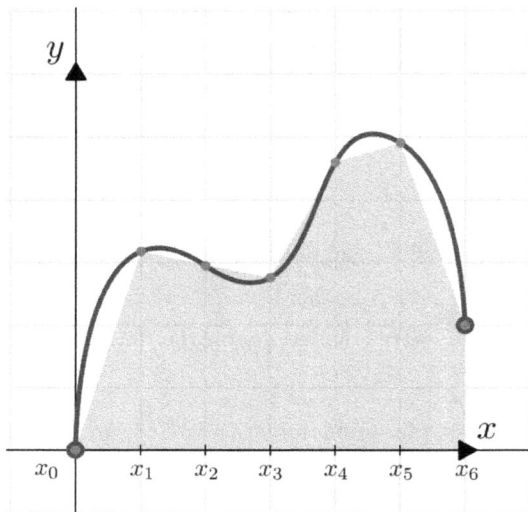

*Figure 14.7: Approximating the area under a function with successive trapezoids*

As the trapezoid's area is given by $h\frac{a+b}{2}$, the area under the curve in $[x_{i-1}, x_i]$ is approximated by

$$(x_i - x_{i-1})\frac{f(x_i) + f(x_{i-1})}{2} = \frac{b-a}{2n}\left(f(x_i) + f(x_{i-1})\right)$$

and we have the approximation

$$\int_a^b f(x)dx \approx \frac{b-a}{2n}\sum_{i=1}^n \left(f(x_{i-1}) + f(x_i)\right)$$

$$= \frac{b-a}{2n}\left(f(x_0) + f(x_n)\right) + \frac{b-a}{n}\sum_{i=1}^{n-1} f(x_i), \quad x_i = a + i\frac{b-a}{n}.$$

(14.11)

This is the trapezoidal rule. It might seem complicated, but (14.11) is just a weighted sum of the $f(x_i)$ values.

Its rate of convergence is quadratic, as stated by the following theorem.

---

**Theorem 14.2.4** *(Trapezoidal rule)*

*Let $f : [a, b] \to \mathbb{R}$ be a twicely differentiable function, and let*

$$I_n := \frac{1}{n} \sum_{i=1}^{n} \frac{f(x_{i-1}) + f(x_i)}{2}$$

*be the approximation given by the trapezoidal rule. Then*

$$\left| \int_a^b f(x)dx - I_n \right| = O\left( \frac{(b-a)^3}{n^2} \right).$$

---

There are other methods, for instance, Simpson's rule approximates the function with a piecewise quadratic one. (Instead of a piecewise linear one, like the trapezoidal rule.) Since the approximation is more accurate, the convergence is also faster: Simpson's rule converges at a $O(n^{-4})$ rate. Without going into details, it is given by

$$S_n = \frac{b-a}{3n} \sum_{i=1}^{\lfloor n/2 \rfloor} \left( f(x_{2i-2}) + 4f(x_{2i-i}) + f(x_{2i}) \right), \tag{14.12}$$

with error

$$\left| \int_a^b f(x)dx - S_n \right| = O\left( \frac{(b-a)^5}{n^4} \right),$$

where $x_i$ is again the equidistant partition $x_i = a + i\frac{b-a}{n}$.

The formula (14.12) can be difficult to unpack, but the essence remains the same: we compute the function's values at given points, then take their weighted sum.

## 14.2.5 Implementing the trapezoidal rule

To show you how straightforward the trapezoidal rule is, let's implement it in practice! To keep it simple, we are implementing this as a function that takes another function as its input.

```python
def trapezoidal_rule(f, a, b, n):
    # Define the partition of the interval [a, b]
    partition = [a + i*(b -_a)/n for i in range(n+1)]

    # Evaluate the function at each partition point
    vals = [f(x) for x in partition]

    # Apply the trapezoidal rule
```

```
    I_n = (b - a) / (2 * n) * (vals[0] + vals[-1]) + (b - a) / n * sum(vals[1:-1])

    return I_n
```

This can be made even simpler with NumPy, but I'll leave this to you as an exercise. Let's test it on an example instead!

With the use of the Newton-Leibniz formula (*Theorem 14.1.2*), you can verify that

$$\int_0^1 x^2 dx = \frac{1}{3}.$$

(We even computed this with our bare hands, using lower and upper sums.) After plugging the function lambda x: x**2 into trapezoidal_rule, we can see that this method is indeed correct.

```
import matplotlib.pyplot as plt

with plt.style.context("seaborn-v0_8"):
    plt.figure()
    ns = range(1, 25, 1)
    Is = [trapezoidal_rule(lambda x: x**2, 0, 1, n) for n in ns]
    plt.axhline(y=1/3, color='r', label="the true integral")
    plt.scatter(ns, Is, label="trapezoidal_rule(f, a, b, n)")
    plt.ylim([0.3, 0.52])
    plt.title("the trapezoidal rule")
    plt.show()
```

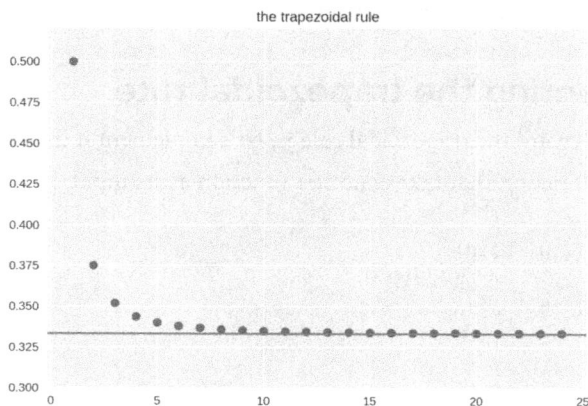

*Figure 14.8: The trapezoidal rule*

## 14.3  Summary

In this chapter, we have learned about integration, one of the technically most challenging subjects so far. Intuitively, the integral of a function describes the signed area under its graph, but mathematically, it is given by the limit

$$\int_a^b f(x)dx = \lim_{n\to\infty} \sum_{k=1}^n (x_i - x_{i-1})f(x_i),$$

where $a = x_0 < x_1 < \dots < x_n = b$ is a partition of the interval $[a, b]$. Of course, we don't often calculate integrals by the definition; we have the Newton-Leibniz formula for that:

$$\int_a^b f(x)dx = F(b) - F(a),$$

where $F$ is the so-called *antiderivative*, satisfying $F'(x) = f(x)$. This is why integration is thought of as the inverse of differentiation.

As one of my professors used to say, symbolic differentiation is easy, numeric differentiation is hard. It's the opposite for integrals: symbolic integration is hard, and numeric integration is easy. We've learned a couple of tricks to pin down the symbolic part, namely the integration by parts formula

$$\int_a^b f'(x)g(x)dx = \left[f(x)g(x)\right]_{x=a}^{x=b} - \int_a^b f(x)g'(x)dx$$

and the integration by substitution formula

$$\int_a^b f\big(g(y)\big)g'(y)dy = \int_{g(a)}^{g(b)} f(x)dx.$$

When symbolic integration is hard (and it's almost always hard), we can resort to numerical methods, such as the Simpson's rule, given by

$$\int_a^b f(x)dx \sim \frac{b-a}{3n} \sum_{i=1}^{\lfloor n/2 \rfloor} \Big( f(x_{2i-2}) + 4f(x_{2i-i}) + f(x_{2i}) \Big).$$

Although integration is quite technical and complicated, its proper exposition is extremely important if you want to understand mathematics. The idea of approximating complex shapes with a sequence of rectangles (as shown in *Figure 14.3*) is the foundation of measure theory, which, in turn, is the foundation of probability theory. (Hell, the idea of approximating complex objects with simpler ones is *the foundation of mathematics*.)

By now, we have mastered differentiation and integration for single-variable functions. However, univariate functions are rare in practice: in machine learning, we often deal with millions or billions of variables. To handle them in practice, we'll generalize all we've learned to higher dimensions. This is the subject of *multivariable calculus*, our next big milestone. Let's get to it!

# 14.4   Problems

**Problem 1.** Use integration by parts to find the following antiderivates.

(a) $\int \sin(x) \cos(x) dx$

(b) $\int x e^x dx$

(c) $\int x^2 e^x dx$

(d) $\int e^x \sin x \, dx$

**Problem 2.** Use integration by substitution to find the following antiderivatives.

(a) $\int x \cos(x^2) dx$

(b) $\int \sin(x) e^{\{x 2\}} dx$

(c) $\int \frac{\sin x}{\cos x}$

(d) $\int x \sqrt{x^2 + 1} dx$

**Problem 3.** Let $f : [a, b] \to \mathbb{R}$ be an integrable function. Show that

$$\left| \int_a^b f(x) dx \right| \leq \int_a^b |f(x)| dx.$$

**Problem 4.** Let $f, g : [a, b] \to \mathbb{R}$ be two integrable functions such that $|f|^2$ and $|g|^2$ are integrable as well. Show that

$$\left| \int_a^b f(x) g(x) dx \right|^2 \leq \int_a^b |f(x)|^2 dx \int_a^b |g(x)|^2 dx.$$

*Hint:* revisit *Chapter 2* about normed spaces, and find an inequality that feels similar to this one.

**Problem 5.** The famous Dirichlet function is defined by

$$D(x) = \begin{cases} 1 & \text{if } x \in \mathbb{Q}, \\ 0 & \text{otherwise.} \end{cases}$$

Is $D(x)$ integrable?

# Join our community on Discord

Read this book alongside other users, Machine Learning experts, and the author himself.

Ask questions, provide solutions to other readers, chat with the author via Ask Me Anything sessions, and much more.

Scan the QR code or visit the link to join the community.

`https://packt.link/math`

# References

[1]  Ramalho, L. (2022). *Fluent Python* (2nd ed.). O'Reilly Media, Inc.

[2]  Rudin, W. (1976). *Principles of Mathematical Analysis* (3rd ed.). McGraw Hill.

[3]  Spivak, M. (2008). *Calculus* (4th ed.). Cambridge University Press.

# PART III

# MULTIVARIABLE CALCULUS

This part comprises the following chapters:

- *Chapter 15, Multivariable Fucntions*
- *Chapter 16, Derivatives and Gradients*
- *Chapter 17, Optimization in Multiple Variables*

# 15

# Multivariable Functions

How different is multivariable calculus from its single-variable counterpart? When I was a student, I had a professor who used to say something like, "multivariable and single-variable functions behave the same, you just have to write more."

Well, this couldn't be further from the truth. Just think about what we are doing in machine learning: training models with gradient descent; that is, finding a configuration of parameters that minimize a parametric function. In one variable (which is not a realistic assumption), we can do this with the derivative, as we saw in *Section 13.2*. How can we extend the derivative to multiple dimensions?

The inputs of multivariable functions are *vectors*. Thus, given a function $f : \mathbb{R}^n \to \mathbb{R}$, we can't just define

$$\frac{df}{d\mathbf{x}}(\mathbf{x}_0) = \lim_{\mathbf{x} \to \mathbf{x}_0} \frac{f(\mathbf{x}_0) - f(\mathbf{x})}{\mathbf{x}_0 - \mathbf{x}}, \quad \mathbf{x}_0, \mathbf{x} \in \mathbb{R}^n$$

to the analogue of *Definition 12.1.1*. Why? Because the division with the vector $\mathbf{x}_0 - \mathbf{x}$ is not defined.

As we'll see, differentiation in multiple dimensions is much more complicated. Think about it: in one dimension, there are only two directions, left and right. This is not true even for two dimensions, with an infinite number of directions at each point.

So, what are multivariable functions anyway?

# 15.1   What is a multivariable function?

We introduced functions in *Chapter 9*, as general mappings between two sets. However, we've only discussed functions that map real numbers to real numbers. Simple scalar-scalar functions are great for conveying ideas, but the world around us is much more complex than what we could describe with them. At the other end of the spectrum, set-set functions are way too general to be useful.

In practice, three categories are special enough to be analyzed mathematically but general enough to describe the patterns in science and engineering: those that

1. map scalars to vectors, that is, $f : \mathbb{R} \to \mathbb{R}^n$,
2. map vectors to scalars, that is, $f : \mathbb{R}^n \to \mathbb{R}$,
3. and those that map vectors to vectors, that is, $f : \mathbb{R}^n \to \mathbb{R}^m$.

The scalar-vector variants are called *curves*, the vector-scalar ones are *scalar fields*, and the vector-vector functions are what we call *vector fields*. This nomenclature looks a bit abstract, so let's see some examples.

**Scalar-vector functions**, or curves to use their more user-friendly name, are the mathematical representations of movement. A space station orbiting around Earth describes a curve. So does the trajectory of a stock in the market.

To give you a concrete example, the scalar-vector function

$$f(t) = \begin{bmatrix} \cos(t) \\ \sin(t) \end{bmatrix}$$

describes the unit circle. This is illustrated by *Figure 15.1*.

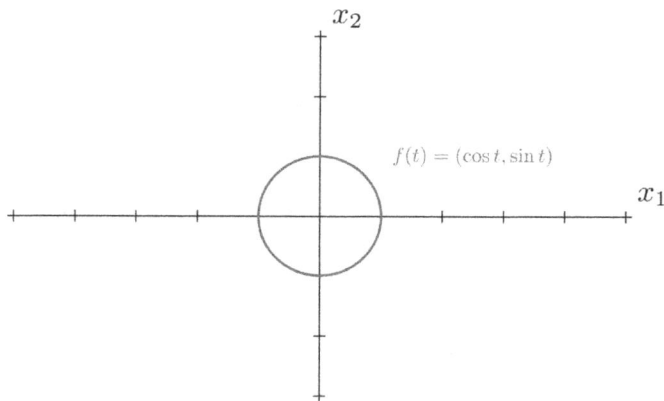

*Figure 15.1: A scalar-vector function, that is, a curve*

Not all curves are closed. For example, the curve

$$g(t) = \begin{bmatrix} \cos(t) \\ \sin(t) \\ t \end{bmatrix}$$

represents a motion that spirals upward, as illustrated by *Figure 15.2*. These curves are called *open*.

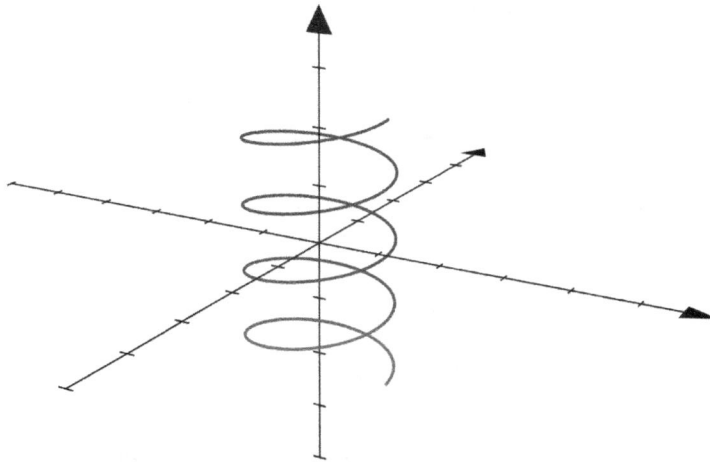

*Figure 15.2: An open curve*

Because of their inherent ability to describe trajectories, scalar-vector functions are essential in mathematics and science. Are you familiar with Newton's second law of motion, stating that force equals mass times acceleration? This is described by the equation, $F = ma$, which is an instance of an *ordinary differential equation*. All of its solutions are curves.

On the surface, scalar-vector functions have little to do with machine learning, but that's not the case. Even though we won't deal with them extensively, they have a serious presence behind the scenes. For instance, gradient descent is a discretized curve, as we saw in *Section 13.3*.

**Vector-scalar** functions will be our focus for the next few chapters. When I write "multivariable function," I'll most often refer to a *vector-scalar* function.

Think about a map of a mountain landscape. This *maps* the height – a scalar – to each coordinate, thereby defining the surface. This is just a function $f : \mathbb{R}^2 \to \mathbb{R}$ in mathematical terms.

Thinking about scalar fields as surfaces is useful for building geometric intuition, giving us a way to visualize them as you can see in *Figure 15.3*. (Note that the surface analog breaks down for dimensions larger than two.)

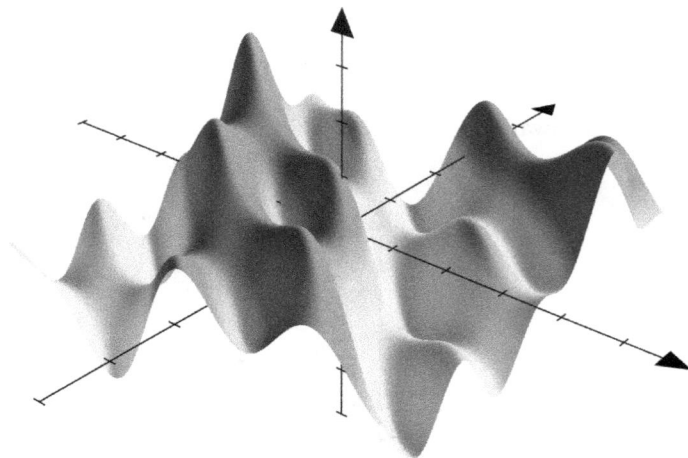

*Figure 15.3: A surface given by a vector-scalar function*

Let's clear up the notation first. If $f : \mathbb{R}^n \to \mathbb{R}$ is a function of $n$ variables, we might write $f(\mathbf{x})$ for an $\mathbf{x} \in \mathbb{R}^n$ or $f(x_1, \dots, x_n)$ for $x_i \in \mathbb{R}$ if we want to emphasize the dependence on its variables. A function of $n$ variables is the same as a function of a single vector variable. I know this seems confusing, but trust me, you'll get used to it in no time.

To give a concrete example for a vector-scalar function, let's consider *pressure*. Pressure is the ratio of the magnitude of the force and the area of the surface of contact:

$$p = \frac{F}{A}.$$

This can be thought of as a function of two variables: $p(x, y) = x/y$.

To illustrate how problematic things can become in multiple dimensions, consider the pressure around $(0, 0)$. Although we haven't talked about the limits of multivariable functions yet, what do you think

$$\lim_{(x,y)\to(0,0)} \frac{x}{y}$$

should be?

Based on how we defined limits for single-variable functions (see *Definition 11.2.1*),

$$\lim_{n \to \infty} \frac{x_n}{y_n}$$

must match for all possible choices for $x_n$ and $y_n$. This is not the case. Consider $x_n = \alpha^2/n$ and $y_n = \alpha/n$ for any $\alpha$ real number. With this choice, we have

$$\lim_{n \to \infty} \frac{x_n}{y_n} = \frac{\alpha^2/n}{\alpha/n} = \alpha.$$

Thus, the above limit is not defined. All we did here is approach zero along slightly different trajectories, yet the result is a total mess. In one variable, we have to flex our intellectual muscles to produce such examples; in multiple variables, a simple $x/y$ will do the trick.

**Vector-vector** functions are called *vector fields*. For example, consider our solar system, modeled by $\mathbb{R}^3$. Each point is affected by a gravitational force, which is a vector. Thus, the gravitational pull can be described by a $f : \mathbb{R}^3 \to \mathbb{R}^3$ function, hence the name vector field.

Although they are often hidden in the background, vector fields play an essential role in machine learning. Remember when we discussed why gradient descent works in *Section 13.3?* (At least in one variable.) All the differential equations we have encountered there are equivalent to vector fields.

Why? Consider the differential equation $x' = f(x)$. If $x(t)$ describes the trajectory of a moving object, then its derivative $x'(t)$ is its speed. Thus, we can interpret the equation $x'(t) = f(x(t))$ as prescribing the speed of our object at every position. It's not that spectacular when our object is moving in one dimension, but if the trajectory $x : \mathbb{R} \to \mathbb{R}^2$ describes a motion on the plane, the function $f : \mathbb{R}^2 \to \mathbb{R}^2$ can be visualized neatly.

For example, consider the population dynamics of a simple predator-prey system. Predators feed on the prey, thus, their numbers can grow in the abundance of food. In turn, over-consumption decreases the prey population, causing a famine among the predators and decreasing their numbers. This leads to a growth in the prey, and the cycle starts over again.

If $x_1(t)$ and $x_2(t)$ are the size of the prey and predator populations, respectively, then their dynamics are described by the famous Lotka-Volterra equations:

$$x_1' = x_1 - x_1 x_2$$
$$x_2' = x_1 x_2 - x_2.$$

If we represent the trajectory as the scalar-vector function

$$\mathbf{x} : \mathbb{R} \to \mathbb{R}^2, \quad \mathbf{x}(t) = \begin{bmatrix} x_1(t) \\ x_2(t) \end{bmatrix},$$

then the derivative

$$\mathbf{x}'(t) = \begin{bmatrix} x_1'(t) \\ x_2'(t) \end{bmatrix}$$

is given by the vector-vector function

$$f : \mathbb{R}^2 \to \mathbb{R}^2, \quad f(x_1, x_2) = \begin{bmatrix} x_1 - x_1 x_2 \\ x_1 x_2 - x_2 \end{bmatrix}.$$

$f$ can be visualized by drawing a vector onto each point of the plane, as illustrated by *Figure 15.4.*

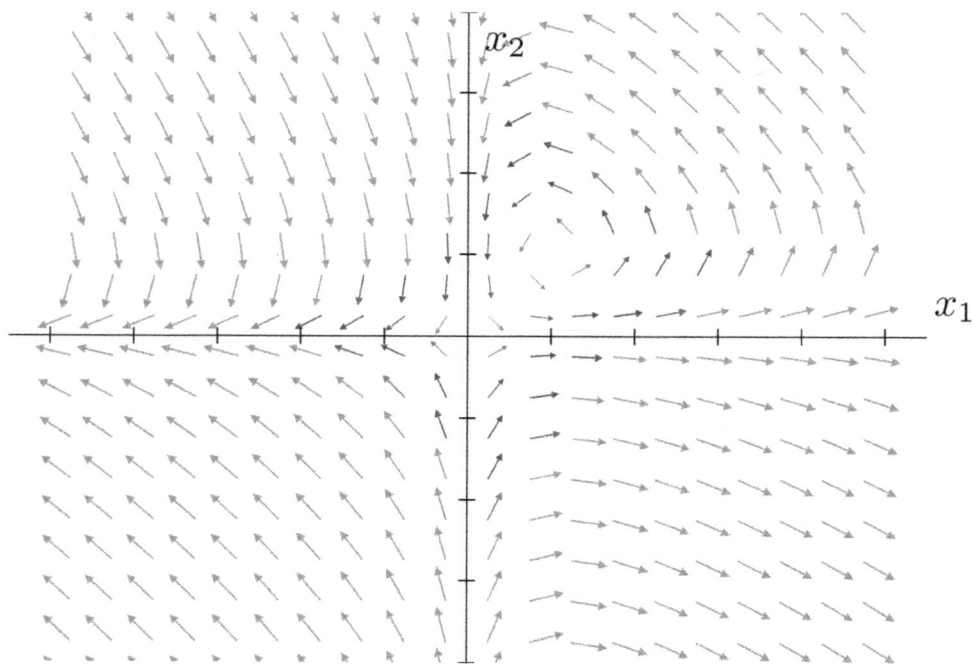

*Figure 15.4: The vector field given by the Lotka-Volterra equations*

Vector fields have serious applications in machine learning. As we shall see soon, the multivariable derivative (called *gradient*) defines a vector field.

Moreover, as indicated by the single-variable case (see *Section 13.3*), the gradient descent algorithm will be the discretized trajectory determined by the vector field of the gradient.

Now that we understand what multivariable functions are, let's see a special case. You know how we roll: examples are essential, and we always start with them whenever possible. This time, we'll put *linear functions* under the microscope.

## 15.2 Linear functions in multiple variables

One of the most important functions in mathematics is the linear function. In one variable, it takes the form $l(x) = ax + b$, where $a$ and $b$ are arbitrary real numbers.

We've seen linear functions several times already. For instance, *Theorem 12.1.1* gives that differentiation is equivalent to finding the best linear approximation.

Linear functions, that is, functions of the form

$$f(x_1, \ldots, x_n) = b + \sum_{i=1}^{n} a_i x_i, \quad b, a_i \in \mathbb{R}$$

are as important in multiple variables as in one.

To build up a deep understanding, we'll take a look at the simplest case: a line on the two-dimensional plane.

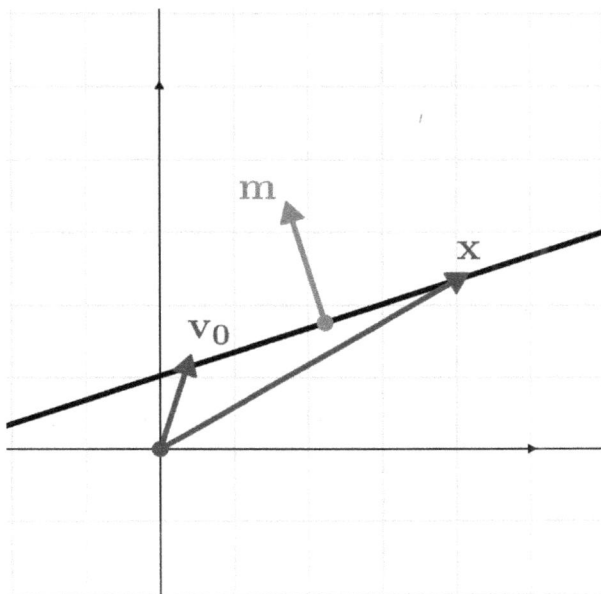

*Figure 15.5: A line on the plane*

Given its normal vector $\mathbf{m} = (m_1, m_2)$ and its arbitrary point $\mathbf{v}_0$, the vector $\mathbf{x}$ is on the line if and only if $\mathbf{m}$ and $\mathbf{x} - \mathbf{v}_0$ is orthogonal, that is, if

$$\langle \mathbf{m}, \mathbf{x} - \mathbf{v}_0 \rangle = 0 \tag{15.1}$$

holds. (15.1) is called the *normal vector equation of the line.*

By using the bilinearity of the inner product and writing out $\langle \mathbf{m}, \mathbf{x} \rangle$ in terms of their coordinates, we can simplify (15.1). Assuming that $m_2 \neq 0$, that is, the line is not parallel to the $x_2$ axis, a quick calculation yields

$$x_2 = -\frac{m_1}{m_2} x_1 + \frac{1}{m_2} \langle \mathbf{m}, \mathbf{v}_0 \rangle.$$

This is a linear function of the single variable $x_1$ in its full glory. The coefficient $-\frac{m_1}{m_2}$ describes the slope, while $\frac{1}{m_2} \langle \mathbf{m}, \mathbf{v}_0 \rangle$ describes the intercept.

In other words, linear functions are equivalent to vector equations of the form (15.1), at least in one variable.

What happens if we apply the same argument in higher dimensional spaces? In $\mathbb{R}^{n+1}$, the normal vector equation

$$\langle \mathbf{m}, \mathbf{x} - \mathbf{v}_0 \rangle = 0, \quad \mathbf{m}, \mathbf{x}, \mathbf{v}_0 \in \mathbb{R}^{n+1} \tag{15.2}$$

defines a *hyperplane*, that is, an $n$-dimensional plane. (One dimension less than the embedding plane, which is $\mathbb{R}^{n+1}$ in our case.) Unraveling (15.2), we obtain

$$x_{n+1} = \frac{1}{m_{n+1}} \langle \mathbf{m}, \mathbf{v}_0 \rangle - \sum_{i=1}^{n} \frac{m_i}{m_{n+1}} x_i.$$

Thus, the general form of a linear function in $n$ variables

$$f(x_1, \ldots, x_n) = b + \sum_{i=1}^{n} a_i x_i, \quad b, a_i \in \mathbb{R}$$

originates from the normal vector equation of the $n$-dimensional plane, embedded in the $(n + 1)$-dimensional space.

This can also be written in the vectorized form

$$f(\mathbf{x}) = b + \langle \mathbf{a}, \mathbf{x} \rangle$$
$$= \mathbf{a}^T \mathbf{x} + b, \quad \mathbf{a}, \mathbf{x} \in \mathbb{R}^n, \quad b \in \mathbb{R}, \tag{15.3}$$

which is how we'll mostly use it in the future. (Note that when looking at the matrix representation of a vector $\mathbf{u} \in \mathbb{R}^n$, we always use the column form $\mathbb{R}^{n \times 1}$. Moreover, $\mathbf{a}$ is *not* the normal vector of the plane.)

Before we move on to study the inner workings of multivariable calculus, I want to emphasize how seriously multiple dimensions complicate things in machine learning.

## 15.3   The curse of dimensionality

First, let's talk about optimization. If all else fails, optimizing a single-variable function $f : [a, b] \to \mathbb{R}$ can be as simple as partitioning $[a, b]$ into a grid of $n$ points, evaluating the function at each point, then finding the minima/maxima.

We cannot do this in higher dimensions. To see why, consider ResNet18, the famous convolutional network architecture. It has precisely 11,689,512 parameters. Thus, training is equivalent to optimizing a function of a whopping 11,689,512-variable function. If we were to construct a grid with just two points along every dimension, we would have $2^{11689512}$ points to evaluate the function at. For comparison, the number of atoms in our observable universe is around $10^{82}$. A number that is dwarfed by the size of our grid. Thus, grid search is currently impossible on such an enormous grid. We are forced to devise clever algorithms that can tackle the size and complexity of large dimensional spaces.

Another issue is that, in high dimensions, a strange thing starts to happen with balls. Recall that, by definition, the n-dimensional ball of radius $r$ around the point $\mathbf{x}_0 \in \mathbb{R}^n$ is defined by

$$B_n(r, \mathbf{x}_0) := \{\mathbf{x} \in \mathbb{R}^n : \|\mathbf{x} - \mathbf{x}_0\| < r\},$$

and we denote its volume by $V_n(r)$. (The volume depends only on the radius and the dimension, not the center.)

It turns out that

$$V_n(r) = \frac{\pi^{\frac{n}{2}}}{\Gamma\left(1 + \frac{n}{2}\right)} r^n,$$

where $\Gamma(z)$ is the famous Gamma function (https://en.wikipedia.org/wiki/Gamma_function), the generalization of the factorial.

The volume formula might seem complicated because of the Gamma function, the $\pi$, and all the other terms, but let's focus on the core of the issue. What happens if we slice off an $\varepsilon$-wide shell from the unit ball?

It turns out that the volume of the unit ball is concentrated around its outer shell, as shown by the volume formula:

$$\lim_{n \to \infty} \frac{V_n(1 - \varepsilon)}{V_n(1)} = \lim_{n \to \infty} (1 - \varepsilon)^n = 0.$$

Heuristically, this means that if you randomly select a point from the unit ball, its distance from the center will be close to 1 in high dimensions.

In other words, distance doesn't behave as you would intuitively expect. Another way of looking at the issue would be to study the effects of taking one step in each possible direction, starting from the origin and arriving at the point

$$\mathbf{1} = (1, 1, \ldots, 1) \in \mathbb{R}^n,$$

something like what *Figure 15.6* illustrates in the three-dimensional case.

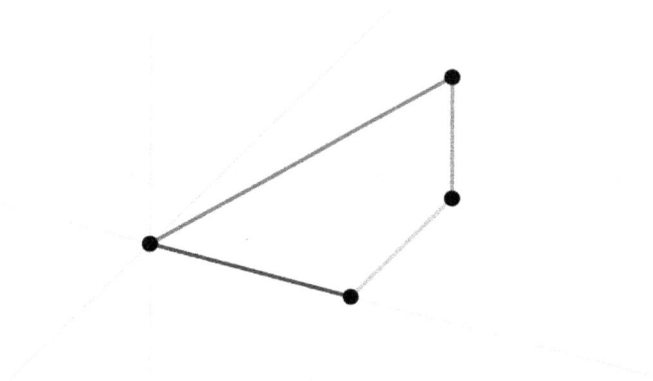

*Figure 15.6: Taking a step in each direction in three dimensions*

The Euclidean distance we have traveled is

$$\|\mathbf{1}\| = \sqrt{\sum_{i=1}^{n} 1} = \sqrt{n},$$

which goes to infinity as the number of dimensions grows. That is, the diagonal of the unit cube is *really big*.

These two phenomena can cause significant headaches in practice. More parameters result in more expressive models but also make training much more difficult. This is called the *curse of dimensionality*.

## 15.4 Summary

In this chapter, we have dipped our toe into the ocean of multivariable functions. The very moment we add more dimensions, the complexity shoots up.

For instance, we have three classes:

1. scalar-vector $f : \mathbb{R} \to \mathbb{R}^n$,
2. vector-scalar $f : \mathbb{R}^n \to \mathbb{R}$,
3. and vector-vector functions $f : \mathbb{R}^n \to \mathbb{R}^m$.

All of them are essential in machine learning. Feature transformations, like layers in neural networks, are vector-vector functions. Loss landscapes are given by vector-scalar functions, but training is done by following along a (discretized) scalar-vector function, also known as a curve.

Besides more complicated notations, we also have the curse of dimensionality to deal with. This is why optimizing functions of millions of variables is hard: not only does the parameter space get insanely large, but the concept of distance also begins to break down.

Now that we've built some intuition about multivariable functions and familiarity with the notation, it's time to dive deep. How can we do calculus in higher dimensions? Let's see in the next chapter!

# Join our community on Discord

Read this book alongside other users, Machine Learning experts, and the author himself.

Ask questions, provide solutions to other readers, chat with the author via Ask Me Anything sessions, and much more.

Scan the QR code or visit the link to join the community.

`https://packt.link/math`

# 16

# Derivatives and Gradients

Now that we understand why multivariate functions and high-dimensional spaces are more complex than the single-variable case we studied earlier, it's time to see how to do things in the general case.

To recap quickly, our goal in machine learning is to *optimize functions with millions of variables*. For instance, think about a neural network $N(\mathbf{x}, \mathbf{w})$ trained for binary classification, where

- $\mathbf{x} \in \mathbb{R}^n$ is the input data,
- $\mathbf{w} \in \mathbb{R}^m$ is the vector compressing all of the weight parameters,
- and $N(\mathbf{x}, \mathbf{w}) \in [0, 1]$ is the prediction, representing the probability of belonging to the positive class.

In the case of, say, binary cross-entropy loss, we have the loss function

$$L(\mathbf{w}) = - \sum_{k=1}^{d} y_i \log N(\mathbf{x}_i, \mathbf{w}),$$

where $\mathbf{x}_i$ is the $i$-th data point with ground truth $y_i \in \{0, 1\}$. See, I told you that we have to write *much* more in multivariable calculus. (We'll talk about binary cross-entropy loss in *Chapter 20*.)

Training the neural network is the same as finding a global minimum of $L(\mathbf{w})$, if it exists. We have already seen how we can do optimization in a single variable:

- figure out the direction of increase by calculating the derivative,
- take a small step,

- then iterate.

For this to work in multiple variables, we need to generalize the concept of the derivative. We can quickly discover the issue: since division with a vector is not defined, the difference quotient

$$\frac{f(\mathbf{x}) - f(\mathbf{y})}{\mathbf{x} - \mathbf{y}}$$

makes no sense when $f : \mathbb{R}^n \to \mathbb{R}$ is a function of $n$ variables and $\mathbf{x}, \mathbf{y} \in \mathbb{R}^n$ are $n$-dimensional vectors.

How can we make sense of it, then? This is what we'll learn in the following chapter.

## 16.1    Partial and total derivatives

Let's take a look at multivariable functions more closely! For the sake of simplicity, let $f : \mathbb{R}^2 \to \mathbb{R}$ be our function of two variables. To emphasize the dependence on the individual variables, we often write

$$f(x_1, x_2), \quad x_1, x_2 \in \mathbb{R}.$$

Here's the trick: by fixing one of the variables, we obtain the two single-variable functions! That is, if $x_1 \in \mathbb{R}^2$ is fixed, we have $x \mapsto f(x_1, x)$, and if $x_2 \in \mathbb{R}^2$ is fixed, we have $x \mapsto f(x, x_2)$, both of which are well-defined univariate functions. Think about this as slicing the function graph with a plane parallel to the $x - z$ or the $y - z$ axes, as illustrated by *Figure 16.1*. The part cut out by the plane is a single-variable function.

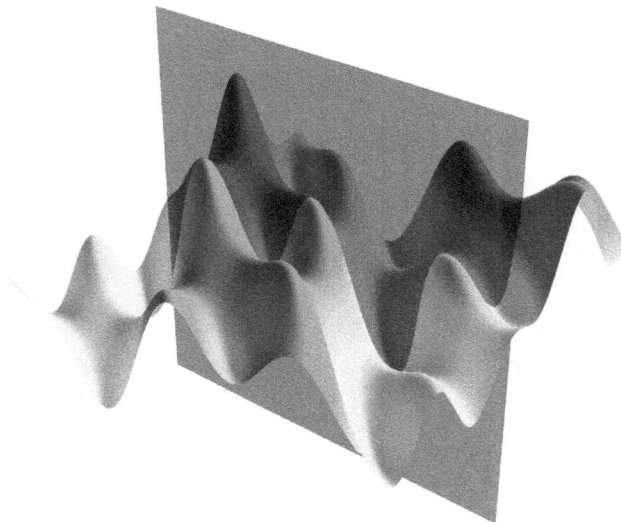

*Figure 16.1: Slicing the surface with the $x - z$ plane*

We can define the derivative of these functions by the limit of difference quotients. These are called the *partial derivatives*:

$$\frac{\partial f}{\partial x_1}(x_1, x_2) = \lim_{x \to x_1} \frac{f(x, x_2) - f(x_1, x_2)}{x - x_1},$$

$$\frac{\partial f}{\partial x_2}(x_1, x_2) = \lim_{x \to x_2} \frac{f(x_1, x) - f(x_1, x_2)}{x - x_2}.$$

(Keep in mind that $x_1$ signifies the variable in $\frac{\partial f}{\partial x_1}$, but an actual scalar value in the argument of $\frac{\partial f}{\partial x_1}(x_1, x_2)$. This can be quite confusing, but you'll soon learn to make sense of it.)

The definition is similar for general multivariable functions; we just have to write much more. There, the partial derivative of $f : \mathbb{R}^n \to \mathbb{R}$ at the point $\mathbf{x} = (x_1, \ldots, x_n)$ with respect to the $i$-th variable is defined by

$$\frac{\partial f}{\partial x_i}(x_1, \ldots, x_n) = \lim_{x \to x_i} \frac{f(x_1, \ldots, \overset{\overset{\text{$i$-th variable}}{\frown}}{x}, \ldots, x_n) - f(x_1, \ldots, x_i, \ldots, x_n)}{x - x_i}. \tag{16.1}$$

One of the biggest challenges in multivariable calculus is to manage the ever-growing notational complexity. Just take a look at the difference quotient above:

$$\frac{f(x_1, \ldots, x, \ldots, x_n) - f(x_1, \ldots, x_i, \ldots, x_n)}{x - x_i}.$$

This is not the prettiest to look at, and this kind of notational complexity can pile up fast. Fortunately, linear algebra comes to the rescue! Not only can we compact the variables into the vector $\mathbf{x} = (x_1, \ldots, x_n)$, we can use the standard basis

$$\mathbf{e}_i = (0, \ldots, 0, \underbrace{1}_{i\text{-th component}}, 0, \ldots, 0)$$

to write the difference quotients as

$$\frac{f(\mathbf{x} + h\mathbf{e}_i) - f(\mathbf{x})}{h}, \quad h \in \mathbb{R}.$$

Thus, (16.1) can be compacted. With this newly found form, we are ready to make a concise and formal definition for partial derivatives.

---

**Definition 16.1.1 (Partial derivatives)**

Let $f : \mathbb{R}^n \to \mathbb{R}$ be a function of $n$ variables. The partial derivative of $f$ at the point $\mathbf{x} = (x_1, \dots, x_n)$ with respect to the $i$-th variable is defined by

$$\frac{\partial f}{\partial x_i}(\mathbf{x}) = \lim_{h \to 0} \frac{f(\mathbf{x} + h e_i) - f(\mathbf{x})}{h}.$$

If the above limit exists, we say that $f$ is *partially differentiable* with respect to the $i$-th variable $x_i$.

---

The partial derivative is again a vector-scalar function. Because of this, it is often written as $\frac{\partial}{\partial x_i} f$, reflecting on the fact that the symbol $\frac{\partial}{\partial x_i}$ can be thought of as a *function* that maps functions to functions. I know, this is a bit abstract, but you'll get used to it quickly.

As usual, there are several alternative notations for the partial derivatives. Among others, the symbols

- $f_{x_i}(\mathbf{x})$,
- $D_i f(\mathbf{x})$,
- $\partial_i f(\mathbf{x})$

denote the $i$-th partial derivative of $f$ at $\mathbf{x}$. For simplicity, we'll use the old-school $\frac{\partial f}{\partial x_i}(\mathbf{x})$.

It's best to start with a few examples to illustrate the concept of partial derivatives.

**Example 1.** Let

$$f(x_1, x_2) = x_1^2 + x_2^2.$$

To calculate, say, $\partial f / \partial x_1$, we fix the second variable and treat $x_2$ as a constant. Formally, we obtain the single-variable function

$$f^1(x) := x^2 + x_2^2, \quad x_2 \in \mathbb{R},$$

whose derivative gives the first partial derivative:

$$\frac{\partial f}{\partial x_1}(x_1, x_2) = \frac{df^1}{dx}(x_1) = 2x_1.$$

Similarly, we get that

$$\frac{\partial f}{\partial x_2}(x_1, x_2) = 2x_2.$$

Once you are comfortable with the mental gymnastics of fixing variables, you'll be able to perform partial differentiation without writing out all the intermediate steps.

**Example 2.** Let's see a more complicated example. Define

$$f(x_1, x_2) = \sin(x_1^2 + x_2).$$

By fixing $x_2$, we obtain a composite function. Thus the chain rule is used to calculate the first partial derivative:

$$\frac{\partial f}{\partial x_1}(x_1, x_2) = 2x_1 \cos(x_1^2 + x_2).$$

Similarly, we obtain that

$$\frac{\partial f}{\partial x_2}(x_1, x_2) = \cos(x_1^2 + x_2).$$

(I highly advise you to carry out the above calculations step by step as an exercise, even if you understand all the intermediate steps.)

**Example 3.** Finally, let's see a function that is partially differentiable in one variable but not in the other. Define the function

$$f(x_1, x_2) = \begin{cases} -1 & \text{if } x_2 < 0, \\ 1 & \text{otherwise.} \end{cases}$$

As $f(x_1, x_2)$ does not depend on $x_1$, we can see that by fixing $x_2$, the resulting function is constant. Thus,

$$\frac{\partial f}{\partial x_1}(x_1, x_2) = 0$$

holds everywhere. However, in $x_2$, there is a discontinuity at 0; thus, $\frac{\partial f}{\partial x_2}$ is undefined there.

## 16.1.1 The gradient

If a function is partially differentiable in every variable, we can compact the derivatives together in a single vector to form the *gradient*.

> **Definition 16.1.2 (The gradient)**
>
> Let $f : \mathbb{R}^n \to \mathbb{R}$ be a function that is partially differentiable in all of its variables. Then, its *gradient* is defined by the (column) vector
>
> $$\nabla f(\mathbf{x}) := \begin{bmatrix} \frac{\partial}{\partial x_1} f(\mathbf{x}) \\ \frac{\partial}{\partial x_2} f(\mathbf{x}) \\ \vdots \\ \frac{\partial}{\partial x_n} f(\mathbf{x}) \end{bmatrix} \in \mathbb{R}^{n \times 1}.$$

A few remarks are in order. First, the symbol $\nabla$ is called *nabla*, a symbol that was conceived to denote gradients.

Second, the gradient can be thought of as a vector-vector function. To see that, consider the already familiar function $f(x_1, x_2) = x_1^2 + x_2^2$. The gradient of $f$ is

$$\nabla f(x_1, x_2) = \begin{bmatrix} 2x_1 \\ 2x_2 \end{bmatrix},$$

or

$$\nabla f(\mathbf{x}) = 2\mathbf{x}$$

in vectorized form. We can visualize this by drawing the vector $\nabla f(x_1, x_2)$ at each point $(x_1, x_2) \in \mathbb{R}^2$.

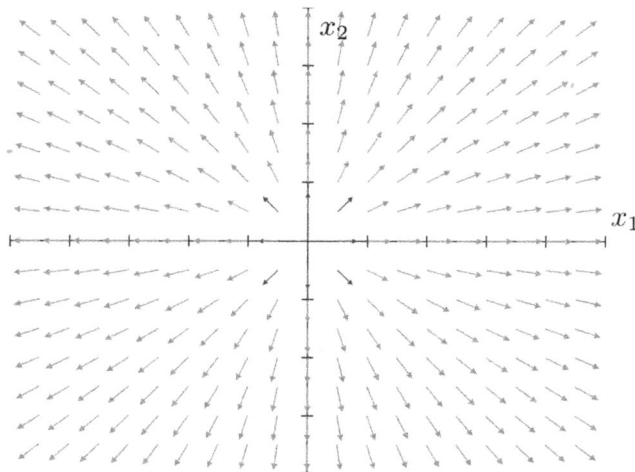

*Figure 16.2: The vector field given by the gradient of $x_1^2 + x_2^2$*

Thus, you can think about $\nabla f$ as a vector-vector function $\nabla f : \mathbb{R}^n \to \mathbb{R}^n$. The gradient at a given point $\mathbf{x}$ is obtained by evaluating this function, yielding $(\nabla f)(\mathbf{x})$.

For clarity, the parentheses are omitted, arriving at the all familiar notation $\nabla f(\mathbf{x})$.

## 16.1.2 Higher order partial derivatives

The partial derivatives of a vector-scalar function $f : \mathbb{R}^n \to \mathbb{R}$ are vector-scalar functions themselves. Thus, we can perform partial differentiation one more time!

If they exist, the second order partial derivatives are defined by

$$\frac{\partial^2 f}{\partial x_i \partial x_j}(\mathbf{a}) := \frac{\partial}{\partial x_i}\left[\frac{\partial f}{\partial x_j}(\mathbf{a})\right], \tag{16.2}$$

where $\mathbf{a} \in \mathbb{R}^n$ is an arbitrary vector. (When the second partial differentiation takes place with respect to the same variable, (16.2) is abbreviated by $\frac{\partial^2 f}{\partial x_i^2}(\mathbf{a})$.)

The definition begs the question: is the order of differentiation interchangeable? That is, does

$$\frac{\partial^2 f}{\partial x_i \partial x_j}(\mathbf{a}) = \frac{\partial^2 f}{\partial x_j \partial x_i}(\mathbf{a})$$

hold? The answer is quite surprising: the order is interchangeable under some mild assumptions, but not in the general case. There is a famous theorem about it which we won't prove, but it's essential to know.

**Theorem 16.1.1** *Let $f : \mathbb{R}^n \to \mathbb{R}$ be an arbitrary vector-scalar function and let $\mathbf{a} \in \mathbb{R}^n$. If there is a small ball $B(\varepsilon, \mathbf{a}) \subseteq \mathbb{R}^n$ centered at $\mathbf{a}$ such that $f$ has continuous second-order partial derivatives at all points of $B(\varepsilon, \mathbf{a})$, then*

$$\frac{\partial^2 f}{\partial x_i \partial x_j}(\mathbf{a}) = \frac{\partial^2 f}{\partial x_j \partial x_i}(\mathbf{a})$$

*holds for all $i = 1, \ldots, n$.*

*Theorem 16.1.1* is known as either Schwarz's theorem, Clairaut's theorem, or Young's theorem.

## 16.1.3   The total derivative

Partial derivatives seem to generalize the notion of differentiability for multivariable functions. However, something is missing. Let's revisit the single-variable case for a moment.

Recall that according to *Theorem 12.1.1*, the differentiability of a single-variable function $f : \mathbb{R} \to \mathbb{R}$ at a given point $a$ is equivalent to a local approximation of $f$ by the linear function

$$l(x) = f(a) + f'(a)(x - a).$$

If $x$ is close to $a$, $l(x)$ is also close to $f(x)$. Moreover, this is the best linear approximation we can do around $a$. In a single variable, this is *equivalent* to differentiation.

This gives us an idea: even though difference quotients like $\frac{f(x)-f(y)}{x-y}$ do not exist in multiple variables, the best local approximation with a multivariable linear function does!

Thus, the notion of total differentiability is born.

---

**Definition 16.1.3  (Total differentiability)**

Let $f : \mathbb{R}^n \to \mathbb{R}$ be a function of $n$ variables. We say that $f$ is *totally differentiable* (or sometimes just *differentiable* for short) at $\mathbf{a} \in \mathbb{R}^n$ if there exists a row vector $D_f(\mathbf{a}) \in \mathbb{R}^{1 \times n}$ such that

$$f(\mathbf{x}) = f(\mathbf{a}) + D_f(\mathbf{a})(\mathbf{x} - \mathbf{a}) + o(\|\mathbf{x} - \mathbf{a}\|) \tag{16.3}$$

holds for all $\mathbf{x} \in B(\varepsilon, \mathbf{a})$, where $\varepsilon > 0$ and $B(\varepsilon, \mathbf{a})$ is defined by

$$B(\varepsilon, \mathbf{a}) = \{\mathbf{x} \in \mathbb{R}^n : \|\mathbf{x} - \mathbf{a}\|\} < \varepsilon.$$

(In other words, $B(\varepsilon, \mathbf{a})$ is a ball of radius $\varepsilon > 0$ around $\mathbf{a}$.) When exists, the vector $D_f(\mathbf{a})$ is called the *total derivative* of $f$ at $\mathbf{a}$.

---

Recall that when it is not stated explicitly, we use *column vectors*, because we want to write our linear transformations in the form $A\mathbf{x}$, where $A \in \mathbb{R}^{m \times n}$ and $\mathbf{x} \in \mathbb{R}^{n \times 1}$. Thus, the "dimensionology" of the formula

$$\underbrace{f(\mathbf{x})}_{\in \mathbb{R}^{1 \times 1}} = \underbrace{f(\mathbf{a})}_{\in \mathbb{R}^{1 \times 1}} + \underbrace{D_f(\mathbf{a})}_{\in \mathbb{R}^{1 \times n}} \underbrace{(\mathbf{x} - \mathbf{a})}_{\in \mathbb{R}^{n \times 1}} + o(\|\mathbf{x} - \mathbf{a}\|) \in \mathbb{R}^{1 \times 1}$$

works out. (Don't be fooled, $\mathbb{R}^{1 \times 1}$ is a scalar.)

Let's unravel the notion of total differentiability. The form (16.3) implies that a totally differentiable function $f$ equals to the linear part $f(\mathbf{a}) + D_f(\mathbf{a})(\mathbf{x} - \mathbf{a})$ plus a small error.

The surface given by the linear part is called the *tangent plane*. We can visualize it for functions of two variables.

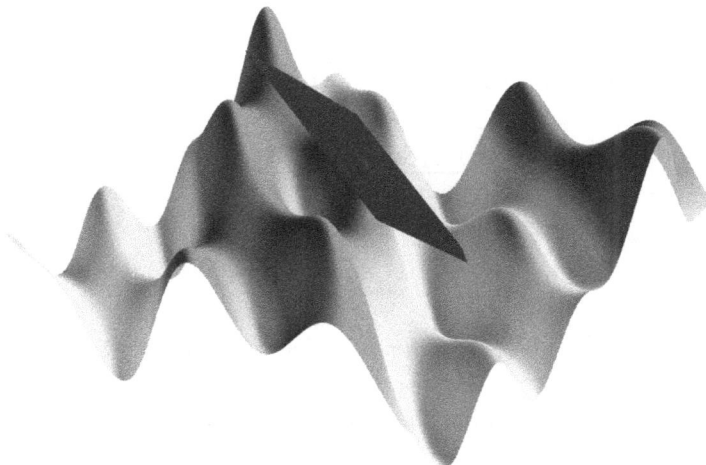

*Figure 16.3: The tangent plane*

Unsurprisingly, the partial and total derivatives share an intimate connection.

**Theorem 16.1.2** *(Total derivative and the partial derivatives)*

*Let $f : \mathbb{R}^n \to \mathbb{R}$ be a function that is totally differentiable at $\mathbf{a} \in \mathbb{R}^n$. Then, all of its partial derivatives exist at $\mathbf{a}$ and*

$$f(\mathbf{x}) = f(\mathbf{a}) + \nabla f(\mathbf{a})^T (\mathbf{x} - \mathbf{a}) + o(\|\mathbf{x} - \mathbf{a}\|) \tag{16.4}$$

*holds for all $\mathbf{a}$ in some $B(\varepsilon, \mathbf{a})$, $\varepsilon > 0$. (That is, $D_f(\mathbf{a}) = \nabla f(\mathbf{a})^T$.)*

In other words, the equation (16.4) gives that the coefficients of the best linear approximation are equal to the partial derivatives.

*Proof.* Because $f$ is totally differentiable at $\mathbf{a}$, the definition gives that $f$ can be written in the form

$$f(\mathbf{x}) = f(\mathbf{a}) + D_f(\mathbf{a})(\mathbf{x} - \mathbf{a}) + o(\|\mathbf{x} - \mathbf{a}\|),$$

where $D_f(\mathbf{a}) = (d_1, \ldots, d_n)$ is the vector that describes the coefficients of the linear part.

Our goal is to show that

$$\lim_{h \to 0} \frac{f(\mathbf{a} + h\mathbf{e}_i) - f(\mathbf{a})}{h} = d_i,$$

where $\mathbf{e}_i$ is the unit (column) vector whose $i$-th component is 1, while the others are 0.

Let's do a quick calculation! Based on what we know, we have

$$\frac{f(\mathbf{a} + h\mathbf{e}_i) - f(\mathbf{a})}{h} = \frac{D_f(\mathbf{a})h\mathbf{e}_i + o(\|h\mathbf{e}_i\|)}{h}$$
$$= D_f(\mathbf{a})\mathbf{e}_i + o(1)$$
$$= d_i + o(1),$$

thus confirming that $\lim_{h \to 0} \frac{f(\mathbf{a}+h\mathbf{e}_i)-f(\mathbf{a})}{h} = d_i$, which is what we had to show.

What's all the hassle with total differentiation, then? *Theorem 16.1.2* tells us that total differentiability is a stronger condition than partial differentiability.

Surprisingly, the other direction is not true: the existence of partial derivatives does not imply total differentiability, as the example

$$f(x, y) = \begin{cases} 1 & \text{if } x = 0 \text{ or } y = 0, \\ 0 & \text{otherwise} \end{cases}$$

illustrates. This function has all its partial derivatives at 0, yet the total derivative does not exist. (You can convince yourself by either drawing a figure, or noting that the function $1 - \mathbf{d}^T\mathbf{x}$ can never be $o(\|\mathbf{x}\|)$, regardless of the choice of $\mathbf{d}$.)

### Remark 16.1.1 (The total derivative as an operator)

Just like for single-variable functions, the total derivative of $f : \mathbb{R}^n \to \mathbb{R}$ is a function $D_f : \mathbb{R}^n \to \mathbb{R}^n$.

At the highest level of abstraction, we can think about the total derivative as an *operator* that maps a vector-scalar function to a vector-vector function:

$$D : (\mathbb{R}^n)^{\mathbb{R}} \to (\mathbb{R}^n)^{\mathbb{R}^n},$$
$$D : f \mapsto D_f,$$

where $A^B$ denotes the set of all functions mapping $A$ to $B$.

You are not *required* to understand this at all, but trust me, the more abstract your thinking is, the more powerful you'll be.

## 16.1.4  Directional derivatives

So far, we have talked about two kinds of derivatives: partial derivatives that describe the rate of change along a fixed axis, and total derivatives that give the best linear approximation of the function at a given point.

Partial derivatives are only concerned with a few particular directions. However, this is not the end of the story in multiple variables. With the standard orthonormal basis vectors $\mathbf{e}_i$, the partial derivatives are defined by

$$\frac{\partial f}{\partial x_i}(\mathbf{a}) = \lim_{h \to 0} \frac{f(\mathbf{a} + h\mathbf{e}_i) - f(\mathbf{a})}{h}, \quad i = 1, 2. \tag{16.5}$$

As we saw earlier, these describe the rate of change along the dimensions. However, the standard orthonormal vectors are just a few special directions.

What about an arbitrary direction $\mathbf{v}$? Can we define the derivative along these? Sure! There is nothing stopping us from replacing $\mathbf{e}_i$ with $\mathbf{v}$ in (16.5). Thus, directional derivatives are born.

**Definition 16.1.4 (Directional derivatives)**

Let $f : \mathbb{R}^n \to \mathbb{R}$ be a function of $n$ variables and let $\mathbf{v} \in \mathbb{R}^n$ be an arbitrary vector. The *directional derivative* of $f$ along $\mathbf{v}$ is defined by the limit

$$\frac{\partial f}{\partial \mathbf{v}} := \lim_{h \to 0} \frac{f(\mathbf{a} + h\mathbf{v}) - f(\mathbf{a})}{h}.$$

Good news: the directional derivatives can be described in terms of the gradient!

**Theorem 16.1.3** *Let $f : \mathbb{R}^n \to \mathbb{R}$ be a function of $n$ variables. If $f$ is totally differentiable at $\mathbf{a} \in \mathbb{R}^n$, then its directional derivatives exist in all directions, and*

$$\frac{\partial f}{\partial \mathbf{v}}(\mathbf{a}) = \nabla f(\mathbf{a})^T \mathbf{v}.$$

*Proof.* Because of the total differentiability, *Theorem 16.2.2* gives that

$$f(\mathbf{x}) = f(\mathbf{a}) + \nabla f(\mathbf{a})^T(\mathbf{x} - \mathbf{a}) + o(\|\mathbf{x} - \mathbf{a}\|)$$

around $\mathbf{a}$. Thus,

$$\frac{f(\mathbf{a} + h\mathbf{v}) - f(\mathbf{a})}{h} = \frac{h\nabla f(\mathbf{a})^T\mathbf{v} + o(h)}{h}$$
$$= \nabla f(\mathbf{a})^T\mathbf{v} + o(1),$$

giving that

$$\frac{\partial f}{\partial \mathbf{v}}(\mathbf{a}) = \lim_{h \to 0} \frac{f(\mathbf{a} + h\mathbf{v}) - f(\mathbf{a})}{h}$$
$$= \lim_{h \to 0} \nabla f(\mathbf{a})^T\mathbf{v} + o(1)$$
$$= \nabla f(\mathbf{a})^T\mathbf{v},$$

as we needed to show.

In other words, *Theorem 16.1.3* gives that no matter the direction $\mathbf{v}$, the directional derivative can be written in terms of the gradient and $\mathbf{v}$. If you think about this for a minute, this is quite amazing: the rates of change along $n$ special directions determine the rate of change in *any* other direction.

## 16.1.5 Properties of the gradient

In one variable, we have learned that if the derivative of $f$ is positive at some $a$, then $f$ is increasing around $a$. (If the derivative is negative, $f$ is decreasing.) If we think about the derivative $f'(a)$ as a one-dimensional vector, then the derivative points towards the direction of increase.

Is this true in higher dimensions? Yes, and this is what makes gradient descent work.

**Theorem 16.1.4** *(The gradient determines the direction of the increase)*

Let $f : \mathbb{R}^n \to \mathbb{R}$ be a function of $n$ variables, and suppose that $f$ is totally differentiable at $\mathbf{a} \in \mathbb{R}^n$.

*Then*

$$\text{argmax}_{\mathbf{v} \in \{\mathbf{x} \in \mathbb{R}^n \,:\, \|\mathbf{x}\| = 1\}} \frac{\partial f}{\partial \mathbf{v}}(\mathbf{a}) = \frac{\nabla f(\mathbf{a})}{\|\nabla f(\mathbf{a})\|}. \tag{16.6}$$

I know, (16.6) is pretty overloaded, so let's unpack it. First, let's start with the mysterious argmax. For a given function $f$,

$$\text{argmax}_{x \in S} f(x)$$

denotes the values that maximize $f$ on the set $S$. As the maximum may not be unique, argmax can yield a set. (The definition of argmin is the same, but with minimum instead of maximum.)

Thus, in English, (16.6) states that the unit direction that maximizes the directional derivative at $\mathbf{a} \in \mathbb{R}^n$ is the normalized gradient. Now we are ready to see the proof!

*Proof.* Do you remember the Cauchy-Schwarz inequality (*Theorem 2.2.1*)? It was a long time ago, so let's recall! In the vector space $\mathbb{R}^n$, the Cauchy-Schwarz inequality tells us that for any $\mathbf{x}, \mathbf{y} \in \mathbb{R}^n$,

$$\mathbf{x}^T \mathbf{y} \leq \|\mathbf{x}\| \|\mathbf{y}\|.$$

Now, as *Theorem 16.1.3* implies, the directional derivatives can be written as

$$\frac{\partial f}{\partial \mathbf{v}}(\mathbf{a}) = \nabla f(\mathbf{a})^T \mathbf{v}.$$

Combined with the Cauchy-Schwarz inequality, we get that

$$\frac{\partial f}{\partial \mathbf{v}}(\mathbf{a}) = \nabla f(\mathbf{a})^T \mathbf{v}$$

$$\leq \|\nabla f(\mathbf{a})\| \|\mathbf{v}\|.$$

By restricting the directions to unit vectors,

$$\frac{\partial f}{\partial \mathbf{v}}(\mathbf{a}) \leq \|\nabla f(\mathbf{a})\| \tag{16.7}$$

follows. Thus, the directional derivatives must be less than or equal to the gradient's norm. (At least, along a direction vector with unit length.)

However, by letting $\mathbf{v}_0 = \nabla f(\mathbf{a})/\|\nabla f(\mathbf{a})\|$, we obtain that

$$
\begin{aligned}
\frac{\partial f}{\partial \mathbf{v}_0}(\mathbf{a}) &= \nabla f(\mathbf{a})^T \mathbf{v}_0 \\
&= \frac{\nabla f(\mathbf{a})^T \nabla f(\mathbf{a})}{\|\nabla f(\mathbf{a})\|} \\
&= \frac{\|\nabla f(\mathbf{a})\|^2}{\|\nabla f(\mathbf{a})\|} \\
&= \|\nabla f(\mathbf{a})\|.
\end{aligned}
$$

Thus, with the choice $\mathbf{v}_0 = \frac{\nabla f(\mathbf{a})}{\|\nabla f(\mathbf{a})\|}$, equality can be attained in (16.7). This means that $\frac{\nabla f(\mathbf{a})}{\|\nabla f(\mathbf{a})\|}$ maximizes the directional derivative at $\mathbf{a}$, which is what we had to prove.

With that, we have the basics of differentiation in multiple variables under our belt. To sum up, we have learned that the difference quotient definition of the derivative does not generalize directly for multiple variables, but we can fix all but one variables to make the difference quotient work, thus obtaining partial derivatives.

On the other hand, the linear approximation definition works in multiple dimensions, but instead of

$$
f(a) + f'(a)(x - a), \quad f : \mathbb{R} \to \mathbb{R}, \quad x, a \in \mathbb{R},
$$

like we had in one variable, we obtain

$$
f(\mathbf{a}) + \nabla f(\mathbf{a})^T (\mathbf{x} - \mathbf{a}), \quad f : \mathbb{R}^n \to \mathbb{R}, \quad \mathbf{x}, \mathbf{a} \in \mathbb{R}^n,
$$

where the analogue of the derivative is the gradient vector $\nabla f(\mathbf{a}) \in \mathbb{R}^n$.

Even when we were studying differentiation in one variable for the first time, I told you that the local linear approximation definition would be useful someday. That time is now, and we are reaping the benefits. Soon, we'll see gradient descent in its full glory.

## 16.2  Derivatives of vector-valued functions

In a single variable, defining higher-order derivatives is easy. We simply have to keep repeating differentiation:

$$
\begin{aligned}
f''(x) &= \left( f'(x) \right)', \\
f'''(x) &= \left( f''(x) \right)',
\end{aligned}
$$

and so on. However, this is not that straightforward with multivariable functions.

So far, we have only talked about gradients, the generalization of the derivative for vector-scalar functions.

As $\nabla f(\mathbf{a})$ is a column vector, the gradient is a vector-vector function $\nabla : \mathbb{R}^n \to \mathbb{R}^n$. We only know how to compute the derivative of vector-scalar functions. It's time to change that!

## 16.2.1 The derivatives of curves

Curves, often describing the solutions of dynamical systems, are one of the most important objects in mathematics. We don't explicitly use them in machine learning, but they are underneath algorithms such as gradient descent. (Where we traverse a discretized curve leading to a local minimum.)

Formally, a curve – that is, a scalar-vector function – is given by a function

$$\gamma : \mathbb{R} \to \mathbb{R}^n, \quad \gamma(t) = \begin{bmatrix} \gamma_1(t) \\ \gamma_2(t) \\ \vdots \\ \gamma_n(t) \end{bmatrix} \in \mathbb{R}^{n(\times 1)},$$

where the $\gamma_i : \mathbb{R} \to \mathbb{R}$ functions are good old single-variable scalar-scalar functions. As the independent variable often represents time, it is customary to denote it with $t$.

We can differentiate $\gamma$ componentwise:

$$\gamma'(t) := \begin{bmatrix} \gamma_1'(t) \\ \gamma_2'(t) \\ \vdots \\ \gamma_n'(t) \end{bmatrix} \in \mathbb{R}^{n(\times 1)}.$$

If we indeed imagine $\gamma(t)$ as a trajectory in space, $\gamma'(t)$ is the tangent vector to $\gamma$ at $t$. Since the differentiation is componentwise, *Theorem 12.1.1* implies that if $\gamma$ is differentiable at some $a \in \mathbb{R}$,

$$\gamma(t) = \gamma(a) + \gamma'(t)^T(t - a) + o(|t - a|) \tag{16.8}$$

there. The equation (16.8) is a true vectorized formula: some components are vectors, and some are scalars. Yet, this is simple and makes perfect sense to us. Hiding the complexities of vectors and matrices is the true power of linear algebra.

It is easy to see that for any two curves $\gamma, \eta : \mathbb{R} \to \mathbb{R}^n$, differentiation is additive, as $(\gamma + \eta)' = \gamma' + \eta'$. What happens when we compose a scalar-vector function with a vector-scalar one?

This situation is commonplace in machine learning. If, say, $L : \mathbb{R}^n \to \mathbb{R}$ describes the loss function and $\gamma : \mathbb{R} \to \mathbb{R}^n$ is our trajectory in the parameter space $\mathbb{R}^n$, the composite function $f(\gamma(t))$ describes the model loss at time $t$. Thus, to compute $(f \circ \gamma)'$, we have to generalize the chain rule.

**Theorem 16.2.1** *(The chain rule for composing scalar-vector and vector-scalar functions)*

*Let $\gamma : \mathbb{R} \to \mathbb{R}^n$ and $f : \mathbb{R}^n \to \mathbb{R}$ be arbitrary functions. If $\gamma$ is differentiable at some $a \in \mathbb{R}$ and $f$ is differentiable at $\gamma(a)$, then $f \circ \gamma : \mathbb{R} \to \mathbb{R}$ is also differentiable at $a$ and*

$$(f \circ \gamma)'(a) = \nabla f(\gamma(a))^T \gamma'(a)$$

*there.*

*Proof.* As $f$ is differentiable at $\gamma(a)$, *Theorem 16.1.2* gives

$$f(\gamma(t)) = f(\gamma(a)) + \nabla f(\gamma(a))^T(\gamma(t) - \gamma(a)) + o(\|\gamma(t) - \gamma(a)\|).$$

Thus,

$$
\begin{aligned}
(f \circ \gamma)'(a) &= \lim_{t \to a} \frac{f(\gamma(t)) - f(\gamma(a))}{t - a} \\
&= \nabla f(\gamma(a))^T \lim_{t \to a} \left[ \frac{\gamma(t) - \gamma(a)}{t - a} + o(1) \right] \\
&= \nabla f(\gamma(a))^T \gamma'(a),
\end{aligned}
$$

which is what we had to prove.

## 16.2.2 The Jacobian and Hessian matrices

Now, our task is to extend the derivative for vector-vector functions, so let $\mathbf{f} : \mathbb{R}^n \to \mathbb{R}^m$ be one. By writing out the output of $\mathbf{f}$ explicitly, we can decompose it into multiple components:

$$\mathbf{f}(\mathbf{x}) = \begin{bmatrix} f_1(\mathbf{x}) \\ \vdots \\ f_m(\mathbf{x}) \end{bmatrix} \in \mathbb{R}^{m(\times 1)}$$

where $f_i : \mathbb{R}^n \to \mathbb{R}$ are vector-scalar functions.

The natural idea is to compute the partial derivatives for $f_i$, compacting them into a matrix. And so we shall!

**Definition 16.2.1 (The Jacobian matrix)**

Let $\mathbf{f} : \mathbb{R}^n \to \mathbb{R}^m$ be an arbitrary vector-vector function, and suppose that

$$\mathbf{f}(\mathbf{x}) = (f_1(\mathbf{x}), \ldots, f_m(\mathbf{x})),$$

where all $f_i : \mathbb{R}^n \to \mathbb{R}$ are (partially) differentiable at some $\mathbf{a} \in \mathbb{R}^n$. The matrix

$$J_{\mathbf{f}}(\mathbf{a}) := \begin{bmatrix} \frac{\partial f_1}{\partial x_1}(\mathbf{a}) & \frac{\partial f_1}{\partial x_2}(\mathbf{a}) & \cdots & \frac{\partial f_1}{\partial x_n}(\mathbf{a}) \\ \frac{\partial f_2}{\partial x_1}(\mathbf{a}) & \frac{\partial f_2}{\partial x_2}(\mathbf{a}) & \cdots & \frac{\partial f_2}{\partial x_n}(\mathbf{a}) \\ \vdots & \vdots & \ddots & \vdots \\ \frac{\partial f_m}{\partial x_1}(\mathbf{a}) & \frac{\partial f_m}{\partial x_2}(\mathbf{a}) & \cdots & \frac{\partial f_m}{\partial x_n}(\mathbf{a}) \end{bmatrix} \in \mathbb{R}^{m \times n}$$

is called the *Jacobian* of $\mathbf{f}$ at $\mathbf{a}$.

In other words, the rows of the Jacobian are the gradients of $f_i$:

$$J_{\mathbf{f}}(\mathbf{a}) = \begin{bmatrix} \nabla f_1(\mathbf{a})^T \\ \nabla f_2(\mathbf{a})^T \\ \vdots \\ \nabla f_m(\mathbf{a})^T \end{bmatrix}.$$

I have good news: the best local linear approximation of $\mathbf{f}$ around $\mathbf{a}$ is given by

$$\mathbf{f}(\mathbf{x}) = \mathbf{f}(\mathbf{a}) + J_{\mathbf{f}}(\mathbf{a})(\mathbf{x} - \mathbf{a}) + o(\|\mathbf{x} - \mathbf{a}\|),$$

if the best local linear approximation exists. Thus, the Jacobian is a proper generalization of the gradient.

We can use the Jacobian to generalize the notion of second-order derivatives for vector-scalar functions: by computing the Jacobian of the gradient, we obtain a special matrix, the analogue of the second derivative.

**Definition 16.2.2 (The Hessian matrix)**

Let $f : \mathbb{R}^n \to \mathbb{R}$ be an arbitrary vector-scalar function, and suppose that all of its second-order partial derivatives exist at $\mathbf{a} \in \mathbb{R}^n$.

The matrix

$$H_f(\mathbf{a}) := \begin{bmatrix} \frac{\partial^2 f}{\partial x_1^2}(\mathbf{a}) & \frac{\partial^2 f}{\partial x_1 \partial x_2}(\mathbf{a}) & \cdots & \frac{\partial^2 f}{\partial x_1 \partial x_n}(\mathbf{a}) \\ \frac{\partial^2 f}{\partial x_2 \partial x_1}(\mathbf{a}) & \frac{\partial^2 f}{\partial x_2^2}(\mathbf{a}) & \cdots & \frac{\partial^2 f}{\partial x_2 \partial x_n}(\mathbf{a}) \\ \vdots & \vdots & \ddots & \vdots \\ \frac{\partial^2 f}{\partial x_n \partial x_1}(\mathbf{a}) & \frac{\partial^2 f}{\partial x_n \partial x_2}(\mathbf{a}) & \cdots & \frac{\partial^2 f}{\partial x_n^2}(\mathbf{a}) \end{bmatrix} \in \mathbb{R}^{n \times n}$$

is called the *Hessian* of $f$ at $\mathbf{a}$.

In other words,

$$H_f(\mathbf{a}) = J_{\nabla f}(\mathbf{a})^T$$

holds by definition. Moreover, if $f$ behaves nicely (for instance, all second-order partial derivatives exist and are continuous), *Theorem 16.1.1* implies that the Hessian is symmetric; that is, $H_f(\mathbf{a}) = H_f(\mathbf{a})^T$.

## 16.2.3   The total derivative for vector-vector functions

One last generalization. (I promise.) Recall that the existence of the gradient (that is, partial differentiability) doesn't imply total differentiability for vector-scalar functions, as the example

$$f(x, y) = \begin{cases} 1 & \text{if } x = 0 \text{ or } y = 0, \\ 0 & \text{otherwise} \end{cases}$$

shows at zero.

This is true for vector-vector functions as well, as the Jacobian is the generalization of the gradient, not the total derivative.

It is best to rip the band-aid off quickly and define the total derivative for vector-vector functions. The definition will be a bit abstract, but trust me, the investment will pay off when talking about the chain rule. (Which is the foundation of backpropagation, the algorithm that makes gradient descent computationally feasible.)

**Definition 16.2.3 (Total differentiability of vector-vector functions)**

Let $\mathbf{f} : \mathbb{R}^n \to \mathbb{R}^m$ be an arbitrary vector-vector function. We say that $f$ is *totally differentiable* (or sometimes just *differentiable* in short) at $\mathbf{a} \in \mathbb{R}^n$ if there exists a matrix $D_{\mathbf{f}}(\mathbf{a}) \in \mathbb{R}^{m \times n}$ such that

$$\mathbf{f}(\mathbf{x}) = \mathbf{f}(\mathbf{a}) + D_{\mathbf{f}}(\mathbf{a})(\mathbf{x} - \mathbf{a}) + o(\|\mathbf{x} - \mathbf{a}\|) \tag{16.9}$$

holds for all $\mathbf{x} \in B(\varepsilon, \mathbf{a})$, where $\varepsilon > 0$ and $B(\varepsilon, \mathbf{a})$ is defined by

$$B(\varepsilon, \mathbf{a}) = \{\mathbf{x} \in \mathbb{R}^n : \|\mathbf{x} - \mathbf{a}\| < \varepsilon\}.$$

(In other words, $B(\varepsilon, \mathbf{a})$ is a ball of radius $\varepsilon > 0$ around $\mathbf{a}$.) When exists, the matrix $D_{\mathbf{f}}(\mathbf{a})$ is called the *total derivative* of $f$ at $\mathbf{a}$.

Notice that *Definition 16.2.3* is almost verbatim to *Definition 16.1.3*, except that the "derivative" is a matrix this time.

You are probably not surprised to hear that its relation with the Jacobian is the same as the gradient and the total derivative in the vector-scalar case.

**Theorem 16.2.2 *(Total derivative and the partial derivatives)***

*Let* $\mathbf{f} : \mathbb{R}^n \to \mathbb{R}^m$ *be a function that is totally differentiable at* $\mathbf{a} \in \mathbb{R}^n$. *Then, all of its partial derivatives exist at* $\mathbf{a}$ *and*

$$D_{\mathbf{f}}(\mathbf{a}) = J_{\mathbf{f}}(\mathbf{a}).$$

The proof is almost identical to the one of *Theorem 16.1.2*, with more complex notations. I strongly recommend you work it out line by line, as this kind of mental gymnastics helps significantly to get used to matrices in practice.

Componentwise, the total derivative can be written as

$$D_{\mathbf{f}}(\mathbf{a}) = \begin{bmatrix} \frac{\partial f_1}{\partial x_1}(\mathbf{a}) & \frac{\partial f_1}{\partial x_2}(\mathbf{a}) & \cdots & \frac{\partial f_1}{\partial x_n}(\mathbf{a}) \\ \frac{\partial f_2}{\partial x_1}(\mathbf{a}) & \frac{\partial f_2}{\partial x_2}(\mathbf{a}) & \cdots & \frac{\partial f_2}{\partial x_n}(\mathbf{a}) \\ \vdots & \vdots & \ddots & \vdots \\ \frac{\partial f_m}{\partial x_1}(\mathbf{a}) & \frac{\partial f_m}{\partial x_2}(\mathbf{a}) & \cdots & \frac{\partial f_m}{\partial x_n}(\mathbf{a}) \end{bmatrix} \in \mathbb{R}^{m \times n}.$$

By introducing the notation

$$\frac{\partial \mathbf{f}}{\partial x_i}(\mathbf{a}) = \begin{bmatrix} \frac{\partial f_1}{\partial x_i}(\mathbf{a}) \\ \frac{\partial f_2}{\partial x_i}(\mathbf{a}) \\ \vdots \\ \frac{\partial f_m}{\partial x_i}(\mathbf{a}) \end{bmatrix} \in \mathbb{R}^{m \times 1},$$

the total derivative $D_{\mathbf{f}}(\mathbf{a})$ can be written in the block-forms

$$D_{\mathbf{f}}(\mathbf{a}) = \begin{bmatrix} \frac{\partial \mathbf{f}}{\partial x_1}(\mathbf{a}) & \frac{\partial \mathbf{f}}{\partial x_2}(\mathbf{a}) & \cdots & \frac{\partial \mathbf{f}}{\partial x_n}(\mathbf{a}) \end{bmatrix}$$

and

$$D_{\mathbf{f}}(\mathbf{a}) = \begin{bmatrix} \nabla f_1(\mathbf{a})^T \\ \nabla f_2(\mathbf{a})^T \\ \vdots \\ \nabla f_m(\mathbf{a})^T \end{bmatrix}.$$

## 16.2.4   Derivatives and function operations

We have generalized the notion of derivatives as far as possible for us. Now it's time to study their relations with the two essential function operations: addition and composition. (As there is no vector multiplication in higher dimensional spaces, the product and ratio of vector-vector functions are undefined.)

Let's start with the simpler one: addition.

**Theorem 16.2.3** *(Linearity of the total derivative)*

Let $\mathbf{f}, \mathbf{g} : \mathbb{R}^n \to \mathbb{R}^m$ *be two vector-vector functions that are differentiable at some* $\mathbf{a} \in \mathbb{R}^n$, *and let* $\alpha, \beta \in \mathbb{R}$ *be two arbitrary scalars.*

*Then,* $\alpha \mathbf{f} + \beta \mathbf{g}$ *is also differentiable at* $\mathbf{a}$ *and*

$$D_{\alpha \mathbf{f} + \beta \mathbf{g}}(\mathbf{a}) = \alpha D_{\mathbf{f}}(\mathbf{a}) + \beta D_{\mathbf{g}}(\mathbf{a})$$

*there.*

*Proof.* Because of the total differentiability, (16.9) implies that

$$\alpha \mathbf{f}(\mathbf{x}) + \beta \mathbf{g}(\mathbf{x}) = \alpha \mathbf{f}(\mathbf{a}) + \beta \mathbf{g}(\mathbf{a})$$
$$+ \big(\alpha D_{\mathbf{f}}(\mathbf{a}) + \beta D_{\mathbf{g}}(\mathbf{a})\big)(\mathbf{x} - \mathbf{a})$$
$$+ o(\|\mathbf{x} - \mathbf{a}\|),$$

which implies

$$D_{\alpha \mathbf{f} + \beta \mathbf{g}}(\mathbf{a}) = \alpha D_{\mathbf{f}}(\mathbf{a}) + \beta D_{\mathbf{g}}(\mathbf{a}).$$

This is what we had to show.

Linearity is always nice, but what we need is the ultimate generalization of the chain rule. We previously saw the special case of composing a scalar-vector and a vector-vector function (see *Theorem 16.2.1*), but we need to go one step further.

The multivariable chain rule is extremely important in machine learning. A neural network is a composite function, with layers acting as components. During gradient descent, we use the chain rule to calculate the derivative of this composition.

**Theorem 16.2.4** *(Multivariable chain rule)*

*Let* $\mathbf{f} : \mathbb{R}^m \to \mathbb{R}^l$ *and* $\mathbf{g} : \mathbb{R}^n \to \mathbb{R}^m$ *be two vector-vector functions. If* $\mathbf{g}$ *is totally differentiable at* $\mathbf{a} \in \mathbb{R}^n$ *and* $\mathbf{f}$ *is totally differentiable at* $\mathbf{g}(\mathbf{a})$, *then* $\mathbf{f} \circ \mathbf{g}$ *is also totally differentiable at* $\mathbf{a}$ *and*

$$D_{\mathbf{f} \circ \mathbf{g}}(\mathbf{a}) = D_{\mathbf{f}}(\mathbf{g}(\mathbf{a}))D_{\mathbf{g}}(\mathbf{a}) \tag{16.10}$$

*holds.*

To our advantage, the derivative of a composed function (16.10) is given by the product of two matrices. Since matrix multiplication can be done lightning fast, this is good news.

We will see two proofs for *Theorem 16.2.4*. One is done with a faster-than-light engine, while the other shows much more by reducing the general case to *Theorem 16.2.1*. Both provide a ton of insight. Let's start with the heavy machinery.

*Proof.* (First method.)

As f is totally differentiable at $\mathbf{g(a)}$, the equation (16.9) implies

$$\mathbf{f(g(x))} = \mathbf{f(g(a))} + D_{\mathbf{f}}(\mathbf{g(a)})\big(\mathbf{g(x)} - \mathbf{g(a)}\big) + o(\|\mathbf{g(x)} - \mathbf{g(a)}\|).$$

In turn, again because of the total differentiability of $\mathbf{g}$ at $\mathbf{a}$, we have

$$\mathbf{g(x)} - \mathbf{g(a)} = D_{\mathbf{g}}(\mathbf{a})(\mathbf{x} - \mathbf{a}) + o(\|\mathbf{x} - \mathbf{a}\|).$$

Thus, we can continue our calculation by

$$\begin{aligned}
\mathbf{f(g(x))} &= \mathbf{f(g(a))} + D_{\mathbf{f}}(\mathbf{g(a)})\big(\mathbf{g(x)} - \mathbf{g(a)}\big) + o(\|\mathbf{g(x)} - \mathbf{g(a)}\|) \\
&= \mathbf{f(g(a))} + D_{\mathbf{f}}(\mathbf{g(a)})D_{\mathbf{g}}(\mathbf{a})(\mathbf{x} - \mathbf{a}) \\
&\quad + \underbrace{D_{\mathbf{f}}(\mathbf{g(a)})\Big[o(\|\mathbf{x} - \mathbf{a}\|) + o(\|\mathbf{g(x)} - \mathbf{g(a)}\|)\Big]}_{=o(\|\mathbf{x-a}\|)},
\end{aligned}$$

showing that $\mathbf{f} \circ \mathbf{g}$ is totally differentiable at $\mathbf{a}$ with total derivative

$$D_{\mathbf{f} \circ \mathbf{g}}(\mathbf{a}) = D_{\mathbf{f}}(\mathbf{g(a)})D_{\mathbf{g}}(\mathbf{a}),$$

which is what we needed to show.

Now, about that second proof.

*Proof.* (Second method.)

Let's unpack $D_{\mathbf{f} \circ \mathbf{g}}(\mathbf{a})$ a bit. Writing out the components of $\mathbf{f} \circ \mathbf{g}$, we have

$$(\mathbf{f} \circ \mathbf{g})(\mathbf{x}) = \begin{bmatrix} (\mathbf{f} \circ \mathbf{g})_1(\mathbf{x}) \\ (\mathbf{f} \circ \mathbf{g})_2(\mathbf{x}) \\ \vdots \\ (\mathbf{f} \circ \mathbf{g})_l(\mathbf{x}) \end{bmatrix} \in \mathbb{R}^l, \quad \mathbf{x} \in \mathbb{R}^n.$$

By definition, the $i$-th row and $j$-th column of $D_{\mathbf{f} \circ \mathbf{g}}(\mathbf{a})$ is

$$\Big(D_{\mathbf{f} \circ \mathbf{g}}(\mathbf{a})\Big)_{i,j} = \frac{\partial (\mathbf{f} \circ \mathbf{g})_i}{\partial x_j}(\mathbf{a}).$$

If you look at it long enough, you'll realize that $\frac{\partial(\mathbf{f}\circ\mathbf{g})_i}{\partial x_j}(\mathbf{a})$ is the derivative of a single variable function. Indeed, the function to be differentiated is the composition of the curve

$$\gamma \: : \: t \mapsto \mathbf{g}(a_1, \ldots, a_{j-1}, t, a_{j+1}, \ldots, a_n)$$

and the vector-scalar function $f_i \: : \: \mathbb{R}^m \to \mathbb{R}$. Thus, the chain rule for the composition of scalar-vector and vector-scalar functions (given by *Theorem 16.2.1*) can be applied:

$$\frac{\partial(\mathbf{f}\circ\mathbf{g})_i}{\partial x_j}(\mathbf{a}) = \nabla f_i(\mathbf{g}(\mathbf{a}))^T \frac{\partial}{\partial x_j}\mathbf{g}(\mathbf{a}),$$

where $\frac{\partial}{\partial x_j}\mathbf{g}(\mathbf{a})$ is the componentwise derivative

$$\frac{\partial}{\partial x_j}\mathbf{g}(\mathbf{a}) = \begin{bmatrix} \frac{\partial g_1(\mathbf{a})}{\partial x_j} \\ \frac{\partial g_2(\mathbf{a})}{\partial x_j} \\ \vdots \\ \frac{\partial g_m(\mathbf{a})}{\partial x_j} \end{bmatrix}.$$

To sum up, we have

$$\frac{\partial(\mathbf{f}\circ\mathbf{g})_i}{\partial x_j}(\mathbf{a}) = \nabla f_i(\mathbf{g}(\mathbf{a}))^T \frac{\partial}{\partial x_j}\mathbf{g}(\mathbf{a})$$

$$= \sum_{k=1}^{m} \frac{\partial f_i}{\partial x_k}(\mathbf{a})\frac{\partial g_k}{\partial x_j}(\mathbf{a}).$$

This is the element in the *i*-th row and *j*-th column of the matrix product $D_{\mathbf{f}}(\mathbf{g}(\mathbf{a}))D_{\mathbf{g}}(\mathbf{a})$, hence

$$D_{\mathbf{f}\circ\mathbf{g}}(\mathbf{a}) = D_{\mathbf{f}}(\mathbf{g}(\mathbf{a}))D_{\mathbf{g}}(\mathbf{a}),$$

which is what we had to show.

With the concept of total derivatives for vector-vector functions and the general chain rule under our belt, we are ready to actually do things with multivariable functions. Thus, our next stop lays the foundations of optimization.

# 16.3  Summary

You know by now: half the success in mathematics is picking the right representations and notations. Although multivariable calculus can seem insanely complex, it's a cakewalk if we have a good understanding of linear algebra. This is why we started our entire journey with vectors and matrices! Going from $f(x_1, \ldots, x_n)$ to $f(\mathbf{x})$ is a big deal.

In this chapter, we have learned that differentiation in multiple dimensions is slightly more complicated than in the single-variable case. First, we have the *partial derivatives* defined by

$$\frac{\partial f}{\partial x_i}(\mathbf{a}) = \lim_{h \to 0} \frac{f(\mathbf{a} + h\mathbf{e}_i) - f(\mathbf{a})}{h}, \quad \mathbf{a} \in \mathbb{R}^n,$$

where $\mathbf{e}_i$ is the vector whose $i$-th component is one, while the others are zero. We can think about $\frac{\partial f}{\partial x_i}$ as the derivative of the single-variable function obtained by fixing all but the $i$-th variable of $f$. Together, the partial derivatives form the gradient:

$$\nabla f(\mathbf{a}) := \begin{bmatrix} \frac{\partial}{\partial x_1} f(\mathbf{a}) \\ \frac{\partial}{\partial x_2} f(\mathbf{a}) \\ \vdots \\ \frac{\partial}{\partial x_n} f(\mathbf{a}) \end{bmatrix} \in \mathbb{R}^{n \times 1}.$$

However, the partial derivatives are not exactly the perfect analogue of the univariate derivatives. There, we learned that the derivative is the best local linear approximation, and this is the version that can be generalized to multiple variables. Thus, we say that $f$ is *totally differentiable* at $\mathbf{a} \in \mathbb{R}^n$ if it can be written in the form

$$f(\mathbf{x}) = f(\mathbf{a}) + \nabla f(\mathbf{a})^T (\mathbf{x} - \mathbf{a}) + o(\|\mathbf{x} - \mathbf{a}\|).$$

In machine learning, one of the most essential tools is the multivariable chain rule

$$D_{f \circ g}(\mathbf{a}) = D_f(\mathbf{g}(\mathbf{a})) D_g(\mathbf{a}),$$

which is used to compute the derivatives in practice. Without the chain rule, we wouldn't have any effective method to compute the gradient. In turn, as the name suggests, the gradient is the cornerstone of *gradient descent*. We already understand the single-variable version, so it's time to dive deep into the general one. See you in the next chapter!

# 16.4 Problems

**Problem 1.** Compute the partial derivatives and the Hessian matrix of the following functions.

(a)   $f(x_1, x_2) = x_1^{3x_2^2} + 2x_1 x_2 + x_2^3$

(b)   $f(x_1, x_2) = e^{x_1^2 - x_2} + \sin(x_1 x_2)$

(c) $f(x_1, x_2) = \ln(x_1^2 + x_2^2) + x_1 e^{x_2}$

(d) $f(x_1, x_2) = \cos(x_1 x_2) + x_1^2 \sin(x_2)$

(e) $f(x_1, x_2) = f(x_1, x_2) = \frac{x_1^2 + x_2^2}{x_1 - x_2}$

**Problem 2.** Compute the Jacobian matrix of the following functions.

(a)

$$\mathbf{f}(x_1, x_2) = \begin{bmatrix} x_1^2 x_2 + e^{x_2} \\ \sin(x_1 x_2) + x_1 e^{x_2} \end{bmatrix}$$

(b)

$$\mathbf{f}(x_1, x_2) = \begin{bmatrix} \ln(x_1^2 + x_2^2) + x_1 x_2 \\ \cos(x_1) + x_2^2 e^{x_1} \end{bmatrix}$$

(c)

$$\mathbf{f}(x_1, x_2) = \begin{bmatrix} x_1^3 - x_2^2 \\ e^{x_1 x_2} + x_1 \cos(x_2) \end{bmatrix}$$

(d)

$$\mathbf{f}(x_1, x_2) = \begin{bmatrix} \tan(x_1 x_2) + x_2^3 \\ \sqrt{x_1^2 + x_2^2} + \sin(x_1) \end{bmatrix}$$

(e)

$$\mathbf{f}(x_1, x_2) = \begin{bmatrix} x_1 e^{x_2} - \ln(1 + x_1^2) \\ x_2^2 \cos(x_1) + x_1 x_2 \end{bmatrix}$$

**Problem 3.** Let $f(x_1, x_2) = x_1 \sqrt{|x_2|}$. Show that $f$ is partially differentiable but not totally differentiable at $(0, 0)$.

# Join our community on Discord

Read this book alongside other users, Machine Learning experts, and the author himself.

Ask questions, provide solutions to other readers, chat with the author via Ask Me Anything sessions, and much more.

Scan the QR code or visit the link to join the community.

https://packt.link/math

# 17

# Optimization in Multiple Variables

Hey! We are at the last checkpoint of our calculus study. What's missing? Gradient descent, of course.

In the previous eight chapters, we lined up all of our ducks in a row, and now it's time to take that shot. First, we'll put multivariable functions to code. Previously, we built a convenient interface in the form of our `Function` class to represent differentiable functions. After the lengthy setup in the previous chapter, we can easily extend it, and with the power of vectorization, we don't even have to change that much. Let's go!

## 17.1   Multivariable functions in code

It's been a long time since we put theory into code. So, let's take a look at multivariable functions!

Last time, we built a `Function` base class with two main methods: one for computing the derivative (`Function.prime`) and one for getting the dictionary of parameters (`Function.parameters`).

This won't be much of a surprise: the multivariate function base class is not much different. For clarity, we'll appropriately rename the `prime` method to `grad`.

```
class MultivariableFunction:
    def __init__(self):
        pass

    def __call__(self, *args, **kwargs):
        pass

    def grad(self):
        pass

    def parameters(self):
        return dict()
```

Let's see a few examples right away. The simplest one is the squared Euclidean norm $f(\mathbf{x}) = \|\mathbf{x}\|^2$, a close relative to the mean squared error function. Its gradient is given by

$$\nabla f(\mathbf{x}) = 2\mathbf{x},$$

thus everything is ready to implement it. As we've used NumPy arrays to represent vectors, we'll use them as the input as well.

```
import numpy as np


class SquaredNorm(MultivariableFunction):
    def __call__(self, x: np.array):
        return np.sum(x**2)

    def grad(self, x: np.array):
        return 2*x
```

Note that SquaredNorm is different from $f(\mathbf{x}) = \|\mathbf{x}\|^2$ in a mathematical sense, as it accepts any NumPy array, not just an $n$-dimensional vector. This is not a problem now, but will be later, so keep that in mind.

Another example can be given with the parametric linear function

$$g(x, y) = ax + by,$$

where $a, b \in \mathbb{R}$ are arbitrary parameters. Let's see how $g(x, y)$ is implemented!

```python
class Linear(MultivariableFunction):
    def __init__(self, a: float, b: float):
        self.a = a
        self.b = b

    def __call__(self, x: np.array):
        """
        x: np.array of shape (2, )
        """
        x = x.reshape(2)
        return self.a*x[0] + self.b*x[1]

    def grad(self, x: np.array):
        return np.array([self.a, self.b]).reshape(2, 1)

    def parameters(self):
        return {"a": self.a, "b": self.b}
```

To check if our implementation works correctly, we can quickly test it out on a simple example.

```python
g = Linear(a=1, b=-1)

g(np.array([1, 0]))
```

```
np.int64(1)
```

Perhaps we might have overlooked this question until now, but trust me, specifying the input and output shapes is of crucial importance. When doing mathematics, we can be flexible in our notation and treat any vector $\mathbf{x} \in \mathbb{R}^n$ as a column or row vector, but painfully, this is not the case in practice.

Correctly keeping track of array shapes is of utmost importance and can save you hundreds of hours. No joke.

# 17.2 Minima and maxima, revisited

In a single variable, we have successfully used the derivatives to find the local optima of differentiable functions.

Recall that if $f : \mathbb{R} \to \mathbb{R}$ is differentiable everywhere, then *Theorem 13.1.4* gives that

(a) $f'(a) = 0$ and $f''(a) > 0$ implies a local minimum.

(b) $f'(a) = 0$ and $f''(a) < 0$ implies a local maximum.

(A simple $f'(a) = 0$ is not enough, as the example $f(x) = x^3$ shows at 0.)

Can we do something similar in multiple variables?

Right from the start, there seems to be an issue: the derivative is not a scalar (thus, we can't equate it to 0).

This is easy to solve: the analogue of the condition $f'(a) = 0$ is $\nabla f(\mathbf{a}) = (0, 0, \ldots, 0)$. For simplicity, the zero vector $(0, 0, \ldots, 0)$ will also be denoted by 0. Don't worry, this won't be confusing; it's all clear from the context. Introducing a new notation for the zero vector would just add more complexity.

We can visualize what happens with the tangent plane at critical points. In a single variable, we have already seen this: as *Figure 17.1* illustrates, $f'(a) = 0$ implies that the tangent line is horizontal.

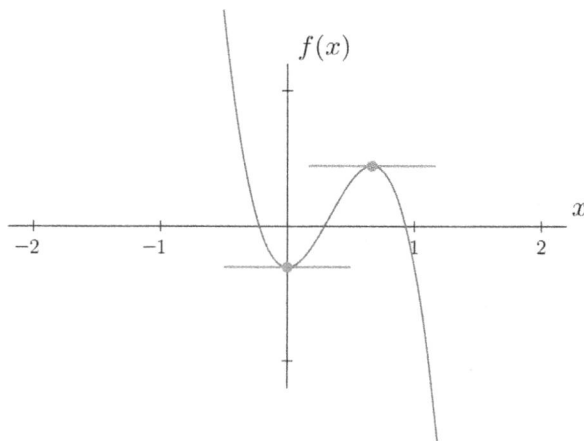

*Figure 17.1: Local extrema in a single variable*

In multiple variables, the situation is similar: $\nabla f(\mathbf{a}) = \mathbf{0}$ implies that the best local linear approximation (16.3) is constant; that is, the tangent *plane* is horizontal. (As visualized by *Figure 17.2*.)

*Figure 17.2: Local extrema in multiple variables*

So, what does $\nabla f(\mathbf{a}) = \mathbf{0}$ imply? Similarly to the single-variable case, an $\mathbf{a} \in \mathbb{R}^n$ is called a *critical point* of $f$ if $\nabla f(\mathbf{a}) = \mathbf{0}$ holds. The similarity doesn't stop at the level of terminologies. We also have three options in multiple variables as well: $\mathbf{a}$ is

1. a local minimum,
2. a local maximum,
3. or neither.

In multiple variables, a non-extremal critical point is called a *saddle point*, because the two-dimensional case bears a striking resemblance to an actual horse saddle, as you are about to see. Saddle points are the high-dimensional analogues of the one-dimensional inflection points. The functions

$$f(x, y) = x^2 + y^2,$$
$$g(x, y) = -(x^2 + y^2),$$
$$h(x, y) = x^2 - y^2$$

at $(0, 0)$ provide an example for all three, as *Figure 17.3*, *Figure 17.4*, and *Figure 17.5* show. (Keep in mind that a local extremum might be global.)

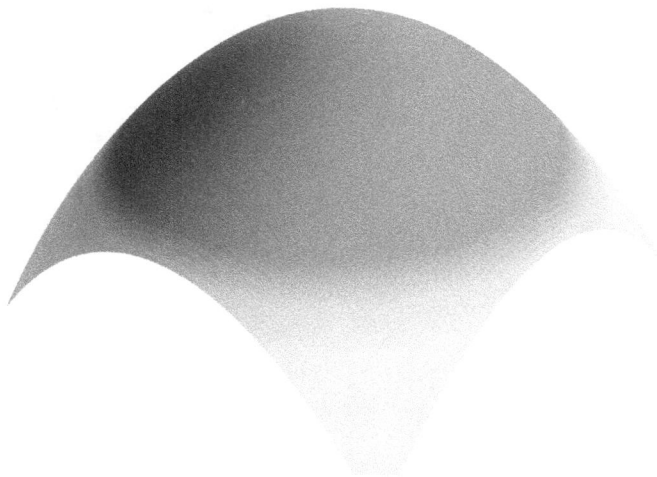

*Figure 17.3: A (local) maximum*

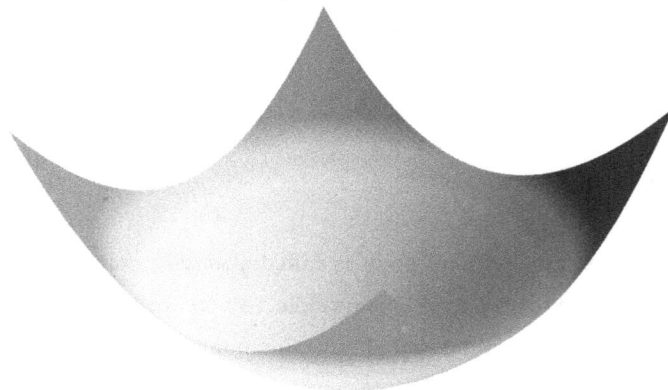

*Figure 17.4: A (local) minima*

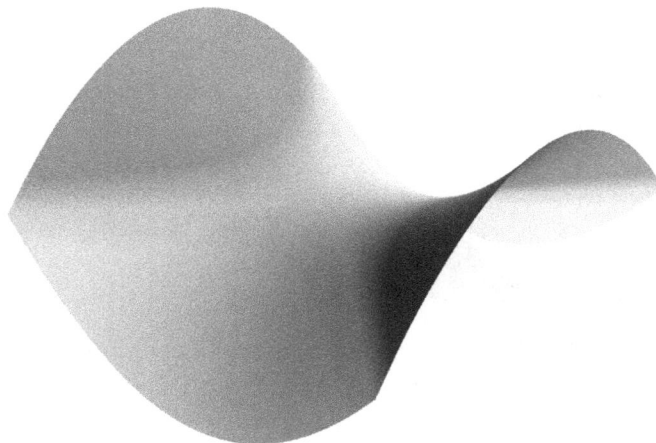

*Figure 17.5: A saddle point*

To put things into order, let's start formulating definitions and theorems.

**Definition 17.2.1 (Critical points)**

Let $f : \mathbb{R}^n \to \mathbb{R}$ be an arbitrary vector-scalar function. We say that $\mathbf{a} \in \mathbb{R}^n$ is a *critical point* of $f$ if either

$$\nabla f(\mathbf{a}) = \mathbf{0}$$

holds, or $f$ is not partially differentiable at $\mathbf{a}$ in at least one variable.

The second case (where $f$ is not differentiable at **a**) is there to handle situations like $f(x, y) = |x| + |y|$.

For the sake of precision, let's define local extrema in multiple dimensions as well.

**Definition 17.2.2 (Local minima and maxima)**

Let $f : \mathbb{R}^n \to \mathbb{R}$ be an arbitrary vector-scalar function and let $\mathbf{a} \in \mathbb{R}^n$ be an arbitrary point.

*(a)* $\mathbf{a}$ is a *local minimum* if there exists an $\varepsilon > 0$ such that

$$f(\mathbf{a}) \leq f(\mathbf{x}), \quad \mathbf{x} \in B(\varepsilon, \mathbf{a}).$$

*(b)* $\mathbf{a}$ is a *strict local minimum* if there exists an $\varepsilon > 0$ such that

$$f(\mathbf{a}) < f(\mathbf{x}), \quad \mathbf{x} \in B(\varepsilon, \mathbf{a}).$$

*(c)* $\mathbf{a}$ is a *local maximum* if there exists an $\varepsilon > 0$ such that

$$f(\mathbf{a}) \geq f(\mathbf{x}), \quad \mathbf{x} \in B(\varepsilon, \mathbf{a}) \setminus \{\mathbf{a}\}.$$

*(d)* $\mathbf{a}$ is a *strict local maximum* if there exists an $\varepsilon > 0$ such that

$$f(\mathbf{a}) > f(\mathbf{x}), \quad \mathbf{x} \in B(\varepsilon, \mathbf{a}) \setminus \{\mathbf{a}\}.$$

As the example of $x^2 - y^2$ shows, a critical point is not necessarily a local extremum, but a local extremum is always a critical point. The next result, which is the analogue of *Definition 17.2.1*, makes this mathematically precise.

**Theorem 17.2.1** *Let $f : \mathbb{R}^n \to \mathbb{R}$ be an arbitrary vector-scalar function, and suppose that $f$ is partially differentiable with respect to all variables at some $\mathbf{a} \in \mathbb{R}^n$.*

*If $f$ has a local extremum at $\mathbf{a}$, then $\nabla f(\mathbf{a}) = \mathbf{0}$.*

*Proof.* This is a direct consequence of *Theorem 13.1.3*, as if $\mathbf{a} = (a_1, \ldots, a_n)$ is a local extremum of the vector-scalar function $f$, then it is a local extremum of the single-variable functions $h \mapsto f(\mathbf{a} + h\mathbf{e}_i)$, where $\mathbf{e}_i$ is the vector whose $i$-th component is 1, while the others are 0.

According to the very definition of the partial derivative given by *Definition 16.1.1*,

$$\frac{d}{dh} f(\mathbf{a} + h\mathbf{e}_i) = \frac{\partial f}{\partial x_i}(\mathbf{a}).$$

Thus, *Theorem 13.1.3* gives that

$$\frac{\partial f}{\partial x_i}(\mathbf{a}) = 0$$

for all $i = 1, \ldots, n$, giving that $\nabla f(\mathbf{a}) = \mathbf{0}$.

So, how can we find the local extrema with the derivative? As we have already suggested, studying the second derivative will help us pinpoint the extrema among critical points. Unfortunately, things are much more complicated in $n$ variables, so let's focus on the two-variable case first.

---

**Theorem 17.2.2** *(The two-variable second derivative test)*

*Let $f : \mathbb{R}^2 \to \mathbb{R}$ be an arbitrary vector-scalar function, and suppose that $f$ is partially differentiable at some $\mathbf{a} \in \mathbb{R}^2$. Also suppose that $\mathbf{a}$ is a critical point, that is, $\nabla f(\mathbf{a}) = \mathbf{0}$.*

*(a) If $\det H_f(\mathbf{a}) > 0$ and $\frac{\partial^2 f}{\partial x_2^2} > 0$, then $\mathbf{a}$ is a local minimum.*

*(b) If $\det H_f(\mathbf{a}) > 0$ and $\frac{\partial^2 f}{\partial x_2^2} < 0$, then $\mathbf{a}$ is a local maximum.*

*(c) If $\det H_f(\mathbf{a}) < 0$, then $\mathbf{a}$ is a saddle point.*

---

We will not prove this, but some remarks are in order. First, as the determinant of the Hessian can be 0, *Theorem 17.2.2* does not cover all possible cases.

It's probably best to see a few examples, so let's revisit the previously seen functions

$$f(x, y) = x^2 + y^2,$$
$$g(x, y) = -(x^2 + y^2),$$
$$h(x, y) = x^2 - y^2.$$

All three have a critical point at $\mathbf{0}$, so the Hessians can provide a clearer picture. The Hessians are given by the matrices

$$H_f(x, y) = \begin{bmatrix} 2 & 0 \\ 0 & 2 \end{bmatrix}, \quad H_g(x, y) = \begin{bmatrix} -2 & 0 \\ 0 & -2 \end{bmatrix}, \quad H_h(x, y) = \begin{bmatrix} 2 & 0 \\ 0 & -2 \end{bmatrix}.$$

For functions of two variables, *Theorem 17.2.2* says that it is enough to study $\det H_f(\mathbf{a})$ and $\frac{\partial^2 f}{\partial y^2}(\mathbf{a})$.

In the case of $f(x, y) = x^2 + y^2$, we have $H_f(0, 0) = 4$ and $\frac{\partial^2 f}{\partial y^2}(0, 0) = 2$, giving that $\mathbf{0}$ is a local minimum of $f(x, y) = x^2 + y^2$.

Similarly, we can conclude that $\mathbf{0}$ is a local maximum of $g(x, y) = -(x^2 + y^2)$ (which shouldn't surprise you, as $g = -f$).

Finally, for $h(x, y) = x^2 - y^2$, the second derivative test confirms that $\mathbf{0}$ is indeed a saddle point.

So, what's up with the general case? Unfortunately, just studying the determinant of the Hessian matrix is not enough. We need to bring in the heavy-hitters: eigenvalues. (See *Definition 6.0.1*.) Here is the second derivative test in its full glory.

---

**Theorem 17.2.3** *(The multivariable second derivative test)*

Let $f : \mathbb{R}^n \to \mathbb{R}$ *be an arbitrary vector-scalar function, and suppose that $f$ is partially differentiable with respect to all variables at some* $\mathbf{a} \in \mathbb{R}^n$. *Also suppose that $\mathbf{a}$ is a critical point, that is,* $\nabla f(\mathbf{a}) = \mathbf{0}$.

*(a) If all the eigenvalues of $H_f(\mathbf{a})$ are positive, then $\mathbf{a}$ is a local minimum.*

*(b) If all the eigenvalues of $H_f(\mathbf{a})$ are negative, then $\mathbf{a}$ is a local maximum.*

*(c) If all the eigenvalues of $H_f(\mathbf{a})$ are either positive or negative, then $\mathbf{a}$ is a saddle point.*

---

That's right: if any of the eigenvalues are 0, then the test is inconclusive. You might recall from linear algebra that in practice, computing the eigenvalues is not as fast as computing the second-order derivatives, but there are plenty of numerical methods (like the QR-algorithm, as we saw in *Section 7.5*).

To sum it up, the method of optimizing (differentiable) multivariable functions is a simple two-step process:

1. find the critical points by solving the equation $\nabla f(\mathbf{x}) = 0$,
2. then use the second derivative test to determine which critical points are extrema.

Do we use this method in practice to optimize functions? No. Why? Most importantly, because computing the eigenvalues of the Hessian for a vector-scalar function with millions of variables is extremely hard. Why is the second derivative test so important? Because understanding the behavior of functions around their extremal points is essential to truly understand gradient descent. Believe it or not, this is the key behind the theoretical guarantees for convergence.

Speaking of gradient descent, now is the time to dig deep into the algorithm that powers neural networks.

# 17.3 Gradient descent in its full form

Gradient descent is one of the most important algorithms in machine learning. We have talked about this a lot, although up until this point, we have only seen it for single-variable functions (which is, I admit, not the most practical use case).

However, now we have all the tools we need to talk about gradient descent in its general form. Let's get to it!

Suppose that we have a differentiable vector-scalar function $f : \mathbb{R}^n \to \mathbb{R}$ that we want to *maximize*. This can describe the return on investment of an investing strategy, or any other quantity. Calculating the gradient and finding the critical points is often not an option, as solving the equation $\nabla f(\mathbf{x}) = \mathbf{0}$ can be computationally unfeasible. Thus, we resort to an iterative solution.

The algorithm is the same as for single-variable functions (as seen in *Section 13.2*):

1. Start from a random point.
2. Calculate its gradient.
3. Take a step towards its direction.
4. Repeat until convergence.

This is called *gradient ascent*. We can formalize it in the following way.

---

**Algorithm 17.3.1** *(The gradient ascent algorithm)*

**Step 1.** *Initialize the starting point* $\mathbf{x}_0 \in \mathbb{R}^n$ *and select a learning rate* $h \in (0, \infty)$.

**Step 2.** *Let*

$$\mathbf{x}_{n+1} := \mathbf{x}_n + h\nabla f(\mathbf{x}_n).$$

**Step 3.** *Repeat* **Step 2.** *until convergence.*

---

If we want to *minimize* $f$, we might as well *maximize* $-f$. The only effect of this is a sign change for the gradient. In this form, the algorithm is called *gradient descent*, and this is the version that's widely used to train neural networks.

---

**Algorithm 17.3.2** *(The gradient descent algorithm)*

**Step 1.** *Initialize the starting point* $\mathbf{x}_0 \in \mathbb{R}^n$ *and select a learning rate* $h \in (0, \infty)$.

**Step 2.** *Let*

$$\mathbf{x}_{n+1} := \mathbf{x}_n - h\nabla f(\mathbf{x}_n).$$

**Step 3.** *Repeat* **Step 2.** *until convergence.*

---

After all of this setup, implementing gradient descent is straightforward.

```python
def gradient_descent(
    f: MultivariableFunction,
    x_init: np.array,                  # the initial guess
    learning_rate: float = 0.1,        # the learning rate
    n_iter: int = 1000,                # number of steps
):
    x = x_init

    for n in range(n_iter):
        grad = f.grad(x)
        x = x - learning_rate*grad

    return x
```

Notice that it is almost identical to the single-variable version in *Section 13.2*. To see if it works correctly, let's test it out on the squared Euclidean norm function, implemented by SquaredNorm earlier!

```python
squared_norm = SquaredNorm()
local_minimum = gradient_descent(
    f=squared_norm,
    x_init=np.array([10.0, -15.0])
)

local_minimum
```

```
array([ 1.23023192e-96, -1.84534788e-96])
```

There is nothing special to it, really. The issues with multivariable gradient descent are the same as what we discussed with the single-variable version: it can get stuck in local minima, it is sensitive to our choice of learning rate, and the gradient can be computationally hard to calculate in high dimensions.

# 17.4   Summary

Although this chapter was short and sweet, we took quite a big step by dissecting the fine details of gradient descent in high dimensions. The chapter's brevity is a testament to the power of vectorization: same formulas, code, and supercharged functionality. It's quite unbelievable, but the simple algorithm

$$\mathbf{x}_{n+1} = \mathbf{x}_n - h\nabla f(\mathbf{x}_n)$$

is behind most of the neural network models. Yes, even state-of-the-art ones.

This lies on the same theoretical foundations as the univariate case, but instead of checking the positivity of the second derivatives, we have to study the full Hessian matrix $H_f$. To be more precise, we have learned that a critical point $\nabla f(\mathbf{a}) = \mathbf{0}$ is

1. a local minimum if all the eigenvalues of $H_f(\mathbf{a})$ are positive,
2. and a local maximum if all the eigenvalues of $H_f(\mathbf{a})$ are negative.

Deep down, this is the reason why gradient descent works. And with this, we have finished our study of calculus, both in single and multiple variables.

Take a deep breath and relax a bit. We are approaching the final stretch of our adventure: our last stop is the theory of probability, the thinking paradigm that is behind predictive modeling. For instance, the most famous loss functions, like the mean-squared error or the cross-entropy, are founded upon probabilistic concepts. Understanding and taming uncertainty is one of the biggest intellectual feats of science, and we are about to undertake this journey ourselves.

See you in the next chapter!

# 17.5   Problems

**Problem 1.** Let $\mathbf{y} \in \mathbb{R}^n$ be an arbitrary vector. The general version of the famous mean-squared error is defined by

$$\mathrm{MSE}(\mathbf{x}) = \frac{1}{n} \sum_{i=1}^{n} (x_i - y_i)^2.$$

Compute its gradient and implement it using the `MultivariateFunction` base class!

**Problem 2.** Let $f : \mathbb{R}^2 \to \mathbb{R}$ be the function defined by

$$f(x, y) = (2x^2 - y)(y - x^2).$$

Does $f$ have a local extremum in $\mathbf{x} = (0, 0)$?

**Problem 3.** Use the previously implemented `gradient_descent` function to find the minimum of

$$f(x, y) = \sin(x + y) + x^2 y^2.$$

Experiment with various learning rates and initial values!

**Problem 4.** In the problem section of Chapter 13, we saw the improved version of gradient descent, called *gradient descent with momentum*. We can do the same in multiple variables: define

$$\mathbf{d}_{n+1} = \alpha \mathbf{d}_n - h f'(\mathbf{x}_n),$$

$$\mathbf{x}_{n+1} = \mathbf{x}_n + \mathbf{d}_n,$$

where $\mathbf{d}_0 = 0$ and $\mathbf{x}_0$ is arbitrary. Implement it!

# Join our community on Discord

Read this book alongside other users, Machine Learning experts, and the author himself.

Ask questions, provide solutions to other readers, chat with the author via Ask Me Anything sessions, and much more.

Scan the QR code or visit the link to join the community.

`https://packt.link/math`

# References

[1] Rudin, W. (1976). *Principles of Mathematical Analysis* (3rd ed.). McGraw Hill.

[2] Rudin, W. (1986). *Real and Complex Analysis* (3rd ed.). McGraw Hill.

[3] Spivak, M. (2008). *Calculus* (4th ed.). Cambridge University Press.

# PART IV
# PROBABILITY THEORY

This part comprises the following chapters:

# 18

# What is Probability?

When going about our lives, we almost always think in binary terms. A statement is either true or false. An outcome has either occurred or not.

In practice, we rarely have the comfort of certainty. We have to operate with incomplete information. When a scientist observes the outcome of an experiment, can they verify their hypothesis with 100% certainty? No. Because they do not have complete control over all the variables (such as the weather or the alignment of stars), the observed effect might be unintentional. Each result will either strengthen or weaken our belief in the hypothesis, but none will provide ultimate proof.

In machine learning, our job is not simply to provide a prediction about some class label but to formulate a mathematical model that summarizes our knowledge about the data in a way that conveys information about the degree of our certainty in the prediction.

So, fitting a parametric function $f : \mathbb{R}^n \to \mathbb{R}^m$ to model the relation between the data and the variable to be predicted is not enough. We will need an entirely new vocabulary to formulate such models. We need to think in terms of *probabilities*.

## 18.1   The language of thinking

First, let's talk about how we think. On the most basic level, our knowledge about the world is stored in propositions. In a mathematical sense, a proposition is a declaration that is either true or false. (In binary terms, true is denoted by 1 and false is denoted by 0.)

*"The sky is blue."*

*"There are infinitely many prime numbers."*

*"1 + 1 = 3."*

*"I got the flu."*

Propositions are often abbreviated as variables such as $A$ = "it's raining outside".

Determining the truth value of a given proposition using evidence and reasoning is called *inference*. To be able to formulate valid arguments and understand how inference works, we'll take a quick visit to the world of mathematical logic.

## 18.1.1  Thinking in absolutes

So, we have propositions such as $A$ = "it's raining outside" or $B$ = "the sidewalk is wet". We need more expressive power: propositions are building blocks, and we want to combine them, yielding more complex propositions. (We'll review the fundamentals of mathematical logic here, but check out *Appendix A* for more.)

We can formulate complex propositions from simpler ones with *logical connectives*. Consider the proposition "if it is raining outside, then the sidewalk is wet". This is the combination of $A$ and $B$, strung together by the implication connective.

There are four essential connectives:

- NOT ($\neg$), also known as negation,
- AND ($\wedge$), also known as conjunction,
- OR ($\vee$), also known as disjunction,
- THEN ($\rightarrow$), also known as implication.

Connectives are defined by the truth values of the resulting propositions. For instance, if $A$ is true, then $\neg A$ is false; if $A$ is false, then $\neg A$ is true. Denoting true by 1 and false by 0, we can describe connectives with *truth tables*. Here is the one for negation:

| $A$ | $\neg A$ |
|---|---|
| 0 | 1 |
| 1 | 0 |

AND ($\wedge$) and OR ($\vee$) connect two propositions. $A \wedge B$ is true if both $A$ and $B$ are true, while $A \vee B$ is true if either one is.

| $A$ | $B$ | $A \wedge B$ | $A \vee B$ |
|-----|-----|--------------|------------|
| 0 | 0 | 0 | 0 |
| 0 | 1 | 0 | 1 |
| 1 | 0 | 0 | 1 |
| 1 | 1 | 1 | 1 |

The implication connective THEN ($\rightarrow$) formalizes the deduction of a conclusion $B$ from a premise $A$. By definition, $A \rightarrow B$ is true if $B$ is true or both $A$ and $B$ are false. An example: IF *"it's raining outside"*, THEN *"the sidewalk is wet"*.

| $A$ | $B$ | $A \rightarrow B$ |
|-----|-----|-------------------|
| 0 | 0 | 1 |
| 0 | 1 | 1 |
| 1 | 0 | 0 |
| 1 | 1 | 1 |

Note that $A \rightarrow B$ does not imply $B \rightarrow A$. This common logical fallacy is called *affirming the consequent*, and we've all fallen victim to it at some point in our lives. To see a concrete example: if *"it's raining outside"*, then *"the sidewalk is wet"*, but not the other way around. The sidewalk can be wet for other reasons, such as someone spilling a barrel of water.

Connectives correspond to set operations. Why? Let's take a look at the *formal* definition of set operations.

**Definition 18.1.1 (The (reasonably) formal definition of set operations and relations)**

Let $A$ and $B$ be two sets.

*(a)* The union of $A$ and $B$ is defined by

$$A \cup B := \{x : (x \in A) \vee (x \in B)\},$$

that is, $A \cup B$ contains all elements that are in $A$ or $B$.

*(b)* The intersection of $A$ and $B$ is defined by

$$A \cap B := \{x : (x \in A) \wedge (x \in B)\},$$

that is, $A \cap B$ contains all elements that are in $A$ and $B$.

*(c)* We say that $A$ is a subset of $B$, that is, $A \subseteq B$ if

$$(x \in A) \rightarrow (x \in B)$$

is true for all $x \in A$.

*(d)* The complement of $A$ with respect to an $\Omega \supset A$ is defined by

$$\Omega \setminus A := \{x \in \Omega : \neg(x \in A)\},$$

that is, $\Omega \setminus A$ contains all elements that are in $\Omega$, but not in $A$.

If you carefully read through the definitions, you can see how connectives and set operations relate. $\wedge$ is intersection, $\vee$ is union, $\neg$ is the complement, and $\rightarrow$ is the subset relation. This is illustrated by *Figure 18.1*. (I've slightly abused the notation here, as statements such as $A \wedge B \iff A \cap B$ are mathematically incorrect. $A$ and $B$ cannot be a proposition and a set at the same time, thus the equivalence is not precise. )

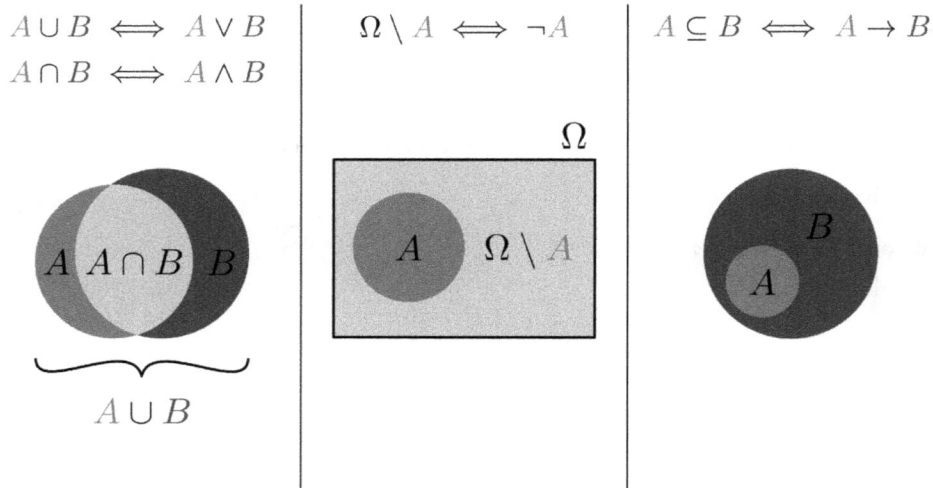

Figure 18.1: Connectives and set operations

Why is this important? Because probability operates on sets, and sets play the role of propositions. We'll see this later, but first, let's dive deep into how mathematical logic formalizes scientific thinking.

Let's refine the inference process of mathematical logic. A proposition is either true or false, fair and square. How can we determine that in practice? For example, how do we find the truth value of the proposition "there are infinitely many prime numbers"?

By using evidence and deduction. Like Sherlock Holmes solving a crime by connecting facts, we rely on knowledge of the form *"if A, then B"*. Our knowledge about the world is stored in true implications. For example:

- *"If it is raining, then the sidewalk is wet."*
- *"If ABC is a right triangle, then $A^2 + B^2 = C^2$."*
- *"If a system is closed, then its entropy cannot decrease."*

As we have seen, the implication can be translated into the language of set theory (as all the other connectives). While $\wedge$ corresponds to intersection and $\vee$ to union, the implication is the subset relation. Keep this in mind, as it's going to be important.

During inference, we use implications in the following way:

1. If $A$, then $B$.
2. $A$.
3. Therefore, $B$.

This is called the *modus ponens*. If it sounds abstract, here is a concrete example:

1. If it is raining, the sidewalk is wet.
2. It is raining.
3. Therefore, the sidewalk is wet.

Thus, we can infer the state of the sidewalk without looking at it. This is bigger than it sounds: modus ponens is a cornerstone of scientific thinking. We would still be living in caves without it. Modus ponens enables us to build robust skyscrapers of knowledge.

However, it's not all perfect. Classical deductive logic might help to prove the infinity of prime numbers, but it fails spectacularly when confronted with inference problems outside the realms of mathematics and philosophy.

Classical logic has a fatal flaw: it is unable to deal with uncertainty. Think about the simple proposition *"it is raining outside"*. If we are unable to actually observe the weather but have some indirect evidence (such as the fact that the sidewalk is wet, or the sky is cloudy, or it's autumn out there), *"it is raining outside"* is probable but not certain.

We need a tool to measure the truth value on a $0 - 1$ scale. This is where probabilities come in.

## 18.1.2 Thinking in probabilities

In a mathematical sense, probability is a function that assigns a numerical value between zero and one to various *sets* that represent events. (You can think of events as propositions.) Events are subsets of the event space, often denoted with the capital Greek letter omega ($\Omega$). This is illustrated in *Figure 18.2*.

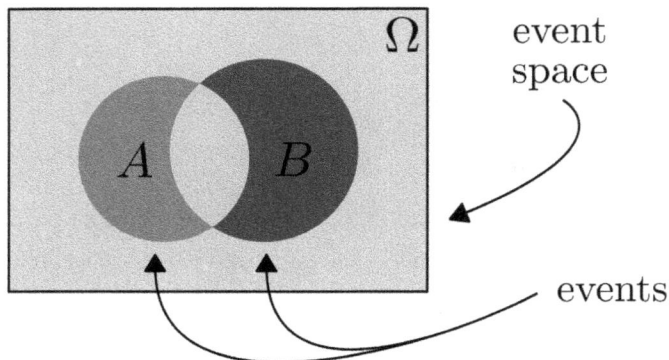

*Figure 18.2: Events and the event space*

This sounds quite abstract, so let's see a simple example: rolling a fair six-sided dice. We can encode all possible outcomes with the event space $\Omega = \{1, 2, 3, 4, 5, 6\}$. Events such as $A =$ "the outcome is even" or $B =$ "the outcome is larger than 3" are represented by the sets

$$A = \{2, 4, 6\},$$
$$B = \{4, 5, 6\}.$$

As the dice is fair, the probability of each outcome is the same:

$$P(\{1\}) = \cdots = P(\{6\}) = \frac{1}{6}.$$

There are two properties that make such a function $P$ a proper measure of probability:

1. the probability of the event space is one,
2. and the probability of the union of disjoint events is the sum of probabilities.

In our dice-rolling example, this is translated to, for instance,

$$P(\text{the outcome is even}) = P(\{2, 4, 6\})$$
$$= P(\{2\}) + P(\{4\}) + P(\{6\})$$
$$= \frac{1}{6} + \frac{1}{6} + \frac{1}{6}$$
$$= \frac{1}{2}.$$

We'll talk about these properties extensively in the next section. As logical connectives can be represented in the language of set theory, set operations translate the semantics of logic into probabilities. Intersection is joint occurrence of events. Union is the occurrence of either one.

$$P(A \cap B): \text{ probability that } A \text{ AND } B \text{ occurs}$$
$$P(A \cup B): \text{ probability that } A \text{ OR } B \text{ occurs}$$

*Figure 18.3: The probabilities of intersection and union*

In this way, we are able to build models involving uncertainty and develop a calculus to work with said models. In the Tower of Babel that is mathematics, *statistics* deals with the modeling part, and *probability theory* deals with the calculus part.

Even technically well-trained engineers conflate modeling and working with models. For instance, when we talk about flipping fair coins, the probability of heads and tails are both 1/2. Even when we are absolutely sure about the model but have ten heads in a row, most would immediately jump to the conclusion that our coin is biased.

To make sure we are not making this mistake, first, we are going to learn what probability is.

# 18.2 The axioms of probability

In the previous section, we have talked about probability as an extension of mathematical logic. Just like formal logic, probability has its axioms, which we need to understand to work with probability models. Now, we are going to seek the answer to a fundamental question: what is the mathematical model of probability and how do we work with it?

Probabilities are defined in the context of experiments and outcomes. To talk about probabilities, we need to define what we assign probabilities to. Formally speaking, we denote the probability of the event $A$ by $P(A)$. First, we'll talk about what events are.

## 18.2.1 Event spaces and $\sigma$-algebras

Let's revisit the six-sided example from the previous section. There are six different mutually exclusive outcomes (that is, events that cannot occur at the same time), and together they form the *event space*, denoted by $\Omega$:

$$\Omega := \{1, 2, 3, 4, 5, 6\}.$$

In general, the event space is the collection of all mutually exclusive outcomes. It can be any set.

What kind of events can we assign probabilities to? Obviously, the individual outcomes come to mind. However, we can think of events such as *"the result is an odd number"*, *"the result is 2 or 6"*, or *"the result is not 1"*. Following this logic, our expectations are that for any two events $A$ and $B$,

- *"A or B"*,
- *"A and B"*,
- and *"not A"*

are events as well. These can be translated to the language of set theory and are formalized by the notion of $\sigma$-algebras.

---

**Definition 18.2.1 ($\sigma$-algebras)**

Let $\Omega$ be an event space. A collection of its subsets $\Sigma \subseteq 2^\Omega$ is called an $\sigma$-algebra over $\Omega$ if the following properties hold:

*(a)* $\Omega \in \Sigma$. (That is, the set of all outcomes is an event.)

*(b)* For all $A \in \Sigma$, the set $\Omega \setminus A$ is also an element of $\Sigma$. (That is, $\sigma$-algebras are closed to complements.)

*(c)* For all $A_1, A_2, \cdots \in \Sigma$, the set $\cup_{n=1}^\infty A_n$ is also an element of $\Sigma$. (That is, $\sigma$-algebras are closed to unions.)

Since events are modeled by sets, logical concepts such as *"and"*, *"or"*, and *"not"* can be translated into set operations:

- the joint occurrence of events $A$ and $B$ is equivalent to $A \cap B$,
- "$A$ or $B$" is equivalent to $A \cup B$,
- and "not $A$" is equivalent to $\Omega \setminus A$.

An immediate consequence of the definition is that, for any events $A_1, A_2, \cdots \in \Sigma$, their intersection $\cap_{n=1}^{\infty} A_n$ is also a member of $\Sigma$. Indeed, as De Morgan's laws (*Theorem C.2.2*) suggest,

$$\Omega \setminus \left( \cap_{n=1}^{\infty} A_n \right) = \cup_{n=1}^{\infty} \left( \Omega \setminus A_n \right).$$

Since $\Omega \setminus (\Omega \setminus A) = A$, we have

$$\cap_{n=1}^{\infty} A_n = \Omega \setminus \left( \Omega \setminus (\cap_{n=1}^{\infty} A_n) \right)$$
$$= \Omega \setminus \cup_{n=1}^{\infty} \left( \underbrace{\Omega \setminus A_n}_{\in \Sigma} \right).$$

Thus, the defining properties of $\sigma$-algebras guarantee that $\cap_{n=1}^{\infty} A_n$ is indeed an element of $\Sigma$. Another immediate consequence of the definition is that since $\Omega \in \Sigma$, the empty set $\varnothing$ is also a member of $\Sigma$.

At first glance, $\sigma$-algebras seem a bit abstract. As usual, a bit of abstraction now will pay us huge dividends later in our studies. To bring this concept closer, here is a summary of $\sigma$-algebras in English:

- The set of all possible outcomes is an event.
- For any event, it not occurring is an event as well.
- For any events, their joint occurrence is an event as well.
- For any events, at least one of them occurring is an event as well.

Now that we have the formal definition under our belt, let's see the first example.

**Example 1.** Rolling a six-sided dice. There, the $\sigma$-algebra is simply the power set of the event space:

$$\Omega = \{1, 2, 3, 4, 5, 6\}, \quad \Sigma = 2^{\Omega}.$$

(Recall that the power set of $A$ is the set $2^A$ containing all subsets of $A$, as defined in *Definition C.1.1*.) Even though this is one of the simplest examples, it will serve as a prototype and a building block for constructing more complicated event spaces.

**Example 2.** Tossing a coin $n$ times. A single toss has two possible outcomes: heads or tails. For simplicity, we are going to encode heads with 0 and tails with 1. Since we are tossing the coin $n$ times, the result of an experiment will be an $n$-long sequence of ones and zeros, like this: $(0, 1, 1, 1, \ldots, 0, 1)$. Thus, the complete event space is $\Omega = \{0, 1\}^n$.

(We are not talking about probabilities just yet, but feel free to spend some time figuring out how to assign them to these events. Don't worry if this is not clear; we will go through it in detail.)

As in the previous example, the $\sigma$-algebra $2^\Omega$ is a good choice. This covers all events that we need, for instance, *"the number of tails is $k$"*.

In practice, $\sigma$-algebras are rarely given explicitly. Sure, for simple cases such as the above, it is possible.

What about cases where the event spaces are not countable? For instance, suppose that we are picking a random number between 0 and 1. Then, $\Omega = [0, 1]$, but selecting $\Sigma = 2^{[0,1]}$ is extremely problematic. Recall that we want to assign a probability to every event in $\Sigma$. The power set $2^{[0,1]}$ is so large that *very* strange things can occur. In certain scenarios, we can cut up sets into a finite number of pieces and reassemble *two identical copies of the set* from its pieces. (If you are interested in more, check out the Banach-Tarski paradox.)

To avoid weird things like the above-mentioned, we need another way to describe $\sigma$-algebras.

## 18.2.2  Describing $\sigma$-algebras

Let's start with a simple yet fundamental property of $\sigma$-algebras that we'll soon use to give a friendly description of $\sigma$-algebras.

**Theorem 18.2.1**  *(Intersection of event algebras.)*

*Let $\Omega$ be a sample space, and let $\Sigma_1$ and $\Sigma_2$ be two $\sigma$-algebras over it. Then $\Sigma_1 \cap \Sigma_2$ is also a $\sigma$-algebra.*

*Proof.* As we saw in the definition of $\sigma$-algebras (*Definition 18.2.1*), there are three properties we need to verify to show that $\Sigma_1 \cap \Sigma_2$ is indeed a $\sigma$-algebra. This is very simple to check, so I suggest taking a shot by yourself first before reading my explanation.

*(a)* As both $\Sigma_1$ and $\Sigma_2$ are $\sigma$-algebras, $\Omega \in \Sigma_1$ and $\Omega \in \Sigma_2$ both hold. Thus, by definition of the intersection, $\Omega \in \Sigma_1 \cap \Sigma_2$.

*(b)* Let $A \in \Sigma_1 \cap \Sigma_2$. As both of them are $\sigma$-algebras, $\Omega \setminus A \in \Sigma_1$ and $\Omega \setminus A \in \Sigma_2$. Thus, $\Omega \setminus A$ is an element of the intersection as well.

*(c)* Let $A_1, A_2, \cdots \in \Sigma_1 \cap \Sigma_2$ be arbitrary events. We can use the exact same argument as before: since both $\Sigma_1$ and $\Sigma_2$ are $\sigma$-algebras, we have

$$\bigcup_{n=1}^{\infty} A_n \in \Sigma_1 \quad \text{and} \quad \bigcup_{n=1}^{\infty} A_n \in \Sigma_2.$$

So, the union is also a member of the intersection, i.e.,

$$\bigcup_{n=1}^{\infty} A_n \in \Sigma_1 \cap \Sigma_2.$$

With all that, we are ready to describe $\sigma$-algebras with a generating set.

**Theorem 18.2.2** *(Generated $\sigma$-algebras)*

*Let $\Omega$ be an event space and $S \subseteq 2^{\Omega}$ be an arbitrary collection of its sets. Then there is an unique smallest $\sigma$-algebra $\sigma(S)$ that contains $S$.*

(By smallest, we mean that if $\Sigma$ is an $\sigma$-algebra containing $S$, then $\sigma(S) \subseteq \Sigma$.)

*Proof.* Our previous result shows that the intersection of $\sigma$-algebras is also an $\sigma$-algebra. So, let's take all $\sigma$-algebras that contain $S$ and take their intersection. Formally, we define

$$\sigma(S) = \cap\{\Sigma \,:\, \Sigma \text{ is an } \sigma\text{-algebra and } S \subseteq \Sigma\}.$$

By definition, $\sigma(S)$ is clearly the smallest, and it also contains $S$.

It's also clear that the generated $\sigma$-algebra is unique as, if there would be another $\hat{\sigma}(S)$ satisfying the conditions, then due to the construction, $\hat{\sigma}(S) \subseteq \sigma(S)$ and $\sigma(S) \subseteq \hat{\sigma}(S)$, hence $\hat{\sigma}(S) = \sigma(S)$.

Right away, we can use this to precisely construct the $\sigma$-algebra for an extremely common task: picking a number between 0 and 1.

**Example 3.** Selecting a random number between 0 and 1. It is clear that the event space is $\Omega = [0, 1]$. What about the events? In this situation, we want to ask questions such as the probability of a random number $X$ falling between some $a, b \in [0, 1]$. That is, events such as $(a, b), (a, b], [a, b), [a, b]$ (whether or not we want strict inequality regarding $a$ and $b$).

So, a proper $\sigma$-algebra can be given by the algebra generated by events of the form $(a, b]$. That is,

$$\Sigma = \sigma\big(\{(a, b] \,:\, 0 \le a < b \le 1\}\big).$$

This $\Sigma$ has a rich structure. For instance, it contains simple events such as $\{x\}$, where $x \in [0, 1]$, but also more complex ones such as "$X$ is a rational number" or "$X$ is an irrational number". Give yourself a few minutes to see why this is true. Don't worry if you don't see the solution—we'll work this out in the problems section. (If you think this through, you'll also see why we chose intervals of the form $(a, b]$ instead of others such as $(a, b)$ or $[a, b]$.)

## 18.2.3  $\sigma$-algebras over real numbers

From all the examples we have seen so far, it is clear that most commonly, we define probability spaces on $\mathbb{N}$ or on $\mathbb{R}$. When $\Omega \subseteq \mathbb{N}$, the choice of $\sigma$-algebra is clear, as $\Sigma = 2^\Omega$ will always work.

However, as suggested in Example 3 above, selecting $\Sigma = 2^\Omega$ when $\Omega \subseteq \mathbb{R}$ can lead to some weird stuff. Because we are interested in the probability of events such as $[a, b]$, our standard choice is going to be the generated $\sigma$-algebra

$$\mathcal{B}(\mathbb{R}) = \sigma\big(\{(a, b) : a, b \in \mathbb{R}\}\big), \tag{18.1}$$

called the Borel algebra, named after the famous French mathematician Émile Borel. Due to its construction, $\mathcal{B}$ contains all events that are important to us, such as intervals and unions of intervals. Elements of $\mathcal{B}$ are called Borel sets.

Because $\sigma$-algebras are closed to unions, you can see that all types of intervals can be found in $\mathcal{B}(\mathbb{R})$. This is summarized by the following theorem.

**Theorem 18.2.3**

*For all $a, b \in \mathbb{R}$, the sets $[a, b]$, $(a, b]$, $[a, b)$, $(-\infty, a]$, $(-\infty, a)$, $(a, \infty)$, and $[a, \infty)$ are elements of $\mathcal{B}(\mathbb{R})$.*

As an exercise, try to come up with the proof by yourself. One trick to get the ideas flowing is to start drawing some figures. If you can visualize what happens, you'll discover a proof quickly.

*Proof.* In general, for a given set $S$, we can show that it belongs to $\mathcal{B}(\mathbb{R})$ by writing it as the union/intersection/difference of known Borel sets. First, we have

$$(a, \infty) = \bigcup_{n=1}^{\infty} (a, n),$$

so $(a, \infty) \in \mathcal{B}(\mathbb{R})$. With a similar argument, we see that $(-\infty, a) \in \mathcal{B}(\mathbb{R})$.

Next,

$$(-\infty, a] = \mathbb{R} \setminus (a, \infty),$$

$$[a, \infty) = \mathbb{R} \setminus (-\infty, a),$$

so $(-\infty, a], [a, \infty) \in \mathcal{B}(\mathbb{R})$ for all $a$. From these, the sets $[a, b], (a, b], [a, b)$ can be produced by intersections.

Now that we understand what events and $\sigma$-algebras are, we can take our first detailed look at *probability*. In the next section, we will introduce its precise mathematical definition.

## 18.2.4   Probability measures

Let's recap what we have learned so far! In the language of mathematics, experiments with intrinsic uncertainty are described using outcomes, event spaces, and events.

The collection of all possible mutually exclusive outcomes of an experiment is called the *event space*, denoted by $\Omega$. Certain subsets of $\Omega$ are called *events*, to which we want to assign probabilities.

These events form what is known as a $\sigma$-*algebra*, denoted by $\Sigma$. We denote the probability of an event $A$ by $P(A)$.

Intuitively speaking, we have three reasonable expectations about probability:

1. $P(\Omega) = 1$, that is, the probability that at least one outcome occurs is 1. In other words, our event space is a complete description of the experiment.

2. $P(\emptyset) = 0$, that is, the probability that *none* of the outcomes occur is 0. Again, this means that our event space is complete.

3. The probability that either of two events occurs for two mutually exclusive events is the sum of the individual probabilities.

These are formalized by the following definition.

> **Definition 18.2.2 (Probability measures and spaces)** Let $\Omega$ be an event space and $\Sigma$ be an $\sigma$-algebra over $\Omega$. We say that the function
>
> $P : \Sigma \to [0, 1]$ is a *probability measure* on $\Sigma$ if the following properties hold:
>
> *(a)* $P(\Omega) = 1$.
>
> *(b)* If $A_1, A_2, \ldots$ are mutually disjoint events (that is, $A_i \cap A_j = \emptyset$ for all $i \neq j$), then $P(\cup_{n=1}^{\infty} A_n) = \sum_{n=1}^{\infty} P(A_n)$. This property is called the $\sigma$-additivity of probability measures.
>
> Along with the probability measure $P$, the structure $(\Omega, \Sigma, P)$ is said to form a *probability space*.

As usual, let's see some concrete examples first! We are going to continue with the ones we worked out when discussing $\sigma$-algebras.

**Example 1, continued.** Rolling a six-sided dice. Recall that the event space and algebra were defined by

$$\Omega = \{1, 2, 3, 4, 5, 6\}, \quad \Sigma = 2^{\Omega}.$$

If we don't have any extra knowledge about our dice, it is reasonable to assume that each outcome is equally probable. That is, since there are six possible outcomes, we have

$$P(\{1\}) = \cdots = P(\{6\}) = \frac{1}{6}.$$

Notice that, in this case, knowing the probabilities for the individual outcomes is enough to determine the probability of any event. This is due to the ($\sigma$-)additivity of the probability. For instance, the event *"the outcome of the dice roll is an odd number"* is described by

$$P(\{1, 3, 5\}) = P(\{1\}) + P(\{3\}) + P(\{5\}) = \frac{3}{6}.$$

In English, the probability of any event can be written down with the following formula:

$$P(\text{event}) = \frac{\text{favorable outcomes}}{\text{all possible outcomes}}.$$

You might remember this from your elementary and high school studies (depending on the curriculum in your country). This is a useful formula, but there is a caveat: it only works if we assume that each outcome has an equal probability.

In the case when our dice is not uniformly weighted, the occurrences of individual outcomes are not equal. (Just think of a lead dice, where one side is significantly heavier than the others.) For now, we are not going to be concerned with this case. Later, this generalization will be discussed in detail.

**Example 2, continued.** Tossing a coin $n$ times. Here, our event space and algebra was $\Omega = \{0, 1\}^n$ and $\Sigma = 2^{\Omega}$. For simplicity, let's assume that $n = 5$.

What is the probability of a particular result, say HHTTT? Going step by step, the probability that the first toss will be *heads* is $1/2$. That is,

$$P(\text{first toss is heads}) = \frac{1}{2}$$

Since the first toss is independent of the second,

$$P(\text{second toss is heads}) = \frac{1}{2}$$

as well. To combine this and calculate the probability that the *first two* tosses are both heads, we can think in the following way. Among the outcomes where the first toss is heads, exactly half of them will have the second toss heads as well. So, we are looking for the *half of the half*. That is,

$$P(\text{first two tosses are heads}) = P(\text{first toss is heads})P(\text{second toss is heads})$$
$$= \frac{1}{4}.$$

Going further with the same logic, we obtain

$$P(\text{HHTTT}) = \frac{1}{2^5} = \frac{1}{32}.$$

If we look a bit deeper, we will notice that this follows the previously seen "favorable/all" formula. Indeed, as we can see with a bit of combinatorics, there are $2^5$ total possibilities, all of them having equal probability.

Considering this, what is the probability that out of our five tosses, exactly two of them are heads? In the language of sets, we can encode each five-toss experiment as a subset of $\{1, 2, 3, 4, 5\}$, the elements signifying the toss that resulted in heads. (So, for example, $\{1, 4, 5\}$ would encode the outcome HTTHH.) With this, the experiments when there are two heads are exactly the two-element subsets of $\{1, 2, 3, 4, 5\}$.

From our combinatorics studies, we know that the number of subsets with given elements is

$$\binom{n}{k} = \frac{n!}{k!(n-k)!},$$

where $n$ is the size of our set and $k$ is the desired size of the subsets. So, in total, there are $\binom{5}{2}$ number of occurrences with exactly two heads. Thus, following the "favorable/all" formula, we have

$$P(\text{two heads out of five tosses}) = \binom{5}{2}\frac{1}{32} = \frac{10}{32}.$$

One more example, and we are ready to move forward.

**Example 3, continued.** Selecting a random number between 0 and 1. Here, our event space was $\Omega = [0, 1]$, and our $\sigma$-algebra was the generated algebra

$$\Sigma = \sigma\big(\{(a, b] : 0 \leq a < b \leq 1\}\big).$$

Without any further information, it is reasonable to assume that every number can be selected with an equal probability. What does this even mean for an infinite event space such as $\Omega = [0, 1]$? We can't divide 1 into infinitely many equal parts.

So, instead of thinking about individual outcomes, we should start thinking about events. Let's denote our randomly selected number with $X$. If all numbers are "equally likely", what is $P(X \in (0, 1/2])$? Intuitively, given our equally likely hypothesis, this probability should be proportional to the size of $[0, 1/2]$. Thus,

$$P(X \in I) = |I|,$$

where $I$ is some interval and $|I|$ is its length. For instance,

$$\begin{aligned} P(a < X < b) &= P(a \leq X < b) \\ &= P(a < X \leq b) \\ &= P(a \leq X \leq b) \\ &= b - a. \end{aligned}$$

By giving the probabilities on the generating set of the $\sigma$-algebra, the probabilities for all other events can be deduced. For instance,

$$\begin{aligned} P(X = x) &= P(0 \leq X \leq x) - P(0 \leq X < x) \\ &= x - x \\ &= 0. \end{aligned}$$

Thus, the probability of picking a given number is zero. There is an important lesson here: **events with zero probability can happen.** This sounds counterintuitive at first, but based on the above example, you can see that it is true.

## 18.2.5  Fundamental properties of probability

Now that we are familiar with the mathematical model of probability, we can start working with them. Manipulating expressions of probabilities gives us the ability to deal with complex scenarios.

If you recall, probability measures had three simple defining properties (see *Definition 18.2.2*):

*(a)* $P(\Omega) = 1$,

*(b)* $P(\emptyset) = 0$, and

*(c)* $P\left(\bigcup_{n=1}^{\infty} A_n\right) = \sum_{n=1}^{\infty} P(A_n)$, if the events $A_n$ are mutually disjoint.

From these properties, many others can be deduced. For simplicity, here is a theorem summarizing the most important ones.

---

**Theorem 18.2.4**

*Let $(\Omega, \Sigma, P)$ be a probability space and let $A, B \in \Sigma$ be two arbitrary events.*

*(a)* $P(A \cup B) = P(A) + P(B) - P(A \cap B)$.

*(b)* $P(A) = P(A \cap B) + P(A \setminus B)$. Specifically, $P(\Omega \setminus A) + P(A) = 1$.

*(c)* If $A \subseteq B$, then $P(A) \leq P(B)$.

---

The proof of this is so simple that it is left to you as an exercise. All of these follow from the additivity of probability measures with respect to disjoint events. (If you don't see the solution, sketch some Venn diagrams!)

Another fundamental tool is the *law of total probability*, which is used all the time when dealing with more complex events.

---

**Theorem 18.2.5**  *(Law of total probability)*

*Let $(\Omega, \Sigma, P)$ be a probability space and let $A \in \Sigma$ be an arbitrary event. If $A_1, A_2, \cdots \in \Sigma$ are mutually disjoint events (that is, $A_i \cap A_j = \emptyset$ if $i \neq j$) for which $\cup_{n=1}^{\infty} A_n = \Omega$, then*

$$P(A) = \sum_{n=1}^{\infty} P(A \cap A_n). \tag{18.2}$$

*We call mutually disjoint events whose union is the entire event space partitions.*

---

*Proof.* This simply follows from the $\sigma$-additivity of probability measures. Feel free to give the proof a shot by yourself to test your understanding.

If you can't see this, no worries. Here is a brief explanation. Since $A_1, A_2, \ldots$ are mutually disjoint, $A \cap A_1, A \cap A_2, \ldots$ are mutually disjoint as well. Moreover, since $\cup_{n=1}^{\infty} A_n = \Omega$, we also have

$$\bigcup_{n=1}^{\infty} (A_n \cap A) = \left( \bigcup_{n=1}^{\infty} A_n \right) \cap A$$

$$= \Omega \cap A$$

$$= A.$$

Thus, the $\sigma$-additivity of probability measures implies that

$$P(A) = \sum_{n=1}^{\infty} P(A \cap A_n),$$

which is what we had to show.

Let's see an example right away! Suppose that we toss *two* dice. What is the probability that the sum of the results is 7?

First, we should properly describe the probability space. For notational simplicity, let's denote the result of the throws with $X$ and $Y$. What we are looking for is $P(X + Y = 7)$. Modeling the toss with two dice is the simplest if we impose order among them. With this in mind, the event space $\Omega$ is described by the Cartesian product

$$\Omega = \{1, 2, 3, 4, 5, 6\} \times \{1, 2, 3, 4, 5, 6\}$$

$$= \{ (i, j) : i, j \in \{1, 2, 3, 4, 5, 6\} \},$$

and the outcomes are tuples of the form $(i, j)$. (That is, the tuple $(i, j)$ encodes the elementary event $\{X = i, Y = j\}$.) Since the tosses are independent of each other,

$$P(X = i, Y = j) = \frac{1}{6} \cdot \frac{1}{6} = \frac{1}{36}.$$

(When it is clear, we omit the brackets of the event $\{X = i, Y = j\}$.)

Since the first throw falls between 1 and 6, we can partition the event space by forming

$$A_n := \{X = n\}, \quad n = 1, \ldots, 6.$$

Thus, the law of total probability gives

$$P(X + Y = 7) = \sum_{n=1}^{6} P(\{X + Y = 7\} \text{ and } \{X = n\}).$$

However, if we know that $X + Y = 7$ and $X = n$, we know that $Y = 7 - n$ must hold as well. So, continuing the calculation above,

$$\begin{aligned} P(X + Y = 7) &= \sum_{n=1}^{6} P(\{X + Y = 7\} \text{ and } \{X = n\}) \\ &= \sum_{n=1}^{6} P(X = n, Y = 7 - n) \\ &= \sum_{n=1}^{6} \frac{1}{36} \\ &= \frac{1}{6}. \end{aligned}$$

So, the law of total probability helps us deal with complex events by decomposing them into simpler ones. We have seen this pattern dozens of times now, and once again, it proves to be essential.

As yet another consequence of $\sigma$-additivity, we can calculate the probability of an increasing sequence of events by taking the limit.

**Theorem 18.2.6** *(Lower continuity of probability measures)*

*Let $(\Omega, \Sigma, P)$ be a probability space and let $A_1 \subseteq A_2 \subseteq \cdots \in \Sigma$ be an increasing sequence of events. Then,*

$$P(\cup_{n=1}^{\infty} A_n) = \lim_{n \to \infty} P(A_n) \tag{18.3}$$

*holds. This property is called the lower continuity of probability measures.*

*Proof.* Since the events are increasing, that is, $A_{n-1} \subseteq A_n$, we can write $A_n$ as

$$A_n = A_{n-1} \cup (A_n \setminus A_{n-1}),$$

where $A_{n-1}$ and $A_n \setminus A_{n-1}$ are disjoint.

Thus,

$$\bigcup_{n=1}^{\infty} A_n = \bigcup_{n=1}^{\infty} (A_n \setminus A_{n-1}), \quad A_0 := \emptyset,$$

which gives

$$
\begin{aligned}
P(\cup_{n=1}^{\infty} A_n) &= \sum_{n=1}^{\infty} P(A_n \setminus A_{n-1}) \\
&= \lim_{N \to \infty} \sum_{n=1}^{N} P(A_n \setminus A_{n-1}) \\
&= \lim_{N \to \infty} \sum_{n=1}^{N} \left[ P(A_n) - P(A_{n-1}) \right] \\
&= \lim_{N \to \infty} P(A_N),
\end{aligned}
$$

where we used that $P(\emptyset) = 0$.

We can state an analogue of the above theorem for a decreasing sequence of events.

**Theorem 18.2.7** *(Upper continuity of probability measures)*

*Let $(\Omega, \Sigma, P)$ be a probability space and let $A_1 \supseteq A_2 \supseteq \cdots \in \Sigma$ be a decreasing sequence of events. Then,*

$$P(\cap_{n=1}^{\infty} A_n) = \lim_{n \to \infty} P(A_n) \tag{18.4}$$

*holds. This property is called the upper continuity of probability measures.*

*Proof.* For simplicity, let's denote the infinite intersection by $A := \cap_{n=1}^{\infty} A_n$.

By defining $B_n := A_1 \setminus A_n$, we have $\cup_{n=1}^{\infty} B_n = A_1 \setminus A$. Since $A_n$ is decreasing, $B_n$ is increasing, so we can apply *Theorem 18.2.6* to obtain

$$
\begin{aligned}
P(A_1 \setminus A) &= \lim_{n \to \infty} P(A_1 \setminus A_n) \\
&= P(A_1) - \lim_{n \to \infty} P(A_n).
\end{aligned}
$$

Since $P(A_1 \setminus A) = P(A_1) - P(A)$, we obtain $P(A) = \lim_{n \to \infty} P(A_n)$.

Now that we have a mathematical definition of a probabilistic model, it is time to take a step toward the space where machine learning is done: $\mathbb{R}^n$.

## 18.2.6 Probability spaces on $\mathbb{R}^n$

In machine learning, every data point is an elementary outcome, located somewhere in the Euclidean space $\mathbb{R}^n$. Because of this, we are interested in modeling experiments there.

How can we define a probability space on $\mathbb{R}^n$? As we did with the real line in *Section 18.2.1*, we describe a convenient $\sigma$-algebra by generating. There, we can use the higher dimensional counterpart of the $(a, b)$ intervals: $n$-dimensional spheres. For this, we define the set

$$B(\mathbf{x}, r) := \{\mathbf{y} \in \mathbb{R}^n : \|\mathbf{x} - \mathbf{y}\| < r\},$$

where $\mathbf{x}$ is the center of the sphere, $r > 0$ is its radius, and $\|\cdot\|$ denotes the usual Euclidean norm. (The $B$ denotes the word *ball*. In mathematics, $n$-dimensional spheres are often called balls.) Similar to the real line, the Borel algebra is defined by

$$\mathcal{B}(\mathbb{R}^n) := \sigma\big(\{B(\mathbf{x}, r) : \mathbf{x} \in \mathbb{R}^n, r > 0\}\big). \tag{18.5}$$

As we saw on the real line (see *Section 18.2.3*), the structure of $\mathcal{B}(\mathbb{R}^n)$ is richer than what the definition suggests at first glance. Here, the analogue of interval is a rectangle, defined by

$$\begin{aligned}(\mathbf{a}, \mathbf{b}) &= (a_1, b_1) \times \cdots \times (a_n, b_n) \\ &= \{\mathbf{x} \in \mathbb{R}^n : a_i < x_i < b_i, i = 1, \ldots, n\},\end{aligned}$$

where $A \times B$ is the Cartesian product. (See *Definition C.3.1*.) Similarly, we can define $[a, b], (a, b], [a, b)$, and others.

**Theorem 18.2.8**

*For any* $\mathbf{a}, \mathbf{b} \in \mathbb{R}^n$, *the sets* $[\mathbf{a}, \mathbf{b}], [\mathbf{a}, \mathbf{b}), (\mathbf{a}, \mathbf{b}], (\mathbf{a}, \infty), [\mathbf{a}, \infty), (-\infty, \mathbf{a}), (-\infty, \mathbf{b}]$ *are elements of* $\mathcal{B}(\mathbb{R}^n)$.

*Proof.* The proof goes along the same line as the counterpart for $\mathcal{B}(\mathbb{R})$. As such, it is left as an exercise to you.

As a hint, first, we can show that $(\mathbf{a}, \mathbf{b})$ can be written as a countable union of balls. We can also show that this holds true for sets such as

$$\mathbb{R} \times \cdots \times \underbrace{(-\infty, a_i)}_{i\text{-th component}} \times \cdots \times \mathbb{R}$$

as well. From these two, we can write the others as unions/intersections/differences.

As an example, let's throw a few darts at a rectangular wall. Suppose that we are terrible darts players and hitting any point on the wall is equally likely.

We can model this event space with $\Omega = [0,1] \times [0,1] \subseteq \mathbb{R}^2$, representing our wall. What are the possible events? For instance, there is a circular darts board hanging on the wall, and we want to find the probability of hitting it. In this scenario, we can restrict the Borel sets defined by (18.5) to

$$\mathcal{B}(\Omega) := \{A \cap \Omega : A \in \mathcal{B}(\mathbb{R}^n)\}.$$

Now that the event space and algebra is clear, we need to think about assigning probabilities. Our assumption is that hitting any point is equally likely. So, by generalizing the $\frac{\text{favorable outcomes}}{\text{all possible outcomes}}$ formula we have seen in the discrete case, we define the probability measure by

$$P(A) = \frac{\text{volume}(A)}{\text{volume}(\Omega)}.$$

(In two dimensions, we have the area instead of volume.) This is illustrated by *Figure 18.4*.

*Figure 18.4: Probability space of throwing darts at a wall. Source: https://unsplash.com/photos/black-and-white-round-logo-i3WlrO7oAHA*

As we shall see later, this is a special case of *uniform distributions*, one of the most prevalent distributions in probability theory. However, there is a lot to talk about until then. Before we conclude our discussion of the fundamentals of probability, let's discuss how can we interpret them.

## 18.2.7 How to interpret probability

Now that we know how to work with probabilities, it is time to study how can we assign probabilities to real-life events.

First, we are going to take a look at the *frequentist* interpretation, explaining probabilities with relative frequencies. (If you are one of those people who are religious about this question, calm down. We'll discuss the Bayesian interpretation in detail, but it is not time yet.)

Let's go back to the beginning and consider the coin-tossing experiment. If I toss a fair coin 1000 times, how many of them will be heads? Most people immediately answer 500, but this is not correct. There is no right answer, as any number of heads between 0 and 1000 can happen. Of course, most probably it will be around 500, but with a very small probability, there can be zero heads as well.

In general, the probability of an event describes its *relative frequency* among infinitely many attempts. That is,

$$P(\text{event}) \approx \frac{\text{number of occurrences}}{\text{number of attempts}}.$$

When the number of attempts goes toward infinity, the relative frequency of occurrences converges to the true underlying probability. In other words, if $X_i$ quantitatively describes our *i*-th attempt by

$$X_i = \begin{cases} 1 & \text{if the event occurs} \\ 0 & \text{otherwise,} \end{cases}$$

then

$$P(\text{event}) = \lim_{n \to \infty} \frac{X_1 + \cdots + X_n}{n}.$$

We can illustrate this by doing a quick simulation using the coin-tossing example. Don't worry if you don't understand the code; we'll talk about it in detail in the next chapters.

```python
import numpy as np
from scipy.stats import randint


n_tosses = 1000
# coin tosses: 0 for tails and 1 for heads
coin_tosses = [randint.rvs(low=0, high=2) for _ in range(n_tosses)]
averages = [np.mean(coin_tosses[:k+1]) for k in range(n_tosses)]
```

Let's plot the results for some insight:

```python
import matplotlib.pyplot as plt


with plt.style.context("seaborn-v0_8"):
    plt.figure(figsize=(16, 8))
    plt.title("Relative frequency of the coin tosses")
    plt.xlabel("Number of tosses")
    plt.ylabel("Relative frequency")

    # plotting the averages
    plt.plot(range(n_tosses), averages, linewidth=3)  # the averages

    # plotting the true expected value
    plt.plot([-100, n_tosses+100], [0.5, 0.5], c="k")
    plt.xlim(-10, n_tosses+10)
    plt.ylim(0, 1)
    plt.show()
```

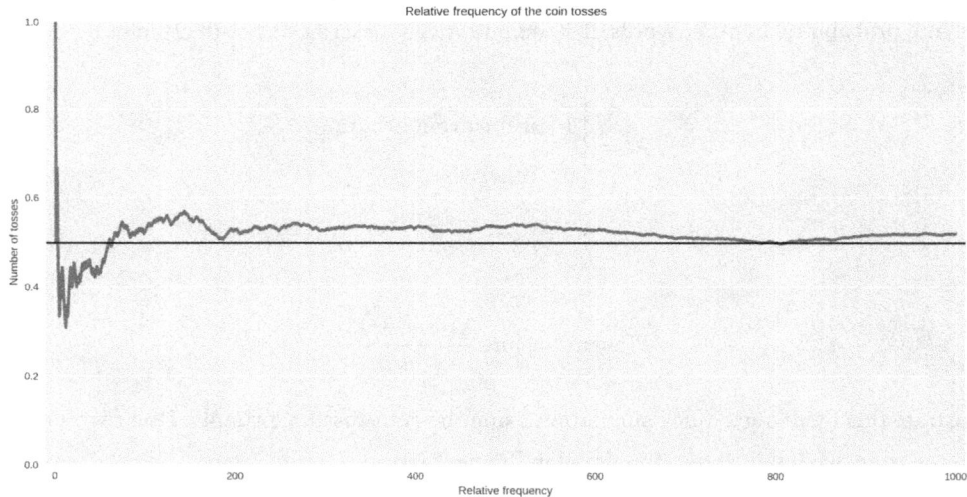

*Figure 18.5: The relative frequency of coin tosses*

The relative frequency quite nicely stabilizes around 1/2, which is the true probability of our fair coin landing on its heads. Is this an accident? No.

We will make all of this mathematically precise when talking about *the law of large numbers* in *Section 20.5*, but first, we'll introduce the Bayesian viewpoint, a probabilistic framework for updating our models given new observations.

# 18.3 Conditional probability

In the previous sections, we learned the foundations of probability. Now we can speak in terms of outcomes, events, and chances. However, in real-life applications, these basic tools are not enough to build useful predictive models.

To illustrate this, let's build a probabilistic spam filter! For every email we receive, we want to estimate the probability $P(\text{email is spam})$. The closer this is to 1, the more likely that we are looking at a spam email.

Based on our inbox, we might calculate the relative frequency of spam emails and obtain that

$$P(\text{email is spam}) \approx \frac{\text{number of our spam emails}}{\text{number of emails in our inbox}}.$$

However, this doesn't help us at all. Based on this, we can randomly discard every email with probability $P(\text{email is spam})$, but that would be a horrible spam filter.

To improve, we need to dig a bit deeper. When analyzing spam emails, we start to notice patterns. For instance, the phrase "act now" can be found almost exclusively in spam. After a quick count, we get that

$$P(\text{email containing "act now" is a spam}) = \frac{\#\text{spam emails with the phrase "act now"}}{\#\text{emails with the phrase "act now"}}$$

$$\approx 0.95.$$

This looks much more useful for our spam filtering efforts. By checking for the presence of the phrase "act now", we can confidently classify an email as spam.

Of course, there is much more to spam filtering, but this example demonstrates the importance of probabilities *conditional* on other events. To put this into mathematical form, we introduce the following definition.

**Definition 18.3.1 (Conditional probability)**

Let $(\Omega, \Sigma, P)$ be a probability space, let $A, B \in \Sigma$ be two events, and suppose that $P(A) > 0$. The *conditional probability* of $B$ given $A$ is defined by

$$P(B \mid A) := \frac{P(A \cap B)}{P(A)}.$$

You can think about $P(B \mid A)$ as restricting the event space to $A$, as illustrated by *Figure 18.6*.

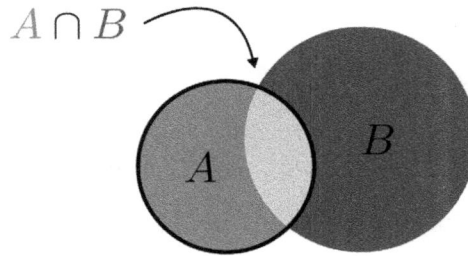

$$P(B \mid A) = \frac{P(A \cap B)}{P(A)}$$

*Figure 18.6: A visual representation of conditional probability*

When there are more conditions, say $A_1$ and $A_2$, the definition takes the form

$$P(B \mid A_1, A_2) = \frac{P(B \cap A_1 \cap A_2)}{P(A_1 \cap A_2)},$$

and so on for even more events.

To bring this concept closer, let's revisit the dice-rolling experiment. Suppose that your friend rolls a six-sided dice and tells you that the outcome is an odd number. Given this information, what is the probability that the result is 3? For simplicity, let's denote the outcome of the roll with $X$. Mathematically speaking, this can be calculated by

$$
\begin{aligned}
P(X = 3 \mid X \in \{1, 3, 5\}) &= \frac{P(X = 3 \text{ and } X \in \{1, 3, 5\})}{P(X \in \{1, 3, 5\})} \\
&= \frac{P(X = 3)}{P(X \in \{1, 3, 5\})} \\
&= \frac{1/6}{1/2} \\
&= \frac{1}{3}.
\end{aligned}
$$

This is the number that we expected.

Although this simple example doesn't demonstrate the usefulness of conditional probability, this is a cornerstone in machine learning. In essence, learning from data can be formulated as estimating $P(\text{label} \mid \text{data})$. We are going to expand on this idea later in this chapter.

# 18.3.1 Independence

The idea behind conditional probability is that observing certain events changes the probability of others. Is this always the case, though?

In probabilistic modeling, recognizing when observing an event doesn't influence another is equally important. This motivates the concept of *independence*.

> **Definition 18.3.2 (Independence of events)**
>
> Let $(\Omega, \Sigma, P)$ be a probability space and let $A, B \in \Sigma$ be two events. We say that $A$ and $B$ are *independent* if
>
> $$P(A \cap B) = P(A)P(B)$$
>
> holds.

Equivalently, this can be formulated in terms of conditional probabilities. By the definition, if $A$ and $B$ are independent, we have

$$
\begin{aligned}
P(B \mid A) &= \frac{P(A \cap B)}{P(A)} \\
&= \frac{P(A)P(B)}{P(A)} \\
&= P(B).
\end{aligned}
$$

To see an example, let's go back to coin tossing and suppose that we toss a coin two times. Let the result of the first and second toss be denoted by $X_1$ and $X_2$, respectively. What is the probability that both of these tosses are heads? As we saw when discussing this example in *Section 18.2.4*, we can see that

$$P(X_1 = \text{ heads and } X_2 = \text{ heads}) = P(X_1 = \text{ heads})P(X_2 = \text{ heads}) = \frac{1}{4}.$$

That is, the two events are independent of each other.

Regarding probability, there are many common misconceptions. One is about the interpretation of independence. Suppose that I toss a fair coin ten times, all of them resulting in heads. What is the probability that my next toss will be heads?

Most would immediately conclude that this must be very small since having eleven heads in a row is highly unlikely. However, once we have the ten results available, we no longer talk about the probability of eleven coin tosses, just the last one! Since the coin tosses are independent of each other, the chance of heads for the eleventh toss (given the results of the previous ten) is still 50%.

This phenomenon is called the *gambler's fallacy*, and I am pretty sure that at some point in your life, you fell victim to it. (I sure did.)

In practical scenarios, working with conditional probabilities might be easier. (For instance, sometimes we can estimate them directly, while the standard probabilities are difficult to gauge.) Because of this, we need tools to work with them.

## 18.3.2   The law of total probability revisited

Remember the law of total probability from *Theorem 18.2.5?* We can use conditional probabilities to put it into a slightly different form.

---

**Theorem 18.3.1**  *(Law of total probability, conditional version)*

*Let* $(\Omega, \Sigma, P)$ *be a probability space and let* $A \in \Sigma$ *be an arbitrary event. If* $A_1, A_2, \cdots \in \Sigma$ *are mutually disjoint events (that is,* $A_i \cap A_j = \varnothing$ *if* $i \neq j$*) for which* $\cup_{n=1}^{\infty} A_n = \Omega$*, then*

$$P(A) = \sum_{k=1}^{\infty} P(A \mid A_k)P(A_k). \tag{18.6}$$

---

*Proof.* The proof is the trivial application of the law of total probability (*Theorem 18.2.5*) and the definition of conditional probabilities: as $P(A \cap A_k) = P(A \mid A_k)P(A_k)$,

$$P(A) = \sum_{k=1}^{\infty} P(A \cap A_k)$$
$$= \sum_{k=1}^{\infty} P(A \mid A_k)P(A_k)$$

holds, which is what we had to show.

Why is this useful for us? Let's demonstrate this with an example. Suppose that we have three urns containing light and dark colored balls.

The first one contains 4 dark, the second one contains 2 light and 2 dark, while the last one contains 1 light and 3 dark balls.

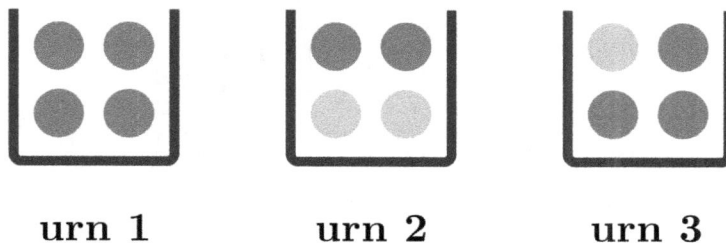

urn 1          urn 2          urn 3

*Figure 18.7: Urns with colored balls*

We randomly pick an urn; however, picking the first one is twice as likely as picking the other two. (That is, we pick the first urn 50% of the time, while the second and the third 25%–25% of the time.) From that urn, we also randomly pick a ball. What is the probability that we select a light ball? Without using the law of total probability, this is difficult to compute.

Let's denote the color of the selected ball by $X$ and suppose that the event $A_n$ describes picking the $n$-th urn. Then, we have

$$P(X = \text{light}) = \sum_{k=1}^{3} P(\{X = \text{light}\} \cap A_k) = \sum_{k=1}^{3} P(X = \text{light} \mid A_k)P(A_k).$$

Without using conditional probabilities, calculating $P(\{X = \text{light}\} \cap A_k)$ is difficult (since we are not picking each urn with equal probability). However, we can simply calculate the conditionals by counting the number of light balls in each urn. That is, we have

$$P(X = \text{light} \mid A_1) = 0$$
$$P(X = \text{light} \mid A_2) = \frac{2}{4}$$
$$P(X = \text{light} \mid A_3) = \frac{1}{4}.$$

Since $P(A_1) = 1/2, P(A_2) = 1/4,$ and $P(A_3) = 1/4$, the probability we are looking for is

$$P(X = \text{light}) = \sum_{k=1}^{3} P(X = \text{light} \mid A_k)P(A_k)$$

$$= 0 \cdot \frac{1}{2} + \frac{2}{4} \cdot \frac{1}{4} + \frac{1}{4} \cdot \frac{1}{4}$$

$$= \frac{3}{16}.$$

Note that because the urns are not selected with equal probability,

$$P(X = \text{light}) \neq \frac{\text{number of light balls}}{\text{number of balls}},$$

as one would naively guess.

Another useful property of the conditional probability is that, due to its definition, we can use it to express the *joint probability* of events:

$$P(A \cap B) = P(B \mid A)P(A).$$

Even though this sounds trivial, there are cases when we can estimate/compute the conditional probability but not the joint probability. In fact, this simple identity can be generalized for an arbitrary number of conditions. This is called the *chain rule*. (Despite its name, it has nothing to do with the chain rule for differentiation.)

**Theorem 18.3.2**  *(The chain rule)*

Let $(\Omega, \Sigma, P)$ *be a probability space and* $A_1, A_2, \cdots \in \Sigma$ *be arbitrary events. Then,*

$$P(A_1 \cap \cdots \cap A_n) = \prod_{k=1}^{n} P(A_k \mid A_1, \ldots, A_{k-1}) \qquad (18.7)$$

*holds.*

*Proof.* First, we notice that $P(A_1 \cap \cdots \cap A_n)$ can be written as

$$P(A_1 \cap \cdots \cap A_n) = \frac{P(A_1 \cap \cdots \cap A_n)}{P(A_1 \cap \cdots \cap A_{n-1})} \frac{P(A_1 \cap \cdots \cap A_{n-1})}{P(A_1 \cap \cdots \cap A_{n-2})} \cdots \frac{P(A_1 \cap A_2)}{P(A_1)} P(A_1),$$

because the terms cancel out each other.

Since

$$\frac{P(A_1 \cap A_k)}{P(A_1 \cap A_{k-1})} = P(A_k \mid A_1, \ldots, A_{k-1}),$$

the chain rule (18.7) follows.

## 18.3.3 The Bayes' theorem

In essence, machine learning is about turning observations into predictive models. Probability theory gives us a language to express our models. For instance, going back to our spam filter example, we can notice that 5% of our emails are spam. However, this is not enough information to filter out spam emails. Upon inspection, we have observed that 95% of emails that contain the phrase "act now" are spam (but only 1% of all the emails contain "act now"). In the language of conditional probabilities, we have concluded that

$$P(\text{spam} \mid \text{contains "act now"}) = 0.95.$$

With this, we can start looking for emails containing the phrase "act now" and discard them with 95% confidence. Is this spam filter effective? Not really, since there can be other frequent keywords in spam mails that we don't check. How can we check this?

For one, we can take a look at the conditional probability $P(\text{contains "act now"} \mid \text{spam})$, describing the frequency of the "act now" keyword among all the spam emails. A low frequency means that we are missing out on other keywords that we can use for filtering.

Generally speaking, we often want to compute/estimate the quantity $P(A \mid B)$, but our observations only allow us to infer $P(B \mid A)$. So, we need a way to reverse the condition and the event. With a bit of algebra, we can do this easily.

**Theorem 18.3.3** *(The Bayes' formula)*

*Let $(\Omega, \Sigma, P)$ be a probability space, $A, B$ be two arbitrary events, and suppose that $P(A), P(B) > 0$. Then,*

$$P(B \mid A) = \frac{P(A \mid B)P(B)}{P(A)} \tag{18.8}$$

*holds.*

*Proof.* By the definition of conditional probability, we have

$$P(B \mid A) = \frac{P(A \cap B)}{P(A)}$$
$$= \frac{P(A \cap B)P(B)}{P(B)P(A)}$$
$$= P(A \mid B)\frac{P(B)}{P(A)},$$

which is what we had to show.

To see how it works in action, let's put it to the test in our spam filtering example. Given the information we know, we have

$$P(\text{spam} \mid \text{contains "act now"}) = 0.95,$$
$$P(\text{contains "act now"}) = 0.01,$$
$$P(\text{spam}) = 0.05.$$

So, according to the Bayes' formula,

$$P(\text{contains "act now"} \mid \text{spam}) = \frac{P(\text{spam} \mid \text{contains "act now"})P(\text{contains "act now"})}{P(\text{spam})}$$
$$= \frac{0.95 \cdot 0.01}{0.05}$$
$$= 0.19.$$

Thus, by filtering only for the phrase "act now", we are missing a *lot* of spam.

We can take the Bayes' formula one step further by combining it with the law of total probability in *Theorem 18.3.1*. (See the equation (18.6).)

**Theorem 18.3.4** *(The Bayes' theorem)*

Let $(\Omega, \Sigma, P)$ be a probability space and let $A, B \in \Sigma$ be arbitrary events. Moreover, let $A_1, A_2, \cdots \in \Sigma$ be a partition of the event space $\Omega$. (That is, the $A_n$-s are pairwise disjoint and their union is the entire event space.) Then,

$$P(B \mid A) = \frac{P(A \mid B)P(B)}{\sum_{n=1}^{\infty} P(A \mid A_n)P(A_n)}$$

holds.

*Proof.* The proof immediately follows from the Bayes' formula (*Theorem 18.3.3*) and the law of total probability (*Theorem 18.3.1.*)

## 18.3.4   The Bayesian interpretation of probability

Historically, probability was introduced as the relative frequency of observed events in *Section 18.2.7*. However, the invention of conditional probabilities and the Bayes' formula enabled another interpretation that slowly became prevalent in statistics and machine learning.

In pure English, the Bayes' formula can be thought of as updating our probabilistic models using new observations. Suppose that we are interested in the event $B$. Without observing anything, we can formulate a probabilistic model by assigning a probability to $B$, that is, estimating $P(B)$. This is what we call the *prior*. However, observing another event $A$ might change our probabilistic model.

Thus, we would like to estimate the *posterior* probability $P(B \mid A)$. We can't do this directly, but thanks to our prior model, we can tell $P(A \mid B)$. The quantity $P(A \mid B)$ is called the *likelihood*. Combining these with the Bayes' formula, we can see that the posterior is proportional to the likelihood and the prior.

$$\underbrace{P(B \mid A)}_{\text{posterior}} = \frac{\overbrace{P(A \mid B)}^{\text{likelihood}}\overbrace{P(B)}^{\text{prior}}}{P(A)}$$

*Figure 18.8: The Bayes' formula, as the product of the likelihood and prior*

Let's see a concrete example that will make the idea clear. Suppose that we are creating a diagnostic test for an exotic disease. How likely is the disease present in a random person?

Without knowing any specifics about the situation, we can only use statistics to formulate the probability model. Let's say that only 2% of the population is affected. So, our probabilistic model is

$$P(\text{infected}) = 0.02, \quad P(\text{healthy}) = 0.98.$$

However, once someone produces a positive test, things change. The goal is to estimate the posterior probability $P(\text{infected} \mid \text{positive})$, a more accurate model.

Since no medical test is perfect, false positives and false negatives can happen. From the manufacturer, we know that it gives true positives 99% of the time, but the chance of a false positive is 5%. In probabilistic terms, we have

$$P(\text{positive} \mid \text{infected}) = 0.99,$$

$$P(\text{positive} \mid \text{healthy}) = 0.05.$$

With these, the Bayes' theorem gives us

$$
\begin{aligned}
P(\text{infected} \mid \text{positive}) &= \frac{P(\text{positive} \mid \text{infected})P(\text{infected})}{P(\text{positive} \mid \text{infected})P(\text{infected}) + P(\text{positive} \mid \text{healthy})P(\text{healthy})} \\
&= \frac{0.99 \cdot 0.02}{0.99 \cdot 0.02 + 0.05 \cdot 0.98} \\
&\approx 0.29.
\end{aligned}
$$

So, the chance of being infected upon producing a positive test is surprisingly 29% (given these specific true and false positive rates).

These probabilistic thinking principles are also valid for machine learning. If we abstract away the process of learning from data, we are essentially 1) making observations, 2) updating our models given the new observations, and 3) starting the process all over again. The Bayes' theorem gives a concrete tool for the job.

## 18.3.5 The probabilistic inference process

As we have seen before, probability theory is the extension of mathematical logic. So far, we have discussed how logical connectives correspond to set operations and how probability generalizes the truth value by adding the component of uncertainty. What about the probabilistic inference process? Can we generalize classical inference and use probabilistic reasoning to construct arguments? Yes.

To illustrate, let's start with a story. It's 6:00 AM. The alarm clock is blasting but you are having a hard time getting out of bed. You don't feel well. Your muscles are weak, and your head is exploding. After a brief struggle, you manage to call a doctor and list all the symptoms. Your sore throat makes speaking painful.

*"It's probably just the flu,"* they say.

Interactions like this are everyday occurrences. Yet, we hardly think about the reasoning process behind them. After all, you could have been hungover. Similarly, if the police find a murder weapon at your house, they'll suspect that you are the killer. The two are related but not the same. For instance, the murder weapon could have been planted.

The bulk of humanity's knowledge is obtained in this manner: we collect evidence, then build hypotheses.

How do we infer the underlying cause from observing the effect? Most importantly, how can we avoid fooling ourselves into false conclusions?

Let's focus on *"muscle fatigue, headache, sore throat → flu"*. This is certainly not true in an absolute sense, as these symptoms resemble how you would feel after shouting and drinking excessively during a metal concert, which is far from the flu. Yet, a positive diagnosis of flu is plausible. Given the evidence at hand, our belief is increased in the hypothesis.

Unfortunately, classical logic cannot deal with plausible, only with the absolute. Probability theory solves this problem by measuring plausibility on a $0 - 1$ scale, instead of being stuck at the extremes. Zero is impossible. One is certain. All the values in between represent degrees of uncertainty.

Let's put this into mathematical terms!

How can we establish a probabilistic link between cause and effect? In classical logic, events are interesting in the context of other events. Before, implication and modus ponens provided the context. Translated to the language of probability, the question is the following: What is the probability of $B$, given that $A$ is observed? The answer: conditional probabilities.

Why does conditional probability generalize the concept of implication? It's easier to draw a picture, so consider the two extreme cases in *Figure 18.9*. (Recall that implication corresponds to the subset relation, as we saw earlier.)

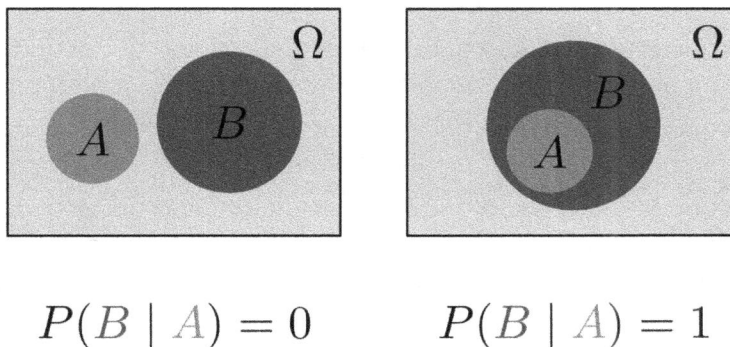

$$P(B \mid A) = 0 \qquad P(B \mid A) = 1$$

*Figure 18.9: Conditional probability as logical implication*

Essentially, $P(B \mid A) = 1$ means that $A \to B$ is true, while $P(B \mid A) = 0$ means that it is not. We can take this analogy further: a small $P(B \mid A)$ means that $A \to B$ is likely to be false, and a large $P(B \mid A)$ means that it is likely to be true.

This is illustrated by *Figure 18.10*.

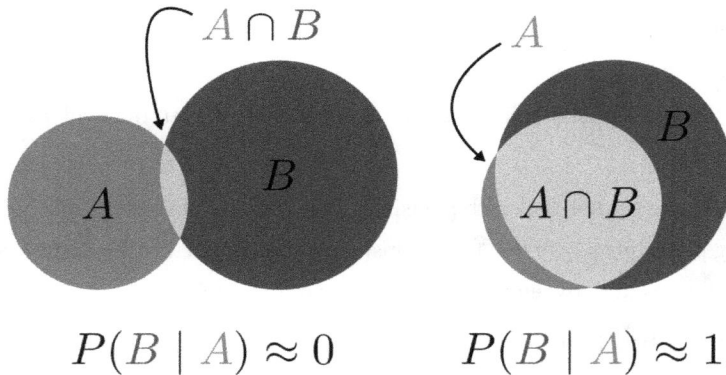

$$P(B \mid A) \approx 0 \qquad P(B \mid A) \approx 1$$

Figure 18.10: Conditional probability as the extension of logical implication

Thus, the *"probabilistic modus ponens"* goes like this:

1. $P(B \mid A) \approx 1$.
2. $A$.
3. Therefore, $B$ is probable.

This is quite a relief, as now we have a solid theoretical justification for most of our decisions. Thus, the diagnostic process that kicked up our investigation makes a lot more sense now:

1. $P(\text{flu} \mid \text{headache, muscle fatigue, sore throat}) \approx 1$.
2. *"Headache and muscle fatigue"*.
3. Therefore, *"flu"* is probable.

However, one burning question remains. How do we know that $P(B \mid A) \approx 1$ holds?

Let's focus on the probabilistic version of *"headache, sore throat, muscle fatigue → flu"*. We know that this is not certain, only plausible. Yet, the reverse implication *"flu → headache, sore throat, muscle fatigue"* is almost certain.

When naively arguing that the evidence implies the hypothesis, we have the opposite in mind. Instead of applying the modus ponens, we use the faulty argument

1. $A \rightarrow B$.
2. $B$.
3. Therefore, $A$.

We have talked about this before: this logical fallacy is called *affirming the consequent*, and it's completely wrong from a purely logical standpoint. However, the Bayes' theorem provides a probabilistic twist.

The proposition $A \rightarrow B$ translates to $P(B \mid A) = 1$, which implies that when $A$ is observed, $B$ occurs as well. Why? Because then we have

$$
\begin{aligned}
P(A \mid B) &= \frac{P(B \mid A)P(A)}{P(B)} \\
&= \frac{P(A)}{P(B)} \\
&\geq P(A).
\end{aligned}
$$

This is good news, as reversing the implication is not totally wrong. Instead, we have the *probabilistic affirming the consequent*:

1. $A \rightarrow B$.
2. $B$.
3. Therefore, $A$ is more probable.

With this, the probabilistic reasoning process makes perfect sense. To recall, the issue with arguments such as *"if you have muscle fatigue, sore throat, and a headache, then you have the flu"* is that the symptoms can be caused by other conditions, and in rare cases, the flu does not carry all of these symptoms.

Yet, this kind of thinking can be surprisingly effective in real-life decision-making. Probability and conditional probability extend our reasoning toolkit with inductive methods in three steps:

1. Generalizes the binary $0 - 1$ truth values to allow the representation of uncertainty.
2. Defines the analogue of *"if A, then B"*-type implications using conditional probability.
3. Provides a method to infer the cause from observing the effect.

These three ideas are seriously powerful, and their inception has enabled science to perform unbelievable feats. (If you are interested in learning more about the relation of probability theory and logic, I recommend you the great book *Probability Theory: The Logic of Science* by E. T. Jaynes.)

There's one more thing I would like to show you. Let's go back to the mid-twentieth century and look at how a TV show shaped probabilistic thinking.

## 18.3.6 The Monty Hall paradox

Before we finish with conditional probability, we'll touch on an important problem. Regarding probability, we often have seemingly contradictory phenomena, going against our intuitive expectations. These are called *paradoxes*. To master probabilistic thinking, we need to resolve them and eliminate common fallacies from our thinking processes. So far, we have already seen the gambler's fallacy when talking about the concept of independence in *Section 18.3.1*. Now, we'll discuss the famous Monty Hall paradox.

In the '60s, there was a TV show in the United States called Let's Make a Deal (`https://en.wikipedia.org/w iki/Let%27s_Make_a_Deal`). As a contestant, you faced three closed doors, one having a car behind it (that you could take home), while the others had nothing. You had the opportunity to open one.

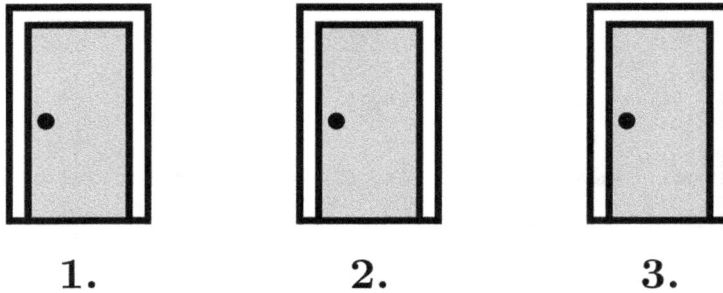

*Figure 18.11: Three closed doors, one of which contains a reward behind it*

Suppose that after selecting door no. 1, Monty Hall — the show host — opens the third door, showing that it was not the winning one. Now, you have the opportunity to change your mind and open door no. 2 instead of the first one. Do you take it?

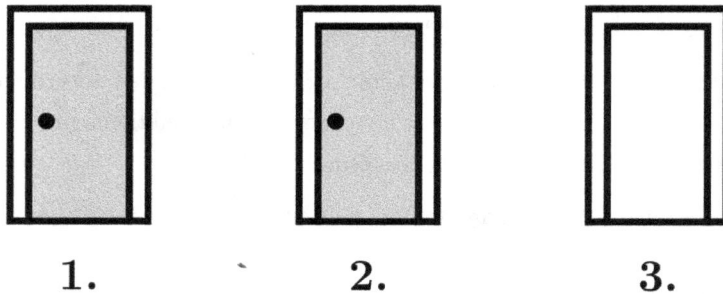

*Figure 18.12: Monty opened the third door for you. Do you switch?*

At first glance, your chances are 50%/50%, so you might not be better off by switching. However, this is not true!

To set things straight, let's do a careful probabilistic analysis. Let $A_i$ denote the event that the prize is behind the $i$-th door, while $B_i$ is the event of Monty opening the $i$-th door. Before Monty opens the third one, our model is

$$P(A_1) = P(A_2) = P(A_3) = \frac{1}{3},$$

and we want to calculate $P(A_1 \mid B_3)$ and $P(A_2 \mid B_3)$.

By thinking from the perspective of the show host, which door would you open? If you know that the prize is behind the 1st door, you open the 2nd and 3rd one with equal probability. However, if the prize is actually behind the 2nd door (and the contestant selected the 1st one), you always open the 3rd one. That is,

$$P(B_3 \mid A_1) = P(B_2 \mid A_1) = \frac{1}{2},$$
$$P(B_3 \mid A_2) = 1.$$

Thus, by applying the Bayes' formula, we have

$$P(A_1 \mid B_3) = \frac{P(B_3 \mid A_1)P(A_1)}{P(B_3)}$$
$$= \frac{1/6}{P(B_3)},$$

and

$$P(A_2 \mid B_3) = \frac{P(B_3 \mid A_2)P(A_2)}{P(B_3)}$$
$$= \frac{1/3}{P(B_3)}.$$

In conclusion, $P(A_2 \mid B_3)$ is twice as large as $P(A_1 \mid B_3)$, from which we deduce

$$P(A_1 \mid B_3) = \frac{1}{3}, \quad P(A_2 \mid B_3) = \frac{2}{3}.$$

So, you should *always* switch doors. Surprising, isn't it? Here, the paradox is that contrary to what we might expect, changing our minds is the better option. With clear probabilistic thinking, we can easily resolve this.

## 18.4 Summary

Phew! We are at the end of an intimidatingly long, albeit extremely essential chapter. Although we've talked about the mathematical details of probability for a couple of dozen pages, the most important takeaway can be summarized in a sentence: probability theory extends our reasoning toolkit by handling uncertainty. Instead of measuring the truthiness of a proposition on a true-or-false binary scale, it opens up a spectrum between 0 and 1, where 0 represents (almost) impossible, and 1 represents (almost) certain.

Mathematically speaking, probabilistic models are defined by *probability measures and spaces*, that is, structures of the form $(\Omega, \Sigma, P)$, where $\Omega$ is the set of possible elementary outcomes, $\Sigma$ is the collection of events, and $P$ is a probability measure, satisfying

1. $P(\Omega) = 1$
2. and $P(\cup_{n=1}^{\infty} - A_n) = \sum_{n=1}^{\infty} P(A_n)$ for all mutually disjoint $A_n \in \Sigma$,

which are called the Kolmogorov axioms. Thinking in probabilities enables us to reason under uncertainty: if $P(A)$ is the probabilistic version of the truth value of a statement, then the conditional probability

$$P(B \mid A) = \frac{P(A \cap B)}{P(A)}$$

is the probabilistic version of the implication $A \to B$.

However, all the tools we learned are just the tips of a massive iceberg. To build truly beefy and useful models, we need to once more turn qualitative into quantitative, as we did for many advances in science and mathematics. Can you recall the dice rolling experiment, where we used a mysterious variable $X$ to represent the outcome of the roll? Thus, we could talk about events such as "$X = k$", turning a probability space into a sequence of numbers.

It's not a coincidence; it's a method. $X$ is an instance of a *random variable*, the premier object of probability theory and statistics. Random variables translate between abstract probability spaces to numbers and vectors, our old friends. Let's make them a permanent tool in our belt.

## 18.5 Problems

**Problem 1.** Let's roll two six-sided dice! Describe the event space, $\sigma$-algebra, and the corresponding probabilities for this experiment.

**Problem 2.** Let $\Omega = [0, 1]$, and the corresponding $\sigma$-algebra be the generated algebra

$$\Sigma = \sigma\big(\{(a, b] \,:\, 0 \leq a < b \leq 1\}\big).$$

Show that the following events are members of $\Sigma$:

*(a)* $S_1 = \{x\}$ for all $x \in [0, 1]$.

*(b)* $S_2 = \cup_{i=1}^{n}(a_i, b_i)$. (Show that this is also true when the intervals $[\dots]$ are replaced with open and half-open versions $(\dots), (\dots], [\dots)$.)

*(c)* $S_3 = [0, 1] \cap \mathbb{Q}$. (That is, the set of rational numbers in $[0, 1]$.)

*(d)* $S_4 = [0,1] \cap (\mathbb{R} \setminus \mathbb{Q})$. (That is, the set of irrational numbers in $[0,1]$.)

**Problem 3.** Let's roll two six-sided dice. What is the probability that

*(a)* Both rolls are odd numbers?

*(b)* At least one of them is an odd number?

*(c)* None of them are odd numbers?

**Problem 4.** Let $\Omega = \mathbb{R}^2$ be the event space, where we define the open disks

$$D(\mathbf{x}, r) := \{\mathbf{z} \in \mathbb{R}^2 : \|\mathbf{x} - \mathbf{z}\| < r\}, \quad \mathbf{x} = (x_1, x_2) \in \mathbb{R}^2, \quad r > 0,$$

and the open rectangles by

$$R(\mathbf{x}, \mathbf{y}) = (x_1, y_1) \times (x_2, y_2)$$
$$= \{\mathbf{z} = (z_1, z_2) \in \mathbb{R}^2 : x_1 < z_1 < y_1, x_2 < z_2 < y_2\}.$$

Show that the $\sigma$-algebras generated by these sets are the same, that is,

$$\sigma\left(\{D(\mathbf{x}, r) : \mathbf{x} \in \mathbb{R}^2, r > 0\}\right) = \sigma\left(\{R(\mathbf{x}, \mathbf{y}) : \mathbf{x}, \mathbf{y} \in \mathbb{R}^2\}\right).$$

**Problem 5.** Let's consider a variant of the Monty Hall problem. Suppose there are a hundred doors instead of three; only one contains a reward. Upon picking a door, Monty opens ninety-eight other doors, all of which are empty. Should you switch now?

# Join our community on Discord

Read this book alongside other users, Machine Learning experts, and the author himself.

Ask questions, provide solutions to other readers, chat with the author via Ask Me Anything sessions, and much more.

Scan the QR code or visit the link to join the community.

`https://packt.link/math`

# 19

# Random Variables and Distributions

Having a probability space to model our experiments and observations is fine and all, but in almost all of the cases, we are interested in a quantitative measure of the outcome. To give you an example, let's consider an already familiar situation: tossing coins. Suppose that we are tossing a fair coin $n$ times but we are only interested in the number of heads. How do we model the probability space this time?

By taking things one step at a time; first, we construct an event space by enumerating all possible outcomes in a single set, just like we already did in *Section 18.2.1*:

$$\Omega = \{0, 1\}^n, \quad \Sigma = 2^{\Omega}.$$

Since the coin is fair, each outcome $\omega$ has the probability $P(\omega) = \frac{1}{2^n}$. This probability space $(\Omega, \Sigma, P)$ is nice and simple so far. Using the additivity of probability measures (see *Definition 18.2.2*), we can calculate the probability of any event. That is, for any $A \in \Sigma$, we have

$$P(A) = \frac{|A|}{|\Omega|},$$

where $|\cdot|$ denotes the number of elements in a given set.

However, as mentioned, we are only interested in the number of heads. Should we just incorporate this information somewhere in the probability space? Sure, we could do that, but that would couple the elementary outcomes (that is, a series of heads or tails) with the measurements. This can significantly complicate our model.

Instead of overloading this probability space to directly deal with the desired measurements, we can do something much simpler: introduce a function $X : \Omega \to \mathbb{N}$, mapping *outcomes* to *measurements*.

These functions are called *random variables*, and they are at the very center of probability theory and statistics. By collecting data, we are observing random variables, and by fitting predictive models, we approximate them using the observations. Now that we understand why we need them, we are going to make this notion mathematically precise.

# 19.1   Random variables

Hold your horses, though; it's not that simple. Random variables are hard to understand in their general form, so we'll slow down and focus on special cases, taking one step at a time. This is how learning is done most effectively, and we'll follow this path as well.

Let's deal with so-called *discrete* random variables (such as the above example) first, real random variables second, and the general case last.

## 19.1.1   Discrete random variables

Following our motivating example describing the number of heads in $n$ coin tosses, we can create a formal definition.

> **Definition 19.1.1  (Discrete random variables)**
>
> Let $(\Omega, \Sigma, P)$ be a probability space and $\{x_k\}_{k=1}^{\infty}$ be an arbitrary sequence of real numbers. The function $X : \Omega \to \{x_1, x_2, \ldots\}$ is called a *discrete random variable* if the sets
>
> $$S_k = \{\omega \in \Omega : X(\omega) = x_k\}$$
>
> are events for any integer $k \in \mathbb{Z}$ (that is, $S_k \in \Sigma$).

You might ask why we are requiring the sets $\{\omega \in \Omega : X(\omega) = x_k\}$ to be events. It seems like just another technical condition, but this plays an essential role. Ultimately, we are defining random variables because we want to measure the probabilities of our observations. This condition ensures that we can do this.

To simplify our notations, we write

$$P(X = x_k) := P\big(\{\omega \in \Omega \,:\, X(\omega) = x_k\}\big)$$

whenever we talk about these probabilities. Let's see a concrete example!

In the case of the coin tossing above, our random variable is defined by

$$X = \text{number of heads}.$$

Even though we can define $X$ in terms of formulas by

$$X(\omega) = \sum_{k=1}^{n} \omega_k, \quad \omega = (\omega_1, \dots, \omega_n) \in \Omega,$$

this is not needed. Often, such a thing is not even possible. Regarding our random variables, we are not interested in knowing the entire mapping, only in questions such as the probability of $k$ heads among $n$ tosses.

If we record the "timestamps" where the outcome is heads, we can encode each $\omega$ as a subset of $\{1, 2, \dots, n\}$. For instance, if the 1st, 3rd, and 37th tosses are heads and the rest are tails, this is $\{1, 3, 37\}$. To calculate the probability of $k$ heads, we need to count the number of $k$-sized subsets for a set of $n$ elements. This is given by the binomial coefficient $\binom{n}{k}$. So,

$$P(X = k) = \binom{n}{k}\frac{1}{2^n}.$$

We'll see this in detail when talking about the *binomial distribution*, whatever it might be. For now, we are ready to generalize our random variables!

## 19.1.2 Real-valued random variables

What if our measurements are not discrete? For instance, suppose that we have a class of students in front of us. We are interested in the distribution of their body height. So, we pick one student at random and measure their height with our shiny new tool, which is capable of measuring height with perfect precision.

In this case, discrete random variables are not enough, but we can define something similar.

---

**Definition 19.1.2 (Real-valued random variables)**

Let $(\Omega, \Sigma, P)$ be a probability space. The function $X : \Omega \to \mathbb{R}$ is called a *random variable* if the set

$$X^{-1}\big((a, b)\big) := \{\omega \in \Omega : a < X(\omega) < b\}$$

is an event for all $a, b \in \mathbb{R}$. (That is, $X^{-1}\big((a, b)\big) \in \Sigma$ for all $a, b \in \mathbb{R}$.)

---

Let's unwrap this definition. First of all, $X$ is a mapping from the event space $\Omega$ to the set of real numbers $\mathbb{R}$, as illustrated by *Figure 19.1*.

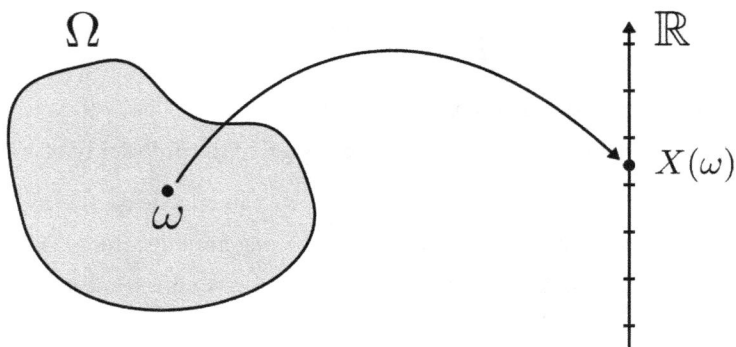

Figure 19.1: A real-valued random variable is a mapping from the event space to the set of real numbers

Similarly to the discrete case, we are interested in the probabilities of events such as $X^{-1}\big((a, b)\big)$. Again, for simplicity, we write

$$P(a < X < b) = P\big(X^{-1}\big((a, b)\big)\big).$$

You can imagine $X^{-1}((a, b))$ as the subset of $\Omega$ that is mapped to $(a, b)$. (In general, sets of the form $X^{-1}(A)$ are called *inverse images*.)

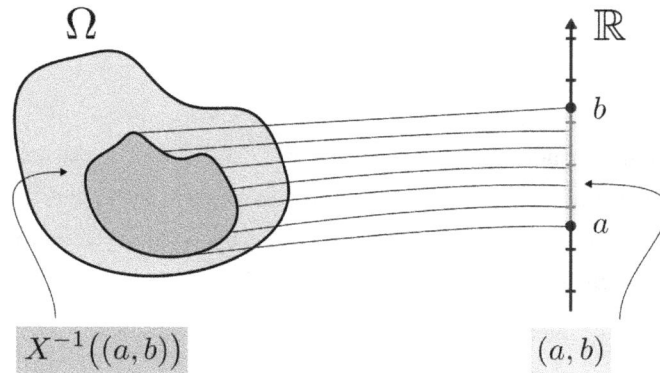

*Figure 19.2: Inverse image of an interval*

Let's see an example right away. Suppose that we are throwing darts at a circular board on the wall. (For simplicity, assume that we are so good that we always hit the board.) As we have seen when discussing $\sigma$-algebras in higher dimensions (*Section 18.2.6*), we can model this by selecting

$$\Omega = B(0, 1)$$
$$= \{\mathbf{x} \in \mathbb{R}^2 : \|\mathbf{x}\| < 1\}$$

and

$$\Sigma = \mathcal{B}\big(B(0, 1)\big)$$
$$= \sigma\big(\{A \cap B(0, 1) : A \in \mathcal{B}(\mathbb{R}^n)\big),$$

while

$$P(A) = \frac{\text{area}(A)}{\text{area}(\Omega)} = \frac{\text{area}(A)}{\pi}.$$

Since dart boards are subdivided into concentric circles, scoring is determined by the distance from the center. So, we might as well define our random variable by

$$X = \text{ distance of the impact point from the center.}$$

$X$ encodes all that we are interested in, in terms of scoring. In general, we have

$$P(X < r) = \begin{cases} 0 & \text{if } r \leq 0, \\ r^2 & \text{if } 0 < r < 1, \\ 1 & \text{otherwise.} \end{cases}$$

What if we have more than one measurement? For instance, in the case of the famous Iris dataset (`https://en.wikipedia.org/wiki/Iris_flower_data_set`) (one that we have seen a few times so far), we have four measurements. Sure, we can just define four random variables, but then we cannot take advantage of all the heavy machinery we have built so far: linear algebra and multivariate calculus.

For this, we will take a look at random variables in the general case.

## 19.1.3   Random variables in general

Let's cut to the chase.

---

**Definition 19.1.3  (Random variables)**

Let $(\Omega_1, \Sigma_1, P_1)$ be a probability space and let $(\Omega_2, \Sigma_2)$ be another event space $\Omega_2$ with $\sigma$-algebra $\Sigma_2$. The function $X : \Omega_1 \to \Omega_2$ is a *random variable* if, for every $E \in \Sigma_2$, the set

$$X^{-1}(E) := \{\omega \in \Omega_1 : X(\omega) \in E\}$$

is a member of $\Sigma_1$. (That is, $X^{-1}(E) \in \Sigma_1$.)

---

In mathematical literature, random variables are usually denoted by either capital Latin letters such as $X, Y$, or Greek letters (mostly starting from $\xi$).

Random variables essentially push probability measures forward from abstract probability spaces to more tractable ones. On the event space $(\Omega_2, \Sigma_2)$, we can define a probability measure $P_2$ by

$$P_2(E) := P_1\big(X^{-1}(E)\big), \quad E \in \Sigma_2,$$

making it possible to transform one probability space to another while keeping the underlying probabilistic model intact.

This general case covers all the mathematical objects we are interested in for machine learning. Staying with the Iris dataset (`https://en.wikipedia.org/wiki/Iris_flower_data_set`), the random variable

$$X \ : \ \text{set of iris flowers} \to \mathbb{R}^4,$$

$$\text{iris flower} \mapsto (\text{petal width, petal length, sepal width, sepal length})$$

describes the generating distribution for the dataset, while for classification tasks, we are interested in approximating the random variable

$$Y \ : \ \text{set of iris flowers} \to \{\text{setosa, versicolor, virginica}\},$$

$$\text{iris flower} \mapsto \text{class label.}$$

Now we will take a deeper look at why random variables are defined this way. This will be a bit technical, so feel free to skip it. It won't adversely affect your ability to work with random variables.

## 19.1.4 Behind the definition of random variables

So, random variables are functions, mapping the probability space onto a measurement space. The only question is, why are the sets $X^{-1}(E)$ so special? Let's revisit one of our motivating examples: picking a random student and measuring their height. We are interested in questions such as the probability of a student having a body height between 155 cm and 185 cm. (If you prefer using the imperial metric system, then 155 cm is roughly 5.09 feet and 185 cm is around 6.07 feet.) Translating this into formulas, we are interested in

$$P(155 \leq X \leq 185) = P\Big(X^{-1}\big([155, 185]\big)\Big).$$

(In the above formula, I wrote the same thing using two different notations.)

So, how is $X^{-1}\big([155, 185]\big)$ an event? To find this out, let's look at *inverse images* in general.

**Definition 19.1.4 (Inverse image of sets with respect to functions)**

Let $f \ : \ E \to H$ be a function between the two sets $E$ and $H$, and let $A \subseteq H$ be an arbitrary set. The *inverse image* of $A$ with respect to the function $f$ is defined by

$$f^{-1}(A) := \{x \in E \ : \ f(x) \in A\}.$$

We like inverse images of sets because they behave nicely under set operations.

This is formalized by the following theorem.

**Theorem 19.1.1** *Let $f : E \to H$ be a function between the two sets $E$ and $H$. For any $A_1, A_2, \cdots \subseteq H$, the following hold:*

*(a)*

$$f^{-1}\left(\bigcup_{n=1}^{\infty} A_n\right) = \bigcup_{n=1}^{\infty} f^{-1}(A_n),$$

*(b)*

$$f^{-1}(A_1 \setminus A_2) = f^{-1}(A_1) \setminus f^{-1}(A_2),$$

*(c)*

$$f^{-1}\left(\bigcap_{n=1}^{\infty} A_n\right) = \bigcap_{n=1}^{\infty} f^{-1}(A_n).$$

*Proof.* (a) We can easily see this by simply writing out the definitions. That is, we have

$$f^{-1}\left(\bigcup_{n=1}^{\infty} A_n\right) = \left\{x \in E : f(x) \in \cup_{n=1}^{\infty} A_n\right\}$$
$$= \bigcup_{n=1}^{\infty} \left\{x \in E : f(x) \in A_n\right\}$$
$$= \bigcup_{n=1}^{\infty} f^{-1}(A_n),$$

which is what we had to show. (If you are not comfortable with working with sets, feel free to review *Appendix C* on introductory set theory.)

*(b)* This can be done in the same manner as *(a)*.

*(c)* The De Morgan laws (*Theorem C.2.2*) imply that

$$H \setminus \left(\bigcup_{n=1}^{\infty} A_n\right) = \bigcap_{n=1}^{\infty} (H \setminus A_n)$$

holds. Combining this with *(a)* and *(b)*, *(c)* follows.

Why is this important? Recall that the Borel sets, our standard $\sigma$-algebra on real numbers (as seen in *Section 18.2.3*), is defined by

$$\mathcal{B} := \sigma\big(\{(-\infty, x] \,:\, x \in \mathbb{R}\}\big). \tag{19.1}$$

These contain all the events that we are interested in regarding the measurements. Combined with our previous result, we can reveal what is not in plain sight about random variables.

**Theorem 19.1.2** *Let $(\Omega, \Sigma, P)$ be a probability space and $X : \Omega \to \mathbb{R}$ be a random variable, and let $A \in \mathcal{B}$, where $\mathcal{B}$ is the Borel algebra defined by (19.1). Then, $X^{-1}(A) \in \Sigma$.*

That is, we can measure the probability of $X^{-1}(A)$ for any Borel set $A$. Without this, our random variables would not be that useful. To make our notations more intuitive, we write

$$P(X \in A) := P\big(X^{-1}(A)\big).$$

In plain English, $P(X \in A)$ is the probability of our measurement $X$ falling into the set $A$.

Now that we understand what all of this means, let's see the simple proof!

*Proof.* This is a simple consequence of the fact that $\mathcal{B}$ is the $\sigma$-algebra generated by sets of the form $(-\infty, x]$, and the inverse images behave nicely under set operations (as *Theorem 19.1.1* suggests).

## 19.1.5 Independence of random variables

When building probabilistic models of the external world, the assumption of independence significantly simplifies the subsequent mathematical analysis. Recall that on a probability space $(\Omega, \Sigma, P)$, the events $A, B \in \Sigma$ are independent if

$$P(A \cap B) = P(A)P(B),$$

or equivalently,

$$P(A \mid B) = P(A).$$

In plain English, observing one event doesn't change our probabilistic belief about the other.

Since a random variable $X$ is described by events of the form $X^{-1}(E)$, we can generalize the notion of independence to random variables.

> **Definition 19.1.5 (Independence of random variables)**
>
> Let $X, Y : \Omega_1 \to \Omega_2$ be two random variables between the probability space $(\Omega_1, \Sigma_1, P)$ and $\sigma$-algebra $(\Omega_2, \Sigma_2)$. We say that $X$ and $Y$ are independent if, for every $A, B \in \Sigma_2$,
>
> $$P(X \in A, Y \in B) = P(X \in A)P(Y \in B)$$
>
> holds.

Again, think about two coin tosses. $X_1$ describes the first coin toss and $X_2$ describes the other. Since the tosses are independent, no observation of the first one reveals any extra information about the second one. This is formalized by the definition above.

On the other hand, to show two *dependent* random variables, consider the following. We'll roll a six-sided dice, and denote the result as $X$. After that, we roll with $X$ pieces of six-sided dice and denote the sum total of their values as $Y$. $X$ and $Y$ are dependent on each other. For instance, consider that $P(X = 1, Y > 7) = 0$, but neither $P(X = 1)$ nor $P(Y > 7)$ are zero.

Independence is an assumption that we often make. When working with sequences of random variables represented by $X_1, X, 2, \dots$, we almost always assume that they are *independent and identically distributed*; that is, *i.i.d.* random variables.

Now that we understand how to work with random variables, it's time to show how to represent them in a compact form.

# 19.2   Discrete distributions

Let's recap what we have learned so far. In probability theory, our goal is to first model real-life scenarios affected by uncertainty and then to analyze them using mathematical tools such as calculus.

For the latter purpose, probability spaces are not easy to work with. A probability measure is a function defined on an $\sigma$-algebra, so we can't really use calculus there.

Random variables bring us one step closer to the solution, but they can also be difficult to work with. Even though a real-valued random variable $X : \Omega \to \mathbb{R}$ maps an abstract probablity space to the set of real numbers, there are some complications. $\Omega$ can be anything, and if you recall, we might not even have a tractable formula for $X$.

For example, if $X$ denotes the lifetime of a lightbulb, we don't have a formula. So, again, we can't use calculus. However, there is a way to represent the information contained by a random variable in a sequence, a vector-scalar function, or a scalar-scalar function.

Enter *probability distributions* and *density functions*.

Consider a simple experiment, such as tossing a fair coin $n$ times and counting the number of heads, denoting it with $X$. As we have seen before in *Definition 19.1.1*, $X$ is a discrete random variable with

$$P(X = k) = \begin{cases} \binom{n}{k} \frac{1}{2^n} & \text{if } k = 0, 1, \dots, n, \\ 0 & \text{otherwise.} \end{cases}$$

However, the sequence $\{P(X = k)\}_{k=0}^n$ fully describes the random variable $X$!

Think about it. As our event space is $\Omega = \{0, 1, \dots, n\}$, any event is of the form $A = \{a_1, a_2, \dots, a_l\} \subset \Omega$ for some $l \le n + 1$. Thus,

$$P(X \in A) = \sum_{i=1}^{l} P(X = a_i),$$

where we used the ($\sigma$-)additivity of probability. The sequence $\{P(X = k)\}_{k=0}^n$ is all the information we need.

As a consequence, instead of working with $X : \Omega \to \mathbb{N}$, we can forget about it and use only $\{P(X = k)\}_{k=0}^n$. Why is this good for us?

Because sequences are awesome. As opposed to the mysterious random variables, we have a lot of tools to work with them. Most importantly, we can represent them in a programming language as an array of numbers. We can't do such a thing with pure random variables.

### Definition 19.2.1 (Probability mass function)

Let $X$ be a real-valued discrete random variable. The function $p_X : \mathbb{R} \to [0, 1]$ defined by

$$p_X(x) = P(X = x), \quad x \in \mathbb{R}$$

is called the *probability mass function* (or PMF in short) of the discrete random variable $X$.

In general, a sequence of real numbers defines a *discrete distribution* if its elements are non-negative and it sums up to one.

---

**Definition 19.2.2 (Discrete probability distribution)**

Let $\{p_k\}_{k=1}^{\infty}$ be a sequence of real numbers. We say that $\{p_k\}$ is a *discrete probability distribution* if

*(a)* $p_k \geq 0$ for all $k$,

*(b)* and $\sum_{k=1}^{\infty} p_k = 1$.

---

**Remark 19.2.1** Note that if the random variable assumes finitely many values (such as in our coin tossing example before), only finitely many values are nonzero in the distribution.

---

As recently hinted, every discrete random variable $X$ defines the distribution $\{P(X = x_k)\}_{k=1}^{\infty}$, where $\{x_1, x_2, \dots\}$ are the possible values that $X$ can take. This also holds in the reverse direction: given a discrete distribution $\mathbf{p} = \{p_k\}_{k=1}^{\infty}$, we can construct a random variable $X$ whose **probability mass function (PMF)** is $\mathbf{p}$.

Thus, the probability mass function of $X$ is also referred to as its distribution. I know, it is a bit confusing, as the word "distribution" is quite overloaded in math. You'll get used to it.

These discrete probability distributions are well suited for performing quantitative analysis, as opposed to the base form of random variables. As an additional benefit, think about how distributions generalize random variables. No matter whether we talk about coin tosses or medical tests, the rate of success is given by the above discrete probability distribution.

Before moving on to discussing the basic properties of discrete distributions, let's see some examples!

## 19.2.1 The Bernoulli distribution

Let's start the long line of examples with the most basic probability distribution possible: the Bernoulli distribution, describing a simple coin-tossing experiment. We are tossing a coin having probability $p$ of coming up heads and probability $1 - p$ of coming up tails. The experiment is encoded in the random variable $X$, which takes the value 1 if the toss results in heads, 0 otherwise:

$$X = \begin{cases} 1 & \text{if the toss results in heads,} \\ 0 & \text{otherwise.} \end{cases}$$

Thus,

$$P(X = k) = \begin{cases} 1 - p & \text{if } k = 0, \\ p & \text{if } k = 1, \\ 0 & \text{otherwise.} \end{cases}$$

When a random variable $X$ is distributed according to this, we write

$$X \sim \text{Bernoulli}(p),$$

where $p \in [0, 1]$ is a parameter of the distribution.

> **Remark 19.2.2 (An alternative form of the Bernoulli distribution)**
>
> There is a clever alternative formulation of the Bernoulli distribution that gets rid of the if-else definition. As $k$ is either zero or one, $P(X = k)$ can be written as
>
> $$P(X = k) = p^k (1 - p)^{1-k}.$$
>
> Keep this form in mind, as it'll be extremely useful later down the line.

It's time to talk about distributions in practice. There are several stats packages for Python, but we'll use the almighty `scipy` (which is not exactly a stats package, but it has an excellent statistical module):

```
from scipy.stats import bernoulli
```

We can generate random values using the `rvs` method of the `bernoulli` object (just like for any other distribution from `scipy`):

```
[bernoulli.rvs(p=0.5) for _ in range(10)]    # ten Bernoulli(0.5)-distributed random
numbers
```

```
[1, 1, 1, 1, 0, 1, 0, 1, 1, 0]
```

In `scipy`, the probability mass function is implemented in the `pmf` method.

We can even visualize the distribution using Matplotlib:

```python
import matplotlib.pyplot as plt

params = [0.25, 0.5, 0.75]


with plt.style.context("seaborn-v0_8"):
    fig, axs = plt.subplots(1, len(params), figsize=(4*len(params), 4), sharey=True)
    fig.suptitle("The Bernoulli distribution")
    for ax, p in zip(axs, params):
        x = range(2)
        y = [bernoulli.pmf(k=k, p=p) for k in x]
        ax.bar(x, y)
        ax.set_title(f"p = {p}")
        ax.set_ylabel("P(X = k)")
        ax.set_xlabel("k")
    plt.show()
```

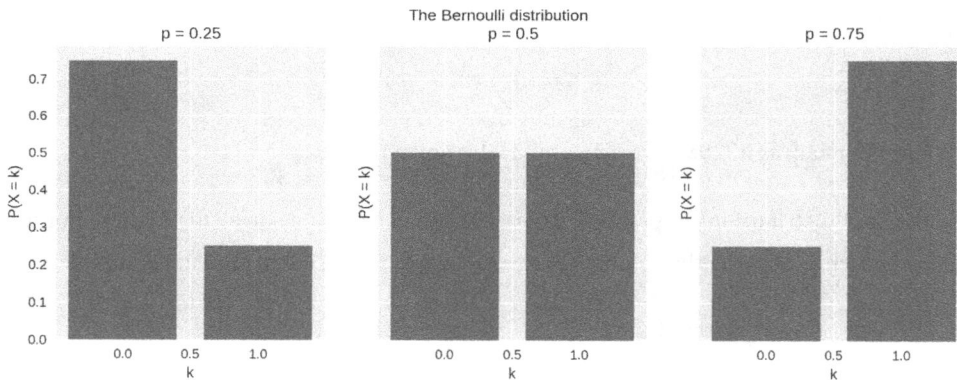

Figure 19.3: The Bernoulli distribution

If you are interested in the details, feel free to check out the SciPy documentation (`https://docs.scipy.org/doc/scipy/reference/generated/scipy.stats.bernoulli.html`) for further methods!

## 19.2.2 The binomial distribution

Let's take our previous coin-tossing example one step further. Suppose that we toss the same coin $n$ times, and $X$ denotes the number of heads out of $n$ tosses. What is the probability of getting exactly $k$ heads?

Say, $n = 5$ and $k = 3$. For example, the configuration 11010 (where 0 denotes tails and 1 denotes heads) has the probability $p^3(1 - p)^2$, as there are three heads and two tails from five independent (*Definition 19.1.5*) tosses.

How many such configurations are available? Selecting the position of the three heads is the same as selecting a three-element subset out of a set of five elements. Thus, there are $\binom{5}{3}$ possibilities. In general, there are $\binom{n}{k}$ possibilities for selecting a $k$-element subset out of a set of $n$ elements.

Combining this, we have

$$P(X = k) = \begin{cases} \binom{n}{k} p^k (1-p)^{n-k} & \text{if } k = 0, 1, \dots, n, \\ 0 & \text{otherwise.} \end{cases}$$

This is called the binomial distribution, one of the most frequently encountered ones in probability and statistics. In notation, we write

$$X \sim \text{Binomial}(n, p),$$

where the $n \in \mathbb{N}$ and $p \in [0, 1]$ are its two parameters. Let's visualize the distribution!

```python
from scipy.stats.distributions import binom

params = [(20, 0.25), (20, 0.5), (20, 0.75)]

with plt.style.context("seaborn-v0_8"):
    fig, axs = plt.subplots(1, len(params), figsize=(4*len(params), 4), sharey=True)
    fig.suptitle("The binomial distribution")
    for ax, (n, p) in zip(axs, params):
        x = range(n+1)
        y = [binom.pmf(n=n, p=p, k=k) for k in x]
        ax.bar(x, y)
        ax.set_title(f"n = {n}, p = {p}")
        ax.set_ylabel("P(X = k)")
        ax.set_xlabel("k")

    plt.show()
```

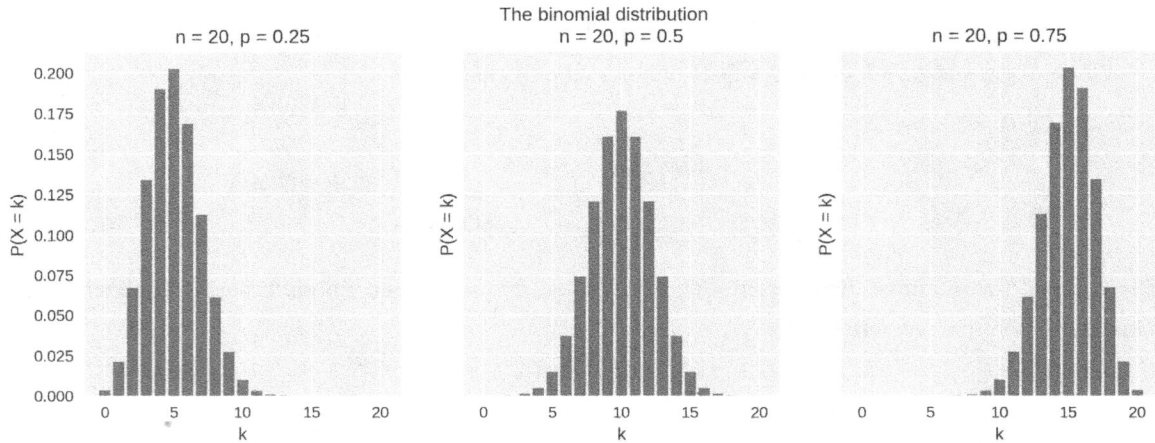

*Figure 19.4: The binomial distribution*

## 19.2.3   The geometric distribution

A bit more coin tossing. We toss the same coin until a heads turn up. Let $X$ denote the number of tosses needed. With some elementary probabilistic thinking, we can deduce that

$$P(X = k) = \begin{cases} (1-p)^{k-1}p & \text{if } k = 1, 2, \dots \\ 0 & \text{otherwise.} \end{cases}$$

(Since if heads turn up first for the $k$-th toss, we tossed $k-1$ tails previously.) This is called the geometric distribution and is denoted as

$$X \sim \text{Geo}(p),$$

with $p \in [0, 1]$ being the only parameter. Similarly, we can plot the histograms to visualize the distribution family:

```python
from scipy.stats import geom


params = [0.2, 0.5, 0.8]


with plt.style.context("seaborn-v0_8"):
```

```
fig, axs = plt.subplots(1, len(params), figsize=(5*len(params), 5), sharey=True)
fig.suptitle("The geometric distribution")
for ax, p in zip(axs, params):
    x = range(1, 20)
    y = [geom.pmf(p=p, k=k) for k in x]
    ax.bar(x, y)
    ax.set_title(f"p = {p}")
    ax.set_ylabel("P(X = k)")
    ax.set_xlabel("k")

plt.show()
```

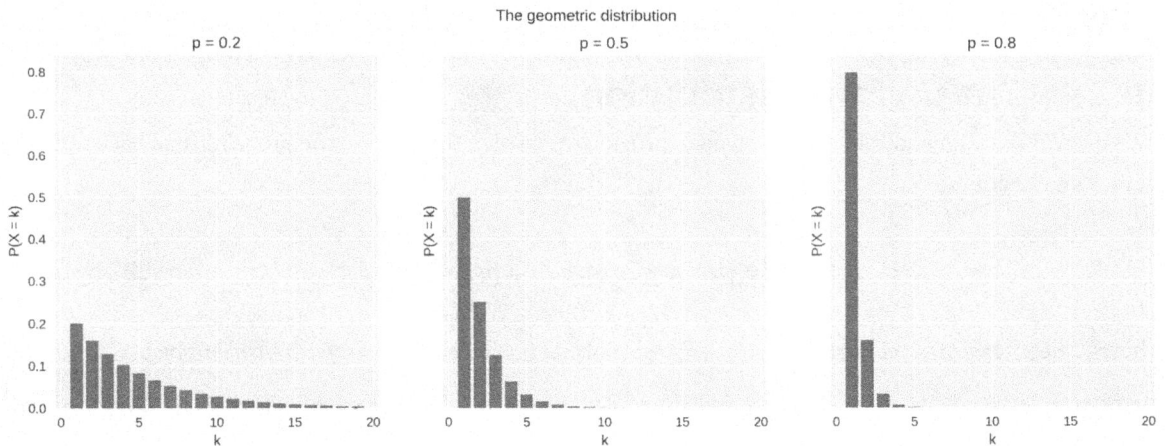

Figure 19.5: The geometric distribution

Note that *none* of the probabilities $P(X = k)$ are zero, but as $k$ grows, they become extremely small. (The closer $p$ is to 1, the faster the decay.)

It might not be immediately obvious that $\sum_{k=1}^{\infty}(1-p)^{k-1}p = 1$. To do that, we'll apply a magic trick. (You know. Paraphrasing the famous Arthur C. Clarke quote, "Any sufficiently advanced mathematics is indistinguishable from magic.")

In fact, for an arbitrary $x \in (-1, 1)$, the astounding identity

$$\sum_{k=0}^{\infty} x^k = \frac{1}{1-x} \tag{19.2}$$

holds.

This is the famous geometric series. Using (19.2), we have

$$\sum_{k=1}^{\infty} P(X = k) = \sum_{k=1}^{\infty} (1-p)^{k-1} p$$

$$= p \sum_{k=0}^{\infty} (1-p)^k$$

$$= p \frac{1}{1-(1-p)}$$

$$= 1.$$

Using the geometric series is one of the most common tricks up a mathematician's sleeve. We'll use this, for instance, when talking about expected values for certain distributions.

## 19.2.4  The uniform distribution

Let's discard the coin and roll a six-sided dice instead. We've seen this before: the probability of each outcome is the same, that is,

$$P(X = 1) = P(X = 2) = \cdots = P(X = 6) = \frac{1}{6},$$

where $X$ denotes the outcome of the roll. This is a special instance of the *uniform distribution*.

In general, let $A = \{a_1, a_2, \ldots, a_n\}$ be a finite set. The discrete random variable $X : \Omega \to A$ is *uniformly distributed on $A$*, that is,

$$X \sim \text{Uniform}(A),$$

if

$$P(X = a_1) = P(X = a_2) = \cdots = P(X = a_n) = \frac{1}{n}.$$

Note that $A$ must be a finite set: no discrete uniform distribution exists on infinite sets. When we have an uniform distribution on $\{1, 2, \ldots, n\}$, we often abbreviate it as $\text{Uniform}(n)$.

Here is the probability mass function for rolling a six-sided dice. Not the most exciting one, I know:

```
from scipy.stats import randint
```

```
with plt.style.context("seaborn-v0_8"):
    fig = plt.figure(figsize=(10, 5))
    plt.title("The uniform distribution")

    x = range(-1, 9)
    y = [randint.pmf(k=k, low=1, high=7) for k in x]
    plt.bar(x, y)
    plt.ylim(0, 1)
    plt.ylabel("P(X = k)")
    plt.xlabel("k")

    plt.show()
```

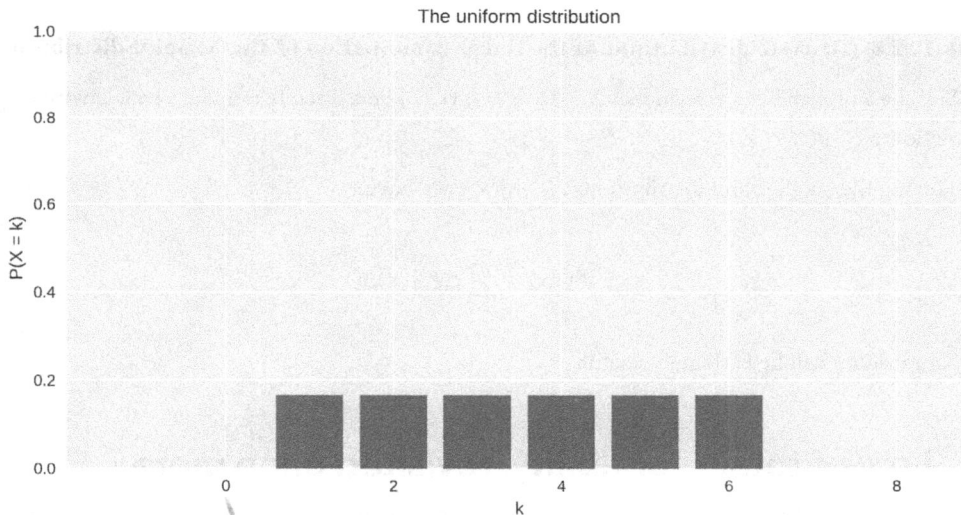

*Figure 19.6: The (discrete) uniform distribution*

## 19.2.5   The single-point distribution

We've left the simplest one till last: the single-point distribution. For that, let $a \in \mathbb{R}$ be an arbitrary real number. We say that the random variable $X$ is distributed according to $\delta(a)$ if

$$P(X = x) = \begin{cases} 1 & \text{if } x = a, \\ 0 & \text{otherwise.} \end{cases}$$

That is, $X$ assumes $a$ with probability 1. Their corresponding cumulative distribution function is

$$F_X(x) = \begin{cases} 1 & \text{if } x \geq a, \\ 0 & \text{otherwise,} \end{cases}$$

which is a simple step function with a single jump.

Trust me, explicitly naming such a simple distribution is immensely useful. There are two main reasons that come to mind. First, the single-point distribution often arises as the limit distribution of sequences of random variables.

Second, every discrete distribution can be written in terms of single-point distributions. This is not absolutely necessary for you to understand right now, but it'll be essential on a more advanced level.

---

**Remark 19.2.3 (Discrete distributions as the linear combination of single-point distributions)**

Let $(\Omega, \Sigma, P)$ be a probability space and let $X : \Omega \to \{x_1, x_2, \dots\}$ be a discrete random variable with probability mass function $p_i = P(X = x_i)$.

By introducing the single-point distributions $X_i \sim \delta(x_i)$, we have

$$F_X(x) = \sum_{i=1}^{\infty} p_i F_{X_i}(x).$$

This decomposition can be extremely useful.

---

## 19.2.6   Law of total probability, revisited once more

With the help of discrete random variables, we can dress the law of total probability (*Theorem 18.3.1*) in new clothes.

---

**Theorem 19.2.1 *(Law of total probability, discrete random variable version)***

*Let $(\Omega, \Sigma, P)$ be a probability space and let $A \in \Sigma$ be an arbitrary event. If $X : \Omega \to \{x_1, x_2, \dots\}$ is a discrete random variable, then*

$$P(A) = \sum_{k=1}^{\infty} P(A \mid X = x_k) P(X = x_k). \tag{19.3}$$

*Proof.* For any discrete random variable $X : \Omega \to \{x_1, x_2, \dots\}$, the events $\{X = x_k\}$ partition the event space: they are mutually disjoint, and their union gives $\Omega$. Thus, the law of total probability can be applied, obtaining

$$P(A) = \sum_{k=1}^{\infty} P(A, X = x_k)$$
$$= \sum_{k=1}^{\infty} P(A \mid X = x_k)P(X = x_k),$$

which is what we had to prove.

In other words, we can study events in the context of discrete random variables. This is extremely useful in practice. (Soon, we'll see that it's not only for the discrete case.)

Let's put (19.3) to work right away.

## 19.2.7    Sums of discrete random variables

Since discrete probability distributions are represented by sequences, we can use a wide array of tools from mathematical analysis to work with them. (This was the whole reason behind switching random variables to distributions.) As a consequence, we can easily describe more complex random variables by constructing them from simpler ones.

For instance, consider rolling two dice, where we are interested in the distribution of the sum. So, we can write this as the sum of random variables $X_1$ and $X_2$, denoting the outcome of the first and second toss, respectively. We know that

$$P(X_i = k) = \begin{cases} \frac{1}{6} & \text{if } k = 1, 2, \dots, 6, \\ 0 & \text{otherwise} \end{cases}$$

for $i = 1, 2$. Using (19.3) and the fact that the two outcomes are independent, we have

$$P(X_1 + X_2 = k) = \sum_{l=1}^{6} P(X_1 + X_2 = k \mid X_2 = l)P(X_2 = l)$$
$$= \sum_{l=1}^{6} P(X_1 = k - l)P(X_2 = l)$$

If this looks familiar, it is not an accident.

What you see here is the famous convolution operation in action.

---

**Definition 19.2.3 (Discrete convolution)**

Let $a = \{a_k\}_{k=-\infty}^{\infty}$ and $b = \{b_k\}_{k=-\infty}^{\infty}$ be two arbitrary sequences. Their *convolution* is defined by

$$a * b := \left\{ \sum_{l=-\infty}^{\infty} a_{k-l} b_l \right\}_{k=-\infty}^{\infty}.$$

---

That is, the $k$-th element of the sequence $a * b$ is defined by the sum $\sum_{l=-\infty}^{\infty} a_{k-l} b_l$. This might be hard to imagine, but thinking about the probabilistic interpretation makes the definition clear. The random variable $X_1 + X_2$ can assume the value $k$ if $X_1 = k - l$ and $X_2 = l$, for all possible $l \in \mathbb{Z}$.

---

**Remark 19.2.4 Remark 19.2.4 (Switching up the indices)**

Due to symmetry,

$$\sum_{l=-\infty}^{\infty} a_{k-l} b_l = \sum_{l=-\infty}^{\infty} a_l b_{k-l}.$$

Thus, an alternative definition of $a * b$

$$a * b = \left\{ \sum_{l=-\infty}^{\infty} a_l b_{k-l} \right\}_{k=-\infty}^{\infty}.$$

This trick is often extremely useful, as when $a_k$ and $b_k$ is explicitly given, sometimes $\sum_{l=-\infty}^{\infty} a_l b_{k-l}$ is simpler to calculate than $\sum_{l=-\infty}^{\infty} a_{k-l} b_l$, and vice versa.

---

Convolution is supported by NumPy, so with its help, we can visualize the distribution of our $X_1 + X_2$:

```python
import numpy as np

dist_1 = [0, 1/6, 1/6, 1/6, 1/6, 1/6, 1/6]
dist_2 = [0, 1/6, 1/6, 1/6, 1/6, 1/6, 1/6]
sum_dist = np.convolve(dist_1, dist_1)

with plt.style.context("seaborn-v0_8"):
    plt.figure(figsize=(10, 5))
    plt.bar(range(0, len(sum_dist)), sum_dist)
    plt.title("Distribution of X1 + X2")
    plt.ylabel("P(X1 + X1 = k)")
    plt.xlabel("k")
    plt.show()
```

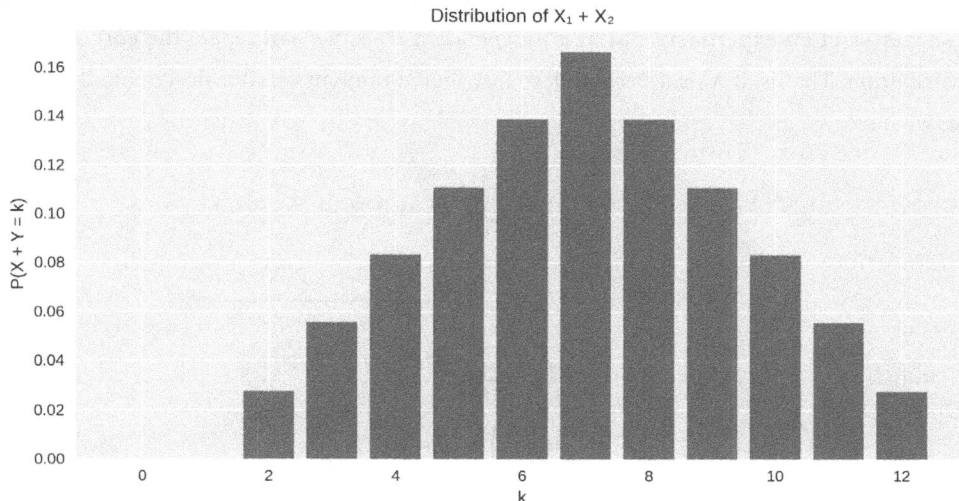

*Figure 19.7: Distribution of the sum of two random variables*

Let's talk about the general case. The pattern is clear, so we can formulate a theorem.

**Theorem 19.2.2** *(Sums of discrete random variables.)*

*If* $X, Y : \Omega \rightarrow \mathbb{Z}$ *are both integer-valued random variables, then the distribution of* $X + Y$ *is given by the convolution of the respective distributions:*

$$P(X + Y = k) = \sum_{l=-\infty}^{\infty} P(X = k - l)P(Y = l),$$

*that is,*

$$p_{X+Y} = p_X * p_Y.$$

*Proof.* The proof is a straightforward application of the law of total probability (19.3):

$$P(X + Y = k) = \sum_{l=-\infty}^{\infty} P(X + Y = k \mid Y = l)P(Y = l)$$

$$= \sum_{l=-\infty}^{\infty} P(X = k - l)P(Y = l)$$

$$= (p_X * p_Y)(k),$$

which is what we had to prove.

Another example of random variable sums is the binomial distribution itself. Instead of thinking about the number of successes of an experiment out of $n$ independent tries, we can model the core experiment as a Bernoulli distribution. That is, if $X_i$ is a Bernoulli($p$) distributed random variable describing the success of the $i$-th attempt, we have

$$
\begin{aligned}
P(X_1 + \cdots + X_n = k) &= \sum_{i_1 + \cdots + i_n = k} P(X_1 = i_1, \ldots, X_n = i_n) \\
&= \sum_{i_1 + \cdots + i_n = k} \underbrace{P(X_1 = i_1) \ldots P(X_n = i_n)}_{X_1, \ldots, X_n \text{ are independent}} \\
&= \sum_{i_1 + \cdots + i_n = k} p^k (1 - p)^{n-k} \\
&= \binom{n}{k} p^k (1 - p)^{n-k},
\end{aligned}
$$

where the sum $\sum_{i_1 + \cdots + i_n = k}$ traverses all tuples $(i_1, \ldots, i_n) \in \{0, 1\}^n$ for which $i_1 + \cdots + i_n = k$. (As there are $\binom{n}{k}$ of such tuples, we have $\sum_{i_1 + \cdots + i_n = k} p^k (1 - p)^{n-k} = \binom{n}{k} p^k (1 - p)^{n-k}$ in the last step.)

# 19.3   Real-valued distributions

So far, we have talked about discrete random variables, that is, random variables with countably many values. However, not all experiments/observations/measurements are like this. For instance, the height of a person is a random variable that can assume a continuum of values.

To give a tractable example, let's pick a number $X$ from $[0, 1]$, with each one having an "equal chance." In this context, equal chance means that

$$
P(a < X \leq b) = |b - a|.
$$

Can we describe $X$ with a single real function? As in the discrete case, we can try

$$
F(x) = P(X = x),
$$

but this wouldn't work. Why?

Because for each $x \in X$, we have $P(X = x) = 0$. That is, picking a particular number $x$ has zero probability. Instead, we can try $F_X(x) = P(X \leq x)$, which is

$$F_X(x) = \begin{cases} 0 & \text{if } x \leq 0, \\ x & \text{if } 0 < x \leq 1, \\ 1 & \text{otherwise.} \end{cases}$$

We can plot this for visualization:

```
from scipy.stats import uniform
X = np.linspace(-0.5, 1.5, 100)
y = uniform.cdf(X)


with plt.style.context('seaborn-v0_8'):
    plt.figure(figsize=(10, 5))
    plt.title("The uniform distribution")
    plt.plot(X, y)
    plt.show()
```

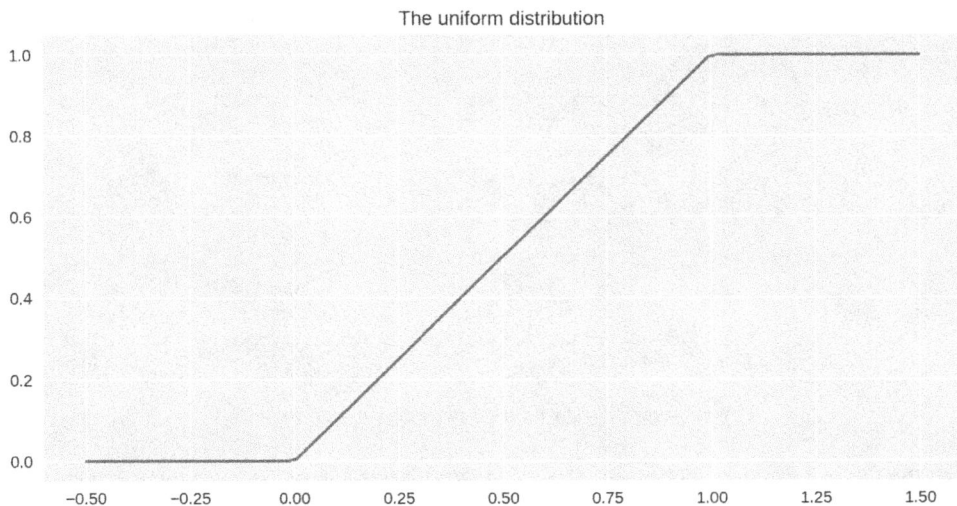

*Figure 19.8: The uniform distribution*

In the following section, we will properly define and study this object in detail for all real-valued random variables.

## 19.3.1   The cumulative distribution function

What we have seen in our motivating example is an instance of a *cumulative distribution function*, or CDF in short. Let's jump into the formal definition right away.

> **Definition 19.3.1 (Cumulative distribution function)**
>
> Let $X$ be a real-valued random variable. The function defined by
>
> $$F_X(x) := P(X \leq x) \tag{19.4}$$
>
> is called the *cumulative distribution function* (CDF) of $X$.

Again, let's unpack this. Recall that in the definition of real-valued random variables (*Definition 19.1.2*), we have used the inverse images $X^{-1}((a, b))$.

Something similar is going on here. $P(X \leq x)$ is the abbreviation for $P\left(X^{-1}((-\infty, x])\right)$, which we are too lazy to write. Similarly to $X^{-1}((a, b)))$, you can visualize $X^{-1}((-\infty, x]))$ by pulling the interval $(-\infty, x]$ back to $\Omega$ using the mapping $X$.

Sets of the form $X^{-1}((-\infty, x])$ are called the *level sets* of $X$.

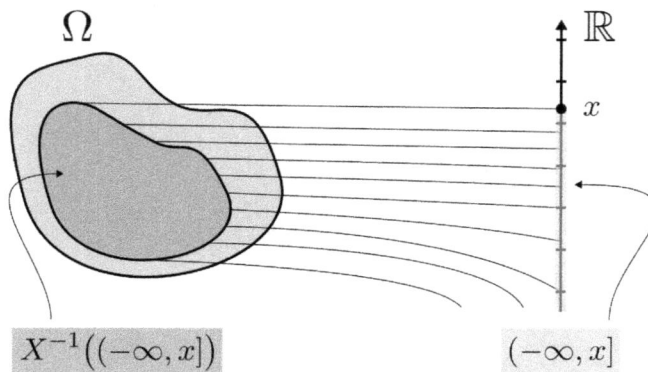

*Figure 19.9: The level set of a random variable*

According to the Oxford English Dictionary, the word *cumulative* means *"increasing or increased in quantity, degree, or force by successive additions."* For discrete random variables, using $P(X = k)$ was enough, but since real random variables are more nuanced, we have to use the cumulative probabilities $P(X \leq x)$ to meaningfully describe them.

Why do we like to work with distribution functions? Because they condense all the relevant information about a random variable in a real function.

For instance, we can express probabilities like

$$P(a < X \leq b) = F_X(b) - F_X(a).$$

To give an example, let's revisit the introduction, where we were selecting a random number between zero and one. There, the random variable $X$ with CDF

$$F_X(x) = \begin{cases} 0 & \text{if } x \leq 0, \\ x & \text{if } 0 < x \leq 1, \\ 1 & \text{otherwise.} \end{cases} \tag{19.5}$$

is said to be *uniformly distributed over* $[0, 1]$, or $X \sim \text{Uniform}(0, 1)$ in short. We'll see a ton of examples later, but keep note of this, as the uniform distribution will be our textbook example throughout this section.

## 19.3.2   Properties of the distribution function

Cumulative distribution functions have three properties that characterize them: they are always non-decreasing, right-continuous (whatever that might be), and their limits are 0 and 1 toward $-\infty$ and $\infty$ respectively. You might have guessed some of this from the definition, but here is the formal theorem that summarizes this.

**Theorem 19.3.1** *(Properties of CDFs)*

*Let $X$ be a real-valued random variable with CDF $F_X$. Then, $F_X$ is*

*(a) non-decreasing (that is, $x \leq y$ implies $F_X(x) \leq F_X(y)$),*

*(b) right-continuous (that is, $\lim_{x \to x_0+} F_X(x) = F_X(x_0)$, or in other words, taking the right limit is interchangeable with $F_X$),*

*(c) and the limits*

$$\lim_{x \to -\infty} F_X(x) = 0, \quad \lim_{x \to \infty} F_X(x) = 1$$

*hold.*

*Proof.* The proofs are relatively straightforward. *(a)* follows from the fact that if $x < y$, then we have

$$X^{-1}\big((-\infty, x]\big) \subseteq X^{-1}\big((-\infty, y]\big).$$

In other words, the event $X \leq x$ is a subset of $X \leq y$. Thus, due to the monotonicity of probability measures, we have $P(X \leq x) \leq P(X \leq y)$.

*(b)* Here, we need to show that $\lim_{x \to x_0+} P(X \leq x) = P(X \leq x_0)$. For this, note that for any $x_n \to x_0$ with $x_n > x_0$, the event sequence $\{\omega \in \Omega : X(\omega) \leq x_n\}$ is decreasing, and

$$\cap_{n=1}^{\infty} X^{-1}\big((-\infty, x_n]\big) = X^{-1}\big((-\infty, x_0]\big).$$

Because of the upper continuity of probability measures (see *Theorem 18.2.7*), the right continuity of $F_X$ follows.

*(c)* Again, this follows from the fact that

$$\cap_{n=1}^{\infty} X^{-1}\big((-\infty, n]\big) = \emptyset$$

and

$$\cup_{n=1}^{\infty} X^{-1}\big((-\infty, n]\big) = \Omega.$$

Since $P(\emptyset) = 0$ and $P(\Omega) = 1$, the statement follows from the upper and lower continuity of probability measures. (See *Theorem 18.2.6* and *Theorem 18.2.7*.)

### Remark 19.3.1 (Alternative definition of CDF-s)

In the literature, you can sometimes find that instead of (19.4), the CDF of $X$ is defined by

$$F_X^*(x) := P(X < x),$$

that is, $X < x$ instead of $X \leq x$. This doesn't change the big picture, but some details are slightly different. For instance, this change makes $F_X$ *left-continuous* instead of right-continuous. These minute details matter if you dig really deep, but in machine learning, we'll be fine without thinking too much about them.

*Theorem 19.3.1* is true the other way around: if you give me a non-decreasing right-continuous function $F(x)$ with $\lim_{x \to -\infty} F(x) = 0$ and $\lim_{x \to \infty} F(x) = 1$, I can construct a random variable such that its distribution function matches $F(x)$.

## 19.3.3 Cumulative distribution functions for discrete random variables

The discrete and real-valued case is not entirely disjoint: in fact, discrete random variables have cumulative distribution functions as well. (But not the other way around; that is, real-valued random variables cannot be described with sequences.)

Say, if $X$ is a discrete random variable taking the values $x_1, x_2, \ldots$, then its CDF is

$$F_X(x) = \sum_{x_i \leq x} P(X = x_i),$$

which is a piecewise continuous function. For example, *Figure 19.10* illustrates the CDF of the binomial distribution.

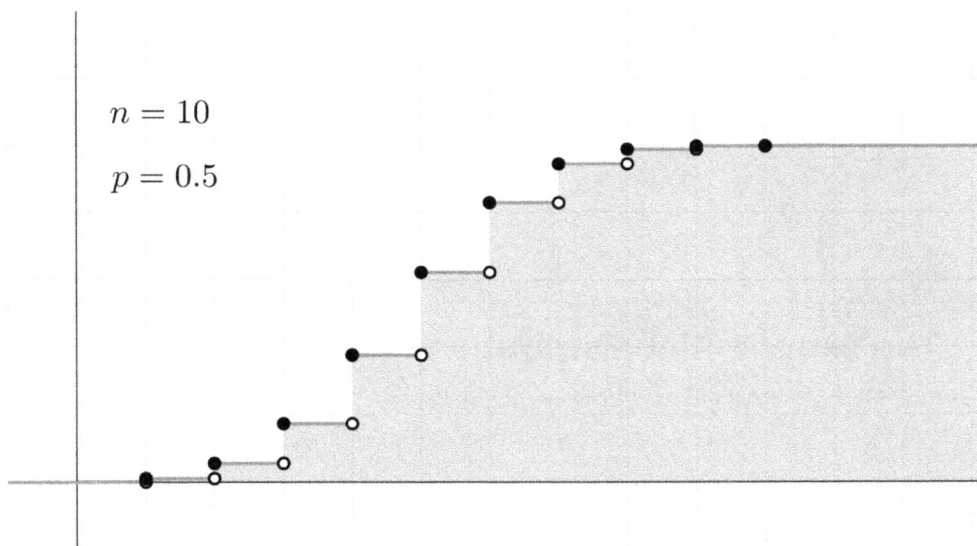

*Figure 19.10: The CDF of Binomial(10, 0.5)*

The strength or probability lies in its ability to translate real-world phenomena into coin tosses, dice rolls, dart throws, lightbulb lifespans, and many more. This is possible because of distributions. Distributions are the ribbons stringing together a vast bundle of random variables.

Let's meet some of the most important ones!

## 19.3.4   The uniform distribution

We have already seen a special case of the uniform distribution: selecting a random number from the interval $[0, 1]$, such that all outcomes are "equally likely." The general uniform distribution captures the same concept, except on an arbitrary interval $[a, b]$ for any $a < b$. That is, the random variable $X$ is uniformly distributed on the interval $[a, b]$, or $X \sim \text{Uniform}(a, b)$ in symbols, if

$$P(\alpha < X \le \beta) = \frac{1}{b - a} \Big| [a, b] \cap (\alpha, \beta] \Big|$$

for all $\alpha < \beta$, where $\big| [c, d] \big|$ denotes the length of the interval $[c, d]$,

In other words, the probability of our random number falling into a given interval is proportional to the interval's length. This is how the condition "equally likely" makes sense: as there are uncountably many possible outcomes, the probability of each individual outcome is zero, but equally long intervals have an equal chance.

In line with the definition, the distribution function of $X$ is

$$F_X(x) = \begin{cases} 0 & \text{if } x \le a, \\ \frac{x-a}{b-a} & \text{if } a < x \le b, \\ 1 & \text{otherwise.} \end{cases}$$

## 19.3.5   The exponential distribution

Let's turn our attention toward a different problem: lightbulbs. According to some mysterious (and probably totally inaccurate) lore, lightbulbs possess the so-called memoryless property. That is, their expected lifespan is the same at *any* point in their life.

To put this into a mathematical form, let $X$ be a random variable denoting the lifespan of a given lightbulb. The memoryless property states that if the lightbulb has already lasted $s$ seconds, then the probability of lasting another $t$ is the same as in the very first moment of its life. That is,

$$P(X > t + s \mid X > s) = P(X > t).$$

Expanding the left side, we have

$$P(X > t + s \mid X > s) = \frac{P(X > t + s, X > s)}{P(X > s)}$$
$$= \frac{P(X > t + s)}{P(X > s)},$$

as $\{X > t + s\} \cap \{X > s\} = \{X > t + s\}$. Thus, the memoryless property implies that

$$P(X > t + s) = P(X > t)P(X > s). \tag{19.6}$$

If we think about the probabilities as a function $f(t) = P(X > t)$, (19.6) can be viewed as a functional equation. And a famous one at that. Without going into the painful details, the only continuous solution is the exponential function $f(t) = e^{at}$, where $a \in \mathbb{R}$ is a parameter.

As we are talking about the lifespan of a lightbulb here, the probability of it lasting forever is zero. That is,

$$\lim_{t \to \infty} P(X > t) = 0$$

holds. Thus, as

$$\lim_{t \to \infty} e^{at} = \begin{cases} 0 & \text{if } a < 0, \\ 1 & \text{if } a = 0, \\ \infty & \text{if } a > 0, \end{cases}$$

only the negative parameters are valid in our case. This characterizes the exponential distribution. In general, $X \sim \exp(\lambda)$ for a $\lambda > 0$ if

$$F_X(x) = \begin{cases} 0 & \text{if } x < 0, \\ 1 - e^{-\lambda x} & \text{if } x \geq 0. \end{cases}$$

Let's plot this for visualization!

```
from scipy.stats import expon
X = np.linspace(-0.5, 10, 100)
params = [0.1, 1, 10]
ys = [expon.cdf(X, scale=1/l) for l in params]

with plt.style.context('seaborn-v0_8'):
    plt.figure(figsize=(10, 5))
```

```
for l, y in zip(params, ys):
    plt.plot(X, y, label=f"lambda = {l}")

plt.title("The exponential distribution")
plt.legend()
plt.show()
```

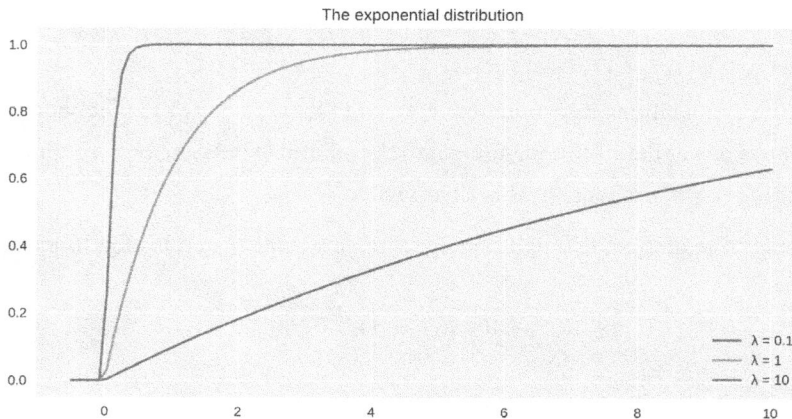

*Figure 19.11: The exponential distribution*

The exponential distribution is extremely useful and frequently encountered in real-life applications. For instance, it models the requests incoming to a server, customers standing in a queue, buses arriving at a bus stop, and many more.

We'll talk more about special distributions in later chapters, and we'll add quite a few others as well.

## 19.3.6  The normal distribution

You have probably seen the bell curve at one point in your life, as it is used to describe a wide range of statistical phenomena. Salaries, prices, height, intelligence: they all seem to follow the same symmetric bell-shaped distribution.

This is described by the famous normal distribution: we say that $X$ is normally distributed, or $X \sim \mathcal{N}(\mu, \sigma^2)$, if

$$F_X(x) = \frac{1}{\sigma\sqrt{2\pi}} \int_{-\infty}^{x} e^{-\frac{(t-\mu)^2}{2\sigma^2}} \, dt,$$

where $\mu, \sigma \in \mathbb{R}$. The parameter $\mu$ is called the *mean* of $X$, while $\sigma^2$ is its *variance* and $\sigma$ is its *standard deviation*. (We'll see more about these quantities when talking about the expected value and variance in *Chapter 20*.)

Let's see the plot the inner part first, which you know as the famous bell curves:

```python
from scipy.stats import norm
X = np.linspace(-10, 10, 1000)
sigmas = [0.5, 1, 2, 3]
ys = [norm.pdf(X, scale=sigma) for sigma in sigmas]


with plt.style.context('seaborn-v0_8'):
    plt.figure(figsize=(10, 5))

    for sigma, y in zip(sigmas, ys):
        plt.plot(X, y, label=f"sigma = {sigma}")

    plt.title("The bell curves")
    plt.savefig("bell_curve.png", dpi=300)
    plt.legend()
    plt.show()
```

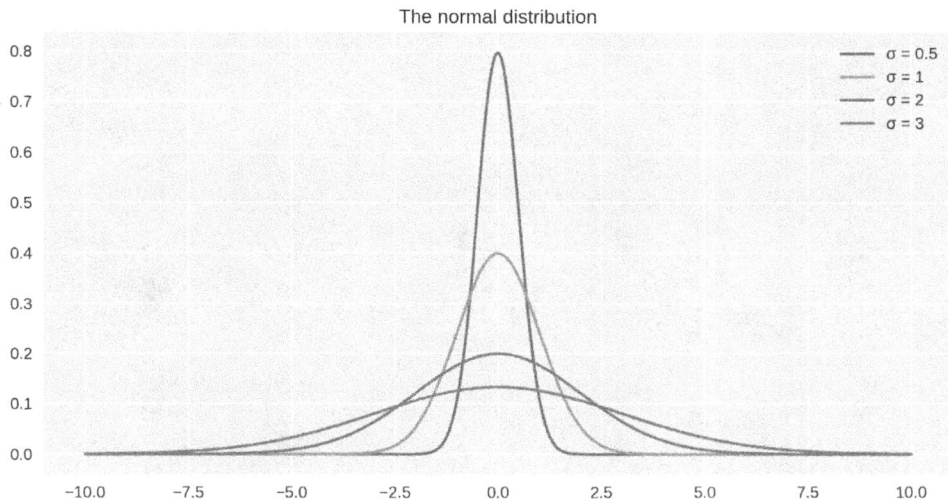

*Figure 19.12: The bell curves*

Surprisingly, no closed expression exists for its CDF

$$F_X(x) = \frac{1}{\sigma \sqrt{2\pi}} \int_{-\infty}^{x} e^{-\frac{(t-\mu)^2}{2\sigma^2}} \, dt.$$

No, it's not that mathematicians were not smart enough to figure it out; it provably doesn't exist. In the ancient days, statisticians used to read out its values from statistical tables, located in massive tomes.

Now, let's plot $F_X$:

```python
X = np.linspace(-10, 10, 1000)
sigmas = [0.5, 1, 2, 3]
ys = [norm.cdf(X, scale=sigma) for sigma in sigmas]


with plt.style.context('seaborn-v0_8'):
    plt.figure(figsize=(10, 5))

    for sigma, y in zip(sigmas, ys):
        plt.plot(X, y, label=f"sigma = {sigma}")

    plt.title("The normal distribution")
    plt.legend()

    plt.show()
```

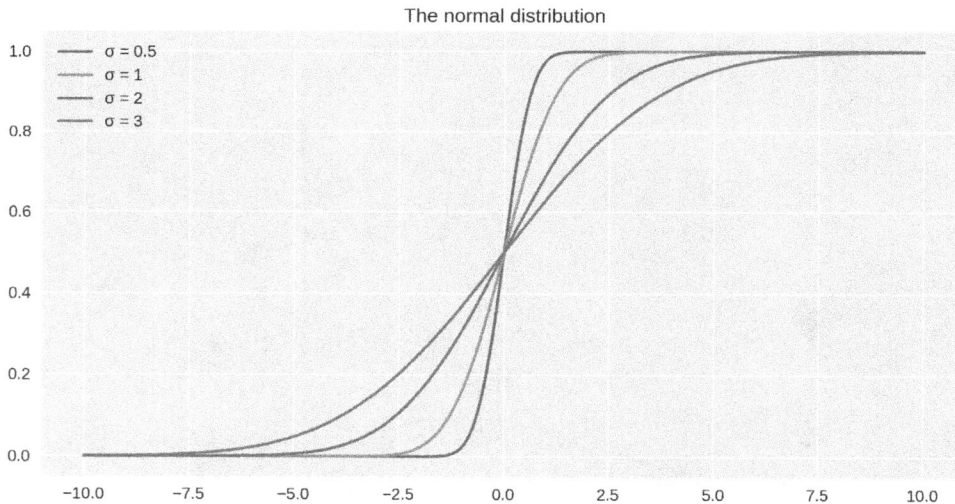

*Figure 19.13: The normal distribution*

Normal distribution is the single most important one in statistics, and we'll see it appearing everywhere, not just in practice but in theory as well.

To sum up, distributions are the lifeblood of probability theory, and distributions can be represented with cumulative distribution functions.

However, CDFs have a significant drawback: it's hard to express the probability of more complex events with them. Later, we'll see several concrete examples of where CDFs fail.

Without going into details, one example points toward multidimensional distributions. (I hope that their existence and importance are not surprising to you.) There, the distribution functions can be used to express the probability of rectangle-shaped events, but not, say, spheres.

To be a bit more precise, if $X, Y \sim \text{Uniform}(0, 1)$, then the probability

$$P(X^2 + Y^2 < 1)$$

cannot be directly expressed in terms of the two-dimensional CDF $F_{X,Y}(x, y)$ (whatever that may be). Fortunately, this is not our only tool. Recall the $e^{-\frac{(x-\mu)^2}{2\sigma^2}}$ part inside the CDF of the normal distribution? This is a special instance of *density functions*, which is what we'll learn about in the next section.

## 19.4 Density functions

Distribution functions are not our only tool to describe real-valued random variables. If you have studied probability theory from a book/lecture/course written by a non-mathematician, you have probably seen a function such as

$$p(x) = \frac{1}{\sqrt{2\pi}} e^{-\frac{x^2}{2}}$$

referred to as "probability" at some point. Let me tell you, this is definitely *not* a probability. I have seen this mistake so much that I decided to write short X/Twitter threads properly explaining probabilistic concepts, from which this book was grown out of. So, I take this issue to heart.

Here is the problem with cumulative distribution functions: they represent global information about local objects. Let's unpack this idea. If $X$ is a real-valued random variable, the CDF

$$F_X(x) = P(X \le x)$$

describes the probability of $X$ being smaller than a given $x$. But what if we are interested in what happens *around* $x$? Say, in the case of the uniform distribution (19.5), we have

$$\begin{aligned} P(X = x) &= \lim_{\varepsilon \to 0} P(x - \varepsilon < X \le x) \\ &= \lim_{\varepsilon \to 0} \left( F_X(x) - F_X(x - \varepsilon) \right) \\ &= \lim_{\varepsilon \to 0} \varepsilon \\ &= 0. \end{aligned}$$

(We used *Theorem 18.2.7* when taking the limit.)

Thus, as we have already seen, the probability of picking a particular point is zero. Contrary to the discrete case, $P(X = x)$ tells us nothing about how the distribution of $X$ behaves around $x$.

And the worst thing is, this is the same for a wide array of distributions. For instance, you can check it manually for the exponential distribution.

Isn't this strange? The probability of individual outcomes is zero for both the uniform and exponential distribution, yet the distributions themselves couldn't be more different. Let's examine the problem from another perspective. By definition,

$$P(a < X \le b) = F_X(b) - F_X(a)$$

holds. Does this look familiar to you? Increments of $F_X$ on the right, probabilities on the left. Where have we seen *increments* before?

In the fundamental theorem of calculus (*Theorem 14.1.2*), that's where. That is, if $F_X$ is differentiable and its derivative is $F_X'(x) = f_X(x)$, then

$$\int_a^b f_X(x)dx = F_X(b) - F_X(a). \tag{19.7}$$

The function $f_X(x)$ seems to be what we are looking for: it represents the local behavior of $X$ around $x$. But instead of describing the probability, it describes its rate of change. This is called a *probability density function*.

By turning this argument around, we can *define* density functions using (19.7). Here is the mathematically precise version.

**Definition 19.4.1 (Density functions)**

Let $(\Omega, \Sigma, P)$ be a probability space, and $X : \Omega \to \mathbb{R}$ be a real-valued random variable. The function $f_X : \mathbb{R} \to \mathbb{R}$ is called the *probability density function* (PDF) of $X$, if it is integrable, and

$$\int_a^b f_X(x)dx = F_X(b) - F_X(a) \tag{19.8}$$

holds for all $a, b \in \mathbb{R}$.

Again, (19.8) is the Newton-Leibniz formula (*Theorem 14.1.2*) in disguise.

The following theorem makes this connection precise.

---

**Theorem 19.4.1** *(The density function as derivative)*

*Let X be a real-valued random variable. If the cumulative distribution function $F_X(x)$ is everywhere differentiable,*
*then*

$$f_X(x) = \frac{d}{dx} F_X(x)$$

*is a density function for X.*

---

*Proof.* This is just a simple application of the fundamental theorem of calculus (*Theorem 14.1.2*). If the derivative
indeed exists, then

$$\int_a^b \frac{d}{dx} F_X(x) dx = F_X(b) - F_X(a),$$

which means that $f_X(x) = \frac{d}{dx} F_X(x)$ is indeed a density function.

---

**Remark 19.4.1 (Density functions are not unique)**

Note that density functions are *not* unique. If $X$ is a random variable with density $f_X$, then, say, modifying $f_X$
at a single point still functions as a density function for $X$.

To be more precise, define

$$f_X^*(x) = \begin{cases} f_X(x) & \text{if } x \neq 0, \\ f_X(0) + 1 & \text{if } x = 0. \end{cases}$$

You can check by hand that $f_X^*$ is still a density for $X$, yet $f_X \neq f_X^*$.

---

One more thing before we move on. Recall that discrete random variables are characterized by probability
mass functions (*Definition 19.2.1*). Mass functions and densities are two sides of the same coin.

The probability mass function is analogous to the density function, yet we don't have terminology for random
variables with the latter. We'll fix this now.

---

**Definition 19.4.2 (Continuous random variables)**

Let $(\Omega, \Sigma, P)$ be a probability space, and $X : \Omega \to \mathbb{R}$ be a real-valued random variable. We say that $X$ is
*continuous* if it has a probability density function.

Discrete and continuous random variables are the backbones of probability theory: the most interesting random variables are falling into either of these two classes. (Later in the chapter, we'll see that there are more types, but these two are the most important.)

Now we are ready to get our hands dirty and see some density functions in practice.

## 19.4.1 Density functions in practice

After all this introduction, let's see a few concrete examples. So far, we have seen two real-valued non-discrete distributions: uniform and exponential.

**Example 1.** Let's start with $X \sim \text{Uniform}(0,1)$. Can we apply *Theorem 19.4.1* directly? Not without a little snag. Or two, to be more precise.

Why? Because the distribution function

$$F_X(x) = \begin{cases} 0 & \text{if } x \leq 0, \\ x & \text{if } 0 < x \leq 1, \\ 1 & \text{if } x > 1 \end{cases}$$

is not differentiable at $x = 0$ and $x = 1$. However, it is differentiable everywhere else, and its derivative

$$F_X'(x) = \begin{cases} 0 & \text{if } x < 0, \\ 1 & \text{if } 0 < x < 1, \\ 0 & \text{if } x > 1 \end{cases}$$

is indeed a density function. (You can check this by hand.) This density is patched together from the derivative of $F_X(x)$ on the intervals $(-\infty, 0)$, $(0, 1)$, and $(1, \infty)$.

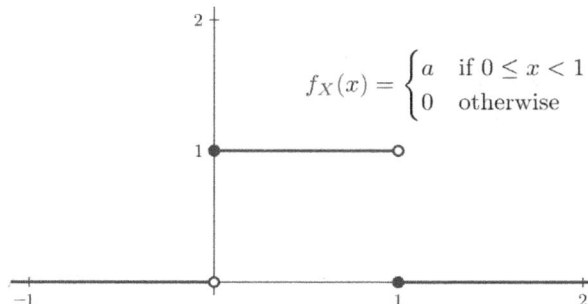

$$f_X(x) = \begin{cases} a & \text{if } 0 \leq x < 1 \\ 0 & \text{otherwise} \end{cases}$$

Figure 19.14: Density function of the uniform distribution on $[0, 1]$

**Example 2.** In the case of the exponentially distributed random variable $Y \sim \exp(\lambda)$, the function

$$f_Y(x) = \begin{cases} 0 & \text{if } x < 0, \\ \lambda e^{-\lambda x} & \text{if } x \geq 0 \end{cases}$$

is a proper density function, which we obtained by differentiating $F_Y(x)$ whenever possible. Again, the density $f_X(x)$ is patched together from the derivatives on the intervals $(-\infty, 0)$ and $(0, \infty)$.

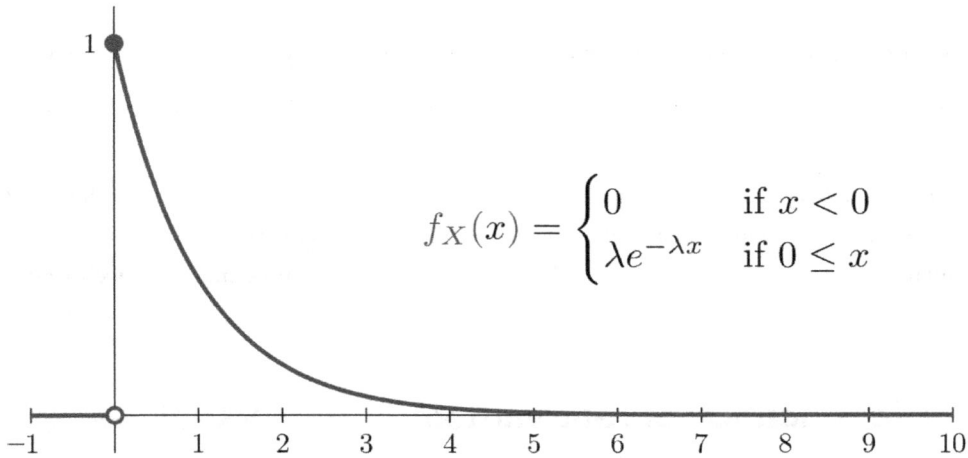

$$f_X(x) = \begin{cases} 0 & \text{if } x < 0 \\ \lambda e^{-\lambda x} & \text{if } 0 \leq x \end{cases}$$

*Figure 19.15: Density function of the* $\exp(1)$ *distribution*

**Example 3.** Now, I am going to turn everything upside down. Let $Z \sim \text{Bernoulli}(1/2)$, which is a discrete random variable with probability mass function

$$p_Z(0) = p_Z(1) = \frac{1}{2},$$

and cumulative distribution function

$$F_Z(x) = \begin{cases} 0 & \text{if } x < 0, \\ \frac{1}{2} & \text{if } 0 \leq x < 1, \\ 1 & \text{if } x \geq 1. \end{cases}$$

Like the uniform and exponential distributions, this CDF is also differentiable except for a few points (which are 0 and 1).

Thus, it is natural to guess that, like before, we can patch its derivatives together to obtain a density function. However, there is a bigger snag: the derivative of $F_Z$ is zero, at least wherever it exists. It turns out that $Z$ *does not* have a density function at all!

What's the issue? I'll tell you: the *jump discontinuities* of $F_Z(x)$ at $x = 0$ and $x = 1$. Although the CDFs of the uniform and exponential distributions were not differentiable at finitely many points, they did not include any jump discontinuities.

We are not going to dive deep into the details, but the gist is: if there is a jump discontinuity in the CDF, the density function does not exist.

> **Remark 19.4.2 (The non-existence of density despite the lack of jump discontinuities)**
>
> Unfortunately, the reverse direction of "jump discontinuity in the CDF $\implies$ no PDF exists" is not true, I repeat, *not true*.
>
> We can find random variables whose cumulative distribution functions are continuous, but their density does not exist. One famous example is the Cantor function (https://en.wikipedia.org/wiki/Cantor_function), also known as the Devil's staircase. (Only follow this link if you are brave enough or well-trained in real analysis, which is the same.)

## 19.4.2   Classification of real-valued random variables

So far, we have been focusing on two special kinds of real-valued random variables: discrete random variables (*Definition 19.1.1*) and continuous ones (*Definition 19.4.2*).

We've seen all kinds of objects describing them. Every real-valued random variable has a cumulative distribution function (*Definition 19.3.1*), but while discrete ones are characterized by probability mass functions (*Definition 19.2.1*), the continuous ones are by density functions (*Definition 19.4.1*).

Are these two all that's out there?

No. There are mixed cases. For instance, consider the following example. We are selecting a random number from $[0, 1]$, but we add a little twist to the picking process. First, we toss a fair coin, and if it comes up heads, we pick 0. Otherwise, we pick uniformly between zero and one.

To describe this weird process, let's introduce two random variables: let $X$ be the final outcome and $Y$ be the outcome of the coin toss. Then, using the conditional version of the law of total probability (see *Theorem 18.3.1*), we have

$$P(X \leq x) = P(X \leq x \mid Y = \text{heads})P(Y = \text{heads})$$
$$+ P(X \leq x \mid Y = \text{tails})P(Y = \text{tails}).$$

As

$$P(X \le x \mid Y = \text{heads}) = \begin{cases} 0 & \text{if } x < 0, \\ 1 & \text{if } x \ge 1, \end{cases}$$

and $P(X \le x \mid Y = \text{tails}) = F_{\text{Uniform}(0,1)}(x)$, we ultimately have

$$F_X(x) = \begin{cases} 0 & \text{if } x < 0, \\ \frac{x+1}{2} & \text{if } 0 \le x < 1, \\ 1 & \text{if } x \ge 1. \end{cases}$$

Ultimately, $F_X$ is the *convex combination* of two cumulative distribution functions. (A convex combination is a linear combination where the coefficients are positive and their sum is 1.)

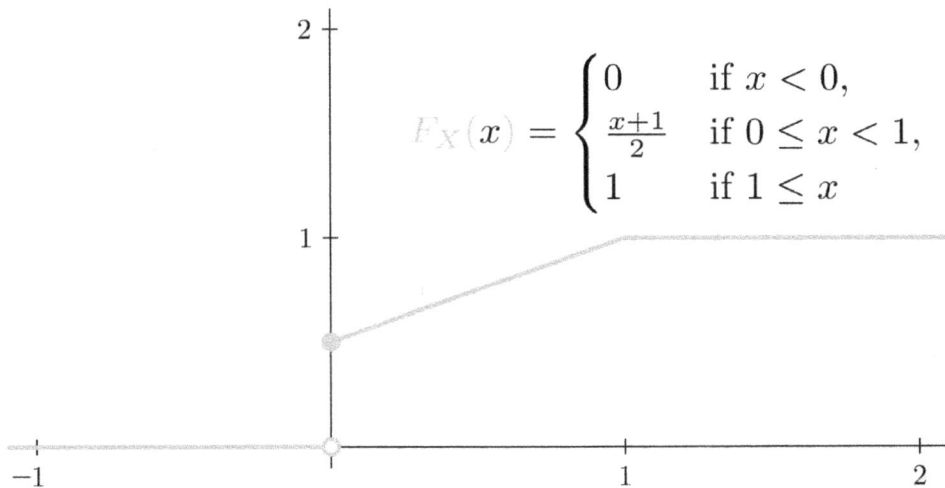

*Figure 19.16: CDF of the mixed distribution X*

Thus, the random variable $X$ is not discrete nor continuous. So, what is it?

It's time to add order to chaos! In this section, we are going to provide a complete classification for our real-valued random variables. This is a beautiful, albeit advanced, topic so feel free to skip it on a first read.

Let's start at a seemingly distant topic: subsets of $\mathbb{R}$ that are so small that they practically vanish. Since $\mathbb{R}$ is a one-dimensional object, we are usually talking about *length* here, but let's forget that terminology and talk about *measure* instead. We'll denote the measure of a set $A \subsetneq \mathbb{R}$ by $\lambda(A)$, whatever that might be.

We are not going too deep into the details and will keep on using the notion of measure intuitively. For instance, the measure of an interval $[a, b]$ is $\lambda([a, b]) = b - a$.

Our measure $\lambda$ has some fundamental properties, for instance,

1. $\lambda(\emptyset) = 0$,
2. $\lambda(A) \leq \lambda(B)$ if $A \subseteq B$,
3. and $\lambda(\cup_{k=1}^{\infty} A_k) = \sum_{k=1}^{\infty} \lambda(A_k)$ if $A_i \cap A_j = \emptyset$.

This almost behaves like a probability measure, with one glaring exception: $\lambda(\mathbb{R}) = \infty$. This is not an accident.

What is the measure of a finite set $\{a_1, \ldots, a_n\}$? Intuitively, it is zero, and from this example, we'll conjure up the concept of *sets of zero measure*.

**Theorem 19.4.2** *(Sets of zero measure)*

*Let $A \subseteq \mathbb{R}$ be an arbitrary set. Suppose that for any arbitrarily small $\varepsilon > 0$, there exists a union of intervals $E = \cup_{k=1}^{\infty}(a_i, b_i)$ such that*

*(a) $\lambda(E) < \varepsilon$,*

*(b) and $A \subseteq E$,*

*then, $\lambda(A) = 0$.*

*Proof.* As $A \subseteq E$, $\lambda(A) \leq \lambda(E) < \varepsilon$. This means that $\lambda(A)$ is smaller than any positive real number, thus it must be zero.

Let's see some examples.

**Example 1.** A set of a single element has zero measure. As any $\{a\}$ can be covered by the interval $(a - \varepsilon, a + \varepsilon)$ for some $\varepsilon > 0$. As $\lambda\big((a - \varepsilon, a + \varepsilon)\big) = 2\varepsilon$, the conditions of *Theorem 19.4.2* apply, thus $\lambda(\{a\}) = 0$.

**Example 2.** A finite set has zero measure. To see this, let $A = \{a_1, \ldots, a_n\}$ be our finite set. The system of intervals

$$E = \bigcup_{k=1}^{n} \left( a_k - \frac{\varepsilon}{2n}, a_k + \frac{\varepsilon}{2n} \right), \quad \varepsilon > 0$$

will do the job, as the intervals are mutually disjoint for a small enough $\varepsilon$, thus

$$\lambda(E) = \sum_{k=1}^{n} \lambda\left(\left(a_k - \frac{\varepsilon}{2n}, a_k + \frac{\varepsilon}{2n}\right)\right)$$

$$= \sum_{k=1}^{n} \frac{\varepsilon}{n}$$

$$= \varepsilon.$$

**Example 3.** A countable set has zero measure. For any $A = \{a_1, a_2, \dots\}$, the system of intervals

$$E = \bigcup_{k=1}^{\infty} \left(a_k - \frac{\varepsilon}{2^{k+1}}, a_k + \frac{\varepsilon}{2^{k+1}}\right), \quad \varepsilon > 0$$

work perfectly, as

$$\lambda(E) \le \sum_{k=1}^{\infty} \frac{\varepsilon}{2^k} = \varepsilon.$$

For instance, as the set of integers and rational numbers are both countable, $\lambda(\mathbb{Z}) = \lambda(\mathbb{Q}) = 0$.

Overall, sets of zero measure are true to their name: they are small. (They are not necessarily countable though.) Why are these important? We'll see this in the next section.

> **Remark 19.4.3 (Density functions are not unique, take two)**
>
> Do you recall *Remark 19.4.1*, where we saw that changing the density function of $X$ at a single point is also a density for $X$?
>
> Turns out that you can actually modify $f_X$ at an entire set of measure zero. Say,
>
> $$f_X^*(x) = \begin{cases} f_X(x) & \text{if } x \notin \mathbb{Q}, \\ 0 & \text{if } x \in \mathbb{Q} \end{cases}$$
>
> is still a density function for $X$. Unfortunately, we don't have the tools to show this, as it would require moving beyond the good old Riemann integral, which is way beyond our scope.

The main difference between a discrete and a continuous random variable is the set where they live. Fundamentally, they are both real-valued random variables, but the range of a discrete variable is a set of measure zero.

Let's introduce the concept of *singular random variables* to make this notion precise.

---

**Definition 19.4.3 (Singular random variables)**

Let $(\Omega, \Sigma, P)$ be a probability space and $X : \Omega \to \mathbb{R}$ be a real-valued random variable. We say that $X$ is *singular* if its range $X(\Omega) = \{X(\omega) : \omega \in \Omega\} \subseteq \mathbb{R}$ is a set of zero measure, that is,

$$\lambda\big(X(\Omega)\big) = 0$$

holds.

---

All discrete random variables are singular, but not the other way around. For instance, the Cantor function (https://en.wikipedia.org/wiki/Cantor_function) is a good example.

Why are singular random variables so special? Because every distribution can be written as the sum of a singular and a continuous one! Here is the famous Lebesgue decomposition theorem.

---

**Theorem 19.4.3 *(Lebesgue's decomposition theorem)***

*Let $(\Omega, \Sigma, P)$ be a probability space and $X : \Omega \to \mathbb{R}$ be a real-valued random variable. Then, there exists a singular random variable $X_s$ and a continuous random variable $X_c$ such that*

$$F_X = \alpha F_{X_s} + \beta F_{X_c},$$

*where $\alpha + \beta = 1$, and $F_X$, $F_{X_s}$, $F_{X_c}$ are the corresponding cumulative distribution functions.*

---

We are not going to prove this here but the gist is this: there are singular random variables, continuous ones, and their sum.

# 19.5 Summary

With the introduction of random variables, we learned to represent abstract probability spaces as random variables, mapping a sufficiently expressive collection of events to the real numbers. Instead of $\sigma$-algebras and probability measures, now we can deal with numbers. As I told you, *"The strength or probability lies in its ability to translate real-world phenomena into coin tosses, dice rolls, dart throws, lightbulb lifespans, and many more."*

Most common random variables come in two forms: discrete or continuous, meaning that either it can be described with a probability mass function

$$\left\{ P(X = x_k) \right\}_{k=1}^{\infty},$$

or with a density function $f_X$, satisfying

$$P(a \le X \le b) = \int_a^b f_X(x)dx.$$

Translating experiments to distributions is the secret sauce of probability theory and statistics. For instance, the time between call center calls, bus arrivals, earthquakes, and insurance claims are all modeled with the exponential distribution, a mathematical object we can work with.

I know that learning takes a lifetime, but we must wrap this book up at some point. There is one more concept left that I want to tell you about: the expected value, enabling us to measure the statistical properties of our distributions. See you in the next chapter!

# 19.6 Problems

**Problem 1.** Let $X$ and $Y$ be two independent random variables, and let $a, b \in \mathbb{R}$ be two arbitrary constants. Show that $X - a$ and $Y - b$ are also independent from each other.

**Problem 2.** Let $X$ be a continuous random variable. Show that $P(X = x) = 0$ for any $x \in \mathbb{R}$.

**Problem 3.** Let $X \sim \text{Bernoulli}(p)$ and $Y \sim \text{Binomial}(n, p)$. Calculate the probability distribution of $X + Y$.

**Problem 4.** Let $X \sim \text{Bernoulli}(p)$ be the result of a coin toss. We select a random number $Y$ from $[0, 2]$ based on the result of the toss: if $X = 0$, we pick a number from $[0, 1]$ using the uniform distribution, but if $X = 1$, we pick a number from $[1, 2]$, once more using the uniform distribution. Find the cumulative distribution function of $Y$. Does $Y$ have a density function? If yes, find it.

# Join our community on Discord

Read this book alongside other users, Machine Learning experts, and the author himself.

Ask questions, provide solutions to other readers, chat with the author via Ask Me Anything sessions, and much more.

Scan the QR code or visit the link to join the community.

`https://packt.link/math`

# 20

# The Expected Value

In the last chapter, we learned about probability distributions, the objects that represent probabilistic models as sequences or functions. After all, there is the entire field of calculus to help us deal with functions, so they open up a wide array of mathematical tools.

However, we might not need all the information available. Sometimes, simple descriptive statistics such as mean, variance, or median suffice. Even in machine learning, loss functions are given in terms of them. For instance, the famous mean-squared error

$$\text{MSE}(\mathbf{x}, \mathbf{y}) = \frac{1}{n} \sum_{i=1}^{n} (f(x_i) - y_i)^2, \quad \mathbf{x}, \mathbf{y} \in \mathbb{R}^n$$

is the variance of the prediction error. Deep down, these familiar quantities are rooted in probability theory, and we'll devote this chapter to learning about them.

# 20.1  Discrete random variables

Let's play a simple game. I toss a coin, and if it comes up heads, you win $1 . If it is tails, you lose $2 .

Up until now, we were dealing with questions like the probability of winning. Say, for the coin toss, whether you win or lose, we have

$$P(\text{heads}) = P(\text{tails}) = \frac{1}{2}.$$

Despite the equal chances of winning and losing, should you play this game? Let's find out.

After $n$ rounds, your earnings can be calculated by the number of heads times $1 minus the number of tails times $2 . If we divide total earnings by $n$, we obtain your average winnings per round. That is,

$$
\begin{aligned}
\text{your average winnings} &= \frac{\text{total winnings}}{n} \\
&= \frac{1 \cdot \#\text{heads} - 2 \cdot \#\text{tails}}{n} \\
&= 1 \cdot \frac{\#\text{heads}}{n} - 2 \cdot \frac{\#\text{tails}}{n},
\end{aligned}
$$

where #heads and #tails denote the number of heads and tails respectively.

Recall the frequentist interpretation of probability from *Section 18.2.7*? According to our intuition, we should have

$$
\begin{aligned}
\lim_{n\to\infty} \frac{\#\text{heads}}{n} &= P(\text{heads}) = \frac{1}{2}, \\
\lim_{n\to\infty} \frac{\#\text{tails}}{n} &= P(\text{tails}) = \frac{1}{2}.
\end{aligned}
$$

This means that if you play long enough, your average winnings per round is

$$
\begin{aligned}
\text{your average winnings} &= 1 \cdot P(\text{heads}) - 2 \cdot P(\text{tails}) \\
&= -\frac{1}{2}.
\end{aligned}
$$

So, as you are losing half a dollar per round on average, you definitely shouldn't play this game.

Let's formalize this argument with a random variable. Say, if $X$ describes your winnings per round, we have

$$P(X = 1) = P(X = -2) = \frac{1}{2},$$

so the average winnings can be written as

$$\text{average value of } X = 1 \cdot P(X = 1) - 2 \cdot P(X = -2)$$
$$= -\frac{1}{2}.$$

With a bit of a pattern matching, we find that for a general discrete random variable $X$, the formula looks like

$$\text{average value of } X = \sum_{\text{value}} (\text{value}) \cdot P(X = \text{value}).$$

And from this, the definition of *expected value* is born.

---

**Definition 20.1.1 (The expected value of discrete random variables)**

Let $(\Omega, \Sigma, P)$ be a probability space, and $X : \Omega \to \{x_1, x_2, \dots\}$ be a discrete random variable. The *expected value* of $X$ is defined by

$$\mathbb{E}[X] := \sum_{k} x_k P(X = x_k).$$

(Note that if $X$ assumes finitely many values, the sum only contains a finite number of terms.)

---

In English, the expected value describes the average value of a random variable in the long run. The expected value is also called the *mean* and is often denoted by $\mu$. Instead of using random variables, we'll often use the expected value symbol by plugging in distributions, like $\mathbb{E}[\text{Bernoulli}(p)]$. Although this is mathematically not precise, 1) it is simpler in certain cases, 2) and the expected value only depends on the distribution anyway.

It's time for examples.

**Example 1.** *Expected value of the Bernoulli distribution.* (See the definition of the Bernoulli distribution in *Section 19.2.1*.) Let $X \sim \text{Bernoulli}(p)$. Its expected value is quite simple to compute, as

$$\mathbb{E}[X] = 0 \cdot P(X = 0) + 1 \cdot P(X = 1) =$$
$$= 0 \cdot (1 - p) + 1 \cdot p$$
$$= p.$$

We've seen this before: the introductory example with the simple game is the transformed Bernoulli distribution $3 \cdot \text{Bernoulli}(1/2) - 2$.

**Example 2.** *Expected value of the binomial distribution.* (See the definition of the binomial distribution in Section 19.2.2.) Let $X \sim \text{Binomial}(n, p)$. Then

$$
\begin{aligned}
\mathbb{E}[X] &= \sum_{k=0}^{n} k P(X = k) \\
&= \sum_{k=0}^{n} k \binom{n}{k} p^k (1 - p)^{n-k} \\
&= \sum_{k=0}^{n} k \frac{n!}{k!(n - k)!} p^k (1 - p)^{n-k}.
\end{aligned}
$$

The plan is the following: absorb that $k$ with the fraction $\frac{n!}{k!(n-k)!}$, and adjust the sum such that its terms form the probability mass function for $\text{Binomial}(n - 1, p)$. As $n - k = (n - 1) - (k - 1)$, we have

$$
\begin{aligned}
\mathbb{E}[X] &= \sum_{k=0}^{n} k \frac{n!}{k!(n - k)!} p^k (1 - p)^{n-k} \\
&= np \sum_{k=1}^{n} \frac{(n - 1)!}{(k - 1)!((n - 1) - (k - 1))!} p^{k-1} (1 - p)^{(n-1)-(k-1)} \\
&= np \sum_{k=0}^{n-1} \frac{(n - 1)!}{k!(n - 1 - k)!} p^k (1 - p)^{(n-1-k)} \\
&= np \sum_{k=0}^{n-1} P(\text{Binomial}(n - 1, p) = k) \\
&= np.
\end{aligned}
$$

This computation might not look like the simplest, but once you get familiar with the trick, it'll be like second nature for you.

**Example 3.** *Expected value of the geometric distribution.* (See the definition of the geometric distribution in Section 19.2.3.) Let $X \sim \text{Geo}(p)$. We need to calculate

$$
\mathbb{E}[X] = \sum_{k=1}^{\infty} k(1 - p)^{k-1} p.
$$

Do you remember the geometric series (19.2)? This is almost it, except for the $k$ term, which throws a monkey wrench into our gears. To fix that, we'll use another magic trick. Recall that

$$
\frac{1}{1 - x} = \sum_{k=0}^{\infty} x^k.
$$

Now, we are going to *differentiate* the geometric series, thus obtaining

$$\frac{d}{dx}\frac{1}{1-x} = \frac{d}{dx}\sum_{k=0}^{\infty} x^k$$

$$= \sum_{k=0}^{\infty} \frac{d}{dx} x^k$$

$$= \sum_{k=1}^{\infty} kx^{k-1},$$

where we used the linearity of the derivative and the pleasant analytic properties of the geometric series. Mathematicians would scream upon the sight of switching the derivative and the infinite sum, but don't worry, everything here is correct as is. (Mathematicians are really afraid of interchanging limits. Mind you, for a good reason!)

On the other hand,

$$\frac{d}{dx}\frac{1}{1-x} = \frac{1}{(1-x)^2},$$

thus

$$\sum_{k=1}^{\infty} kx^{k-1} = \frac{1}{(1-x)^2}.$$

Combining all of these, we finally have

$$\mathbb{E}[X] = \sum_{k=1}^{\infty} k(1-p)^{k-1}p$$

$$= p\sum_{k=1}^{\infty} k(1-p)^{k-1}$$

$$= p\frac{1}{p^2} = \frac{1}{p}.$$

**Example 4.** *Expected value of the constant random variable.* Let $c \in \mathbb{R}$ be an arbitrary constant, and let $X$ be the random variable that assumes the value $c$ everywhere. As $X$ is a discrete random variable, its expected value is simply

$$\mathbb{E}[X] = c \cdot P(X = c) = c.$$

I know, this example looks silly, but it can be quite useful. When it is clear, we abuse the notation by denoting the constant $c$ as the random variable itself.

## 20.1.1   The expected value in poker

One more example before we move on. I was a mediocre no-limit Texas hold'em player a while ago, and the first time I heard about the expected value was years before I studied probability theory.

According to the rules of Texas hold'em, each player holds two cards on their own, while five more shared cards are dealt. The shared cards are available for everyone, and the player with the strongest hand wins.

*Figure 20.1* shows how the table looks before the last card (the river) is revealed.

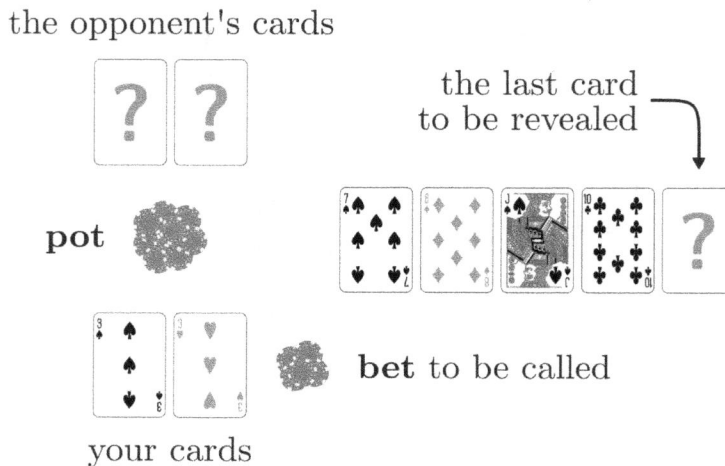

Figure 20.1: The poker table before the river card

There is money in the pot to be won, but to see the river, you have to call the opponent's bet. The question is, should you? Expected value to the rescue.

Let's build a probabilistic model. We would win the pot with certain river cards but lose with all the others. If $X$ represents our winnings, then

$$P(X = \text{pot}) = \frac{\#\text{winning cards}}{\#\text{remaining cards}},$$

$$P(X = -\text{bet}) = \frac{\#\text{losing cards}}{\#\text{remaining cards}}.$$

Thus, the expected value is

$$\mathbb{E}[X] = \text{pot} \cdot P(X = \text{pot}) - \text{bet} \cdot P(X = -\text{bet})$$

$$= \text{pot} \cdot \frac{\#\text{winning cards}}{\#\text{remaining cards}} - \text{bet} \cdot \frac{\#\text{losing cards}}{\#\text{remaining cards}}.$$

When is the expected value positive? With some algebra, we obtain that $\mathbb{E}[X] > 0$ if and only if

$$\frac{\#\text{winning cards}}{\#\text{losing cards}} > \frac{\text{bet}}{\text{pot}},$$

which is called *positive pot odds*. If this is satisfied, making the bet is the right call. You might lose a hand with positive pot odds, but in the long term, your winnings will be positive.

Of course, pot odds are extremely hard to determine in practice. For instance, you don't know what others hold, and counting the cards that would win the pot for you is not possible unless you have a good read on the opponents. Poker is much more than just math. Good players choose their bet specifically to throw off their opponents' pot odds.

Now that we understand the idea behind the expected value, let's move on to the general case!

## 20.2 Continuous random variables

So far, we have only defined the expected value for discrete random variables. As $\mathbb{E}[X]$ describes the average value of $X$ in the long run, it should exist for continuous random variables as well.

The interpretation of the expected value was simple: outcome times probability, summed over all potential values. However, there is a snag with continuous random variables: we don't have such a mass distribution, as the probabilities of individual outcomes are zero: $P(X = x) = 0$. Moreover, we can't sum uncountably many values.

What can we do?

Wishful thinking. This is one of the most powerful techniques in mathematics, and I am not joking.

Here's the plan. We'll pretend that the expected value of a continuous random variable is well-defined, and let our imagination run free. Say goodbye to mathematical precision, and allow our intuition to unfold. Instead of the probability of a given outcome, we can talk about $X$ landing in a small interval. First, we divide up the set of real numbers into really small parts. To be more precise, let $x_0 < x_1 < \ldots < x_n$ be a granular partition of the real line. If the partition is refined enough, we should have

$$\mathbb{E}[X] \approx \sum_{k=1}^{n} x_k P(x_{k-1} < X \leq x_k). \tag{20.1}$$

The probabilities in (20.1) can be expressed in terms of the CDF:

$$\sum_{k=1}^{n} x_k P(x_{k-1} < X \leq X_k) = \sum_{k=1}^{n} x_k \big(F_X(x_k) - F_X(x_{k-1})\big).$$

These increments remind us of the difference quotients. We don't quite have these inside the sum, but with a "fancy multiplication with one," we can achieve this:

$$\sum_{k=1}^{n} x_k \left( F_X(x_k) - F_X(x_{k-1}) \right) = \sum_{k=1}^{n} x_k (x_k - x_{k-1}) \frac{F_X(x_k) - F_X(x_{k-1})}{x_k - x_{k-1}}.$$

If the $x_i$-s are close to each other (and we can select them to be arbitrarily close), the difference quotients are close to the derivative of $F_X$, which is the density function $f_X$. Thus,

$$\frac{F_X(x_k) - F_X(x_{k-1})}{x_k - x_{k-1}} \approx f_X(x_k) \quad \sum_{k=1}^{n} x_k (x_k - x_{k-1}) \frac{F_X(x_k) - F_X(x_{k-1})}{x_k - x_{k-1}} \quad \approx \sum_{k=1}^{n} x_k (x_k - x_{k-1}) f_X(x_k).$$

This is a Riemann-sum, defined by (14.7)! Hence, the last sum is close to a Riemann-integral:

$$\sum_{k=1}^{n} x_k (x_k - x_{k-1}) f_X(x_k) \approx \int_{-\infty}^{\infty} x f_X(x) dx.$$

Although we were not exactly precise in our argument, all of the above can be made mathematically correct. (But we are not going to do it here, as it is not relevant to us.) Thus, we finally obtain the formula of the expected value for continuous random variables.

---

**Definition 20.2.1 (The expected value of continuous random variables)**

Let $(\Omega, \Sigma, P)$ be a probability space, and $X : \Omega \to \mathbb{R}$ be a continuous random variable. The *expected value* of $X$ is defined by

$$\mathbb{E}[X] := \int_{-\infty}^{\infty} x f_X(x) dx.$$

---

As usual, let's see some examples first.

**Example 1.** *Expected value of the uniform distribution.* (See the definition of the uniform distribution in *Section 19.3.4.*) Let $X \sim \text{Uniform}(a, b)$. Then

$$\mathbb{E}[X] = \int_{-\infty}^{\infty} x \frac{1}{b - a} dx$$

$$= \frac{1}{b - a} \int_{a}^{b} x \, dx$$

$$= \left[ \frac{1}{2(b - a)} x^2 \right]_{x=a}^{x=b}$$

$$= \frac{a + b}{2},$$

which is the midpoint of the interval $[a, b]$, where the Uniform$(a, b)$ lives.

**Example 2.** *Expected value of the exponential distribution.* (See the definition of the exponential distribution in *Section 19.3.5.*) Let $X \sim \exp(\lambda)$. Then, we need to calculate the integral

$$\mathbb{E}[X] = \int_0^\infty x\lambda e^{-\lambda x}\, dx.$$

We can do this via integration by parts (*Theorem 14.2.2*): by letting $f(x) = x$ and $g'(x) = \lambda e^{-\lambda x}$, we have

$$\begin{aligned}
\mathbb{E}[X] &= \int_0^\infty x\lambda e^{-\lambda x}\, dx \\
&= \underbrace{\left[ -xe^{-\lambda x} \right]_{x=0}^{x=\infty}}_{=0} + \int_0^\infty e^{-\lambda x}\, dx \\
&= \left[ -\frac{1}{\lambda} e^{-\lambda x} \right]_{x=0}^{x=\infty} \\
&= \frac{1}{\lambda}.
\end{aligned}$$

# 20.3 Properties of the expected value

As usual, the expected value has several useful properties. Most importantly, the expected value is linear with respect to the random variable.

---

**Theorem 20.3.1** *(Linearity of the expected value)*

Let $(\Omega, \Sigma, P)$ be a probability space, and let $X, Y : \Omega \to \mathbb{R}$ be two random variables. Moreover, let $a, b \in \mathbb{R}$ be two scalars. Then

$$\mathbb{E}[aX + bY] = a\mathbb{E}[X] + b\mathbb{E}[Y]$$

holds.

---

We are not going to prove this theorem here, but know that linearity is an essential tool. Do you recall the game that we used to introduce the expected value for discrete random variables? I toss a coin, and if it comes up heads, you win \$1. Tails, you lose \$2. If you think about it for a minute, this is the

$$X = 3 \cdot \text{Bernoulli}(1/2) - 2$$

distribution, and as such,

$$
\begin{aligned}
\mathbb{E}[X] &= \mathbb{E}[3 \cdot \text{Bernoulli}(1/2) - 2] \\
&= 3 \cdot \mathbb{E}[\text{Bernoulli}(1/2)] - 2 \\
&= 3 \cdot \frac{1}{2} - 2 \\
&= -\frac{1}{2}.
\end{aligned}
$$

Of course, linearity goes way beyond this simple example. As you've gotten used to this already, linearity is a crucial property in mathematics. We love linearity.

**Remark 20.3.1** Notice that *Theorem 20.3.1* did not say that $X$ and $Y$ have to be both discrete or both continuous. Even though we have only defined the expected value in such cases, there is a general definition that works for *all* random variables.

The snag is, it requires a familiarity with measure theory, falling way outside of our scope. Suffice to say, the theorem works as is.

If the expected value of a sum is the sum of the expected values, does the same apply to the product? Not in general, but fortunately, this works for independent random variables. (See *Definition 19.1.5* for the definition of independent random variables.)

**Theorem 20.3.2** *(Expected value of the product of independent random variables)*

*Let $(\Omega, \Sigma, P)$ be a probability space, and let $X, Y : \Omega \to \mathbb{R}$ be two independent random variables.*

*Then*

$$
\mathbb{E}[XY] = \mathbb{E}[X]\mathbb{E}[Y]
$$

*holds.*

This property is extremely useful, as we'll see in the next section, where we'll talk about variance and covariance.

One more property that'll help us to calculate the expected value of *functions* of the random variable, such as $X^2$ or $\sin X$:

**Theorem 20.3.3** *(Law of the unconscious statistician)*

*Let $(\Omega, \Sigma, P)$ be a probability space, let $X : \Omega \to \mathbb{R}$ be a random variable, and let $g : \mathbb{R} \to \mathbb{R}$ be an arbitrary function.*

*(a) If $X$ is discrete with possible values $x_1, x_2, \dots$, then*

$$\mathbb{E}[g(X)] = \sum_n g(x_n) P(X = x_n).$$

*(b) If $X$ is continuous with the probability density function $f_X(x)$, then*

$$\mathbb{E}[g(X)] = \int_{-\infty}^{\infty} g(x) f_X(x) dx.$$

Thus, calculating $\mathbb{E}[X^2]$ for a continuous random variable can be done by simply taking

$$\mathbb{E}[X^2] = \int_{-\infty}^{\infty} x^2 f_X(x) dx,$$

which will be used all the time.

# 20.4   Variance

Plainly speaking, the expected value measures the average value of the random variable. However, even though both $\text{Uniform}(-1, 1)$ and $\text{Uniform}(-100, 100)$ have zero expected value, the latter is much more spread out than the former. Thus, $\mathbb{E}[X]$ is not a good descriptor of the random variable $X$.

To add one more layer, we measure the average deviation from the expected value. This is done via the *variance* and the *standard deviation*.

**Definition 20.4.1  (Variance and standard deviation)**

Let $(\Omega, \Sigma, P)$ be a probability space, let $X : \Omega \to \mathbb{R}$ be a random variable, and let $\mu = \mathbb{E}[X]$ be its expected value. The *variance* of $X$ is defined by

$$\text{Var}[X] := \mathbb{E}\big[(X - \mu)^2\big],$$

while its *standard deviation* is defined by

$$\text{Std}[X] := \sqrt{\text{Var}[X]}.$$

Take note that in the literature, the expected value is often denoted by $\mu$, while the standard deviation is denoted by $\sigma$. Together, they form two of the most important descriptors of a random variable.

*Figure 20.2* shows a visual interpretation of the mean and standard deviation in the case of a normal distribution. The mean shows the average value, while the standard deviation can be interpreted as the average deviation from the mean. (We'll talk about the normal distribution in detail later, so don't worry if it is not yet familiar to you.)

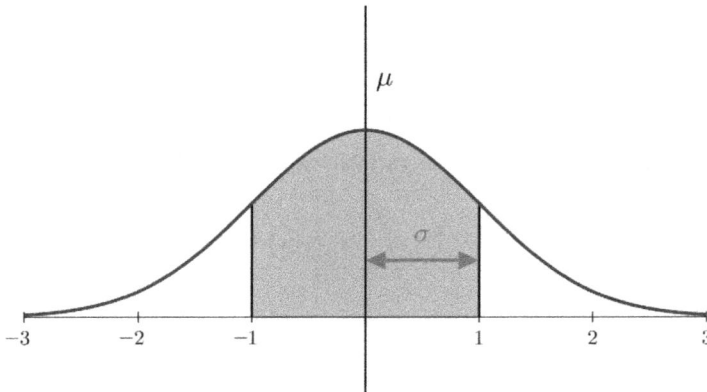

*Figure 20.2: Mean ($\mu$) and standard deviation ($\sigma$) of the standard normal distribution*

The usual method of calculating variance is not taking the expected value of $(X - \mu)^2$, but taking the expected value of $X^2$ and subtracting $\mu^2$ from it. This is shown by the following proposition.

**Proposition 20.4.1** *Let $(\Omega, \Sigma, P)$ be a probability space, and let $X : \Omega \to \mathbb{R}$ be a random variable.*

*Then*

$$\text{Var}[X] = \mathbb{E}[X^2] - \mathbb{E}[X]^2.$$

*Proof.* Let $\mu = \mathbb{E}[X]$. Because of the linearity of the expected value, we have

$$
\begin{aligned}
\mathrm{Var}[X] &= \mathbb{E}\big[(X - \mu)^2\big] \\
&= \mathbb{E}[X^2 - 2\mu X + \mu^2] \\
&= \mathbb{E}[X]^2 - 2\mu\mathbb{E}[X] + \mu^2 \\
&= \mathbb{E}[X]^2 - 2\mu^2 + \mu^2 \\
&= \mathbb{E}[X^2] - \mu^2 \\
&= \mathbb{E}[X^2] - \mathbb{E}[X]^2,
\end{aligned}
$$

which is what we had to show.

Is the variance linear as well? No, but there are some important identities regarding scalar multiplication and addition.

---

**Theorem 20.4.1** *(Variance and the linear operations)*

*Let $(\Omega, \Sigma, P)$ be a probability space, and let $X : \Omega \to \mathbb{R}$ be a random variable.*

*(a) Let $a \in \mathbb{R}$ be an arbitrary constant. Then*

$$
\mathrm{Var}[aX] = a^2 \mathrm{Var}[X].
$$

*(b) Let $Y : \Omega \to \mathbb{R}$ be a random variable that is independent from $X$. Then*

$$
\mathrm{Var}[X + Y] = \mathrm{Var}[X] + \mathrm{Var}[Y].
$$

---

*Proof. (a)* Let $\mu_X = \mathbb{E}[X]$. Then we have

$$
\begin{aligned}
\mathrm{Var}[aX] &= \mathbb{E}\big[(aX - a\mu_X)^2\big] = \mathbb{E}\big[a^2(X - \mu_X)^2\big] \\
&= a^2\mathbb{E}\big[(X - \mu_X)^2\big] = a^2\mathrm{Var}[X],
\end{aligned}
$$

which is what we had to show. *(b)* Let $\mu_Y = \mathbb{E}[Y]$. Then, due to the linearity of the expected value, we have

$$
\begin{aligned}
\mathrm{Var}[X + Y] &= \mathbb{E}\big[(X + Y - (\mu_X + \mu_Y))^2\big] \\
&= \mathbb{E}\big[((X - \mu_X) + (Y - \mu_Y))^2\big] \\
&= \mathbb{E}\big[(X - \mu_X)^2\big] + 2\mathbb{E}\big[(X - \mu_X)(Y - \mu_Y)\big] + \mathbb{E}\big[(Y - \mu_Y)^2\big].
\end{aligned}
$$

592       The Expected Value

Now, as $X$ and $Y$ are independent, $\mathbb{E}[XY] = \mathbb{E}[X]\mathbb{E}[Y]$. Thus, due to the linearity of the expected value,

$$
\begin{aligned}
\mathbb{E}\big[(X - \mu_X)(Y - \mu_Y)\big] &= \mathbb{E}\big[XY - X\mu_Y - \mu_X Y + \mu_X \mu_Y\big] \\
&= \mathbb{E}\big[XY\big] - \mathbb{E}\big[X\mu_Y\big] - \mathbb{E}\big[\mu_X Y\big] + \mu_X \mu_Y \\
&= \mathbb{E}\big[X\big]\mathbb{E}\big[Y\big] - \mathbb{E}\big[X\big]\mu_Y - \mu_X \mathbb{E}\big[Y\big] + \mu_X \mu_Y \\
&= \mu_X \mu_Y - \mu_X \mu_Y - \mu_X \mu_Y + \mu_X \mu_Y \\
&= 0.
\end{aligned}
$$

Thus, continuing the first calculation,

$$
\begin{aligned}
\mathrm{Var}[X + Y] &= \mathbb{E}\big[(X - \mu_X)^2\big] + 2\mathbb{E}\big[(X - \mu_X)(Y - \mu_Y)\big] + \mathbb{E}\big[(Y - \mu_Y)^2\big] \\
&= \mathbb{E}\big[(X - \mu_X)^2\big] + \mathbb{E}\big[(Y - \mu_Y)^2\big],
\end{aligned}
$$

which is what we had to show.

## 20.4.1  Covariance and correlation

Expected value and variance measure a random variable in isolation. However, in real problems, we need to discover relations between separate measurements. Say, $X$ describes the price of a given real estate, while $Y$ measures its size. These are certainly related, but one does not determine the other. For instance, the location might be a differentiator between the prices.

The simplest statistical way of measuring similarity is the *covariance* and *correlation*.

---

**Definition 20.4.2  (Covariance and correlation)**

Let $(\Omega, \Sigma, P)$ be a probability space, let $X, Y : \Omega \to \mathbb{R}$ be two random variables, and let $\mu_X = \mathbb{E}[X], \mu_Y = \mathbb{E}[Y]$ be their expected values and $\sigma_X = \mathrm{Std}[X], \sigma_Y = \mathrm{Std}[Y]$ their standard deviations.

*(a)* The *covariance* of $X$ and $Y$ is defined by

$$
\mathrm{Cov}[X, Y] := \mathbb{E}\big[(X - \mu_X)(Y - \mu_Y)\big].
$$

*(b)* The *correlation* of $X$ and $Y$ is defined by

$$
\mathrm{Corr}[X, Y] := \frac{\mathrm{Cov}[X, Y]}{\sigma_X \sigma_Y}.
$$

---

Similarly to variance, the definition of covariance can be simplified to provide an easier way of calculating its exact value.

**Proposition 20.4.2** *Let $(\Omega, \Sigma, P)$ be a probability space, let $X, Y : \Omega \to \mathbb{R}$ be two random variables, and let $\mu_X = \mathbb{E}[X], \mu_Y = \mathbb{E}[Y]$ be their expected values.*

*Then*

$$\mathrm{Cov}[X, Y] = \mathbb{E}[XY] - \mu_X \mu_Y.$$

*Proof.* This is just a simple calculation. According to the definition, we have

$$
\begin{aligned}
\mathrm{Cov}[X, Y] &= \mathbb{E}\big[(X - \mu_X)(Y - \mu_Y)\big] \\
&= \mathbb{E}\big[XY - X\mu_Y - \mu_X Y + \mu_X \mu_Y\big] \\
&= \mathbb{E}\big[XY\big] - \mathbb{E}\big[X\mu_Y\big] - \mathbb{E}\big[\mu_X Y\big] + \mu_X \mu_Y \\
&= \mathbb{E}\big[XY\big] - \mathbb{E}\big[X\big]\mu_Y - \mu_X \mathbb{E}\big[Y\big] + \mu_X \mu_Y \\
&= \mathbb{E}\big[XY\big] - \mu_X \mu_Y - \mu_X \mu_Y + \mu_X \mu_Y \\
&= \mathbb{E}[XY] - \mu_X \mu_Y,
\end{aligned}
$$

which is what we had to show.

One of the most important properties of covariance and correlation is that they are zero for independent random variables.

**Theorem 20.4.2** *Let $(\Omega, \Sigma, P)$ be a probability space, and let $X, Y : \Omega \to \mathbb{R}$ be two independent random variables.*

*Then, $\mathrm{Cov}[X, Y] = 0$. (And consequently, $\mathrm{Corr}[X, Y] = 0$ as well.)*

The proof follows straight from the definition and *Theorem 20.3.2*, so this is left as an exercise for you.

Take note, as this is extra important: independence implies zero covariance, but zero covariance *does not* imply independence. Here is an example.

Let $X$ be a discrete random variable with the probability mass function

$$P(X = -1) = P(X = 0) = P(X = 1) = \frac{1}{3},$$

and let $Y = X^2$.

The expected value of $X$ is

$$\mathbb{E}[X] = (-1) \cdot P(X = -1) + 0 \cdot P(X = 0) + 1 \cdot P(X = 1)$$
$$= -\frac{1}{3} + 0 + \frac{1}{3}$$
$$= 0,$$

while the law of the unconscious statistician (*Theorem 20.3.3*) gives that

$$\mathbb{E}[Y] = \mathbb{E}[X^2] = 1 \cdot P(X = -1) + 0 \cdot P(X = 0) + 1 \cdot P(X = 1)$$
$$= \frac{1}{3} + 0 + \frac{1}{3}$$
$$= \frac{2}{3},$$

and

$$\mathbb{E}[XY] = \mathbb{E}[X^3] = 0.$$

Thus,

$$\mathrm{Cov}[X, Y] = \mathbb{E}[XY] - \mathbb{E}[X]\mathbb{E}[Y]$$
$$= \mathbb{E}[X^3] - \mathbb{E}[X]\mathbb{E}[X^2]$$
$$= 0 - 0 \cdot \frac{2}{3}$$
$$= 0.$$

However, $X$ and $Y$ are not independent, as $Y = X^2$ is a function of $X$. (I shamelessly stole this example from a brilliant Stack Overflow thread, which you should read here for more on this question: https://stats.stacke xchange.com/questions/179511/why-zero-correlation-does-not-necessarily-imply-independence)

Do you recall that we interpreted the concept of probability as the relative frequency of occurrences? Now that we have the expected value under our belt, we can finally make this precise. Let's look at the famous law of large numbers!

# 20.5    The law of large numbers

We'll continue our journey with a quite remarkable and famous result: the law of large numbers. You have probably already heard several faulty arguments invoking the law of large numbers. For instance, gamblers are often convinced that their bad luck will end soon because of said law. This is one of the most frequently misused mathematical terms, and we are here to clear that up.

We'll do this in two passes. First, we are going to see an intuitive interpretation, then add the technical but important mathematical details. I'll try to be gentle.

## 20.5.1    Tossing coins...

First, let's toss some coins again. If we toss coins repeatedly, what is the relative frequency of heads in the long run?

We should have a pretty good guess already: the average number of heads should converge to $P(\text{heads}) = p$ as well. Why? Because we saw this when studying the frequentist interpretation of probability in *Section 18.2.7*.

Our simulation showed that the relative frequency of heads does indeed converge to the true probability. This time, we'll carry the simulation a bit further.

First, to formulate the problem, let's introduce the independent random variables $X_1, X_2, \ldots$ that are distributed along Bernoulli$(p)$, where $X_i = 0$ if the toss results in tails, while $X_i = 1$ if it is heads. We are interested in the long-term behavior of

$$\overline{X}_n = \frac{X_1 + \cdots + X_n}{n}.$$

$\overline{X}_n$ is called the *sample average*. We have already seen that the sample average gets closer and closer to $p$ as $n$ grows. Let's see the simulation one more time, before we go any further. (The parameter $p$ is selected to be $1/2$ for the sake of the example.)

```
import numpy as np
from scipy.stats import bernoulli


n_tosses = 1000
idx = range(n_tosses)

coin_tosses = [bernoulli.rvs(p=0.5) for _ in idx]
coin_toss_averages = [np.mean(coin_tosses[:k+1]) for k in idx]
```

And here is the plot.

```python
import matplotlib.pyplot as plt


with plt.style.context("seaborn-v0_8"):
    plt.figure(figsize=(10, 5))
    plt.title("Relative frequency of the coin tosses")
    plt.xlabel("Relative frequency")
    plt.ylabel("Number of tosses")

    # plotting the averages
    plt.plot(range(n_tosses), coin_toss_averages, linewidth=3) # the averages

    # plotting the true expected value
    plt.plot([-100, n_tosses+100], [0.5, 0.5], c="k")
    plt.xlim(-10, n_tosses+10)
    plt.ylim(0, 1)
    plt.show()
```

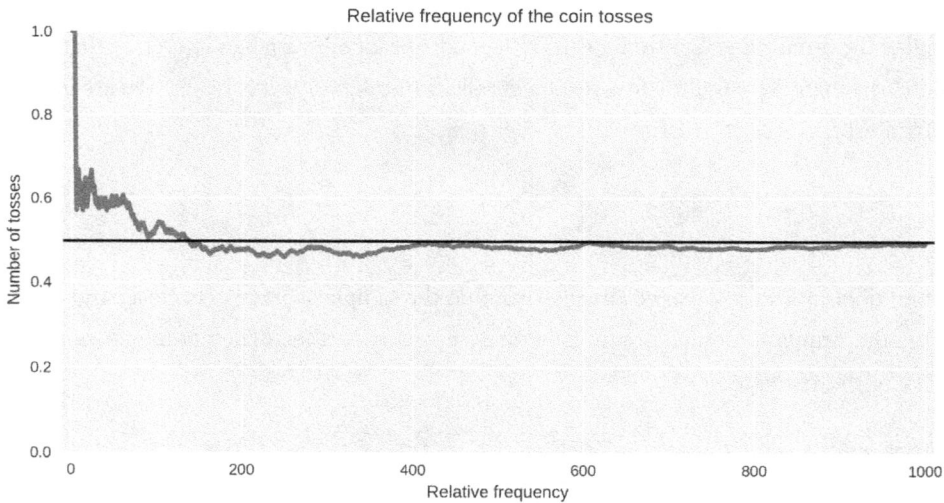

*Figure 20.3: Relative frequency of the coin tosses*

Nothing new so far. However, if you have a sharp eye, you might ask the question: is this just an accident? After all, we are studying the average

$$\overline{X}_n = \frac{X_1 + \cdots + X_n}{n},$$

which is (almost) a binomially distributed random variable! To be more precise, if $X_i \sim$ Bernoulli($p$), then

$$\overline{X}_n \sim \frac{1}{n}\text{Binomial}(n, p).$$

(We saw this earlier when discussing the sums of discrete random variables in *Section 19.2.7*.)

At this point, it is far from guaranteed that this distribution will be concentrated around a single value. So, let's do some more simulations. This time, we'll toss a coin a thousand times to see the distribution of the averages. Quite meta, I know.

```
more_coin_tosses = bernoulli.rvs(p=0.5, size=(n_tosses, n_tosses))
more_coin_toss_averages = np.array([[np.mean(more_coin_tosses[i][:j+1]) for j in idx]
                                    for i in idx])
```

We can visualize the distributions on histograms.

```
with plt.style.context("seaborn-v0_8"):
    fig, axs = plt.subplots(1, 3, figsize=(12, 4), sharey=False)
    fig.suptitle("The distribution of sample averages")
    for ax, i in zip(axs, [5, 100, 999]):
        x = [k/i for k in range(i+1)]
        y = more_coin_toss_averages[:, i]
        ax.hist(y, bins=x)
        ax.set_title(f"n = {i}")

    plt.show()
```

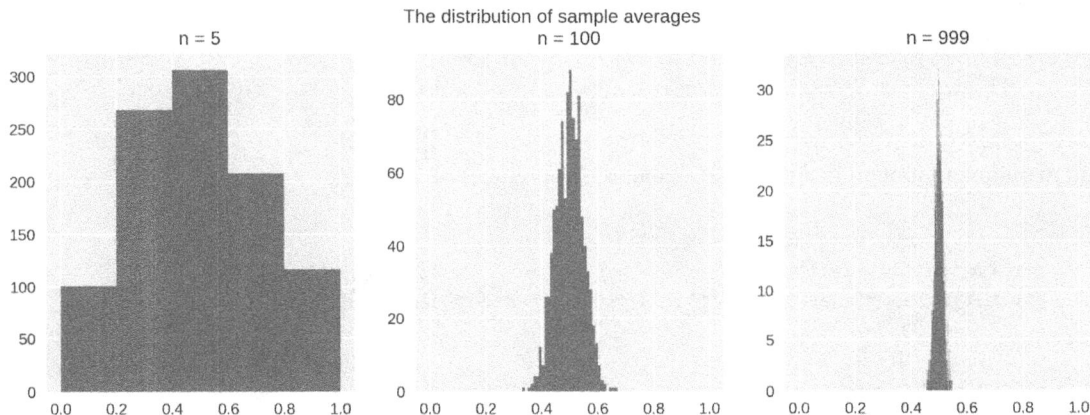

*Figure 20.4: Sample average distributions of coin tosses*

In other words, the probability of $X_n$ falling far from $p$ becomes smaller and smaller. For any small $\varepsilon$, we can formulate the probability of "$\overline{X}_n$ falling farther from $p$ than $\varepsilon$" as $P(|\overline{X}_n - p| > \varepsilon)$.

Thus, mathematically speaking, our guess is

$$\lim_{n \to \infty} P(|\overline{X}_n - p| > \varepsilon) = 0.$$

Again, is this just an accident, and were we just lucky to study an experiment where this is true? Would the same work for random variables other than Bernoulli ones? What will the sample averages converge to? (If they converge at all.)

We'll find out.

## 20.5.2  ...rolling dice...

Let's play dice. To keep things simple, we are interested in the average value of a roll in the long run. To build a proper probabilistic model, let's introduce random variables!

A single roll is uniformly distributed on $\{1, 2, \ldots, 6\}$, and each roll is independent from the others. So, let $X_1, X_2, \ldots$ be independent random variables, each distributed according to Uniform($\{1, 2, \ldots, 6\}$).

How does the sample average $\overline{X}_n$ behave? Simulation time. We'll randomly generate 1000 rolls, then explore how $\overline{X}_n$ behaves.

```
from scipy.stats import randint


n_rolls = 1000
idx = range(n_rolls)

dice_rolls = [randint.rvs(low=1, high=7) for _ in idx]
dice_roll_averages = [np.mean(dice_rolls[:k+1]) for k in idx]
```

Again, to obtain a bit of an insight, we'll visualize the averages on a plot.

```
with plt.style.context("seaborn-v0_8"):
    plt.figure(figsize=(10, 5))
    plt.title("Sample averages of rolling a six-sided dice")

    # plotting the averages
    plt.plot(idx, dice_roll_averages, linewidth=3) # the averages

    # plotting the true expected value
```

```
plt.plot([-100, n_rolls+100], [3.5, 3.5], c="k")

plt.xlim(-10, n_rolls+10)
plt.ylim(0, 6)
plt.show()
```

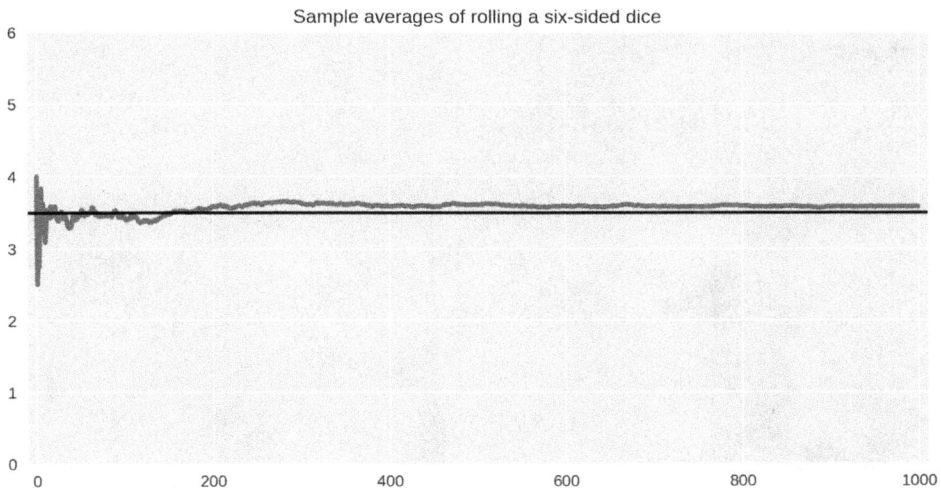

*Figure 20.5: Sample averages of rolling a six-sided dice*

The first thing to note is that these are suspiciously close to 3.5. This is *not* a probability, but the expected value:

$$\mathbb{E}[X_1] = \mathbb{E}[X_2] = \cdots = 3.5.$$

For Bernoulli($p$) distributed random variables, the expected value coincides with the probability $p$. However, this time, $\overline{X}_n$ does not have a nice and explicit distribution like in the case of coin tosses, where the sample averages were binomially distributed. So, let's roll some more dice to estimate how $\overline{X}_n$ is distributed.

```
more_dice_rolls = randint.rvs(low=1, high=7, size=(n_rolls, n_rolls))
more_dice_roll_averages = np.array([[np.mean(more_dice_rolls[i][:j+1]) for j in idx]
                                    for i in idx])
```

```
with plt.style.context("seaborn-v0_8"):
    fig, axs = plt.subplots(1, 3, figsize=(12, 4), sharey=False)
    fig.suptitle("The distribution of sample averages")
    for ax, i in zip(axs, [5, 100, 999]):
        x = [6*k/i for k in range(i+1)]
        y = more_dice_roll_averages[:, i]
        ax.hist(y, bins=x)
        ax.set_title(f"n = {i}")

    plt.show()
```

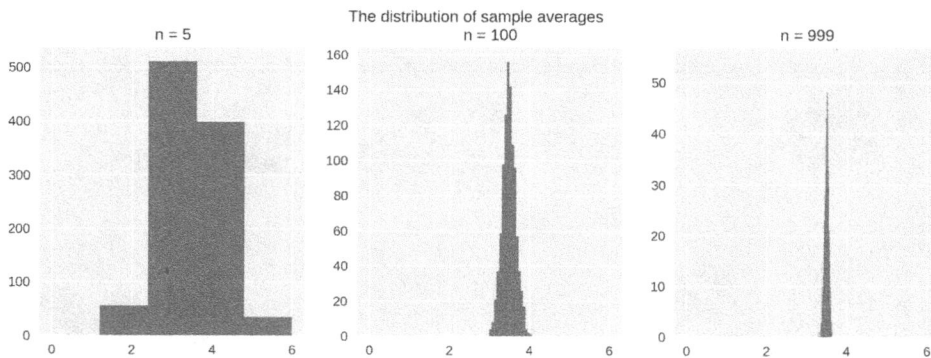

*Figure 20.6: Sample average distributions of dice rolls*

It seems like, once more, the distribution of $\overline{X}_n$ is concentrated around $\mathbb{E}[X_1]$. Our intuition tells us that this is not an accident; that this phenomenon is true for a wide range of random variables.

Let me spoil the surprise: this is indeed the case, and we'll see this now.

## 20.5.3 ...and all the rest

This time, let $X_1, X_2, \ldots$ be a sequence of independent and identically distributed (i.i.d.) random variables. Not coin tosses, not dice rolls, but any distribution. We saw that the sample average $\overline{X}_n$ seems to converge to the joint expected value of the $X_i$-s:

$$^{''}\overline{X}_n \rightarrow \mathbb{E}[X_1]^{''}$$

Note the quotation marks: $\overline{X}_n$ is not a number but a random variable. Thus, we can't (yet) speak about convergence.

In mathematically precise terms, what we saw previously is that for large enough $n$-s, the sample average $\overline{X}_n$ is highly unlikely to fall far from the joint expected value $\mu = \mathbb{E}[X_1]$; that is,

$$\lim_{n \to \infty} P(|\overline{X}_n - \mu| > \varepsilon) = 0 \tag{20.2}$$

holds for all $\varepsilon > 0$.

The limit (20.2) seems hard to prove right now even in the simple case of coin tossing. There, $\overline{X}_n \sim \frac{1}{n}\text{Binomial}(n, p)$, thus

$$P(|\overline{X}_n - \mu| > \varepsilon) = 1 - \sum_{k=\lfloor n(p-\varepsilon) \rfloor}^{\lfloor n(p+\varepsilon) \rfloor} \binom{n}{k} p^k (1-p)^{n-k},$$

where the symbol $\lfloor x \rfloor$ denotes the largest integer that is smaller than $x$. This does not look friendly at all. (I leave the verification as an exercise.)

Thus, our plan is the following.

1. Find a way to estimate $P(|\overline{X}_n - \mu| > \varepsilon)$ in a way that is independent from the distribution of the $X_i$-s.
2. Use the upper estimate to show $\lim_{n \to \infty} P(|\overline{X}_n - \mu| > \varepsilon) = 0$.

Let's go.

## 20.5.4   The weak law of large numbers

First, the upper estimates. There are two general inequalities that'll help us to deal with $P(|\overline{X}_n - \mu| \geq \varepsilon)$.

**Theorem 20.5.1** *(Markov's inequality)*

*Let $(\Omega, \Sigma, P)$ be a probability space and let $X : \Omega \to [0, \infty)$ be a nonnegative random variable. Then*

$$P(X \geq t) \leq \frac{\mathbb{E}[X]}{t}$$

*holds for any $t \in (0, \infty)$.*

*Proof.* We have to separate the discrete and the continuous cases. The proofs are almost identical, so I'll only do the discrete case here, while the continuous is left for you as an exercise to test your understanding.

So, let $X : \Omega \to \{x_1, x_2, ...\}$ be a discrete random variable (where $x_k \geq 0$ for all $k$), and $t \in (0, \infty)$ be an arbitrary positive real number.

Then

$$E[X] = \sum_{k=1}^{\infty} x_k P(X = x_k)$$

$$= \sum_{k:x_k<t} x_k P(X = x_k) + \sum_{k:x_k\geq t} x_k P(X = x_k),$$

where the sum $\sum_{k:x_k<t}$ only accounts for $k$-s with $x_k < t$, and similarly, $\sum_k : x_k \geq t$ only accounts for $k$-s with $x_k \geq t$.

As the $x_k$-s are nonnegative by assumption, we can estimate $E[X]$ from below by omitting one of them. Thus,

$$E[X] = \sum_{k:x_k<t} x_k P(X = x_k) + \sum_{k:x_k\geq t} x_k P(X = x_k)$$

$$\geq \sum_{k:x_k\geq t} x_k P(X = x_k)$$

$$\geq t \sum_{k:x_k\geq t} P(X = x_k)$$

$$= tP(X \geq t),$$

from which Markov's inequality

$$P(X \geq t) \leq \frac{E[X]}{t}$$

follows.

The law of large numbers is only one step away from Markov's inequality. This last step is so useful that it deserves to be its own theorem. Meet the famous inequality of Chebyshev.

---

**Theorem 20.5.2** *(Chebyshev's inequality)*

*Let $(\Omega, \Sigma, P)$ be a probability space and let $X : \Omega \to \mathbb{R}$ be a random variable with finite variance $\sigma^2 = \mathrm{Var}[X] < \infty$ and expected value $E[X] = \mu$.*

*Then*

$$P(|X - \mu| \geq t) \leq \frac{\sigma^2}{t^2}$$

*holds for all $t \in (0, \infty)$.*

*Proof.* As $|X - \mu|$ is a nonnegative random variable, we can apply *Theorem 20.5.1* to obtain

$$P(|X - \mu| \geq t) = P(|X - \mu|^2 \geq t^2)$$
$$\leq \frac{\mathbb{E}[|X - \mu|^2]}{t^2}.$$

However, as $\mathbb{E}[|X - \mu|^2] = \text{Var}[X] = \sigma^2$, we have

$$P(|X - \mu| \geq t) \leq \frac{\mathbb{E}[|X - \mu|^2]}{t^2} = \frac{\sigma^2}{t^2}$$

which is what we had to show.

And with that, we are ready to precisely formulate and prove the law of large numbers. After all this setup, the (weak) law of large numbers is just a small step away. Here it is in its full glory.

**Theorem 20.5.3** *(The weak law of large numbers)*

*Let $X_1, X_2, \ldots$ be a sequence of independent and identically distributed random variables with finite expected value $\mu = \mathbb{E}[X_1]$ and variance $\sigma^2 = \text{Var}[X_1]$, and let*

$$\overline{X}_n = \frac{X_1 + \cdots + X_n}{n}$$

*be their sample average. Then*

$$\lim_{n \to \infty} P(|\overline{X}_n - \mu| \geq \varepsilon) = 0$$

*holds for any $\varepsilon > 0$.*

*Proof.* As the $X_i$-s are independent, the variance of the sample average is

$$\text{Var}[\overline{X}_n] = \text{Var}\left[\frac{X_1 + \cdots + X_n}{n}\right]$$
$$= \frac{1}{n^2}\text{Var}[X_1 + \cdots + X_n]$$
$$= \frac{1}{n^2}\left(\text{Var}[X_1] + \cdots + \text{Var}[X_n]\right)$$
$$= \frac{n\sigma^2}{n^2} = \frac{\sigma^2}{n}.$$

Now, by using Chebyshev's inequality from *Theorem 20.5.2*, we obtain

$$P(|\overline{X}_n - \mu| \geq \varepsilon) \leq \frac{\text{Var}[\overline{X}_n]}{\varepsilon^2} = \frac{\sigma^2}{n\varepsilon^2}.$$

Thus,

$$0 \leq \lim_{n\to\infty} P(|\overline{X}_n - \mu| \geq \varepsilon)$$

$$\leq \lim_{n\to\infty} \frac{\sigma^2}{n\varepsilon^2} = 0,$$

hence

$$\lim_{n\to\infty} P(|\overline{X}_n - \mu| \geq \varepsilon) = 0,$$

which is what we needed to show.

*Theorem 20.5.3* is not all that can be said about the sample averages. There is a stronger result, showing that the sample averages do in fact converge to the mean with probability 1.

## 20.5.5 The strong law of large numbers

Why is *Theorem 20.5.3* called the "weak" law? Think about the statement

$$\lim_{n\to\infty} P(|\overline{X}_n - \mu| \geq \varepsilon) = 0 \tag{20.3}$$

for a moment. For a given $\omega \in \Omega$, this doesn't tell us *anything* about the convergence of a concrete sample average

$$\overline{X}_n(\omega) = \frac{X_1(\omega) + \cdots + X_n(\omega)}{n},$$

it just tells us that in a probabilistic sense, $\overline{X}_n$ is concentrated around the joint expected value $\mu$. In a sense, (20.3) is a weaker version of

$$P(\lim_{n\to\infty} \overline{X}_n = \mu) = 1,$$

hence the terminology *weak* law of large numbers.

Do we have a stronger result than *Theorem 20.5.3*? Yes, we do.

---

**Theorem 20.5.4** *(The strong law of large numbers)*

*Let $X_1, X_2, \ldots$ be a sequence of independent and identically distributed random variables with finite expected value $\mu = \mathbb{E}[X_1]$ and variance $\sigma^2 = \mathrm{Var}[X_1]$, and let*

$$\overline{X}_n = \frac{X_1 + \cdots + X_n}{n}$$

*be their sample average. Then*

$$P(\lim_{n \to \infty} \overline{X}_n = \mu) = 1.$$

---

We are not going to prove this, just know that the sample average will converge to the mean with probability one.

---

**Remark 20.5.1 (Convergence of random variables)**

What we have seen in the weak and strong laws of large numbers are not unique to sample averages. Similar phenomena can be observed in other cases, thus, these types of convergences have their own exact definitions.

If $X_1, X_2, \ldots$ is a sequence of random variables, we say that

*(a)* $X_n$ converges *in probability* towards $X$ if

$$\lim_{n \to \infty} P(|X_n - X| \geq \varepsilon) = 0$$

for all $\varepsilon > 0$. Convergence in probability is denoted by $X_n \xrightarrow{P} X$.

*(b)* $X_n$ converges *almost surely* towards $X$ if

$$P(\lim_{n \to \infty} X_n = X) = 1$$

holds. Almost sure convergence is denoted by $X_n \xrightarrow{\text{a. s.}} X$.

Thus, the weak and strong laws of large numbers state that in certain cases, the sample averages converge to the expected value both in probability and almost surely.

---

# 20.6 Information theory

If you have already trained a machine learning model in your practice, chances are you are already familiar with the mean-squared error

$$\text{MSE}(\mathbf{x}, \mathbf{y}) = \frac{1}{n} \sum_{i=1}^{n} (f(x_i) - y_i)^2, \quad \mathbf{x}, \mathbf{y} \in \mathbb{R}^n,$$

where $f : \mathbb{R}^n \to \mathbb{R}$ represents our model, $\mathbf{x} \in \mathbb{R}^n$ is the vector of one-dimensional observations, and $\mathbf{y} \in \mathbb{R}^n$ is the ground truth. After learning all about the expected value, this sum should be familiar: if we assume a probabilistic viewpoint and let $X$ and $Y$ be the random variables describing the data, then the mean-squared error can be written as the expected value

$$\text{MSE}(\mathbf{x}, \mathbf{y}) = \mathbb{E}\left[(f(X) - Y)^2\right].$$

However, the mean-squared error is not suitable for classification problems. For instance, if the task is to classify the object of an image, the output is a discrete probability distribution for each sample. In this situation, we could use the so-called *cross-entropy*, defined by

$$H[\mathbf{p}, \mathbf{q}] = - \sum_{i=1}^{n} p_i \log q_i$$

where $\mathbf{p} \in \mathbb{R}^n$ denotes the one-hot encoded vector of the class label for a single data sample, and $\mathbf{q} \in \mathbb{R}^n$ is the class label prediction, forming a probability distribution. (One-hot encoding is the process where we represent a finite set of possible class labels, such as $\{a, b, c\}$ as zero-one vectors, like

$$a \longleftrightarrow (1, 0, 0),$$
$$b \longleftrightarrow (0, 1, 0),$$
$$c \longleftrightarrow (0, 0, 1).$$

We do this because it's easier to work with vectors and matrices than with strings.)

Not that surprisingly, $H[\mathbf{p}, \mathbf{q}]$ is also an expected value, but it's much more than that: it quantifies the information content of the distribution $\mathbf{q}$ compared to the ground truth distribution $\mathbf{q}$.

But what is information in a mathematical sense? Let's dive in.

## 20.6.1  Guess the number

Let's start with a simple game. I have thought of an integer between 0 and 7, and your job is to find out which one by asking yes-no questions.

*Figure 20.7: Which number am I thinking of?*

One possible strategy is to guess the numbers one by one. In other words, the sequence of your questions are:

- Is it 0?
- Is it 1?
  $\vdots$
- Is it 7?

Although this strategy works, it is not an effective one. Why? Consider the average number of questions. Let the random variable $X$ denote the number I have picked. As $X$ is uniformly distributed — that is, $P(X = k) = 1/8$ for all $k = 0, \dots, 7$ — the probability of asking exactly $k$ questions is

$$P(\#\text{questions} = k) = P(X = k - 1) = \frac{1}{8}$$

as well. Thus, the number of questions needed is also uniformly distributed on $\{1, \dots, 8\}$, thus

$$\mathbb{E}[\#\text{questions}] = \sum_{k=1}^{8} kP(\#\text{questions} = k)$$
$$= \frac{1}{8} \sum_{k=1}^{8} k$$
$$= \frac{1}{8} \frac{8 \cdot 9}{2} = \frac{9}{2},$$

where we have used that $\sum_{k=1}^{n} = \frac{n(n+1)}{2}$.

Can we do better than this? Yes. In the previous sequential strategy, each question has a small chance of hitting, and a large chance of eliminating, only one potential candidate. It's easy to see that the best would be to eliminate half the search space with each question.

Say, the number I thought of is 2. By asking *"is the number larger than 3"?*, the answer trims out four of the candidates.

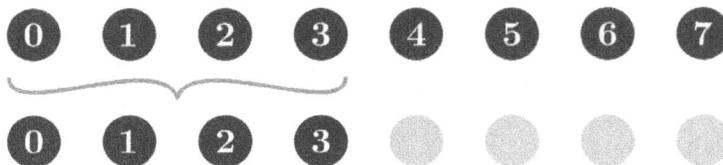

*Figure 20.8: The search space after the question is the number larger than 3?*

Each subsequent question cuts the remaining possibilities in half. In the case of $X = 2$, the three questions are the following:

- Is $X \geq 4$? *(no)*
- Is $X \geq 2$? *(yes)*
- Is $X \geq 3$? *(no)*

This is the so-called *binary search*, illustrated by *Figure 20.9*.

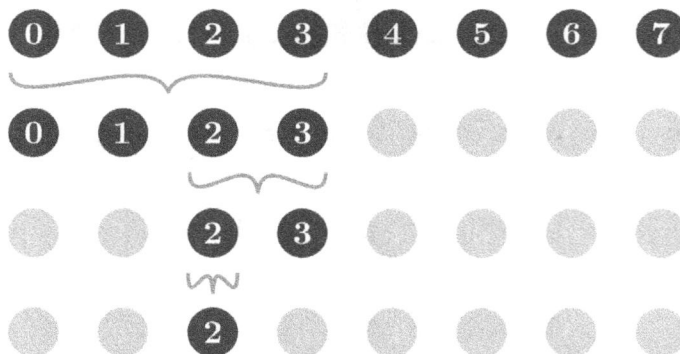

*Figure 20.9: Figuring out the answer with binary search*

If we write down the answers for our three consecutive questions ($X \geq 4$, $X \geq 2$, $X \geq 3$) as a zero-one sequence, we obtain 010. If this looks familiar, it's not an accident: 010 is 2 in binary. In fact, all of the answers can be coded using their binary form:

$$0 = 000_2, \quad 1 = 001_2, \quad 2 = 010_2, \quad 3 = 011_2$$
$$4 = 100_2, \quad 5 = 101_2, \quad 6 = 110_2, \quad 7 = 111_2.$$

Thus, we can reformulate our three questions:

1. Is 1 the 1st digit of $X$ in binary?
2. Is 1 the 2nd digit of $X$ in binary?
3. Is 1 the 3rd digit of $X$ in binary?

As the above example shows, guessing the number is equivalent to finding the binary representation of the objects to be guessed. Each digit represents exactly one *bit* of information. (In this case, the representation is the actual binary form.) Binary codings have an additional perk: we no longer have to sequentially go through the questions, we can ask them simultaneously. From now on, instead of questions, we'll talk about binary representations (codings) and their bits.

Notice that the number of bits is the same for each outcome of $X$. Thus, as this strategy always requires three bits, their average number is three as well:

$$\mathbb{E}[\#\text{bits}] = \sum_{k=0}^{7} 3 \cdot P(X = k) = \sum_{k=0}^{7} 3 \cdot \frac{1}{8} = 3.$$

Can we do better than the three questions on average? No. I invite you to come up with your arguments, but we'll see this later.

Where does the number three in the above come from? In general, if we have $2^k$ possible choices, then $\log_2 2^k = k$ questions will be enough to find the answer. (As each question cuts the set of possible answers in half.) In other words, we have

$$\mathbb{E}[\#\text{bits}] = \sum_{k=0}^{7} P(X = k) \log_2 2^3$$

$$= \sum_{k=0}^{7} P(X = k) \log_2 P(X = k)^{-1}.$$

Thus, the value $\log_2 P(X = k)^{-1}$ is the number of bits needed to represent $k$ in our coding. In other words,

$$\mathbb{E}[\#\text{bits}] = \mathbb{E}[\log_2 P(X = k)^{-1}].$$

Let's get a bit ahead of ourselves: this is the famous *entropy* of the random variable $X$, and the quantity $\log_2 P(X = k)^{-1}$ is the so-called *information content* of the event $X = k$.

However, at this point, these concepts are quite unclear. What does $\log_2 P(X = k)^{-1}$ have to do with information? Why can't we represent $k$ better than $\log_2 P(X = k)^{-1}$ bits? We'll see the answers soon.

## 20.6.2   Guess the number 2: Electric Boogaloo

Let's play the guessing game again but with a twist this time. Now, I have picked a number from $\{0, 1, 2\}$, and you have to guess which one. The catch is, I am twice as likely to select 0 than the others.

In probabilistic terms, if $X$ denotes the number I picked, then $P(X = 0) = 1/2$, while $P(X = 1) = P(X = 2) = 1/4$.

What is the best strategy? There are two key facts to recall:

- good questions cut the search space in half,
- and asking questions is equivalent to finding a binary encoding of the outcomes.

However, as we are looking for the encoding that is optimal *on average*, cutting the search space in half with each digit is *not* meant in a numeric way. Rather, in a probabilistic one. Thus, if 0 is indeed twice as likely, representing $0, 1, 2$ by

$$0 \sim 0,$$
$$1 \sim 01,$$
$$2 \sim 10,$$

the average number of bits is

$$
\begin{aligned}
\mathbb{E}[\#\text{bits}] &= P(X = 0) \cdot 1 + P(X = 1) \cdot 2 + P(X = 2) \cdot 2 \\
&= P(X = 0) \log_2 2 + P(X = 1) \log_2 4 + P(X = 2) \log_2 4 \\
&= \sum_{k=0}^{2} P(X = k) \log_2 P(X = k)^{-1} \\
&= \frac{3}{2}.
\end{aligned}
$$

Once more, we have arrived at the familiar logarithmic formula. We are one step closer to grasping the meaning of the mysterious quantity $\log_2 P(X = k)^{-1}$. The smaller it is, the more questions we need; equivalently, the more bits we need to represent $k$ within our encoding to avoid information loss.

So, what are these mysterious quantities exactly?

## 20.6.3   Information and entropy

It is time to formulate the general problem. Suppose that our random variable $X$ assumes a number from the set $\{1, 2, \ldots, N\}$, each with probability $p_k = P(X = k)$. Upon repeatedly observing $X$, what is the average information content of our observations?

According to what we've learned, we are looking for the quantity

$$\mathbb{E}[I(X)] = -\sum_{k=1}^{N} p_k \log p_k,$$

where $I : \mathbb{N} \to \mathbb{R}$ denotes the information $I(k) = -\log p_k$. Previously, we have seen two special cases where $I(k)$ is the average number of questions needed to guess $k$. (Equivalently, the information is the average number of bits in $k$ using the optimal encoding.)

What is the information in general?

Let's look for $I$ as an unknown function of the probabilities: $I(x) = f\big(P(X = x)\big)$. What can $f$ be? There are two key properties of that'll lead us to the answer. First, the more probable an event is, the less information content there is. (Recall the previous example, where the most probable outcome required the least amount of bits in our binary representation.)

Second, as a function of the probabilities, the information is additive: $f(pq) = f(p) + f(q)$. Why? Suppose that I have picked two numbers, independently from each other, and now you have to guess those two. You can do this sequentially, applying the optimal strategy to the first one, then the second one.

In mathematical terms, $f(p)$ is

- continuous,
- strictly increasing, that is, $f(p) < f(q)$ for any $p < q$,
- and additive, that is, $f(pq) = f(p) + f(q)$ for any $p, q$.

I'll spare you the mathematical details, but with a bit of calculus magic, we can confidently conclude that the only option is $f(p) = -\log_a p$, where $a > 1$. Seemingly, information depends on the base, but as

$$\log_b x = \frac{\log_a x}{\log_a b},$$

the choice of base only influences the information and entropy up to a multiplicative scaling factor. Thus, using the natural logarithm is the simplest choice.

So, here is the formal definition at last.

---

**Definition 20.6.1 (Information)**

Let $X$ be a discrete random variable with probability mass function $\{P(X = x_k)\}_k$.

The information of the event $X = x_k$ is defined by

$$I(x_k) := -\log P(X = x_k) = \log P(X = x_k)^{-1}.$$

---

(Note that whenever the base of log is not indicated, we are using the natural base $e$.) To emphasize the dependency of the information on $X$, we'll sometimes explicitly denote the connection by $I_X(x_k)$.

Armed with the notion of information, we are ready to define entropy, the average amount of information per observation. This quantity is named after Claude Shannon, who essentially founded information theory in his epic paper *"A Mathematical Theory of Communication."*

---

**Definition 20.6.2 (Shannon entropy)**

Let $X$ be a discrete random variable with probability mass function $\{P(X = x_k)\}_k$.

The entropy of $X$ is defined by

$$H[X] := \mathbb{E}[I(X)]$$

$$= -\sum_{k=1}^{\infty} P(X = x_k)\log P(X = x_k).$$

---

Even though $H[X]$ is called the *Shannon entropy*, we'll just simply refer to it as *entropy*, unless an explicit distinction is needed.

One of the first things we can notice is that $H[X] \geq 0$. This is shown in the following proposition.

---

**Proposition 20.6.1 (The nonnegativity of entropy)**

Let $X$ be an arbitrary discrete random variable. Then $H[X] \geq 0$.

---

*Proof.* By definition,

$$H[X] = \sum_k P(X = x_k)\log P(X = x_k)^{-1}.$$

First, suppose that $P(X = x_k) \neq 0$ for all $k$. Then, as $0 < P(X = x_k) \leq 1$, the information is nonnegative:

$\log P(X = x_k)^{-1} \geq 0$. Hence, as all terms in the defining sum are nonnegative, $H[X]$ is nonnegative as well.

If $P(X = x_k) = 0$ for some $k$, then, as $\lim_{x \to 0+} x \log x = 0$, the expression $0 \cdot \log 0$ is taken to be 0. Thus, $H[X]$ is still nonnegative.

Computing the entropy in practice is hard, as we have to evaluate sums that involve logarithms. However, there are a few special cases that shed some much needed light on the concept of entropy. Let's look at them!

**Example 1.** *The discrete uniform distribution.* (See the definition of the discrete uniform distribution in *Section 19.2.4.*) Let $X \sim \text{Uniform}(\{1, \dots, n\})$. Then

$$
\begin{aligned}
H[X] &= -\sum_{k=1}^{n} \frac{1}{n} \log \frac{1}{n} \\
&= \sum_{k=1}^{n} \frac{1}{n} \log n \\
&= \log n.
\end{aligned}
$$

By now, we have an intuitive understanding of entropy as the average amount of information per observation. Take a wild guess: how does the entropy of the uniform distribution compare amongst all other distributions concentrated on $\{1, 2, \dots, n\}$? Is it above or below average? Is it perhaps minimal or maximal?

We'll reveal the answer by the end of this chapter, but take a minute to ponder this question before moving on to the next example.

**Example 2.** *The single-point distribution.* (See the definition of the single-point distribution in *Section 19.2.5*) Let $X \sim \delta(a)$. Then

$$
H[X] = -1 \cdot \log 1 = 0.
$$

In other words, as the event $X = a$ is certain, no information is gained upon observing $X$. Now think back to the previous example. As $X \sim \delta(k)$ is concentrated on $\{1, 2, \dots, n\}$ for all $k = 1, 2, \dots, n$, give the previous question one more thought.

Let's see a partial answer in the next example.

**Example 3.** *The Bernoulli distribution.* (See the definition of the Bernoulli distribution in *Section 19.2.1*). Let $X \sim \text{Bernoulli}(p)$. Then, it is easy to see that

$$
H[X] = -p \log p - (1 - p) \log(1 - p).
$$

Which value of $p$ maximizes the entropy? To find the maxima of $H[X]$, we can turn to the derivatives. (Recall how the derivative and second derivative can be used for optimization, as claimed by *Theorem 13.1.4.*)

Thus, let $f(p) \doteq H[X] = -p \log p - (1-p) \log(1-p)$. Then,

$$f'(p) = -\log p + \log(1-p) = \log \frac{1-p}{p},$$

$$f''(p) = -\frac{1}{p} - \frac{1}{1-p}.$$

By solving $f'(p) = 0$, we obtain that $p = 1/2$, which is the only potential extrema of $f(p)$. As $f''(1/2) = -4 < 0$, we see that $p = 1/2$ is indeed a local maximum. Let's plot $f(p)$ to obtain a visual confirmation as well.

```python
def bernoulli_entropy(p):
    return -p*np.log(p) - (1 - p)*np.log(1 - p)

X = np.linspace(0.001, 0.999, 100)
y = bernoulli_entropy(X)
with plt.style.context('seaborn-v0_8'):
    plt.figure(figsize=(8, 8))
    plt.xlabel("p")
    plt.ylabel("H[X]")
    plt.title("The entropy of Bernoulli(p)")
    plt.plot(X, y)
    plt.show()
```

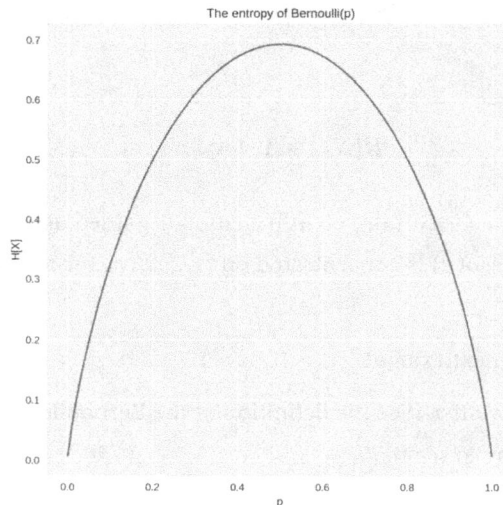

Figure 20.10: *The entropy of the Bernoulli distribution*

For $p = 1/2$, that is, where the entropy of Bernoulli($p$) is maximal, we have a uniform distribution on the two-element set $\{0, 1\}$. On the other hand, for $p = 0$ or $p = 1$, where the entropy is minimal, Bernoulli($p$) is a single-point distribution.

As every random variable on $\{0, 1\}$ is Bernoulli-distributed, we seem to have a partial answer to our question: the uniform distribution maximizes entropy, while single-point ones minimize it.

As the following theorem indicates, this is true in general as well.

> **Theorem 20.6.1** *(The uniform distribution and maximal entropy)*
>
> *Let $E = \{x_1, \ldots, x_n\}$ be a finite set, and let $X : \Omega \to E$ be a random variable that assumes values in E. Then,*
>
> $$H[X] \le H[\text{Uniform}(E)],$$
>
> *and $H[X] = H[\text{Uniform}(E)]$ if and only if X is uniformly distributed on E.*

We are not going to show this here, but there are several proofs out there. For instance, Bishop's classic *Pattern Recognition and Machine Learning* uses the Lagrange multiplier method to explicitly find the maximum of the multivariable function $f(p_1, \ldots, p_n) = -\sum_{k=1}^{n} p_k \log p_k$; feel free to check it out for the details.

What if we don't restrict our discrete random variable to a finite set? In that case, the Shannon entropy has no upper limit. In the problem set of this chapter, you'll see that the entropy of the geometric distribution is

$$H[\text{Geo}(p)] = -\frac{p \log p + (1 - p) \log(1 - p)}{p}.$$

It is easy to see that $\lim_{p \to 0} H[\text{Geo}(p)] = \infty$. Let's plot this!

```python
def geom_entropy(p):
    return -(p*np.log(p) + (1 - p)*np.log(1 - p))/p

X = np.linspace(1e-16, 1-1e-16, 1000)
y = geom_entropy(X)


with plt.style.context('seaborn-v0_8'):
    plt.figure(figsize=(8, 8))
    plt.xlabel("p")
    plt.ylabel("H[X]")
    plt.title("The entropy of Geo(p)")
    plt.plot(X, y)
    plt.show()
```

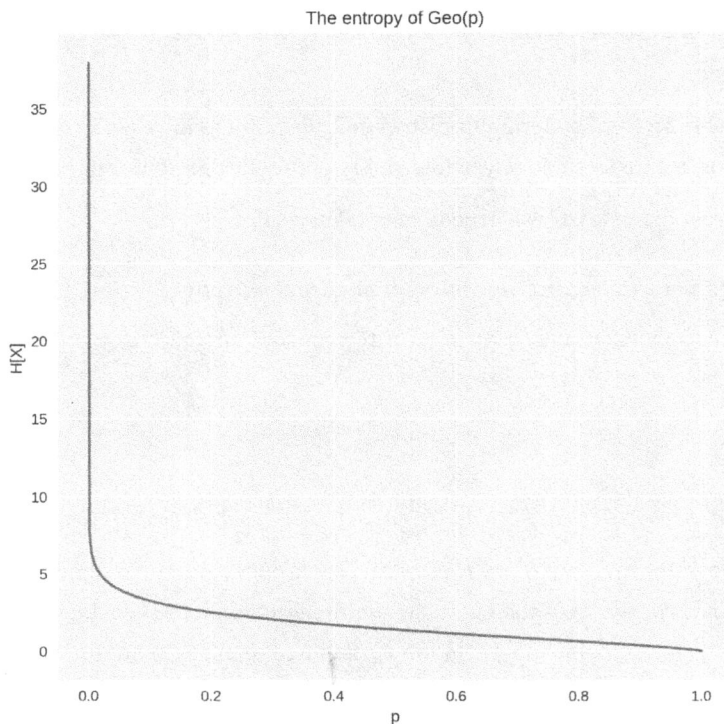

*Figure 20.11: The entropy of the geometric distribution*

Thus, the Shannon entropy can assume any nonnegative value.

## 20.6.4   Differential entropy

So far, we have only defined the entropy for discrete random variables.

Does it translate to continuous ones as well? Yes. The formula $\mathbb{E}[-\log f_X(X)]$ can be directly applied for continuous random variables, yielding the so-called *differential entropy*. Here is the formal definition.

> **Definition 20.6.3  (Differential entropy)**
>
> Let $X$ be a continuous random variable. The *differential entropy* of $X$ is defined by the formula
>
> $$H[X] := -\int_{-\infty}^{\infty} f_X(x) \log f_X(x) dx,$$
>
> where $f_X$ denotes the probability density function of $X$.

Now comes the surprise. Can we derive the formula from the Shannon entropy? We are going to approach the problem like we did when we defined the expected value for continuous random variables in *Section 20.2*: approximate the continuous random variable with a discrete one, then see where the Shannon entropy converges.

Thus, let $X : \Omega \to \mathbb{R}$ be a continuous random variable, and let $[a, b] \subseteq \mathbb{R}$ be a (large) interval, so large that $P(X \notin [a, b])$ is extremely small. We'll subdivide $[a, b]$ into $n$ equal parts by

$$x_k = a + \frac{k(b - a)}{n}, \quad k = 0, 1, \ldots, n,$$

and define the approximating random variable $X^{(n)}$ by

$$X^{(n)}(\omega) := \begin{cases} x_k & \text{if } x \in (x_{k-1}, x_k] \text{ for some } k = 1, 2, \ldots, n, \\ 0 & \text{otherwise.} \end{cases}$$

This way, the entropy of $X^{(n)}$ is given by

$$H[X^{(n)}] = -\sum_{k=1}^{n} P(X^{(n)} = x_k) \log P(X^{(n)} = x_k).$$

However, due to how we defined $X^{(n)}$,

$$P(X^{(n)} = x_k) = P(x_{k-1} < X \le X_k) = \int_{x_{k-1}}^{x_k} f_X(x)dx,$$

where $f_X$ is the density function of $X$. Now, the mean value theorem for definite integrals (*Theorem 14.1.3*) gives that there is a $\xi_k \in [x_{k-1}, x_k]$ such that

$$\int_{x_{k-1}}^{x_k} f_X(x)dx = (x_k - x_{k-1})f_X(\xi_k) = \frac{f_X(\xi_k)}{n},$$

thus, in conclusion,

$$P(X^{(n)} = x_k) = \frac{f_X(\xi_k)}{n}.$$

(Recall that as the partition $x_0 < x_1 < \ldots < x_n$ is equidistant, $x_k - x_{k-1} = 1/n$.)

Now, using $P(X^{(n)} = x_k) = \frac{f_X(\xi_k)}{n}$, we obtain

$$H[X^{(n)}] = -\sum_{k=1}^{n} P(X^{(n)} = x_k) \log P(X^{(n)} = x_k)$$

$$= -\sum_{k=1}^{n} \frac{f_X(\xi_k)}{n} \log \frac{f_X(\xi_k)}{n}$$

$$= -\sum_{k=1}^{n} \frac{f_X(\xi_k)}{n} \log f_X(\xi_k) + \log n \sum_{k=1}^{n} \frac{f_X(\xi_k)}{n}.$$

Both of these terms are Riemann-sums, approximating the integral of the functions inside. If $n$ is large, and the interval $[a, b]$ is big enough, then

$$-\sum_{k=1}^{n} \frac{f_X(\xi_k)}{n} \log f_X(\xi_k) \approx -\int_{-\infty}^{\infty} f_X(x) \log f_X(x) dx = h[X],$$

and

$$\sum_{k=1}^{n} \frac{f_X(\xi_k)}{n} \approx \int_{-\infty}^{\infty} f_X(x) dx = 1,$$

implying

$$H[X^{(n)}] \approx h[X] + \log n.$$

This is quite surprising, as one would expect $H[X^{(n)}]$ to converge towards $h(X)$. This is not the case. In fact,

$$\lim_{n \to \infty} \left( H[X^{(n)}] - \log n \right) = h[X]$$

holds.

It's time for the examples.

**Example 1.** *The uniform distribution.* (See the definition of the uniform distribution in *Section 19.3.4*.) Let $X \sim \text{Uniform}(a, b)$. Then,

$$h[X] = -\int_{a}^{b} \frac{1}{b-a} \log \frac{1}{b-a} dx$$

$$= \log(b - a),$$

which is similar to the discrete uniform case. However, there is one notable difference: $h(X)$ is negative when $b - a < 1$. This is in stark contrast with the Shannon entropy, which is always nonnegative.

**Example 2.** *The normal distribution.* (See the definition of the normal distribution in *Section 19.3.6*.) Let $X \sim \mathcal{N}(\mu, \sigma^2)$. Then,

$$h[X] = -\int_{-\infty}^{\infty} \frac{1}{\sigma\sqrt{2\pi}} e^{-\frac{(x-\mu)^2}{2\sigma^2}} \log\left(\frac{1}{\sigma\sqrt{2\pi}} e^{-\frac{(x-\mu)^2}{2\sigma^2}}\right) dx$$

$$= \log\left(\sigma\sqrt{2\pi}\right) \underbrace{\int_{-\infty}^{\infty} \frac{1}{\sigma\sqrt{2\pi}} e^{-\frac{(x-\mu)^2}{2\sigma^2}} dx}_{=1} + \underbrace{\int_{-\infty}^{\infty} \frac{(x-\mu)^2}{2\sigma^2} \frac{1}{\sigma\sqrt{2\pi}} e^{-\frac{(x-\mu)^2}{2\sigma^2}} dx}_{=\frac{1}{2\sigma^2}\text{Var}[X]=\frac{1}{2}}$$

$$= \frac{1}{2}\left(1 + \log(\sigma^2 2\pi)\right).$$

Depending on the value of $\sigma$, the value of $h[X]$ can be negative here as well.

Previously, we have seen that for discrete distributions on a given finite set, the uniform distribution maximizes entropy, as *Theorem 20.6.1* claims.

What is the analogue of *Theorem 20.6.1* for continuous distributions? Take a wild guess. If we let $X$ be any continuous distribution, then, as we have seen,

$$h[\text{Uniform}(a, b)] = \log(b - a),$$

which can assume any real number. Similarly to the discrete case, we have to make restrictions; this time, we'll fix the variance. Here is the result.

**Theorem 20.6.2** *(Maximizing the differential entropy)*

Let $X$ be a continuous random variable with variance $\sigma^2$. Then

$$h[X] \leq h[\mathcal{N}(0, \sigma^2)],$$

and $h[X] = h[\mathcal{N}(0, \sigma^2)]$ if and only if $X \sim \mathcal{N}(\mu, \sigma^2)$.

Again, we are not going to prove this. You can check Bishop's *Pattern Recognition and Machine Learning* for more details.

## 20.7 The Maximum Likelihood Estimation

I am an evangelist for simple ideas. Stop me any time you want, but whichever field I was in, I've always been able to find a small set of mind-numbingly simple ideas making the entire shebang work. (Not that you could interrupt me, as this is a book. Joke's on you!)

Let me give you a concrete example that's on my mind. What do you think enabled the rise of deep learning, including neural networks with billions of parameters? Three ideas as simple as ABC:

- that you can optimize the loss function by going against its gradient (no matter the number of parameters),
- that you can efficiently compute the gradient with a clever application of the chain rule and matrix multiplication,
- and that we can perform matrix operations blazingly fast on a GPU.

Sure, there's a great tower of work built upon these ideas, but these three lie at the very foundation of machine learning today. Ultimately, these enable you to converse with large language models. To have your car cruise around town while you read the newspaper. To predict the exact shape of massive amino-acid chains called proteins, responsible for building up every living thing. (Including you.)

Gradient descent, backpropagation, and high-performance linear algebra are on the practical side of the metaphorical machine learning coin. If we conjure up a parametric model, we can throw some extremely powerful tools at it.

But where do our models come from?

As I've said, there is a small set of key ideas that go a long way. We are about to meet one: the maximum likelihood estimation.

## 20.7.1  Probabilistic modeling 101

As a self-proclaimed evangelist of simple ideas, I'll start with a *simple* example to illustrate a *simple* idea.

Pick up a coin and toss it a few times, recording each outcome. The question is, once more, simple: what's the probability of heads? We can't just immediately assume $p = 1/2$, that is, a fair coin. For instance, one side of our coin could be coated with lead, resulting in a bias. To find out, let's perform some statistics. (Rolling up my sleeves, throwing down my gloves.)

Mathematically speaking, we can model coin tosses with the Bernoulli distribution (*Section 19.2.1*):

$$P(X = 1) = p, \quad P(X = 0) = 1 - p,$$

where

- $X$ is the random variable representing the outcome of a single toss,
- $X = 1$ for heads and $X = 0$ for tails,
- and $p \in [0, 1]$ is the probability of heads.

That's just the model. We're here to estimate the parameter $p$, and this is what we have statistics for.

Tossing up the coin $n$ times yields the zero-one sequence $x_1, x_2, \dots, x_n$, where each $x_i$ is a realization of a Bernoulli-distributed random variable $X_i \sim$ Bernoulli$(p)$, independent of each other.

As we saw previously when discussing the law of large numbers (*Theorem 20.5.4*), one natural idea is to compute the sample mean to estimate $p$, which is coincidentally the expected value of $X$. To move beyond empirical estimates, let's leverage that, this time, we have a probabilistic model.

The key question is this: which parameter $p$ is the most likely to produce our sample?

In the language of probability, this question is answered by maximizing the *likelihood* function

$$\text{LLH}(p; x_1, \dots, x_n) = \prod_{i=1}^{n} P(X_i = x_i \mid p),$$

where $P(X_i = x_i \mid p)$ represents the probability of observing $x_i$ given a fixed parameter $p$. The larger the $\text{LLH}(p; x_1, \dots, x_n)$, the more likely the parameter $p$ is. In other words, our estimate of $p$ is going to be

$$\hat{p} = \text{argmax}_{p \in [0,1]} \text{LLH}(p; x_1, \dots, x_n).$$

Let's find it.

In our concrete case, $P(X_i = x_i \mid p)$ can be written as

$$P(X_i = x_i \mid p) = \begin{cases} p & \text{if } x_i = 1, \\ 1 - p & \text{if } x_i = 0. \end{cases}$$

Algebra doesn't welcome if-else type functions, so with a clever mathematical trick, we write $P(X_i = x_i \mid p)$ as

$$P(X_i = x_i \mid p) = p^{x_i}(1 - p)^{1 - x_i},$$

making the likelihood function be

$$\text{LLH}(p; x_1, \dots, x_n) = \prod_{i=1}^{n} p^{x_i}(1 - p)^{1 - x_i}.$$

(We'll often write $\text{LLH}(p)$ to minimize notational complexity.)

This is still not easy to optimize, as it is composed of the product of exponential functions. So, here's another mathematical trick: take the logarithm to turn the product into a sum.

As the logarithm is increasing, it won't change the optima, so we're good to go:

$$\log \text{LLH}(p) = \log \prod_{i=1}^{n} p^{x_i}(1-p)^{1-x_i}$$

$$= \sum_{i=1}^{n} \log \left[ p^{x_i}(1-p)^{1-x_i} \right]$$

$$= \sum_{i=1}^{n} \left[ \log p^{x_i} + \log(1-p)^{1-x_i} \right]$$

$$= \log p \sum_{i=1}^{n} x_i + \log(1-p) \sum_{i=1}^{n}(1-x_i).$$

Trust me, this is much better. According to the second derivative test (*Theorem 13.1.4*), we can find the maxima by

1. solving $\frac{d}{dp} \log \text{LLH}(p) = 0$ to find the critical point $\hat{p}$,
2. then showing that $\hat{p}$ is a maximum because $\frac{d^2}{dp^2} \log \text{LLH}(p) < 0$.

Let's get to it.

As $\frac{d}{dp} \log p = \frac{1}{p}$ and $\frac{d}{dp} \log(1-p) = -\frac{1}{1-p}$, we have

$$\frac{d}{dp} \log \text{LLH}(p; x_1, \ldots, x_n) = \frac{1}{p} \sum_{i=1}^{n} x_i - \frac{1}{1-p} \sum_{i=1}^{n}(1-x_i).$$

Solving $\frac{d}{dp} \log \text{LLH}(p; x_1, \ldots, x_n) = 0$ yields a single solution

$$\hat{p} = \frac{1}{n} \sum_{i=1}^{n} x_i.$$

(Pick up a pen and paper and calculate the solution yourself.) Regarding the second derivative, we have

$$\frac{d^2}{dp^2} \log \text{LLH}(p) = -\frac{1}{p^2} \sum_{i=1}^{n} x_i - \frac{1}{(1-p)^2} \sum_{i=1}^{n}(1-x_i),$$

which is uniformly negative. Thus, $\hat{p} = \frac{1}{n} \sum_{i=1}^{n} x_i$ is indeed a (local) maximum. Yay!

In this case, the maximum likelihood estimate is identical to the sample mean. Trust me, this is one of the rare exceptions. Think of it as validating the sample mean: we've obtained the same estimate through different trains of thought, so it must be good.

## 20.7.2 Modeling heights

Let's continue with another example. The coin-tossing example demonstrated the discrete case. It's time to move into the continuous domain!

This time, we are measuring the heights of a high school class, and we want to build a probabilistic model of it. A natural idea is to assume the heights to come from a normal distribution $X \sim \mathcal{N}(\mu, \sigma^2)$. (Check *Section 19.3.6* for the normal distribution.)

Our job is to estimate the expected value $\mu$ and the variance $\sigma^2$. Let's go, maximum likelihood!

To make the problem mathematically precise, we have the measurements $x_1, \ldots, x_n$, coming from independent and identically distributed random variables $X_i \sim \mathcal{N}(\mu, \sigma^2)$. However, there's a snag: as our random variables are continuous,

$$P(X_1 = x_1, \ldots, X_n = x_n \mid \mu, \sigma^2) = \prod_{i=1}^{n} P(X_i = x_i \mid \mu, \sigma^2) = 0.$$

(As all terms of the product are zero.) How can we define the likelihood function, then? No worries: even though we don't have a mass function, we have density! Thus, the likelihood function defined by

$$\mathrm{LLH}(\mu, \sigma; x_1, \ldots, x_n) = \prod_{i=1}^{n} f_{X_i}(x_i)$$

$$= \prod_{i=1}^{n} \frac{1}{\sigma \sqrt{2\pi}} e^{-\frac{(x_i - \mu)^2}{2\sigma^2}},$$

where $f_{X_i}(x)$ is the probability density function of $X_i$.

Let's maximize it. The idea is similar: take the logarithm, find the critical points, then use the second derivative test. Here we go:

$$\log \mathrm{LLH}(\mu, \sigma) = n \log \frac{1}{\sigma \sqrt{2\pi}} - \frac{1}{\sigma^2} \sum_{i=1}^{n} (x_i - \mu)^2.$$

Just the usual business from now on. The derivatives are given by

$$\frac{\partial}{\partial \mu} \log \mathrm{LLH}(\mu, \sigma) = \frac{2}{\sigma^2} \sum_{i=1}^{n} (x_i - \mu),$$

$$\frac{\partial}{\partial \sigma} \log \mathrm{LLH}(\mu, \sigma) = -\frac{n}{\sigma} - \frac{2}{\sigma^3} \sum_{i=1}^{n} (x_i - \mu)^2.$$

With a bit of number-crunching (that you should attempt to carry out by yourself), we get that $\frac{\partial}{\partial \mu} \log \text{LLH}(\mu, \sigma) = 0$ implies

$$\mu = \frac{1}{n} \sum_{i=1}^{n} x_i,$$

and $\frac{\partial}{\partial \sigma} \log \text{LLH}(\mu, \sigma)$ implies

$$\sigma = \frac{1}{n} \sum_{i=1}^{n} (x_i - \mu)^2.$$

We won't do the second derivative test here, but trust me: it's a maximum, leaving us with the estimates

$$\hat{\mu} = \frac{1}{n} \sum_{i=1}^{n} x_i,$$

$$\hat{\sigma} = \frac{1}{n} \sum_{i=1}^{n} (x_i - \hat{\mu})^2.$$

Again, the sample mean and variance. Think of it this way: defaulting to the sample mean and variance is the simplest thing to do, yet even clever methods like the maximum likelihood estimation yield them as parameter estimates.

After working out the above two examples in detail, we are ready to abstract away the details and introduce the general problem.

### 20.7.3 The general method

We've seen how maximum likelihood estimation works. Now, it's time to construct the abstract mathematical framework.

**Definition 20.7.1 (The likelihood function)**

Let $P_\theta$ be a probability distribution parametrized by the parameter $\theta \in \mathbb{R}^k$, and let $x_1, \ldots, x_n \in \mathbb{R}^d$ be an independent realization of the probability distribution. (That is, the samples are coming from independent and identically distributed random variables $X_1, \ldots, X_n$, distributed according to $P_\theta$.)

The *likelihood function* of $\theta$ given the sample $x_1, \ldots, x_n$ is defined by

(a)

$$\text{LLH}(\theta; x_1, \ldots, x_n) := \prod_{i=1}^{n} P_\theta(X_i = x_i)$$

if $P_\theta$ is discrete, and

*(b)*

$$\text{LLH}(\theta; x_1, \dots, x_n) := \prod_{i=1}^{n} f_\theta(x_i)$$

if $P_\theta$ is continuous, where $f_\theta$ is the probability density function of $P_\theta$.

We've already seen two examples of the likelihood function: for the Bernoulli-distribution Bernoulli($p$), given by

$$\text{LLH}(p; x_1, \dots, x_n) = \prod_{i=1}^{n} p^{x_i}(1-p)^{1-x_i},$$

and for the normal distribution $\mathcal{N}(\mu, \sigma)$, given by

$$\text{LLH}(\mu, \sigma; x_1, \dots, x_n) = \prod_{i=1}^{n} \frac{1}{\sigma\sqrt{2\pi}} e^{-\frac{(x_i-\mu)^2}{2\sigma^2}}.$$

Intuitively, the likelihood function $\text{LLH}(\theta; x_1, \dots, x_n)$ expresses the probability of our observation $x_1, \dots, x_n$ if the parameter $\theta$ is indeed true. The maximum likelihood estimate is the parameter $\hat{\theta}$ that maximizes this probability; that is, under which the observation is the most likely.

**Definition 20.7.2 (The maximum likelihood estimate)**

Let $P_\theta$ be a probability distribution parametrized by the parameter $\theta \in \mathbb{R}^k$, and let $x_1, \dots, x_n \in \mathbb{R}^d$ be an independent realization of the probability distribution.

The *maximum likelihood estimate* of $\theta$ is given by

$$\hat{\theta} = \text{argmax}_{\theta \in \mathbb{R}^k} \text{LLH}(\theta; x_1, \dots, x_n).$$

In both examples, we used the logarithm to turn the product into a sum. Use it once and it's a trick; use it (at least) twice and it's a method. Here's the formal definition.

**Definition 20.7.3 (The log-likelihood function)**

Let $P_\theta$ be a probability distribution parametrized by the parameter $\theta \in \mathbb{R}^k$, and let $x_1, \dots, x_n \in \mathbb{R}^d$ be an independent realization of the probability distribution.

The *log-likelihood function* of $\theta$ given the sample $x_1, \ldots, x_n$ is defined by

$$\log \text{LLH}(\theta; x_1, \ldots, x_n),$$

where $\text{LLH}(\theta; x_1, \ldots, x_n)$ is the likelihood function.

In a classical statistical setting, the maximum likelihood estimation is done via

1. pulling a parametric probabilistic model out from the mathematician's hat,
2. massaging the (log-)likelihood function until we obtain an analytically manageable form,
3. and solving $\nabla \text{LLH} = 0$ (or $\nabla \log \text{LLH} = 0$) to obtain the parameter estimate.

Statistics can be extremely powerful under specific circumstances, but let's face it: the above method has quite a few weaknesses. First, constructing a tractable probabilistic model is a challenging task, burdened by the experts' inherent bias. (It's no accident that I indirectly compared the modeling process to pulling a rabbit out of a hat.) Moreover, the more complex the model, the more complex the likelihood function is. Which, in turn, increases the complexity of our optimization problem.

Why did we spend quite a few pages learning this ancient technique, then?

Because its idea is fundamental in machine learning, we'll arrive at (somewhere near the) state of the art by breaking down its barriers one by one. Is modeling hard? Let's construct a function with BILLIONS of parameters that'll do the job. Is optimization computationally intensive? Fear not. We have clusters of GPUs at our disposal.

## 20.7.4 The German tank problem

One more example before finishing up, straight from World War II. Imagine you are an Allied intelligence officer tasked to estimate the size of a German armored division. (That is, to guess the number of tanks.)

There was no satellite imagery back in the day, so there's only a little to go on, except for a tiny piece of information: the serial numbers of the enemy's destroyed tanks. What can we do with these?

Without detailed knowledge of the manufacturing process, we can assume that the tanks are labeled sequentially as they roll out from the factory. We also don't know how the tanks are distributed between the battlefields.

These two pieces of knowledge (or lack of knowledge, to be more precise) translate to a simple probabilistic model: encountering an enemy tank is the same as drawing from the distribution $\text{Uniform}(N)$, where $N$ is the total number of tanks. Thus, if $x_1, \ldots, x_n$ are the serial numbers of destroyed tanks, we can use the maximum likelihood method to estimate $N$.

Let's do it. The likelihood function for the discrete uniform distribution Uniform($N$) is given by

$$\text{LLH}(N) = \prod_{i=1}^{n} P(X_i = x_i),$$

where the probability $P(X_i = x_i)$ has a quite peculiar form:

$$P(X_i = x_i) = \begin{cases} \frac{1}{N} & \text{if } x_i \in \{1, \dots, N\}, \\ 0 & \text{otherwise.} \end{cases}$$

Keeping in mind that $x_1, \dots, x_n \leq N$ (as no observed serial number can be larger than the total number of tanks), we have

$$\text{LLH}(N) = \begin{cases} \frac{1}{N^n} & \text{if } N < \max\{x_1, \dots, x_n\}, \\ 0 & \text{otherwise.} \end{cases}$$

Ponder on this a minute: the larger the $N$, the smaller the LLH($N$) is. Thus, the maximum likelihood estimate is the smallest possible choice

$$\hat{N} = \max\{x_1, \dots, x_n\}.$$

In other words, our guess about the number of tanks is the largest serial number we've encountered.

What do you think about this estimate? I won't lie; I am not a big fan. The German tank problem highlights the importance of modeling assumptions in statistics. The final estimate $\hat{N}$ is the outcome of our choice of Uniform($N$). Common wisdom in machine learning is "garbage in, garbage out." It is true for modeling as well.

# 20.8  Summary

In this chapter, we have learned about the concept of the expected value. Mathematically speaking, the expected value is defined by

$$\mathbb{E}[X] = \sum_k x_k P(X = x_k)$$

for discrete random variables and

$$\mathbb{E}[X] = \int_{-\infty}^{\infty} x f_X(x) dx$$

for continuous ones. Although these formulas involve possibly infinite sums and integrals, the underlying meaning is simple: $\mathbb{E}[X]$ represents the average outcome of $X$, weighted by the underlying probability distribution.

According to the law of large numbers, the expected value also describes a long-term average: if the independent and identically distributed random variables $X_1, X_2, \ldots$ describe the outcomes of a repeated experiment — say, betting a hand in poker — then the sample average converges to the joint expected value, that is,

$$\lim_{n \to \infty} \frac{1}{n} \sum_{i=1}^{n} X_i = \mathbb{E}[X_1]$$

holds with probability 1. In a sense, the law of large numbers allows you to glimpse into the future and see what happens if you make the same choice. In the case of poker, if you only make bets with a positive expected value, you'll win in the long run.

In machine learning, the LLN also plays an essential role. Check out the mean-squared error

$$\text{MSE}(\mathbf{x}, \mathbf{y}) = \frac{1}{n} \sum_{i=1}^{n} (f(x_i) - y_i)^2, \quad \mathbf{x}, \mathbf{y} \in \mathbb{R}^n$$

once more. If the number of samples ($n$) is in the millions, computing the gradient of this sum is not feasible. However, the mean-squared error is the sample average of the prediction errors; thus, it's enough to sample a smaller amount. This is the core principle behind stochastic gradients, an idea that makes machine learning on a large scale feasible.

With this chapter, our journey comes to a close. Still, there's so much to learn; I could probably write this book until the end of time. Sadly, we have to stop somewhere. Now, instead of giving a summary of all that's in the book, let's talk about the most important message: learning never ends.

It's a spiral that you continue to ascend, meeting familiar landscapes from higher and higher vantage points. If you keep going, you'll know what I'm talking about.

If you lead an intellectually challenging life, you'll also find that knowledge is like keeping a dozen leaky cups full of water. If your focus shifts from one, it'll empty faster than you think. In other words, you'll lose it if you don't use it. This is completely normal. The good news is, if you already have a good foundation, refilling the cup can be done quickly. Sometimes, simply glancing at a page from a book you read long ago can do the trick.

This is how I know that it's not goodbye. If you have found the book useful and continue down the rabbit hole that is machine learning, we'll meet again with probability one. You just have to keep going.

## 20.9  Problems

**Problem 1.** Let $X, Y : \Omega \to \mathbb{R}$ be two random variables.

*(a)* Show that if $X \geq 0$, then $\mathbb{E}[X] \geq 0$.

*(b)* Show that if $X \geq Y$, then $\mathbb{E}[X] \geq \mathbb{E}[Y]$.

**Problem 2.** Let $X : \Omega \to \mathbb{R}$ be a random variable. Show that if $\mathrm{Var}[X] = 0$, then $X$ assumes only a single value. (That is, the set $X(\Omega) = \{X(\omega) : \omega \in \Omega\}$ has only a single element.)

**Problem 3.** Let $X \sim \mathrm{Geo}(p)$ be a geometrically distributed (*Section 19.2.3*) discrete random variable. Show that

$$H[X] = -\frac{p \log p + (1 - p) \log(1 - p)}{p}.$$

*Hint:* Use that for any $q \in (0, 1)$, $\sum_{k=1}^{\infty} kq^{k-1} = (1 - q)^{-2}$.

**Problem 4.** Let $X \sim \exp(\lambda)$ be an exponentially distributed continuous random variable. Show that

$$h[X] = 1 - \log \lambda.$$

**Problem 5.** Find the maximum likelihood estimation for the $\lambda$ parameter of the exponential distribution.

# Join our community on Discord

Read this book alongside other users, Machine Learning experts, and the author himself.

Ask questions, provide solutions to other readers, chat with the author via Ask Me Anything sessions, and much more.

Scan the QR code or visit the link to join the community.

`https://packt.link/math`

# References

[1] Jaynes, E. T. (2003). *Probability Theory: The Logic of Science*. Cambridge University Press, Cambridge.

[2] SciPy documentation. `https://docs.scipy.org/doc/scipy/`

[3] Shannon, C. E. (1948). A Mathematical Theory of Communication. *The Bell System Technical Journal*, 27, 379–423.

[4] Bishop, C. M. (2006). *Pattern Recognition and Machine Learning* (Vol. 4). Springer.

# PART V
# APPENDIX

This part comprises the following chapters:

- *Appendix A, It's Just Logic*
- *Appendix B, The Structure of Mathematics*
- *Appendix C, Basics of Set Theory*
- *Appendix D, Complex Numbers*

# A

# It's Just Logic

*The rules of logic are to mathematics what those of structure are to architecture.* — *Bertrand Russell*

*"Mathematics is a language,"* one of my professors used to say all the time. *"Learning mathematics starts with building up a basic vocabulary."*

What he forgot to add is that mathematics is the *language of thinking.* I often get asked the question: do you need to know mathematics to be a software engineer/data scientist/random technical professional? My answer is simple. If you regularly have to solve problems in your profession, then mathematics is extremely beneficial to you. You don't *have to* think effectively, but you are better off.

The learning curve of mathematics is steep. You have experienced it yourself, and the difficulty may have deterred you from reaching a familiarity with its fundamentals. I have good news: if we treat learning mathematics as learning a foreign language, we can start by building up a basic vocabulary instead of diving straight into poems and novels. As my professor suggested.

# A.1   Mathematical logic 101

Logic and clear thinking lie at the very foundations of mathematics. But what are those? How would you explain what "logic" is?

Our thinking processes are formalized by the field of *mathematical logic*. In logic, we work with *propositions*, that is, statements that are either true or false. *"It is raining outside." "The sidewalk is wet."* These are both valid propositions.

To be able to reason about propositions effectively, we often denote them with roman capital letters, such as

$$A = \text{"It is raining outside."}$$
$$B = \text{"The sidewalk is wet."}$$

Each proposition has a corresponding *truth value*, which is either *true* or *false*. These are often abbreviated as 1 and 0. Although this seems like no big deal, finding the truth value can be extremely hard. Think about the proposition

$$A = \text{"If the solutions of an algorithm can be}$$
$$\text{verified in polynomial time, it can also be}$$
$$\text{solved in polynomial time."}$$

This is the famous P = NP conjecture, one of the longest-standing unsolved problems in mathematics. The statement is easy to understand, but solving the problem (that is, finding the truth value of the corresponding proposition) has eluded even the smartest minds.

In essence, the entire body of our scientific knowledge lies in propositions whose truth values we have identified. So, how do we do that in practice?

# A.2   Logical connectives

In themselves, propositions are not enough to provide an effective framework for reasoning. Mathematics (and the entirety of modern science) is the collection of complex propositions formulated from smaller building blocks with *logical connectives*. Each connective takes one or more propositions and transforms their truth value.

*"If it is raining outside, then the sidewalk is wet."* This is the combination of two propositions, strung together by the *implication* connective. There are four essential connectives: *negation, disjunction, conjunction,* and *implication*. We will take a close look at each one.

*Negation* flips the truth value of a proposition to its opposite. It is denoted by the mathematical symbol $\neg$: if $A$ is a proposition, then $\neg A$ is its negation. Connectives are defined by *truth tables* that enumerate all the possible truth values of the resulting expression, given its inputs. In writing, this looks complicated, so here is the truth table of $\neg$ to illustrate the concept.

| $A$ | $\neg A$ |
|---|---|
| 0 | 1 |
| 1 | 0 |

When expressing propositions in a natural language, negation translates to the word "not." For instance, the negation of the proposition *"the screen is black"* is *"the screen is not black."* (Not *"the screen is white."*)

Logical *conjunction* is equivalent to grammatical conjunction "and", denoted by the symbol $\wedge$. The proposition $A \wedge B$ is true if and only if both $A$ and $B$ are true. For example, when we say that *"the table is set and the food is ready,"* we mean to convey that both conjuncts are true. Here is the truth table:

| $A$ | $B$ | $A \wedge B$ |
|---|---|---|
| 0 | 0 | 0 |
| 0 | 1 | 0 |
| 1 | 0 | 0 |
| 1 | 1 | 1 |

*Disjunction* is known as "or" in the English language and is denoted by the symbol $\vee$. The proposition $A \vee B$ is true whenever either one is:

| $A$ | $B$ | $A \vee B$ |
|---|---|---|
| 0 | 0 | 0 |
| 0 | 1 | 1 |
| 1 | 0 | 1 |
| 1 | 1 | 1 |

Disjunction is inclusive, unlike the *exclusive or* we frequently use in our natural language. When you say "I am traveling by train or car," both cannot be true. The disjunction connective is not exclusive.

Finally, the *implication* connective ($\rightarrow$) formalizes the deduction of a conclusion $B$ from a premise $A$: "if $A$, then $B$."

The implication is false only when the premise is true and the conclusion is false; otherwise, it is true.

| $A$ | $B$ | $A \to B$ |
|:---:|:---:|:---:|
| 0 | 0 | 1 |
| 0 | 1 | 1 |
| 1 | 0 | 0 |
| 1 | 1 | 1 |

One example would be the famous quote from Descartes: *"I think, therefore I am."* Translated to the language of formal logic, this is simply

$$\text{"I think"} \to \text{"I exist"}.$$

Sentences of the form *"if A, then B"* are called *conditionals*. It's not all just philosophy. Science is the collection of propositions such as *"if X is a closed system, then the entropy of X cannot decrease."* (As the 2nd law of thermodynamics states.)

Most of our scientific knowledge is made of $A \to B$ propositions, and scientific research is equivalent to pursuing the truth value of implications. When solving problems in practice, we rely on theorems (that is, implications) that turn our premises into conclusions.

# A.3   The propositional calculus

If you got the feeling that the connectives are akin to arithmetic operations, you are correct. Connectives yield propositions. Thus connectives can again be applied, resulting in complex expressions such as $\neg(A \lor B) \land C$. Constructing such expressions and deductive arguments is called the *propositional calculus*.

Just like arithmetic operations, expressions made up of propositions and connectives also have identities. Think about the famous algebraic identity

$$(a + b)(a - b) = a^2 - b^2,$$

which is one of the most frequently used symbolic expressions. Such an identity means we can write one thing in another form.

In mathematical logic, we call these *logical equivalences*.

---

**Definition A.3.1 (Logical equivalences)**

The propositions $P$ and $Q$ are *logically equivalent* if they always have the same truth value.

If $P$ and $Q$ are logically equivalent, we write

$$P \equiv Q.$$

---

To show you an example, let's look at our first *theorem*, one that establishes logical equivalences for the conjunction connective.

---

**Theorem A.3.1 *(Properties of conjunction)***

*Let A, B, and C be arbitrary propositions. Then,*

*(a)* $(A \wedge B) \wedge C \equiv A \wedge (B \wedge C)$ *(associativity)*

*(b)* $A \wedge B \equiv B \wedge A$ *(commutativity)*

*(c)* $A \vee (B \wedge C) \equiv (A \vee B) \wedge (A \vee C)$ *(distributivity)*

*(d)* $A \wedge A \equiv A$ *(idempotence)*

---

*Proof.* Showing these properties is done by drawing up their truth tables. We will do this for *(a)*, while the rest is left for you as an exercise. (I highly suggest you do this, as performing a task by yourself is an excellent learning opportunity.)

For the associativity property, the sizable truth table

| $A$ | $B$ | $C$ | $A \wedge B$ | $B \wedge C$ | $(A \wedge B) \wedge C$ | $A \wedge (B \wedge C)$ |
|---|---|---|---|---|---|---|
| 0 | 0 | 0 | 0 | 0 | 0 | 0 |
| 0 | 0 | 1 | 0 | 0 | 0 | 0 |
| 0 | 1 | 0 | 0 | 0 | 0 | 0 |
| 0 | 1 | 1 | 0 | 1 | 0 | 0 |
| 1 | 0 | 0 | 0 | 0 | 0 | 0 |
| 1 | 0 | 1 | 0 | 0 | 0 | 0 |
| 1 | 1 | 0 | 1 | 0 | 0 | 0 |
| 1 | 1 | 1 | 1 | 1 | 1 | 1 |

provides a proof.

A few remarks are in order. First, we should read the truth table from left to right columns. Strictly speaking, we can omit the columns for $A \wedge B$ and $B \wedge C$. However, including them saves the mental gymnastics.

Second, because of the associativity, we can freely write $A \wedge B \wedge C$, as the order of operations is irrelevant.

Finally, note that our first theorem is a premise and a conclusion, connected by the implication connective. If we denote them by

$$P = \text{``}A, B, C \text{ are propositions,''}$$
$$Q = \text{``}(A \wedge B) \wedge C \equiv A \wedge (B \wedge C).\text{''}$$

then the first part of our theorem is just the proposition $P \rightarrow Q$, one that we have proven to be true via laying out the truth table. This shows the immense power of the *propositional calculus* we are building here.

*Theorem A.3.1* has an analogue for disjunction. This is stated below for the sake of completeness, but the proof is left to you as an exercise.

> **Theorem A.3.2** *(Properties of disjunction)*
>
> *Let A, B, and C be arbitrary propositions. Then,*
>
> *(a) $(A \vee B) \vee C \equiv A \vee (B \vee C)$ (associativity)*
>
> *(b) $A \vee B \equiv B \vee A$ (commutativity)*
>
> *(c) $A \wedge (B \vee C) \equiv (A \wedge B) \vee (A \wedge C)$ (distributivity)*
>
> *(d) $A \vee A \equiv A$ (idempotence)*

Just as with arithmetic operations, connectives have order of precedence as well: $\neg, \wedge, \vee, \rightarrow$. This means that, for instance, $((\neg A) \wedge B) \vee C$ can be written as $\neg A \wedge (B \vee C)$.

In our calculus of propositions, one of the most important rules is De Morgan's laws, describing how conjunction and disjunction behave with respect to negation.

> **Theorem A.3.3** *(De Morgan's laws)*
>
> *Let A and B be two arbitrary propositions. Then,*
>
> *(a) $\neg(A \wedge B) \equiv \neg A \vee \neg B$*
>
> *(b) $\neg(A \vee B) \equiv \neg A \wedge \neg B$*
>
> *hold.*

*Proof.* As usual, we can prove De Morgan's laws by laying out the two truth tables

| $A$ | $B$ | $\neg A$ | $\neg B$ | $A \wedge B$ | $\neg(A \wedge B)$ | $\neg A \vee \neg B$ |
|---|---|---|---|---|---|---|
| 0 | 0 | 1 | 1 | 0 | 1 | 1 |
| 0 | 1 | 1 | 0 | 0 | 1 | 1 |
| 1 | 0 | 0 | 1 | 0 | 1 | 1 |
| 1 | 1 | 0 | 0 | 1 | 0 | 0 |

and

| $A$ | $B$ | $\neg A$ | $\neg B$ | $A \vee B$ | $\neg(A \vee B)$ | $\neg A \wedge \neg B$ |
|---|---|---|---|---|---|---|
| 0 | 0 | 1 | 1 | 0 | 1 | 1 |
| 0 | 1 | 1 | 0 | 1 | 0 | 0 |
| 1 | 0 | 0 | 1 | 1 | 0 | 0 |
| 1 | 1 | 0 | 0 | 1 | 0 | 0 |

that verify our claim.

The propositional calculus we have established so far is the mathematical formalization of thinking. One thing is missing, though: deduction, or as Wikipedia puts it, *"the mental process of drawing inferences in which the truth of their premises ensures the truth of their conclusion."* This is given via the famous rule of *modus ponens*.

**Theorem A.3.4** *(Modus ponens)*

*Let A and B be two propositions. If A and A → B are true, then B is true as well.*

*Proof.* Let's take a look at the truth table of → once again:

| $A$ | $B$ | $A \to B$ |
|---|---|---|
| 0 | 0 | 1 |
| 0 | 1 | 1 |
| 1 | 0 | 0 |
| 1 | 1 | 1 |

By looking at its rows, we can see that when $A$ is true and the implication $A \to B$ is true, $B$ is true as well, as the principle of modus ponens indicates.

As modus ponens sounds extremely abstract, here is a concrete example. From common sense, we know that the implication *"if it's raining, then the sidewalk is wet"* is true. If we observe from a roof window that it's indeed raining, we can confidently conclude that the sidewalk is wet, even without looking at it.

In symbolic notation, we can write

$$A \to B, A \vdash B,$$

where the turnstile symbol $\vdash$ essentially reads as "proves." Thus, the modus ponens says that $A \to B$ and $A$ prove $B$.

Modus ponens is how we use our theorems. It is always in the background.

---

**Remark A.3.1 (Reversing the implication)**

This is a great opportunity to point out one of the most frequent logical fallacies: reversing the implication. When debating about a given topic, participants often resort to the faulty argument

$$A \to B, B \vdash A.$$

Of course, this is not true. For instance, consider our favorite example:

$$A = \text{"it's raining outside,"}$$
$$B = \text{"the sidewalk is wet."}$$

Clearly, $A \to B$ holds, but $B \to A$ does not. There are other reasons for a wet sidewalk. For instance, someone accidentally spilled a barrel of water on it.

---

# A.4   Variables and predicates

So, mathematics is about propositions, implications, and their truth values. We have seen that we can formulate propositions and reason about pretty complicated expressions using our propositional calculus. However, the language we have built up so far is not suitable for propositions with *variables*.

For instance, think about the sentence

$$x \text{ is a non-negative real number.}$$

Because the truth value depends on $x$, this is not a well-formed proposition.

Sentences with variables are called *predicates*, and we denote them by emphasizing the dependence on their variables; for instance,

$$P(x) : x \geq 0,$$

or

$$Q(x, y) : x + y \text{ is an even number.}$$

Each predicate has a *domain* from which its variables can be taken. You can think about a predicate $P(x)$ as a function that maps its domain to the set $\{0, 1\}$, representing its truth value. (Although, strictly speaking, we don't have functions available as tools when defining the very foundation of our formal language. However, we are not philosophers or set theorists, so we don't have to be concerned about such details.)

Predicates define *truth sets*, that is, subsets of the domain where the predicate is true. Formally, they are denoted by

$$\{x \in D : P(x)\}, \tag{A.1}$$

where $P(x)$ is a predicate with domain $D$.

Translated to English, (A.1) reads as "all elements $x$ of $D$ for which $P(x)$ is true."

Although we haven't talked about sets before, truth sets probably seem familiar if you have a computer science background. For instance, if you have ever used the Python programming language, you have probably seen expressions like

```python
s = {x for x in range(1, 100) if x % 5 == 0}
```

all the time. These are called *comprehensions*, and they are inspired by the so-called *set-builder notation* given by (A.1).

# A.5   Existential and universal quantification

Predicates are a big step toward properly formalizing mathematical thinking, but we are not quite there yet. To give you an example from machine learning, let's talk about finding the minima of loss functions (that is, training a model).

A point $x$ is said to be the global minimum of a function $f(x)$ if, for all other $y$ in its domain $D$, $f(x) \leq f(y)$ holds. For instance, the point $x = 0$ is a minima of the function $f(x) = x^2$.

How would you express this in our formal language? For one, we could say that

$$\text{for all } y \in D, f(x) \leq f(y) \text{ is true,}$$

where we fix $f(x) = x^2$ and $x = 0$. There are two parts of this sentence: for all $y \in D$, and $f(x) \leq f(y)$ is true. The second one is a predicate:

$$P(y) \,:\, f(x) \leq f(y),$$

where $y \in \mathbb{R}$.

The second part seems new, as we have never seen the words *"for all"* in our formal language before. They express a kind of *quantification* about when the predicate $P(y)$ is true.

In mathematical logic, there are two quantifiers we need to be happy: the universal quantifier *"for all"* denoted by the symbol $\forall$, and the existential quantifier *"there exists"* denoted by $\exists$.

For example, consider the sentence *"all of my friends are mathematicians."* By defining the set $F$ to be set of my friends and the predicate on this domain as

$$M(x) \,:\, x \text{ is a mathematician,}$$

we can formalize our sentence as

$$\forall x \in F, M(x).$$

Remember that the domain of the predicate $M(x)$ is $F$. We could omit that, but it's much more user-friendly this way.

Similarly, *"I have at least one friend who is a mathematician"* translates to

$$\exists x \in F, M(x).$$

When there is a more complex proposition behind the quantifier, we mark its *scope* with parentheses:

$$\forall x \in F, (A(x) \rightarrow (B(x) \wedge C(x))).$$

Note that as $(\forall x \in F, M(x))$ and $(\exists x \in F, M(x))$ have a single truth value, they are propositions, not predicates! Thus, quantifiers turn predicates into propositions. Just like any other propositions, logical connectives can be applied to them.

Among all the operations, negation is the most interesting here. To see why, let's consider the previous example: *"all of my friends are mathematicians."* At first, you might say that its negation is *"none of my friends are mathematicians,"* but that is not correct. Think about it: I can have mathematician friends, as long as not all of them are mathematicians. Thus,

$$\neg(\text{"all of my friends are mathematicians"}) \equiv \text{"I have at least one non-mathematician friend."}$$

In other words (or should I say *symbols*), we have

$$\neg(\forall x \in F, M(x)) \equiv \exists x \in F, \neg M(x).$$

That is, roughly speaking, the negation of $\forall$ is $\exists$ and the negation of $\exists$ is $\forall$.

## A.6   Problems

**Problem 1.** Using truth tables, show that

*(a)* $A \vee \neg A$ is true,

*(b)* and $A \wedge \neg A$ is false.

In other words, $A \vee \neg A$ is a *tautology*, while $A \wedge \neg A$ is a *contradiction*. (We call expressions that are always true *tautologies*, while expressions that are always false are *contradictions*.)

**Problem 2.** Define the exclusive or operation XOR, denoted by $\oplus$, by the truth table

| $A$ | $B$ | $A \oplus B$ |
|-----|-----|--------------|
| 0 | 0 | 0 |
| 0 | 1 | 1 |
| 1 | 0 | 1 |
| 1 | 1 | 0 |

Show that

*(a)* $A \oplus B \equiv (\neg A \wedge B) \vee (A \wedge \neg B)$

*(b)* and $A \oplus B \equiv (\neg A \vee \neg B) \wedge (A \vee B)$

holds.

# Join our community on Discord

Read this book alongside other users, Machine Learning experts, and the author himself.

Ask questions, provide solutions to other readers, chat with the author via Ask Me Anything sessions, and much more.

Scan the QR code or visit the link to join the community.

`https://packt.link/math`

# B

# The Structure of Mathematics

We've come a long way from the start: we've studied propositions, logical connectives, predicates, quantifiers, and all the formal logic. This was to be able to *talk* about mathematics. However, ultimately, we want to *do* mathematics.

As the only exact science, mathematics is built on top of *definitions*, *theorems*, and *proofs*. We precisely define objects, formulate conjectures about them, then prove those with mathematically correct arguments. You can think of mathematics as a colossal building made of propositions, implications, and modus ponens. If one theorem fails, all others that build upon it fail too.

In other fields of science, the *modus operandi* is to hypothesize, experiment, and validate. However, experiments are not enough in mathematics. For instance, think about the famous Fermat numbers, that is, numbers of the form $F_n := 2^{2^n} + 1$. Fermat conjectured them all to be prime numbers, as $F_0$, $F_1$, $F_2$, $F_3$, and $F_4$ are primes.

Five affirmative "experiments" might have been enough to accept the hypothesis as true in certain fields of science. Not in mathematics. In 1732, Euler showed that $F_5 = 4,294,967,297$ is not a prime, as $4,294,967,297 = 641 \times 6,700,417$. (Imagine calculating that in the 18th century, long before the age of computing.)

So far, we've seen some definitions, theorems, and even proofs when talking about mathematical logic. It's time to put them under the magnification glass and see what they are!

# B.1   What is a definition?

Ambiguity is the drawback of natural languages. How would you define, say, the concept of "hot"? Upon several attempts, you would soon discover that no two people have the same definition.

In mathematics, there is no room for ambiguity. Every object and every property must be precisely defined. It's best to look at a good example instead of philosophizing about it.

> **Definition B.1.1 (Divisors)**
>
> Let $b \in \mathbb{Z}$ be an integer. We say that $a \in \mathbb{Z}$ is a *divisor* of $b$ if there exists an integer $k \in \mathbb{Z}$ such that $b = ka$.
>
> The property *"a is a divisor of b"* is denoted by $a \mid b$.

For example, $2 \mid 10$ and $5 \mid 10$, but $7 \nmid 10$. (Crossed symbols mean the negation of the said property.)

In terms of our formal language, the definition of *"a is a divisor of b"* can be written as

$$a \mid b \ : \ \exists k \in \mathbb{Z}, b = ka. \tag{B.1}$$

Don't let the $a \mid b$ notation deceive you; this is a predicate in disguise. We could have denoted $a \mid b$ by

$$\mathrm{divisor}(a, b) \ : \ \exists k \in \mathbb{Z}, b = ka.$$

Although every mathematical definition can be formalized, we'll prefer our natural language because it is much easier to understand. (At least for humans. Not so much for computers.)

Like building blocks, definitions build on top of each other.

(If you have a sharp eye for details, you noticed that even *Definition B.1.1* is built upon other concepts such as numbers, multiplication, and equality. We haven't defined them precisely, just assumed they are there. Since our goal is not to rebuild mathematics from scratch, we'll let this one slide.)

Again, it's best to see an example here. Let's see what *even* and *odd* numbers are!

> **Definition B.1.2 (Even and odd numbers)**
>
> Let $n \in \mathbb{Z}$ be an integer. We say that $n$ is *even* if $2 \mid n$.
>
> In turn, we say that $n$ is *odd* if $2 \nmid n$. (The notation $a \nmid b$ is the negation of the *"a is a divisor of b"* predicate.)

One more time, with our formal language. For an integer $n \in \mathbb{Z}$, the predicates

$$\text{even}(n) \; : \; 2 \mid n$$

and

$$\text{odd}(n) \; : \; 2 \nmid n$$

express the same as *Definition B.1.2*.

These examples are not that exciting, so let's see something more interesting!

---

**Definition B.1.3 (Prime numbers)**

Let $p \in \mathbb{N}$ be a positive integer. We say that $p$ is a *prime number* if

*(a)* $p > 1$,

*(b)* and if $a \mid p$, then $a = 1$ or $a = p$.

---

In other words, primes have no integer divisors other than themselves. The first few primes are 2, 3, 5, 7, 11, 13, 17, and many more. Non-prime integers are called *composite numbers*.

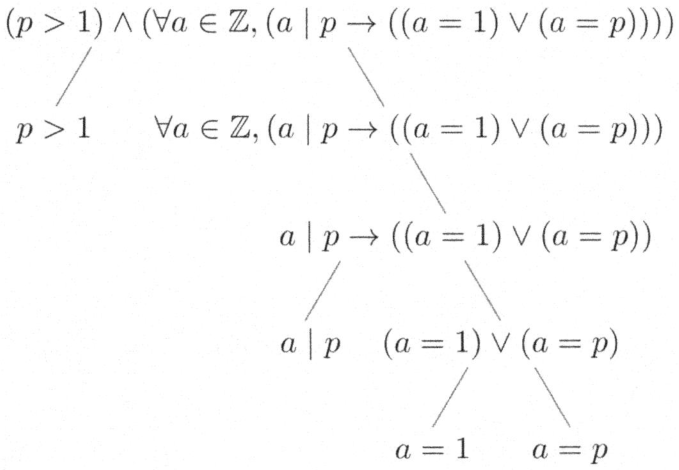

$$(p > 1) \wedge (\forall a \in \mathbb{Z}, (a \mid p \rightarrow ((a = 1) \vee (a = p))))$$

$$p > 1 \qquad \forall a \in \mathbb{Z}, (a \mid p \rightarrow ((a = 1) \vee (a = p)))$$

$$a \mid p \rightarrow ((a = 1) \vee (a = p))$$

$$a \mid p \qquad (a = 1) \vee (a = p)$$

$$a = 1 \qquad a = p$$

*Figure B.1: Definition of primality in predicate logic, decomposed into its parts*

The definition of primality can be written as

$$P(p) \; : \; (p > 1) \wedge (\forall a \in \mathbb{Z}, (a \mid p \rightarrow ((a = 1) \vee (a = p)))).$$

This might look complicated, but we can decompose it into parts, as shown by *Figure B.1*.

Primes play an essential role in our everyday lives! For instance, many mainstream cryptographic methods use large primes to cipher and decipher messages. Without them, you wouldn't be able to initiate financial transactions securely.

Their usefulness is guaranteed by their various properties, established in the form of *theorems*. We'll see a few of them soon enough, but first, let's talk about what theorems *really* are.

# B.2   What is a theorem?

So, a definition is essentially a predicate whose truth set consists of our objects of interest. The whole point of mathematics is to find true propositions involving those objects, most often in the form $A \to B$. Consider the following theorem.

> **Theorem B.2.1**  *(Existence of global minima for convex functions)*
>
> *Let $f : [0, 1] \to \mathbb{R}$ be a function. If $f$ is continuous, then there exists an $x^*$ such that $f$ assumes its minimum at $x^*$ on $[0, 1]$.*
>
> *(That is, for all $x \in [0, 1]$, we have $f(x^*) \le f(x)$.)*

Don't worry if you are unfamiliar with the concepts of continuity and minimum; it's beside the point. The gist is that *Theorem B.2.1* can be written as

$$\forall f \in F, (C(f) \to M(f)),$$

where $F$ denotes the set of all functions $[0, 1] \to \mathbb{R}$, and the predicates $C(f)$ and $M(f)$ are defined by

$$C(f) : f \text{ is continuous on } [0, 1],$$
$$M(f) : \exists x^*, \forall x \in [0, 1], f(x^*) \le f(x).$$

Notice the structure of the theorem: *"Let $x \in A$. If $B(x)$, then $C(x)$."* With the first sentence, we are setting the domains of the predicates $A(x)$ and $B(x)$, and putting a universal quantifier in front of the conditional *"if $B(x)$, then $C(x)$."*

# B.3  What is a proof?

Now that we understand what theorems are, it's time to look at *proofs*. We have just seen that theorems are true propositions. Proofs are deductions that establish the truth of a proposition. Let's see an example instead of talking like a philosopher!

The proof of *Theorem B.2.1* is not within our reach yet, so let's look at something much simpler: the sum of even numbers.

---

**Theorem B.3.1** *(The sum of even numbers)*

*Let $n, m \in \mathbb{Z}$ be two integers. If $n$ and $m$ are even, then $n + m$ is even.*

---

*Proof.* Since $n$ is even, $2 \mid n$. According to *Definition B.1.1*, this means that there exists an integer $k \in \mathbb{Z}$ such that $n = 2k$.

Similarly, as $m$ is also even, there exists an integer $l \in \mathbb{Z}$ such that $m = 2l$. Summing up the two, we obtain

$$n + m = 2k + 2l$$
$$= 2(k + l),$$

giving that $n + m$ is indeed even.

---

If you read the above proof carefully, you might notice that it is a chain of implications and modus ponens. These two form the backbone of our deductive skills. What is proven is set in stone.

Understanding what proofs are is one of the biggest skill gaps in mathematics. Don't worry if you don't get it immediately; this is a deep concept. You'll get used to proofs eventually.

# B.4  Equivalences

The building blocks of mathematics are propositions of the form $A \to B$; at least, this is what I emphasized throughout this chapter.

I was not precise. The proposition $A \to B$ translates to *"if A, then B,"* but sometimes, we know much more. Quite frequently, $A$ and $B$ have the same truth values. In natural language, we express this by saying *"A if and only if B."* (Although this is much rarer than the simple conditional.)

In logic, we express this relation with the *biconditional* connective $\leftrightarrow$, defined by

$$A \leftrightarrow B \equiv (A \to B) \wedge (B \to A).$$

Theorems of the *"if and only if"* type are called *equivalences*, and they play an essential role in mathematics. When proving an equivalence, we must show both $A \to B$ and $B \to A$.

To see an example, let's go back to elementary geometry. As you probably learned in high school, we can describe geometric objects on the plane with *vectors* that are represented by a tuple of two real numbers. This way, geometric properties can be translated into analytic ones, and we can often prove hard theorems by simple calculations.

For instance, let's talk about *orthogonality*, one of the most important concepts in mathematics. Here is how orthogonality is defined for two planar vectors.

**Definition B.4.1 (Orthogonality)**

Let $\mathbf{a}$ and $\mathbf{b}$ be two nonzero vectors on the plane. We say that $\mathbf{a}$ and $\mathbf{b}$ are *orthogonal* if their enclosed angle is $\pi/2$.

Orthogonality is denoted by the $\perp$ symbol; that is, $\mathbf{a} \perp \mathbf{b}$ means that $\mathbf{a}$ and $\mathbf{b}$ are orthogonal.

For the sake of simplicity, we always assume that the enclosed angle is between $0$ and $\pi$. (An angle of $\pi$ radians is 180 degrees, but we'll always use radians.)

However, measuring the angle enclosed by two arbitrary vectors is not as easy as it sounds. We need a tractable formula, and this is where the *dot product* comes in.

**Definition B.4.2 (Dot product of planar vectors)**

Let $\mathbf{a} = (a_1, a_2)$ and $\mathbf{b} = (b_1, b_2)$ be two vectors on the plane. Their dot product $\mathbf{a} \cdot \mathbf{b}$ is defined by

$$\mathbf{a} \cdot \mathbf{b} := |\mathbf{a}||\mathbf{b}| \cos \alpha,$$

where $\alpha$ is the angle enclosed by the two vectors, and $|\cdot|$ denotes the magnitude of a vector.

Dot products give an equivalent definition of orthogonality in the form of an *"if and only if"* theorem.

**Theorem B.4.1** *Let $\mathbf{a} = (a_1, a_2)$ and $\mathbf{b} = (b_1, b_2)$ be two nonzero vectors in the plane. Then, $\mathbf{a}$ and $\mathbf{b}$ are orthogonal if and only if $\mathbf{a} \cdot \mathbf{b} = 0$.*

Let's see the proof of this equivalence!

*Proof.* We have to prove two implications:

*(a)* $\mathbf{a} \perp \mathbf{b} \implies \mathbf{a} \cdot \mathbf{b} = 0$,

*(b)* $\mathbf{a} \cdot \mathbf{b} = 0 \implies \mathbf{a} \perp \mathbf{b}$.

Let's start with *(a)*. If $\mathbf{a} \perp \mathbf{b}$, then their enclosed angle $\alpha$ equals to $\pi/2$. Thus,

$$\mathbf{a} \cdot \mathbf{b} = |\mathbf{a}||\mathbf{b}| \cos \frac{\pi}{2}$$
$$= |\mathbf{a}||\mathbf{b}|0$$
$$= 0,$$

which is what we needed to show.

To prove *(b)*, we have to notice that since $\mathbf{a}$ and $\mathbf{b}$ are nonzero, their magnitudes $|\mathbf{a}|$, $|\mathbf{b}|$ are also nonzero. Thus,

$$\mathbf{a} \cdot \mathbf{b} = |\mathbf{a}||\mathbf{b}| \cos \alpha = 0$$

can only hold if $\cos \alpha = 0$. In turn, this means that $\alpha = \pi/2$; that is, $\mathbf{a} \perp \mathbf{b}$. (Recall that we assumed the enclosed angle $\alpha$ to be between 0 and $\pi$.)

So, we know all about what theorems and proofs are. But how do we find proofs in practice? Let's see the essential techniques.

# B.5 Proof techniques

There is no way around it: proving theorems is hard. Some took the smartest of minds decades, and some conjectures remain unresolved after a century. (That is, they are not proven nor disproven.)

A few basic yet powerful tools can get one push through the difficulties. In the following, we'll look at the three most important ones: proof by induction, proof by contradiction, and the principle of contraposition.

## B.5.1 Proof by induction

How do you climb a set of stairs? Simple. You climb the first step, then climb the next one, and so on.

You might be surprised, but this is something we frequently use in mathematics all the time.

Let's illuminate this by an example.

**Theorem B.5.1** *(Sum of natural numbers)*

*Let $n \in \mathbb{N}$ be an arbitrary integer. Then,*

$$1 + 2 + \cdots + n = \frac{n(n+1)}{2} \tag{B.2}$$

*holds.*

*Proof.* For $n = 1$, the case is clear: the left-hand side of (B.2) evaluates to 1, while the right-hand side is

$$\frac{1(1+1)}{2} = 1.$$

Thus, our proposition holds for $n = 1$, which is called the *base case*.

Here comes the magic, that is, the *induction step*. Let's *assume* that (B.2) holds for a given $n$; that is, we have

$$1 + 2 + \cdots + n = \frac{n(n+1)}{2}.$$

This is what's called the *induction hypothesis*. Using this assumption, we are going to prove that (B.2) holds for $n + 1$ as well. In other words, our goal is to show that

$$1 + 2 + \cdots + n + (n+1) = \frac{(n+1)(n+2)}{2}.$$

Due to our induction hypothesis, we have

$$1 + 2 + \cdots + n + (n+1) = \left[ 1 + 2 + \cdots + n \right] + (n+1)$$
$$= \frac{n(n+1)}{2} + (n+1).$$

Continuing the calculation, we obtain

$$\frac{n(n+1)}{2} + (n+1) = \frac{n(n+1)}{2} + \frac{2(n+1)}{2}$$
$$= \frac{n(n+1) + 2(n+1)}{2}$$
$$= \frac{(n+1)(n+2)}{2},$$

which is what we had to show.

To sum up what happened, let's denote the equation (B.2) by the predicate

$$S(n) : 1 + 2 + \cdots + n = \frac{n(n+1)}{2}.$$

Proof by induction consists of two main steps. First, we establish that the base case $S(1)$ is true. Then, we show that for arbitrary $n$, the implication $S(n) \to S(n+1)$ holds. Starting from the induction step, this implies that $S(n)$ is indeed true for all $n$: the chain of implications

$$S(1) \to S(2),$$
$$S(2) \to S(3),$$
$$S(3) \to S(4),$$
$$\vdots$$

combined with $S(1)$ and the almighty modus ponens (*Theorem A.3.4*) yields the truth of $S(n)$. We took the first step $S(1)$, then proved that we can take the next step from anywhere.

Induction is not simple to grasp, so here is another example. (It is slightly more complex than the previous one.) Follow through with the proof and see if you can identify the marks of induction.

**Theorem B.5.2** *(The fundamental theorem of number theory)*

*Let $n \in \mathbb{Z}$ be an integer and suppose that $n > 1$. Then, $n$ can be uniquely represented as the product of prime numbers; that is, there exists prime numbers $p_1, p_2, \ldots, p_l$ and exponents $k_1, k_2, \ldots, k_l > 1$ such that*

$$n = p_1^{k_1} p_2^{k_2} \cdots p_l^{k_l}. \tag{B.3}$$

*Furthermore, this representation is unique.*

For example, $24 = 2^3 3$, and 24 cannot be written as a different product of primes.

The presence of natural language masks it, but in essence, *Theorem B.5.2* can be translated to the sentence

$$\forall n \in \mathbb{Z}\left[(n > 1) \to (\exists p_1, \ldots, p_l, k_1, \ldots, k_l \in \mathbb{Z}, (\forall i, p_i \text{ is a prime}) \wedge (n = p_1^{k_1} p_2^{k_2} \ldots p_l^{k_l}))\right].$$

For simplicity, we'll only prove the *existence* of the prime factorization, not the unicity.

*Proof.* (Existence.) For $n = 2$, the theorem is trivially true, as 2 is a prime itself.

Now, let $n > 2$ and suppose that (B.3) is true for all integers $m$ that are smaller or equal to $n$. (This is our induction hypothesis.)

Our goal is to show that (B.3) also holds for $n + 1$.

There are two possibilities: either $n + 1$ is a prime or a composite number. If it is a prime, we are done, as $n + 1$ is by itself in the form (B.3). Otherwise, if $n + 1$ is a composite number, we can find a divisor that is not 1 or $n + 1$:

$$n + 1 = ab$$

for some $a, b \in \mathbb{Z}$. Since $a, b \leq n$, we can apply the induction hypothesis! Spelling it out, it means that we can write them as

$$a = p_1^{\alpha_1} \ldots p_l^{\alpha_l},$$
$$b = q_1^{\beta_1} \ldots q_m^{\beta_m},$$

where the $p_i$, $q_i$ are the primes and the $\alpha_i, \beta_i$ are the exponents. Thus,

$$n + 1 = ab$$
$$= p_1^{\alpha_1} \ldots p_l^{\alpha_l} q_1^{\beta_1} \ldots q_m^{\beta_m},$$

which is just (B.3), with a bit more symbols.

Induction is like a power tool in mathematics. It is extremely powerful, and when it is applicable, it'll almost always do the job.

## B.5.2  Proof by contradiction

Sometimes, it is easier to prove theorems by assuming that their conclusion is false, then deduce a contradiction.

Again, it's best to see a quick example. Let's revisit our good old friends, the prime numbers.

**Theorem B.5.3** *There are infinitely many prime numbers.*

*Proof.* Assume that there are finitely many prime numbers: $p_1, p_2, \ldots, p_n$.

Is the integer $p_1 p_2 \ldots p_n + 1$ a prime? If $p_1, p_2, \ldots, p_n$ are all of the prime numbers, it is enough to check if

$$p_i \nmid p_1 p_2 \ldots p_n + 1.$$

This holds indeed, as by definition, $p_1 p_2 \ldots p_n + 1 = p_i k + 1$, where $k$ is simply the product of the prime numbers other than $p_i$.

Since no $p_i$ is a divisor of $p_1 p_2 \ldots p_n + 1$, it must be a prime. We have found a new prime that is not on our list! This means that our assumption (that there are finitely many prime numbers) has led to a contradiction.

Thus, there must be infinitely many prime numbers.

If you have a sharp eye, you probably noticed that the above example is not of the form $A \to B$; it's just a simple proposition:

$$A = \text{"there are infinitely many prime numbers."}$$

In these cases, showing that $\neg A$ is false yields the desired conclusion. However, this technique works for $A \to B$-style propositions as well. (By the way, the existence part of *Theorem B.5.2* can also be shown via contradiction; I'll leave this for you as an exercise.)

## B.5.3  Contraposition

The final technique we will study is *contraposition*, a clever method that puts a twist into the classic $A \to B$-style thinking.

We should get to know the implication connective a bit better to see what it is. As it turns out, $A \to B$ can be written in terms of negation and disjunction.

**Theorem B.5.4** *Let A and B two propositions. Then,*

$$A \to B \equiv \neg A \vee B.$$

*Proof.* The truth table

| $A$ | $B$ | $\neg A$ | $\neg A \vee B$ |
|-----|-----|----------|-----------------|
| 0 | 0 | 1 | 1 |
| 0 | 1 | 1 | 1 |
| 1 | 0 | 0 | 0 |
| 1 | 1 | 0 | 1 |

provides a proof.

Why is this relevant? Simple. Take a look at the following corollary.

**Corollary B.5.1** *(Principle of contraposition)*

*Let A and B be two propositions. Then,*
$$A \to B \equiv \neg B \to \neg A.$$

*Proof.* Theorem B.5.4 implies that

$$A \to B \equiv \neg A \vee B$$
$$\equiv B \vee \neg A$$
$$\equiv \neg B \to \neg A,$$

which is what we had to prove.

Here is a simple proposition about integers to give you a mathematical example.

**Theorem B.5.5** *Let $n \in \mathbb{Z}$ be an integer. If $2 \nmid n$, then $4 \nmid n$.*

*Proof.* We should prove this via contraposition. Thus, assume that $4 \mid n$. This means that

$$n = 4k$$

for some integer $k \in \mathbb{Z}$. However, this implies that

$$n = 2(2k),$$

which shows that $2 \mid n$. Due to the principle of contraposition, $(4 \mid n) \rightarrow (2 \mid n)$ is logically equivalent to $(2 \nmid n) \rightarrow (4 \nmid n)$, which is what we had to prove.

Contraposition is not only useful in mathematics, it is a valuable thinking tool in general. Let's consider our recurring proposition: *"if it is raining outside, then the sidewalk is wet."* We know this to be true, but this also means that *"if the sidewalk is not wet, then it is not raining"* (because, otherwise, the sidewalk would be wet).

You perform these types of arguments every day without even noticing it. Now you have a name for them and can start to apply this pattern consciously.

# Join our community on Discord

Read this book alongside other users, Machine Learning experts, and the author himself.

Ask questions, provide solutions to other readers, chat with the author via Ask Me Anything sessions, and much more.

Scan the QR code or visit the link to join the community.

https://packt.link/math

# C
# Basics of Set Theory

*In other words, general set theory is pretty trivial stuff really, but, if you want to be a mathematician, you need some and here it is; read it, absorb it, and forget it. — Paul R. Halmos*

Although Paul Halmos said the above a long time ago, it has remained quite accurate. Except for one part: set theory is not only necessary for mathematicians, but for computer scientists, data scientists, and software engineers as well.

You might have heard about or studied set theory before. It is hard to see why it is so essential for machine learning, but trust me, set theory *is* the very foundation of mathematics. Deep down, *everything* is a set or a function between sets. (As we saw in *Chapter 9*, even functions are defined as sets.)

Think about the relation of set theory and machine learning like grammar and poetry. To write beautiful poetry, one needs to be familiar with the rules of the language. For example, data points are represented as vectors in vector spaces, often constructed as the Cartesian product of sets. (Don't worry if you are not familiar with Cartesian products, we'll get there soon.) Or, to *really* understand probability theory, you need to be familiar with event spaces, which are systems of sets closed under certain operations.

So, what are sets anyway?

# C.1   What is a set?

On the surface level, a set is just a collection of things. We define sets by enumerating their elements like

$$S = \{\text{red}, \text{green}, \heartsuit\}.$$

Two sets are equal if they have the same elements. Given any element, we can always tell if it is a member of a given set or not. When every element of $A$ is also an element of $B$, we say that $A$ is a *subset* of $B$, or, in notation,

$$A \subseteq B.$$

If $A \subseteq B$ and $A \neq B$, we say that $A$ is a *proper subset* of $B$ and write $A \subset B$. If we have a set, we can define subsets by specifying a property that all of its elements satisfy, for example,

$$\text{even numbers} = \{n \in \mathbb{Z} : n\%2 = 0\}.$$

(The $\%$ denotes the modulo operator.) This latter method is called the set-builder notation, and if you are familiar with the Python programming language, you can see this inspired list comprehensions. There, one would write something like this:

```python
even_numbers = {n for n in range(10) if n%2 == 0}

print(even_numbers)
```

```
{0, 2, 4, 6, 8}
```

We can even describe sets as a collection of other sets, say, the set of all subsets of $A$. This is called the *power set*, a concept so essential that it deserves its own formal definition.

**Definition C.1.1 (Power sets)**

Let $A$ be an arbitrary set. The set defined by

$$2^A := \{B : B \subseteq A\},$$

containing all subsets of $A$, is called the *power set* of $A$.

Both $\emptyset$ and $A$ are elements of the power set $2^A$.

Unfortunately, defining sets as a collection of elements does not work. Without further conditions, it can lead to paradoxes, as the famous Russell paradox shows. (We'll talk about this later in this chapter.) To avoid going down the rabbit hole of set theory, we accept that sets have some proper definition buried within a thousand-page-sized tome of mathematics. Instead of worrying about this, we focus on what we can do with sets.

# C.2 Operations on sets

Describing more complex sets with only these two methods (listing their members or using the set-builder notation) is extremely difficult. To make the job easier, we define *operations* on sets.

## C.2.1 Union, intersection, difference

The most basic operations are the *union*, *intersection*, and *difference*. You are probably familiar with these, as they are encountered frequently as early as high school. Even if you are familiar with them, check out the formal definition next.

---

**Definition C.2.1 (Set operations)**

Let $A$ and $B$ be two sets. We define

(a) their union by $A \cup B := \{x : x \in A \text{ or } x \in B\}$,

(b) their intersection by $A \cap B := \{x : x \in A \text{ and } x \in B\}$,

(c) and their difference by $A \setminus B := \{x : x \in A \text{ and } x \notin B\}$.

---

We can easily visualize these with Venn diagrams, as you can see below.

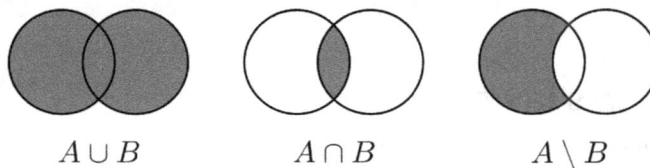

$$A \cup B \qquad A \cap B \qquad A \setminus B$$

Figure C.1: Set operations visualized in Venn diagrams

We can express set operations in plain English as well. For example, $A \cup B$ means *"A or B."* Similarly, $A \cap B$ means *"A and B,"* while $A \setminus B$ is *"A but not B."* When talking about probabilities, these will be useful for translating events to the language of set theory.

These set operations also have a lot of pleasant properties.

For example, they behave nicely with respect to parentheses.

**Theorem C.2.1** *Let A, B, and C be three sets. The union operation is*

*(a) associative, that is, $A \cup (B \cup C) = (A \cup B) \cup C$,*

*(b) commutative, that is, $A \cup B = B \cup A$.*

*Moreover, the intersection operation is also associative and commutative. Finally,*

*(c) the union is distributive with respect to the intersection, that is, $A \cup (B \cap C) = (A \cup B) \cap (A \cup C)$,*

*(d) and the intersection is distributive with respect to the union, that is, $A \cap (B \cup C) = (A \cap B) \cup (A \cap C)$.*

Union and intersection can be defined for an arbitrary number of operands. That is, if $A_1, A_2, \ldots, A_n$ are sets,

$$A_1 \cup \cdots \cup A_n := (A_1 \cup \cdots \cup A_{n-1}) \cup A_n,$$

and similar for the intersection. Note that this is a recursive definition! Because of associativity, the order of parentheses doesn't matter.

The associativity and commutativity might seem too abstract and trivial at the same time. However, this is not the case for all operations, so it is worth emphasizing to get used to the concepts. If you are curious, noncommutative operations are right under our noses. A simple example is string concatenation.

```
a = "string"
b = "concatenation"
a + b == b + a
```

```
False
```

## C.2.2   De Morgan's laws

One of the fundamental rules describes how set difference, union, and intersection behave together regarding set operations. These are called De Morgan's laws.

**Theorem C.2.2** *(De Morgan's laws)*

*Let A, B, and C be three sets. Then*

*(a) $A \setminus (B \cup C) = (A \setminus B) \cap (A \setminus C)$,*

*(b) $A \setminus (B \cap C) = (A \setminus B) \cup (A \setminus C)$.*

*Proof.* For simplicity, we are going to prove this using Venn diagrams. Although drawing a picture is not a "proper" mathematical proof, this is not a problem. We are here to *understand* things, not to get hung up on philosophy.

Here is the illustration.

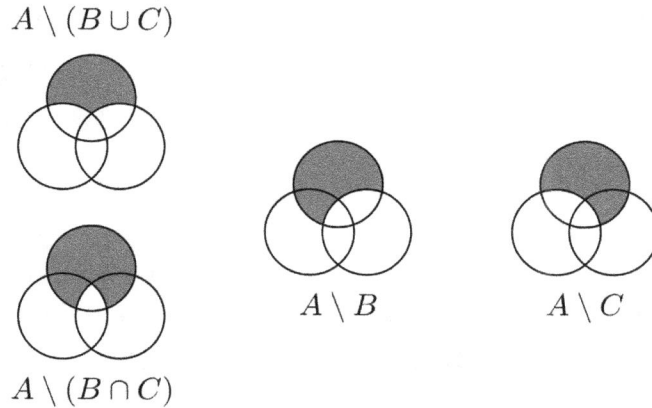

Figure C.2: De Morgan's laws, illustrated on Venn diagrams

Based on this, you can easily see both *(a)* and *(b)*.

Note that De Morgan's laws can be generalized to cover any number of sets. So, for any $\Gamma$ index set,

$$A \setminus (\cap_{\gamma \in \Gamma} B_\gamma) = \cup_{\gamma \in \Gamma} (A \setminus B_\gamma),$$
$$A \setminus (\cup_{\gamma \in \Gamma} B_\gamma) = \cap_{\gamma \in \Gamma} (A \setminus B_\gamma).$$

# C.3   The Cartesian product

One of the most fundamental ways to construct new sets is the Cartesian product.

**Definition C.3.1 (The Cartesian product)**

Let $A$ and $B$ be two sets. Their *Cartesian product* $A \times B$ is defined by

$$A \times B := \{(a, b) : a \in A \text{ and } b \in B\}.$$

The elements of the product are called *tuples*. Note that this operation is not associative nor commutative!

To see this, consider that, for example,

$$\{1\} \times \{2\} \neq \{2\} \times \{1\}$$

and

$$\big(\{1\} \times \{2\}\big) \times \{3\} \neq \{1\} \times \big(\{2\} \times \{3\}\big).$$

The Cartesian product for an arbitrary number of sets is defined with a recursive definition, just like we did with the union and intersection. So, if $A_1, A_2, \dots, A_n$ are sets, then

$$A_1 \times \cdots \times A_n := (A_1 \times \cdots \times A_{n-1}) \times A_n.$$

Here, the elements are tuples of tuples of tuples of..., but to avoid writing an excessive number of parentheses, we can abbreviate it as $(a_1, \dots, a_n)$. When the operands are the same, we usually write $A^n$ instead of $A \times \cdots \times A$. One of the most common examples is the Cartesian plane, which you probably have seen before.

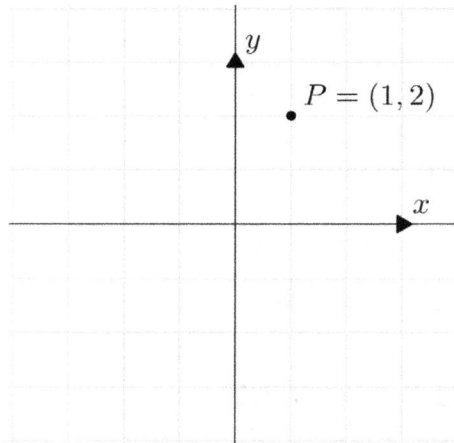

*Figure C.3: The Cartesian plane*

To give a machine-learning-related example, let's take a look at how data is usually presented. We'll focus on the famous *Iris dataset* (`https://scikit-learn.org/stable/auto_examples/datasets/plot_iris_datas et.html`), which is a subset of $\mathbb{R}^4$. In this dataset, the four axes represent *sepal length, sepal width, petal length,* and *petal width.*

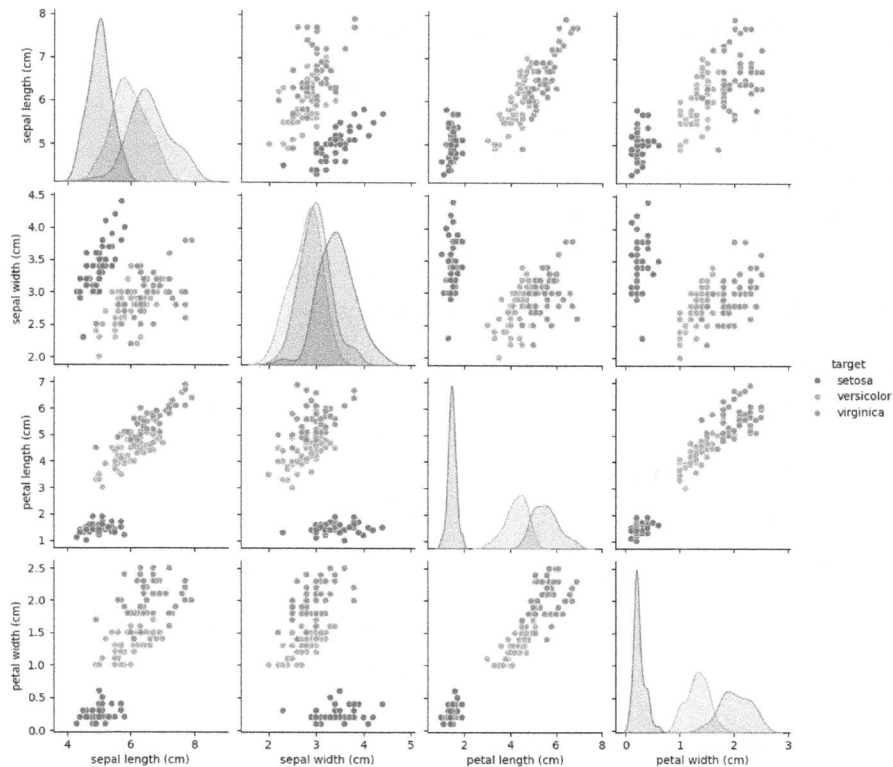

*Figure C.4: The sepal width, plotted against the sepal length in the Iris dataset. Source: scikit-learn documentation*

As the example demonstrates, Cartesian products are useful because they combine related information into a single mathematical structure. This is a recurring pattern in mathematics: building complex things from simpler building blocks and abstracting away the details by turning the result into yet another building block. (As one would do to create complex software as well.)

So far, we've seen the pleasant side of sets. However, they have caused quite a headache for mathematicians upon their first attempts to formalize set theory. We are ready to see why.

# C.4   The cardinality of sets

When it comes to sets, *"How many elements does a set have?"* is a natural question to ask. What you might not expect is how deep of a rabbit hole such an innocent question plunges us into!

Soon, you'll see why. The "size of a set" is formalized by the concept of *cardinality*, denoted by $|A|$; that is, putting the set $A$ between the absolute value signs. Intuitively, $|A|$ seems clear, but let me assure you, it is not.

This is where the true mind-bending begins in mathematics. Sure, for finite sets like $\{4, 6, 42\}$, we can confidently claim that $|\{4, 6, 42\}| = 3$, but what about $|\mathbb{Z}|$, $|\mathbb{Q}|$, $|\mathbb{R}|$, or even $|\mathbb{R}^2|$ ?

Check this: $|\mathbb{Z}| = |\mathbb{Q}|$, but $|\mathbb{Z}| \neq |\mathbb{R}|$, and $|\mathbb{R}| = |\mathbb{R}^2|$. In other words:

- there are "as many" integers as rational numbers,
- but there are "more" real numbers than integers,
- and there are "as many" points on the real line as on the plane.

See, I told you that cardinality is where the crazy stuff starts. (The "as many" and "more" are in quotes because we haven't defined them yet; at least, not in a mathematical sense.)

As with several concepts in mathematics, we won't define cardinality directly. Instead, we'll define how to *compare* the cardinalities of sets, and then anchor special sets that'll serve as points of comparison.

> **Definition C.4.1 (Comparison of cardinality)**
>
> Let $A$ and $B$ be two arbitrary sets.
>
> *(a)* We say that $|A| = |B|$ if there exists a function from $A$ to $B$ that is bijective.
>
> *(b)* We say that $|A| \leq |B|$ if there exists a function from $A$ to $B$ that is injective.
>
> *(c)* We say that $|A| < |B|$ if there exists a function from $A$ to $B$ that is injective but not surjective.

Let's unpack this with a couple of examples.

**Example 1.** Let $A = \{1, 2, 3\}$ and $B = \{-3.2, 12.452, -5.82\}$. Then $|A| = |B|$, because $f : A \to B$,

$$f(1) = -5.83,$$
$$f(2) = -3.2,$$
$$f(3) = 12.452$$

is a bijection.

**Example 2.** Let $A = \{0, 1\}$ and $B = \{2, 3, 4\}$. Then $|A| < |B|$, because $f : A \to B$,

$$f(0) = 2,$$
$$f(1) = 3$$

is injective, but not surjective.

What about more interesting sets, like $\mathbb{N}, \mathbb{Z}, \mathbb{Q}, \mathbb{R}$? This is where things get weird.

The following result is so important that we state it as a theorem.

**Theorem C.4.1** *(The set of natural numbers is infinite)*

*Let $n \in \mathbb{N}$ be any natural number, and define the set $A = \{1, 2, \ldots, n\}$. Then $|A| < \mathbb{N}$.*

*Proof.* The proof is simple, as it's easy to see that the function $f : A \to \mathbb{N}$, defined by $f(a) = a$ is injective but not surjective.

In other words, $\mathbb{N}$ is not a finite set! The set of natural numbers is our first example of an *infinite* set, defining the notion of *countable sets*.

**Definition C.4.2 (Countable sets)**

Let $A$ be an arbitrary set. If $|A| \le \mathbb{N}$, then $A$ is called *countable*.

We can show that a set is countable by enumerating its elements, as every enumeration defines an injective mapping. For instance, $\mathbb{Z}$ is countable, as the function $f : \mathbb{N} \to \mathbb{Z}$ defined by

$$f(n) = \begin{cases} 0 & \text{if } n = 0, \\ k & \text{if } n = 2k \text{ for some } k, \\ -k & \text{if } n = 2k + 1 \text{ for some } k \end{cases}$$

is a bijection. In other words, the sequence

$$0, 1, -1, 2, -2, 3, -3, \ldots$$

is an enumeration of $\mathbb{Z}$. For more examples, see *Chapter 10*.

Regarding countability, there are two essential results: the union and the Cartesian product of countable sets are still countable.

**Theorem C.4.2** *(Union and Cartesian product of countable sets)*

*Let $A_1, A_2, \ldots$ be countable sets.*

*(a) $A_1 \times A_2$ is countable.*

*(b) $\cup_{n=1}^{\infty} A_n$ is countable.*

Among infinite cardinalities, there are two that we frequently encounter: the *countably infinite* and the *continuum*. The countably infinite is the cardinality of the set of natural numbers, denoted by

$$|\mathbb{Z}| = \aleph_0,$$

where $\aleph$ is the Hebrew letter aleph. On the other hand, the continuum is the cardinality of the set of real numbers, denoted by

$$|\mathbb{R}| = c.$$

# C.5   The Russell paradox (optional)

Let's return to a remark I made earlier: naively defining sets as collections of things is not going to cut it. In the following, we are going to see why. Prepare for some mind-twisting mathematics.

Here's a riddle. A barber is *"one who shaves all those, and those only, who do not shave themselves."* Does the barber shave themself? There's no good answer: either yes or no, the definition implies otherwise. This is known as the *barber's paradox*. It's more than a cute little story; it's a paradox that shook the foundations of mathematics.

As we have seen, sets can be made of sets. For instance, $\{\mathbb{N}, \mathbb{Z}, \mathbb{R}\}$ is a collection of the most commonly used number sets. We might as well define the set of all sets, which we'll denote with $\Omega$.

With that, we can use the set-builder notation to describe the following collection of sets:

$$S := \{A \in \Omega \ : \ A \notin A\}.$$

In plain English, $S$ is a collection of sets that are not elements of themselves. Although this is weird, it *looks* valid. We used the property "$A \notin A$" to filter the set of all sets. What is the problem?

For one, we can't decide if $S$ is an element of $S$ or not. If $S \in S$, then by the *defining property*, $S \notin S$. On the other hand, if $S \notin S$, then by the definition, $S \in S$. This is definitely very weird.

We can diagnose the issue by decomposing the set-builder notation. In general terms, it can be written as

$$x \in A \ : \ T(x),$$

where $A$ is some set and $T(x)$ is a property, that is, a true or false statement about $x$.

In the definition $\{A \in \Omega : A \notin A\}$, our abstract property is defined by

$$T(A) := \begin{cases} \text{true} & \text{if } A \notin A, \\ \text{false} & \text{otherwise.} \end{cases}$$

This is perfectly valid, so the problem must be in the other part: the set $\Omega$. It turns out that the *set of all sets* is not a set. So, defining sets as *a collection of things* is not enough. Since sets are at the very foundation of mathematics, this discovery threw a giant monkey wrench into the machine around the late 19th-early 20th century, and it took lots of years and brilliant minds to fix it.

Fortunately, as machine learning practitioners, we don't have to care about such low-level details as the axioms of set theory. For us, it is enough to know that a solid foundation exists somewhere. (Hopefully.)

# Join our community on Discord

Read this book alongside other users, Machine Learning experts, and the author himself.

Ask questions, provide solutions to other readers, chat with the author via Ask Me Anything sessions, and much more.

Scan the QR code or visit the link to join the community.

`https://packt.link/math`

# D

# Complex Numbers

Learning is an upward spiral. Depending on where we are on our journey, we keep revisiting past knowledge, looking at it from a different angle.

Complex numbers is one of those topics to revisit, and understanding them makes us reevaluate certain pieces of knowledge that we take for granted. For example, you were probably taught that −1 does not have a square root. However, after familiarizing yourself with complex numbers, you'll see that there are actually *two* of them; both are complex numbers.

Our primary example here is going to be the quadratic equation

$$x^2 + 1 = 0.$$

To see that it doesn't have any solutions (or *roots*, in other words) among real numbers, we can check that the discriminant $b^2 - 4ac = -4$ is less than 0, but we can also simply plot the graph of $x^2 + 1$.

```python
import numpy as np
import matplotlib.pyplot as plt


X = np.linspace(-3, 3, 1000)
Y = X**2+1
```

```
with plt.style.context("seaborn-v0_8"):
    plt.figure()
    plt.axhline(0, color='black', linewidth=1)
    plt.axvline(0, color='black', linewidth=1)
    plt.plot(X, Y)
    plt.xlim([-3, 3])
    plt.ylim([-5, 10])
    plt.xlabel("x")
    plt.ylabel("y")
    plt.title("The graph of x² + 1")
    plt.show()
```

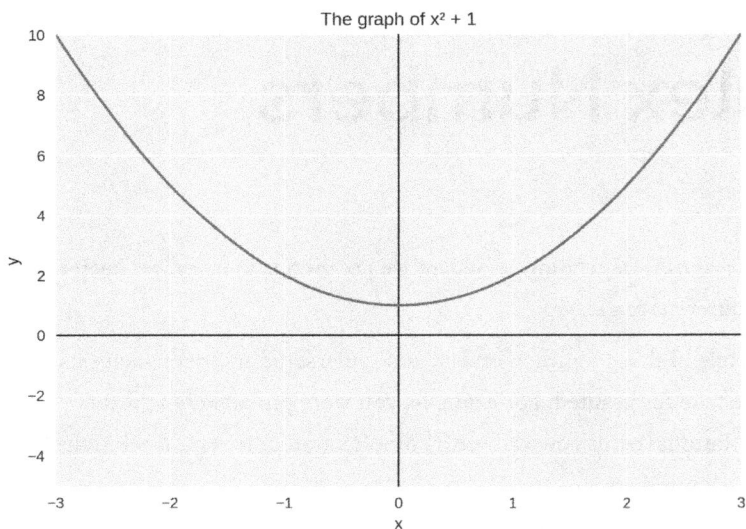

*Figure D.1: Graph of $x^2 + 1$*

It might not seem like a big deal, but a polynomial equation with real coefficients not having real solutions is a huge problem. Lots of essential quantities are described in terms of roots of polynomials, like the eigenvalues of matrices.

So, how can we solve this?

In most cases, we can factor quadratic polynomials into the product of two linear terms. For example,

$$x^2 - 1 = (x - 1)(x + 1),$$

revealing the solutions of the equation $x^2 - 1 = 0$ (which are $x = 1$ and $x = -1$ in this case).

To factorize the polynomial $x^2 - 1$, we used the identity

$$x^2 - a^2 = (x - a)(x + a).$$

What can we do for $x^2 + 1$? Let's get a bit more creative. We can write our polynomial as

$$x^2 + 1 = x^2 - (-1),$$

where in principle, we can use the previous identity. However, there is a problem: no real number satisfies $a^2 = -1$.

The solution that took mathematicians a few centuries to figure out is straightforward: *we imagine such a number*. In mathematics, the most creative abuse of rules can often prove to be most rewarding. So, suppose that our imaginary number $i$ satisfies

$$i^2 = -1.$$

This way, we have

$$x^2 + 1 = x^2 - i^2 = (x - i)(x + i).$$

In other words, the solutions for the notorious quadratic equation $x^2 + 1 = 0$ are $x = i$ and $x = -i$. This $i$ is called an *imaginary number*, and its discovery opened the gates to complex numbers.

So, what are these exotic objects?

# D.1 The definition of complex numbers

Let's jump into the definition right away.

**Definition D.1.1 (Complex numbers)**

The set of numbers written in the form

$$z = a + bi, \quad a, b \in \mathbb{R},$$

where $i^2 = -1$ is satisfied, is called complex numbers. We call $a$ the real part and $b$ the imaginary part of $z$, denoted by

$$\mathrm{Re}(z) = a, \quad \mathrm{Im}(z) = b.$$

If $a + bi$ and $c + di$ are two complex numbers, then we define addition and multiplication by

*(a)*

$$(a + bi) + (c + di) := (a + c) + (b + d)i,$$

*(b)*

$$(a + bi)(c + di) := (ac - bd) + (ad + bc)i.$$

The set of complex numbers is denoted with $\mathbb{C}$. Thus, we write

$$\mathbb{C} := \{a + bi : a, b \in \mathbb{R}\}.$$

According to the definition, addition is straightforward. However, multiplication seems a little bit convoluted. To see why it is defined this way, multiply them term by term, as you do with two polynomials.

An important property of complex numbers is their absolute value, or in other words, their distance from 0.

**Definition D.1.2 (Absolute value of complex numbers)**

Let $z = a + bi$ be a complex number. Its *absolute value* is defined by

$$|z| := \sqrt{a^2 + b^2}.$$

In addition, each complex number has a *conjugate*, which, as we shall see later, corresponds to its mirror image with respect to the real axis.

**Definition D.1.3 (Conjugate of complex numbers)**

Let $z = a + bi$ be a complex number. Its *conjugate* $\overline{z}$ is defined by

$$\overline{z} := a - bi.$$

Note that $z\overline{z} = |z|^2$.

Besides the algebraic representation $z = a + bi$, complex numbers have a rich geometric interpretation, as we are about to see.

## D.2 The geometric representation

We can represent complex numbers in ways other than the one in the definition. If you think about it, each number $z = a + bi$ can be seen as an ordered pair $(a, b)$. These can be visualized as vectors on the Cartesian plane.

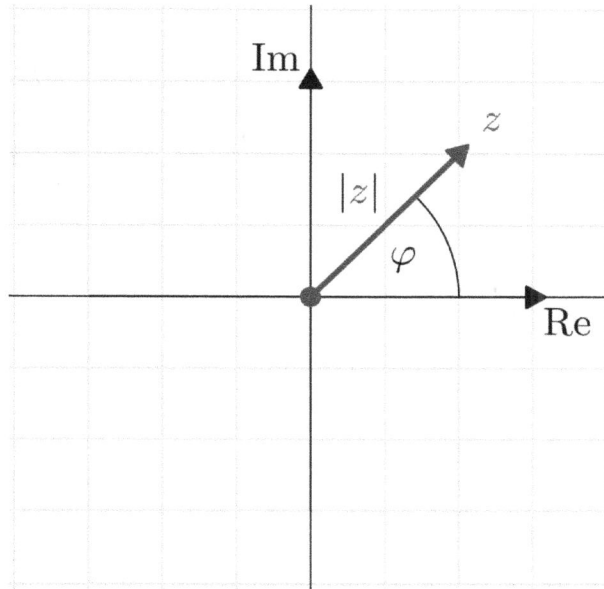

*Figure D.2: Complex numbers as vectors on the Cartesian plane*

The absolute value $|z| = \sqrt{a^2 + b^2}$ of a complex number $z = a + bi$ represents the length of the vector $(a, b)$ from the origin, while conjugation $\overline{z} = a - bi$ corresponds to reflecting the point across the real axis.

This geometric view gives us a new algebraic way to represent complex numbers.

To see why, recall the relation of the unit circle and the trigonometric functions on the plane.

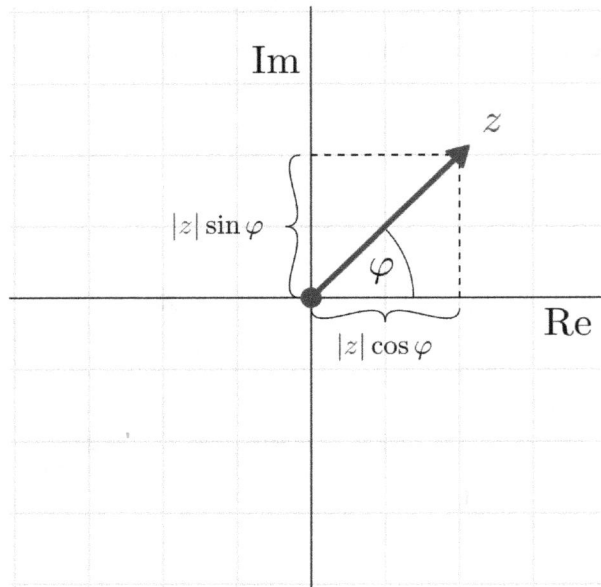

*Figure D.3: Geometric representation of complex numbers*

This means that every complex number with unit absolute value can be written in the form $\cos(\varphi) + i \sin(\varphi)$. From the geometric representation, we can see that every complex number is uniquely determined by its absolute value $|z|$ and the angle $\varphi$. So, we can write all complex numbers in the so-called *polar form*

$$z = r\big( \cos(\varphi) + i \sin(\varphi) \big), \quad r \in [0, \infty), \quad \varphi \in [0, 2\pi),$$

where $r = |z|$ is called the *modulus*, while $\varphi$ is called the *phase* or *argument*.

This geometric representation can also help us make more sense of multiplication. To see this, let's do some algebra first! When multiplying the complex numbers

$$z_1 = r_1\big( \cos(\varphi_1) + i \sin(\varphi_1) \big),$$
$$z_2 = r_2\big( \cos(\varphi_2) + i \sin(\varphi_2) \big)$$

together, we obtain

$$z_1 z_2 = r_1 r_2 \Big[ \big( \cos(\varphi_1) \cos(\varphi_2) - \sin(\varphi_1) \sin(\varphi_2) \big) + i\big( \cos(\varphi_1) \sin(\varphi_2) + \cos(\varphi_2) \sin(\varphi_1) \big) \Big].$$

Do you recognize the real and imaginary parts? These are the famous trigonometric addition formulas. With these, we have

$$z_1 z_2 = r_1 r_2 \big( \cos(\varphi_1 + \varphi_2) + i \sin(\varphi_1 + \varphi_2) \big).$$

This reveals a lot that is unclear from the algebraic definition. Most importantly,

1. the modulus of the product is the product of the moduli,
2. and the argument of the product is the sum of the arguments. (*Moduli* is the plural of modulus.)

In other words, multiplication in the complex plane is equivalent to a scaling and a rotation. All of a sudden, identities like $i^2 = -1$ make much more sense: since the argument of $i$ is $\pi/2$ (or 90 degrees), rotating it anticlockwise by $\pi/2$ yields $-1$.

So, why do we like real numbers? Because all polynomial equations have a solution there. Let's see!

# D.3   The fundamental theorem of algebra

Remember our motivating example? We introduced the imaginary number $i$ so that the equation $x^2 + 1 = 0$ can have a solution. It turned out that complex numbers provide a solution for any polynomial equation.

Let's introduce the set of polynomials with complex coefficients:

$$\mathbb{C}[x] := \left\{ \sum_{k=0}^{n} c_k x^k \ : \ c_k \in \mathbb{C}, n \in \mathbb{N}_0 \right\}.$$

Analogously, $\mathbb{R}[x], \mathbb{Q}[x], \mathbb{Z}[x]$ and $\mathbb{N}[x]$ can be defined as well. The degree of a polynomial (or deg $p$ for short) is the highest power of $x$. (For example, $-3x^8 + \pi x$ has a degree of 8.)

For a given polynomial $p(x)$, solutions of the equation $p(x) = 0$ are called *roots*. Algebraically, it is desirable that for a given set of polynomials over a particular set of numbers, each polynomial has roots there as well. As we have seen, it is not true for $\mathbb{R}[x]$, since $x^2 + 1 = 0$ has no real solutions.

However, this changes for $\mathbb{C}[x]$, as stated by *the fundamental theorem of algebra*.

> **Theorem D.3.1** *(The fundamental theorem of algebra)*
>
> *Every non-constant polynomial $p(x) \in \mathbb{C}[x]$ (i.e., a polynomial with* deg $p \geq 1$*) has at least one root in* $\mathbb{C}$*.*

Although this looks easy to prove, I assure you, it is not. The original algebraic proof is very long and involved, and although there is a shorter version, it requires advanced tools from mathematical analysis.

We can take the fundamental theorem of algebra a bit further. If $p(x)$ is a polynomial with a degree of at least 1 and $x_1$ is its root, then

$$q(x) = \frac{p(x)}{x - x_1}$$

is a polynomial of degree $\deg q = \deg p - 1$. If $q$ is non-constant, the fundamental theorem of algebra once again guarantees that $q$ also has a root in $\mathbb{C}$. Ultimately, the repeated application of the theorem yields that $p(x)$ can be written in the form

$$p(x) = (x - x_1)(x - x_2)\ldots(x - x_n),$$

where $x_1, x_2, \ldots, x_n$ are the *roots* of $p$. Some of them can match, and the number of times a given $x_i$ can be found among its roots is called its *algebraic multiplicity*.

> **Definition D.3.1** *(Algebraic multiplicity of roots)*
>
> Let $p(x) \in \mathbb{C}[x]$ and suppose that
>
> $$p(x) = \prod_{i=1}^{l}(x - x_i)^{p_i}, \quad x_i \neq x_j \quad \text{for } i \neq j.$$
>
> The integer $p_i$ is called the algebraic multiplicity of the root $x_i$.

The fundamental theorem of algebra is called fundamental for a reason. For instance, this is the reason why matrices have eigenvalues, as we learned in *Chapter 6*.

# D.4  Why are complex numbers important?

At first glance, complex numbers might not seem so important in machine learning. Let me assure you that this is not the case: they are absolutely essential. In this section, we will take a quick look forward and see what complex numbers make possible.

For instance, one of the most important applications of complex numbers is the eigenvalue-eigenvector pairs of matrices. For a given matrix $A$, there are special numbers $\lambda$ called eigenvalues, and corresponding vectors $v$ called eigenvectors, such that $Av = \lambda v$ holds. In the language of linear transformations, this means that in the linear subspace spanned by the eigenvectors of $\lambda$, the transformation $A$ is just a stretching.

These are very powerful, as under certain conditions, eigenvalues allow us to simplify the matrix.

Certain square real matrices can be written in the form

$$A = U\Sigma U^{-1}, \quad \Sigma = \begin{pmatrix} \lambda_1 & 0 & \dots & 0 \\ 0 & \lambda_2 & \dots & 0 \\ \vdots & \vdots & \ddots & \vdots \\ 0 & 0 & \dots & \lambda_n \end{pmatrix}$$

is a diagonal matrix composed of the eigenvalues of $A$. Guess what guarantees the existence of eigenvalues: the fundamental theorem of algebra.

By going from real to complex numbers, we obtain a larger degree of freedom and a more powerful set of tools. One such tool is the famous Fourier transform. To give you a practical example, let's talk about audio. Audio data comes in the form of a function $f : \mathbb{R} \to \mathbb{R}$, mapping time to signal intensity. However, in signal processing, it is instrumental to understand the signal in terms of frequencies. Every sound is the superposition of sine waves with various frequencies, and quantifying the contribution of each frequency reveals a lot about the signal.

```python
import matplotlib.pyplot as plt
import numpy as np


def sin(freq, x):
    return np.sin(freq*x)


X = np.linspace(0, 2*np.pi, 1000)
freqs = [1, 2, 3, 4, 5, 6, 7, 8, 9]
y = {freq: sin(freq, X) for freq in freqs}

with plt.style.context("seaborn-v0_8-white"):
    plt.figure(figsize=(10, 10), dpi=100)
    for i, freq in enumerate(freqs):
        plt.subplot(3, 3, i+1)
        plt.plot(X, y[freq])
        plt.title(f"frequency = {freq}")
    plt.show()
```

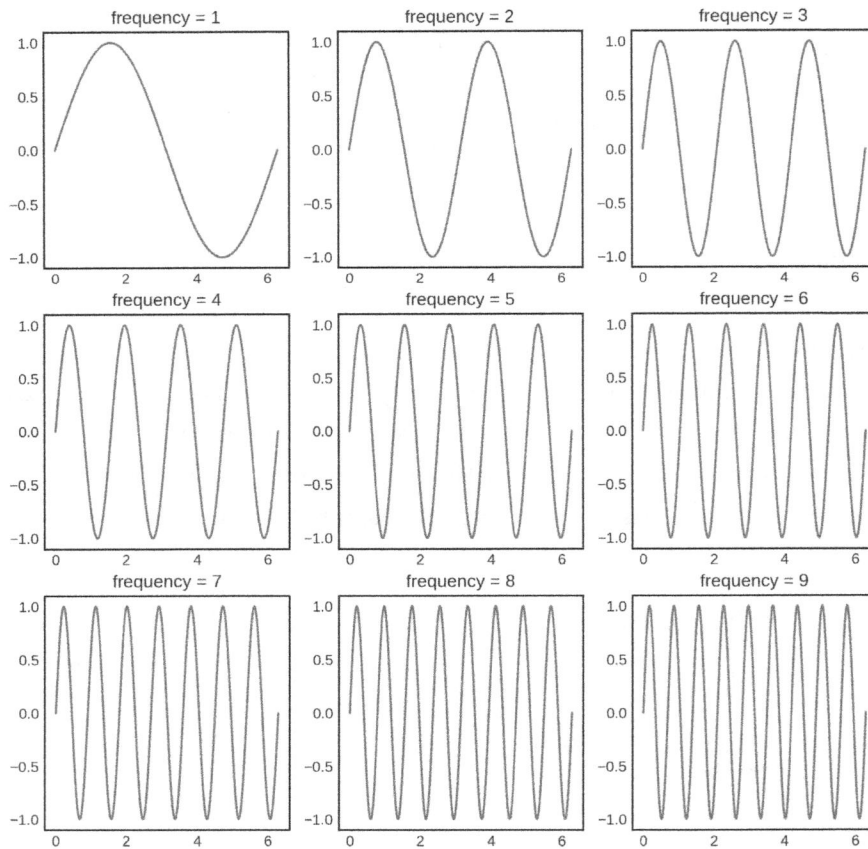

*Figure D.4: Fourier transform*

The thing is, the Fourier transform is a *complex integral.* That is, the transform of $f$ is defined by

$$\hat{f}(\xi) := \int_{-\infty}^{\infty} f(t)\Big(\cos(2\pi t\xi) - i\sin(2\pi t\xi)\Big)dt.$$

Without complex numbers, this tool is not available to us.

Although we will not deal with Fourier transforms in detail, they are an indispensable tool in probability theory. When applied to probability distributions, their convergence properties can be studied easily. For instance, some versions of the *central limit theorem* (https://en.wikipedia.org/wiki/Central_limit_theorem) are proven this way.

# Join our community on Discord

Read this book alongside other users, Machine Learning experts, and the author himself.

Ask questions, provide solutions to other readers, chat with the author via Ask Me Anything sessions, and much more.

Scan the QR code or visit the link to join the community.

https://packt.link/math

# ‹packt›

Subscribe to our online digital library for full access to over 7,000 books and videos, as well as industry leading tools to help you plan your personal development and advance your career. For more information, please visit our website.

## Why subscribe?

- Spend less time learning and more time coding with practical eBooks and Videos from over 4,000 industry professionals
- Improve your learning with Skill Plans built especially for you
- Get a free eBook or video every month
- Fully searchable for easy access to vital information
- Copy and paste, print, and bookmark content

Did you know that Packt offers eBook versions of every book published, with PDF and ePub files available? You can upgrade to the eBook version at packt.com and as a print book customer, you are entitled to a discount on the eBook copy. Get in touch with us at customercare@packtpub.com for more details.

At www.packt.com, you can also read a collection of free technical articles, sign up for a range of free

# Other Books You May Enjoy

If you enjoyed this book, you may be interested in these other books by Packt:

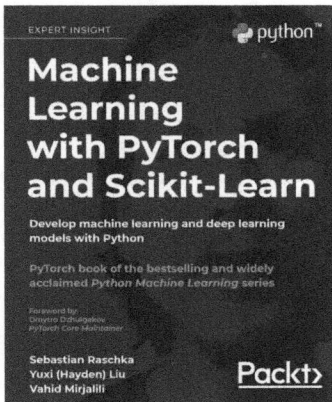

**Machine Learning with PyTorch and Scikit-Learn**

Sebastian Raschka , Yuxi (Hayden) Liu, Vahid Mirjalili

ISBN: 978-1-80181-931-2

- Explore frameworks, models, and techniques for machines to learn from data
- Use scikit-learn for machine learning and PyTorch for deep learning
- Train machine learning classifiers on images, text, and more
- Build and train neural networks, transformers, and boosting algorithms
- Discover best practices for evaluating and tuning models
- Predict continuous target outcomes using regression analysis
- Dig deeper into textual and social media data using sentiment analysis

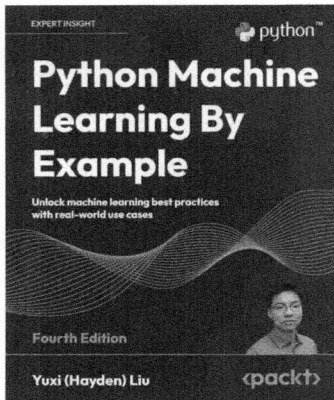

**Python Machine Learning By Example, 4th Edition**

Yuxi (Hayden) Liu

ISBN: 978-1-83508-222-5

- Follow machine learning best practices across data preparation and model development
- Build and improve image classifiers using Convolutional Neural Networks (CNNs) and transfer learning
- Develop and fine-tune neural networks using TensorFlow and PyTorch
- Analyze sequence data and make predictions using RNNs, transformers, and CLIP
- Build classifiers using SVMs and boost performance with PCA
- Avoid overfitting using regularization, feature selection, and more

# Packt is searching for authors like you

If you're interested in becoming an author for Packt, please visit `authors.packtpub.com` and apply today. We have worked with thousands of developers and tech professionals, just like you, to help them share their insight with the global tech community. You can make a general application, apply for a specific hot topic that we are recruiting an author for, or submit your own idea.

# Share your thoughts

Now you've finished *Mathematics of Machine Learning*, we'd love to hear your thoughts! Scan the QR code below to go straight to the Amazon review page for this book and share your feedback or leave a review on the site that you purchased it from.

`https://packt.link/r/1837027870`

Your review is important to us and the tech community and will help us make sure we're delivering excellent quality content.

# Index

# Download a free PDF copy of this book

Thanks for purchasing this book!

Do you like to read on the go but are unable to carry your print books everywhere? Is your eBook purchase not compatible with the device of your choice?

Don't worry; with every Packt book, you now get a DRM-free PDF version of that book at no cost.

Read anywhere, on any device. Search, copy, and paste code from your favorite technical books directly into your application.

The perks don't stop there! You can get exclusive access to discounts, newsletters, and great free content in your inbox daily.

Follow these simple steps to get the benefits:

1. Scan the QR code or visit the link below:

https://packt.link/free-ebook/9781837027873

2. Submit your proof of purchase.
3. That's it! We'll send your free PDF and other benefits to your email address directly.

www.ingramcontent.com/pod-product-compliance
Lightning Source LLC
Chambersburg PA
CBHW081208220326
41598CB00037B/6717